# Joint International Conference SPM-2019-RCWDFM



3rd International Conference "Scanning Probe Microscopy" (SPM), 4th Russia-China Workshop on Dielectric and Ferroelectric Materials (RCWDFM), and International Youth Conference "Functional Imaging of Nanomaterials”

IOP Conference Series: Materials Science and Engineering
Volume 699

Ekaterinburg, Russia
25 - 28 August 2019

ISBN: 978-1-7138-1014-8
ISSN: 1757-8981

**Printed from e-media with permission by:**

Curran Associates, Inc.
57 Morehouse Lane
Red Hook, NY 12571

**Some format issues inherent in the e-media version may also appear in this print version.**



Institute of Physics
Dirac House, Temple Back
Bristol BS1 6BE UK

Phone: 44 1 17 929 7481
Fax: 44 1 17 920 0979

techtracking@iop.org

**Additional copies of this publication are available from:**

Curran Associates, Inc.
57 Morehouse Lane
Red Hook, NY 12571 USA
Phone: 845-758-0400
Fax: 845-758-2633
Email: curran@proceedings.com
Web: www.proceedings.com

# TABLE OF CONTENTS

IOP Conf. Series: Materials Science and Engineering **699** (2019) 011001 doi:10.1088/1757-899X/699/1/011001

# Guest Editorial

These Proceedings contain a collection of papers based on the invited, oral, and poster presentations at the **Joint International Conference SPM-2019-RCWDFM** combining **the 3rd International Conference "Scanning Probe Microscopy"** (SPM), **the 4th Russia-China Workshop on Dielectric and Ferroelectric Materials (RCWDFM)**, and **International Youth Conference "Functional Imaging of Nanomaterials"**, which was held in Ekaterinburg, Russia, August 25-28, 2019.

**Joint International Conference SPM-2019-RCWDFM** covered a wide range of scientific topics.

SPM Topics:
- SPM in materials science
- PFM, MFM, KPFM, SNOM, ESM, SEM, and related techniques
- Tip-enhanced phenomena
- Probe lithography and nano-indentation
- Ferroelectrics, piezoelectrics, and ionic conductors
- Multiferroic phenomena and magnetoelectric coupling on the nanoscale
- Interface and domain engineering
- Biocompatible & organic materials
- 1D and 2D nanostructured materials
- Theory, modelling, and data processing

RCWDFM Topics:
- High-performance piezo-/ferroelectric materials and devices
- Thin films, single crystals, interfaces and nanoscale materials
- Multiferroic materials and devices
- New mechanisms/materials/devices
- Domains, domain walls, and domain engineering

The conference was organized by the School of Natural Sciences and Mathematics and Ural Centre for Shared Use "Modern Nanotechnology" of the Ural Federal University.

The objective of SPM-2019-RCWDFM was to provide an international and interdisciplinary exchange of information on a wide range of topics related to modern state-of-the-art and future development of various Scanning Probe Microscopy techniques in different fields of science, as well as fundamentals and applications of dielectric and ferroelectric materials in order to ensure effective cooperation among scientists and specialists from universities, research institutions and industry.

More than 200 participants from 15 countries: Belarus, Brazil, Canada, China, France, Germany, Ireland, Japan, Korea, Netherlands, Poland, Portugal, Russia, Switzerland, and USA presented their results. The conference comprised 239 presentations, including 42 invited, 47 oral, and 150 poster ones.

The organizers are very grateful to the conference sponsors and exhibitors: Taylor & Francis Group, Crystals journal, NT-MDT Spectrum Instruments, SPECS TII RUS, Ostec ArtTool, OPTEC, INTERTECH Corporation, MTEON, IMC Group, Promenergolab, and Alexander Lukanin.

Vladimir Shur
Chairman of **SPM-2019-RCWDFM**

IOP Conf. Series: Materials Science and Engineering **699** (2019) 011001 doi:10.1088/1757-899X/699/1/011001

# SPM-2019-RCWDFM Committees

**SPM General Chairs**

Prof. Vladimir Shur
Prof. Victor Mironov

**SPM International Program Committee**

Prof. A.V. Ankudinov
Prof. A.A. Bukharaev
Prof. V.A. Bykov
Prof. S.A. Chizhik
Prof. A.L. Gruverman
Prof. A.L. Kholkin
Prof. O.V. Kolosov
Prof. A.V. Latyshev
Prof. V.L. Mironov
Prof. A.A. Saranin
Prof. V.Ya. Shur
Prof. V.V. Shvartsman
Prof. A.P. Volodin
Prof. I.V. Yaminsky

**RCWDFM General Chairs**

Prof. Alexander Sigov
Prof. Zhuo Xu

**RCWDFM International Program Committee**

Prof. I.N. Chugueva
Prof. H.S. Gu
Prof. A.L. Kholkin
Prof. L.N. Korotkov
Prof. H.X. Liu
Prof. W. Ren
Prof. I.V. Shnaidshtein
Prof. V.Ya. Shur
Prof. T.R. Volk
Prof. S.B. Vakhrushev
Prof. X.Y. Wei
Prof. Z.G. Ye

**Local organizing committee**

Prof. Vladimir Shur
Alevtina Shur
Dr. Dmitry Pelegov
Elena Pelegova
Dr. Ekaterina Shishkina
Victoria Pryakhina
Olga Dyachuk

IOP Conf. Series: Materials Science and Engineering **699** (2019) 011001 doi:10.1088/1757-899X/699/1/011001

# SPM-2019-RCWDFM Sponsors and Exhibitors

Taylor & Francis Group
Crystals Journal
NT-MDT Spectrum Instruments
SPECS TII RUS
Ostec-ArtTool
OPTEC
INTERTECH Corporation
MTEON
IMC Group
Promenergolab
Alexander Lukanin

IOP Conf. Series: Materials Science and Engineering **699** (2019) 011001 doi:10.1088/1757-899X/699/1/011001

# List of SPM-2019-RCWDFM Participants

**Belarus**
Melnikova Galina
Petrovskaya Agata

**Brazil**
Araujo Eudes

**Canada**
Ye Zuo-Guang

**China**
Chen Wen
Gong Jihui
Gu Haoshuang
Hao Hua
Hu Yongming
Hu Zhongqiang
Jin Wei
Juan Xiong
Lei Jiang
Li ChangJiao
Li Guorong
Li Jinglei
Li Shenhou
Liang Xuewei
Liu Hanxing
Liu Juyin
Liu Ming
Liu Wenbin
Lyapunov Nikolay
Ma Youqi
Shen Xiaoyuan
Shen Yang
Tang Zhuohua
Wang Chen
Wang Minqiang
Wang Wang
Wang Xusheng
Wang Zhiguang
Wei Wei
Wu Di
Wu Jingen
Xu Zhuo
Yang Wenwen
Yao Xiaogang
Yuan Jianmin
Zhai Jiwei
Zhang Huazhang
Zhang Yong
Zhu Jianguo
Zhu Quanyao
Zhuang Jian

**France**
Neradovskaia Elizaveta

**Germany**
Shvartsman Vladimir

**Ireland**
Tofail Syed
Vasilev Semen

**Japan**
Sawada Chiaki

**Korea**
Hong Seungbum

**Netherlands**
Kozodaev Dmitry

**Poland**
Bauer Joanna

**Portugal**
Kholkin Andrei

**Russia**
Abalmasov Veniamin
Abramov Alexander
Akbaeva Akbaeva
Akhmatkhanov Andrei
Akhmetova Assel
Alekseev Alexander
Alikin Denis
Amangulova Ilyuza
Andreeva Natalia
Ankudinov Alexander
Avilov Vadim
Barabanova Ekaterina
Baruzdina Olga
Bi Dongxue
Bunin Mikhail
Butiaikin Sergey
Bykov Ivan
Bykov Victor
Bystrov Vladimir
Cherkasov Dmitriy
Chernozem Roman
Chezganov Dmitry
Chikova Olga
Chuklanov Anton
Chuvakova Maria
Dokukin Maxim
Dunaevskiy Mikhail
Dyachuk Olga
Eremeev Alexander
Esin Alexander
Fokin Denis
Fomin Lev
Gainutdinov Radmir
Geguzina Galina
Gimadeeva Lubov
Golubev Evgeny
Golubok Alexander
Goncharov Boris
Gorelik Vladimir
Gornostaev Dmitry
Greshnyakov Evgeny
Gunina Ekaterina
Gushchina Ekaterina
Ibragimov Artem
Ivleva Liudmila
Jityaev Igor
Kandaurov Mikhail
Kasimova Valentina
Khannanov Boris
Kokhanchik Lyudmila
Koriazhkina Maria
Korolev Leonid
Korotkov Leonid
Korsakova Elena
Kosobokov Mikhail
Kostritskii Sergey
Kozlowski Alexander
Krasnoborodko Sergey
Krutov Vladislav
Krylov Alexander
Krylova Svetlana
Kunkel Tatiana
Kuznetsov Dmitry
Lebedev Denis
Leesment Stanislav
Linker Eduard
Lisjikh Boris
Lisov Evgeny
Lobov Ivan
Makaev Andrei
Mamin Rinat
Marakhovskiy Mikhail
Maslenikov Igor
Mayorova Yana
Mikhailova Tatiana
Milinskiy Alexey
Mingaliev Evgeny
Minigalieva Ilzira
Minin Maxim
Mironov Viktor
Miruschenko Mikhail
Morozov Ilya
Morozova Anna
Mukhin Ivan
Mushinsky Sergey
Nasedkin Andrey
Nesterenko Lolita
Neudachina Vera
Nuraeva Alla
Panov Dmitry
Pariy Igor
Pashnina Elena
Pavlov Dmitry
Pelegov Dmitry
Pelegova Elena
Petrova Tatiana
Piskunov Oleg
Piskunova Natalia
Podgornov Fyodor
Politova Ekaterina
Polyakov Vyacheslav
Privalova Larisa
Pryakhina Viktoria
Pugachev Alexey
Rassadin Alexander
Rudyk Nikolay
Rumyantsev Evgeny
Ryabova Julia
Savchenkov Evgeny
Saveliev Evgeny
Sazanova Tatiana
Semkin Mikhail
Senkevich Stanislav
Shaginyan Vasily
Shandyba Nikita
Sharov Vladislav
Sherstyuk Natalia
Shikhova Vera
Shishkina Ekaterina
Shkalei Ivan
Shubin Andrei

IOP Conf. Series: Materials Science and Engineering **699** (2019) 011001 doi:10.1088/1757-899X/699/1/011001

Shur Alevtina
Shur Vladimir
Sigov Alexander
Slautin Boris
Smirnova Elena
Sokolov Denis
Solovyeva Svetlana
Tsvetkov Yuri
Turygin Anton
Ushakov Andrei
Vasileva Daria
Vlasov Evgeny
Volchetskaya Ksenia
Volk Tatiana
Vtyurin Aleksander
Vysokikh Yury
Yaminsky Igor
Yuzhakov Vladimir
Zhukov Alexey
Zaytseva Irina
Zhukov Mikhail

**Switzerland**
Kasas Sandor

**USA**
Proksch Roger

# Peer review statement

All papers published in this volume of *IOP Conference Series: Materials Science and Engineering* have been peer reviewed through processes administered by the proceedings Editors. Reviews were conducted by expert referees to the professional and scientific standards expected of a proceedings journal published by IOP Publishing.

IOP Conf. Series: Materials Science and Engineering **699** (2019) 012001 doi:10.1088/1757-899X/699/1/012001

# Mapping hole mobility in PTB7 films at nanoscale

**A M Alekseev[1,2], A T Yedrissov[2,3], B R Ilyassov[2,3], G J Hedley[4], I D W Samuel[5] and S S Kharintsev[6]**

[1]National Research University "MIET", Moscow, 124498, Russia
[2]National Laboratory Astana, Nazarbayev University, Astana, 010000, Kazakhstan
[3]Karaganda State University, Karaganda, 100028, Kazakhstan
[4]University of Glasgow, G12 8QQ Glasgow, UK
[5]University of St. Andrews, KY16 9AJ St. Andrews, UK
[6]Kazan Federal University, Kazan, 420008, Russia

alalrus@gmail.com

**Abstract**. The nanoscale hole mobility in organic semiconducting polymer PTB7 is quantified by using conductive-AFM (C-AFM) measurements in space charge limited (SCLC) regime. The obtained current map of the neat PTB7 film is explained in terms of non-uniform built-in voltage and variations of hole mobility. For mobility estimation, the semi-empirical model of SCLC, known from previous works, was modified and applied. It is found that the values of built-in voltage in C-AFM measurements are usually several times larger than ones derived from macroscopic measurements. It is also shown that value of hole mobility in PTB7 film depends on location and varies in more than two times. These mobility variations are connected with nanoscale film structure revealed by other methods.

## 1. Introduction

One of the most important achievements of the recent progress in synthesis of new materials for organic photovoltaics (OPV) was introduction of the new thiophene based donor with abbreviation PTB7 (Poly [[4,8-bis[(2-ethylhexyl)oxy]benzo[1,2-b:4,5-b']dithiophene-2,6-diyl][3-fluoro-2-[(2 ethylhexyl) carbonyl]thieno[3,4-b]thiophenediyl]]) [1]. The improved processing conditions and device architecture allowed achieving 9.2% [2] and later 10.02% [3] power conversion efficiencies on solar cells with PTB7:PC71BM active layer. Thus, blend PTB7:PC71BM was under intensive investigation since its first introduction and structure and properties of this compound were extensively examined by different methods [4-8]. One of important parameters of organic semiconductors, which influences performance of polymer:fullerene solar cells, is a charge carrier mobility. Mobility can be determined by different methods, in particular, by measurements of the space charge limited current (SCLC) [2, 9-12]. Influence of polymer hole mobility on performance of PTB7:PC71BM solar cell was discussed earlier [10,12]. Beside structure of bulk heterojunction, characterization of the electrical properties of organic semiconductors at nanoscale is one of the important directions towards understanding the physics of organic electronic devices and improving their performance. The use of small electrodes to form local contacts is required for such nanoscale measurements. Thus, Atomic Force Microscopy (AFM) equipped with conductive probe (conductive-AFM or C-AFM) is a promising tool for probing local electrical properties of materials. The quantitative analysis of local current-voltage (I-V) curves obtained by C-AFM is an attractive method of characterization of the local electrical properties, such as charge carrier mobility, quantum efficiency, local open circuit voltage, and short circuit current [8, 13-16].

IOP Conf. Series: Materials Science and Engineering **699** (2019) 012001 doi:10.1088/1757-899X/699/1/012001

However, complicated geometry of the electrodes in C-AFM makes any quantitative analysis much more challenging than just use of standard Mott-Gurney law [10], which is suitable for flat electrodes. The semi-empirical model for local mobility measurements by C-AFM was introduced by O. Reid et al. [16] and it was applied to standard OPV materials such as P3HT and MDMO-PPV.

Here, we use C-AFM for analysis of the local conductivity in PTB7-based films and apply modified semi-empirical model for calculation of the nanoscale mobility and its variations caused by film structure.

## 2. Experimental

PTB7 films were produced by spin-coating from solution in chlorobenzene on glass/ITO/PEDOT:PSS as described elsewhere [7,8]. Different spin-coating rates were used in order to obtain a set of PTB7 films with different thicknesses ranging from 50 to 120 nm. The C-AFM measurements were performed by AFM Smart1000 (AIST-NT) in contact regime. The gold-coated probes CSG10/Au were utilized for C-AFM. The alternating mode of current measurements implemented in software was employed, which implied the lift of the probe from surface when movement from point to point was executed in order to reduce the sample damage. Tip-sample force was estimated from force-distance measurements for each experiment being in the range of 5-10 nN. Cantilever force constant was determined by Sader method implemented into AFM software. Tip-sample area was estimated corresponding to Hertzian contact theory with modulus elasticity of PTB7 taken from recently published data [17]. Gold coated tip radius estimated using SEM was 45 nm. The wavelength of AFM laser was 1300 nm, being far from light adsorption maxima of the films [2,7]. All samples were kept inside nitrogen filled glove-box and taken out just before AFM measurements. Conventional SCLC measurements were performed on devices with structure glass/ITO/PEDOT:PSS/PTB7 film/Au-electrode using potentiostat Autolab PGSTAT302N. The hole mobility calculations were performed corresponding to the unified protocol published recently [11]. Software Nova (NT-MDT) was utilized for drawing local mobility and built-in voltage maps.

## 3. Results and discussions

Nanoscale topography of the neat PTB7 film shown in Figure 1a is very flat and shows very small height variations of less than 10 nm in amplitude. These variations are much smaller than film thickness $L$=102 nm, determined by AFM measurements of the scratch. The current distribution map obtained by C-AFM has nearly no correlation with topography image (Fig. 1). The contrast in current image reveals grain and fiber-like structures, while topography image is rather flat. Current variations in one scan are significant and difference between current level is up to several times in different locations. Thus, even neat PTB7 film demonstrates highly non-uniform electrical properties at nanoscale. In order to understand mechanism of local conductivity, measurements of I-V curves were performed by C-AFM on film surface at the grid of 5x5 points. Figure 2a shows cross-section of obtained data at tip voltage $U_t$=+5V on 0.5x0.5 μm$^2$ area, and Figure 2b demonstrates all 25 I-V curves. It is clearly seen that all the curves are different and, thus, different local electrical characteristics are expected. The average I-V curve plotted in double logarithmic scale represents a straight line with a slope 3.2, which is reduced to

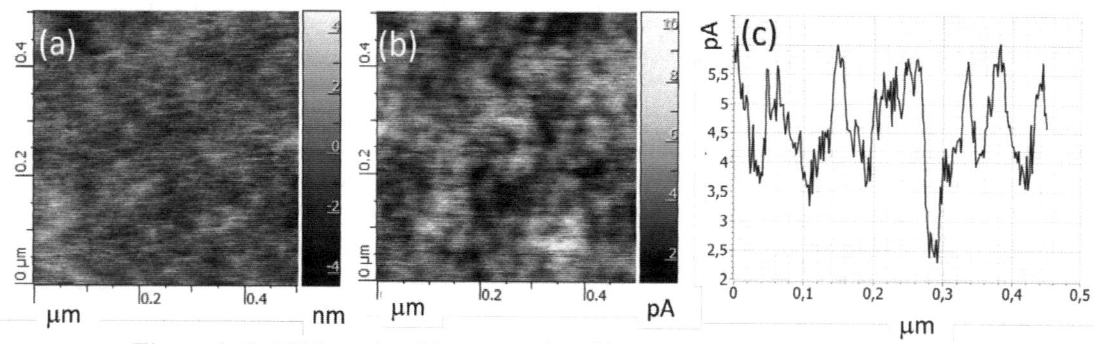


**Figure 1.** C-AFM results: (a) topography, (b) current, (c) cross-section of (b).

IOP Conf. Series: Materials Science and Engineering **699** (2019) 012001 doi:10.1088/1757-899X/699/1/012001

2.0 after taking into account built-in voltage $U_{bi}$ [18], clearly indicating that space charge limited current takes place (Fig. 2c). Corresponding to energy levels of PTB7 and electrodes used, the C-AFM contrast is caused by hole current through Au probe/PTB7/PEDOT:PSS/ITO [7-8]. Only positive voltages on the tip are used for analysis, which correspond to hole injection from tip to polymer. Then obtained results can be used for calculations of hole local mobility in PTB7 film. For this purpose, the protocol of mobility calculation [11], as well as semi-empirical model of SCLC adapted for C-AFM measurements [16], were utilized by us for hole mobility estimation.

First of all, series of PTB7 films with different thicknesses were measured in the same manner by C-AFM and average I-V curve was used for current density $J$ vs. film thickness $L$ plot. From these data the dependence $J \sim L^{-1.7}$ was derived, which lead to modified equation for local mobility [16]:

$$I = A_{eff}\, \alpha\, \varepsilon_0\, \varepsilon\, \mu\, e^{0.89\gamma \left(\frac{U}{L}\right)^{\frac{1}{2}}} \frac{U^2}{L^3}\, \delta \left(\frac{L}{d}\right)^{1.3}, \quad (1)$$

where $\alpha$=8.2 is a parameter from numerical calculations [16], $\varepsilon$=3.5 is material permittivity, $\mu$ is the hole mobility at zero field, $\gamma$ is field dependence parameter of the mobility, $L$ is the sample thickness, $d$ is the tip-sample contact diameter, $\delta$=7.8 is the scaling factor for fitting C-AFM mobility to macroscopic data [16].

Corresponding to our macroscopic SCLC measurements on the same structure, parameter $\gamma$ is equal to zero and we will use it in nanoscale calculations. The voltage $U$ in (1) must take into account the value of built-in voltage $U_{bi}$ [18]: $U = U_t - U_{bi}$. The value of $U_{bi}$ was estimated as 0.25 V for macroscopic data. However, for C-AFM data the same method of estimation applied to local $I$-$V$ curves gives significantly higher values of $U_{bi}$ between 1.2 and 1.8 V with rms value 0.18 V. Such difference between macroscopic and C-AFM results is not clear and can be explained by influence of surface properties, e.g., water or double layers at interfaces as well as some changes of surface properties under the probe when AFM contact mode is used. No correlation between tip-sample force and $U_{bi}$ was found. Mapping of $U_{bi}$ on surface is shown in Figure 3a. Variations of $U_{bi}$ on the surface can be caused by local surface properties as well as non-uniform structure of PEDOT:PSS layer.

Taking into account $U_{bi}(x,y)$ and using Eq. (1), distribution of hole mobility on the $\mu(x,y)$ surface can be obtained. The results of mobility mapping are shown in Figure 3b. Each pixel in Figures 2a and 3 corresponds to the area 100x100 nm$^2$, being roughly a size of current cross-section at PEDOT:PSS as estimated by modelling [16]. One can see that there is no direct correlation between current, $U_{bi}$, and mobility distributions, i.e. despite proportionality between current and mobility in Eq. (1), the influence of non-uniform $U_{bi}(x,y)$ on current distribution is essential. The performed calculations were made manually with limited image resolution. Increased number of measured I-V curves will lead to image similar to Figure 1b; however, it requires automated procedure due to a large data array. The average value of nanoscale mobility is 1.17*10$^{-8}$ m$^2$/(Vs), which is close to macroscopic value of 1.34*10$^{-8}$ m$^2$/(Vs) obtained by regular SCLC on the same structure. At the same time, variations of mobility in more than 2 times from 0.69 to 1.59 10$^{-8}$ m$^2$/(Vs) may be explained by non-uniform structure of PTB7 films, which was revealed earlier by XRD [4,5]. The rms value of the hole mobility is 0.22*10$^{-8}$ m$^2$/(Vs), which is almost 20% from the average mobility.

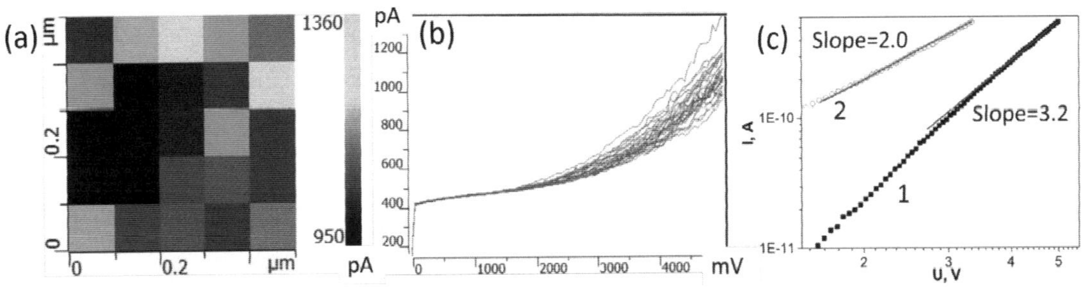


**Figure 2.** (a) Section of array of 25 I-V curves at $U$=+5V; (b) 25 I-V curves; (c) average I-V curve in double logarithmic coordinates: 1- initial voltage scale, 2 – $U_{bi}$ is taken into account.

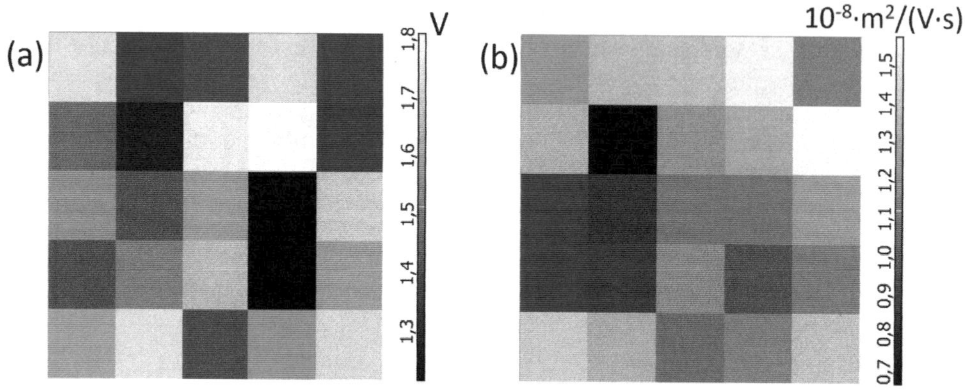


**Figure 3.** Distribution of (a) $U_{bi}$ and (b) hole mobility $\mu$ for the same area as shown in Figure 2a.

Corresponding to XRD results, the PTB7 spin-coated films consist of nanocrystalline areas inside amorphous matrix [4,5]. Since current area between sharp golden tip and flat opposite electrode has gradually increased cross-section, the top layer of film has larger influence on detected current. In top layer of the film, size of ordered areas of PTB7 is closer to current cross-section and then structure influences current, producing contrast in C-AFM image. Thus, contrast in Figure 1b and Figure 2a reflects mainly structure in top layer of PTB7 film. This current contrast at high voltages ($U$-$U_{bi}$>1 V) can be described in terms of non-uniform built-in voltage and hole mobility. The hole mobility variations may also be explained by different amount of crystalline and amorphous polymer inside volume, where charge carrier flow occurs.

## 4. Conclusion

The variations of local current in PTB7 film may be explained by influence of nanoscale structure in top layer of the film. The current image can be decomposed into two images with distributions of local built-in voltage and hole mobility by using semi-empirical equation, which describes local I-V dependences in SCLC regime. Local mobility values on surface of film caused by local structure differ in more than two times and have rms value $0.22*10^{-8}$ $m^2/(Vs)$, which is almost 20% from average mobility.

**Acknowledgements**

This work was supported by Federal Target Program of MES of Russian Federation, contract 14.575.21.0149 (RFMEFI57517X0149). The authors thank D. Evplov (Horiba) and S. Leesment (NT-MDT SI) for their help with software adjustment.

## References

[1] Liang Y, Xu Z, Xia J, Tsai S-T, Wu Y, Li G, Ray C and Yu L 2010 *Adv. Mater.* **22** E135
[2] He Z, Zhong C, Su S, Xu M, Wu H and Cao Y 2012 *Nat. Photon.* **6** 591-5
[3] Ouyang X, Peng R, Ai L, Zhang X and Ge Z 2015 *Nat. Photon.* **9** 520-4
[4] Hammond M R, Kline R J, Herzing A A, Richter L J, Germack D S, Ro H-W, Soles C L, Fischer D A, Xu T, Yu L, Toney M F and DeLongchamp D M 2011 *ACS Nano* **5** 8248-57
[5] Collins B A, Li Z, Tumbleston J R, Gann E, McNeill C R and Ade H 2013 *Adv. Energy Mater.* **3** 65-74
[6] Liu F, Zhao W, Tumbleston J R, Wang C, Gu Y, Wang D, Briseno A L, Ade H and Russell T P 2014 *Adv. Energy Mater.* **4** 1676
[7] Hedley G J, Ward A J, Alekseev A, Howells C T, Martins E R, Serrano L A, Cooke G, Ruseckas A and Samuel I D W 2013 *Nat. Commun.* **4** 2867
[8] Alekseev A, Hedley G J, Al-Afeef A, Ageev O A and Samuel I D W 2015 *J. Mater. Chem. A* **3** 8706
[9] Goh C, Kline R J, McGehee M D, Kadnikova E N and Fréchet J M J 2005 *Appl. Phys. Lett.* **86**

IOP Conf. Series: Materials Science and Engineering **699** (2019) 012001 doi:10.1088/1757-899X/699/1/012001

122110
[10] To C H, Ng A, Dong Q, Djurišić A B, Zapien J A, Chan W K and Surya C 2015 *ACS Appl. Mater. Interf.* **7** 13198
[11] Blakesley J C, Castro F A, Kylberg W, Dibb G F A, Arantes C, Valaski R, Cremona M, Kim J S, and Kim J-S 2014 *Organ. Electron.* **15** 1263
[12] Ebenhoch B, Thomson S A J, Genevicius K, Juška G and Samuel I D W 2015 *Organ. Electron.* **22** 62-8
[13] Alexeev A, Loos J and Koetse M M 2006 *Ultramicrosc.* **106** 191-9
[14] Dang X-D, Tamayo A B, Seo J, Hoven C V, Walker B and Nguyen T-Q 2010 *Adv. Funct. Mater.* **20** 3314
[15] Dang X-D, Mikhailovsky A and Nguyen T-Q 2010 *Appl. Phys. Lett.* **97** 113303
[16] Reid O G, Munechika K, and Ginger D S 2008 *Nano Lett.* **8** 1602
[17] Kim J-H, Noh J, Choi H, Lee J-Y and Kim T-S 2017 *Chem. Mater.* **29** 3954
[18] Malliaras G G, Salem J R, Brock P J and Scott C 1998 *Phys. Rev. B* **58** R13 411

IOP Conf. Series: Materials Science and Engineering **699** (2019) 012002 doi:10.1088/1757-899X/699/1/012002

# Resistive switching phenomena in thin ferroelectric films

**N V Andreeva[1], A Petraru[2], A E Petukhov[3] and A V Batueva[1]**

[1]St. Petersburg Electrotechnical University "LETI", Saint Petersburg, 197376, Russia
[2]Nanoelektronik, Technische Fakultät, Christian-Albrechts-Universität zu Kiel, D-24143 Kiel, Germany
[3]Research Park, St. Petersburg State University, Saint Petersburg, Russia

nvandr@gmail.com

**Abstract**. Switching of the resistance in thin ferroelectric films is a complex phenomenon involving an influence of polarization states, point defects, processes of electron injection from electrodes and peculiarity of polarization charges screening. In order to clarify the mechanism of surface charge screening under ferroelectric polarization switching and its correlation with the local resistive properties in thin ferroelectric films, we studied the changes in the surface potential under dc bias voltage application in $SrTiO_3/La_{0.7}Sr_{0.3}MnO_3/BaTiO_3$ heterostructures using different modes of atomic force microscopy (Kelvin probe, piezoresponse, and conductive) at ambient conditions. It is found that the peculiarity of surface charge screening under polarization reversal gives a strong electrostatic contribution in the measurements of local resistive properties at the ambient conditions and could completely veil the real current distribution over the film surface. Remarkably, scanning tunnelling microscopy measurements under ultra-high vacuum conditions show evidence of oxygen vacancy formation in thin ferroelectric films after poling, revealing an increase of the film thickness due to the chemical expansion.

## 1. Introduction

Point defects in perovskite thin films could significantly affect their structural, electronic, and ferroelectric (FE) properties. Oxygen vacancies (defects in anion sublattice) are the most intensively studied ones in relation to aging and fatigue phenomena in FE [1-3]. The most common models of fatigue involve the mechanisms of the domain wall motion impeding or local polarization disturbance due to the presence of oxygen vacancies. Several plausible studies demonstrate that the polarization driven oxygen vacancy migration plays a key role in resistive switching effects observed in thin FE films [4-8]. It is claimed that oxygen vacancy migration is the sole reason for the resistive switching in all oxide tunnel junctions, including FE ones [9]. Moreover, oxygen vacancies could enable ionic conductivity in perovskite-based solid solution for solid oxide fuel cells [10]. Related with the structural properties, the presence of oxygen vacancies could induce an increase of volume (chemical expansion) [11] in perovskites, giving a significant impact on resistive properties of FE tunnel junction. Besides this, oxygen vacancies together with injected charges and charges from ambience could be considered as a sources of screening charges of the FE polarization at the surface and could strongly influence both the surface potential and polarization reversal behaviour in thin films [12]. Thus, the functional properties of thin ferroelectric films are governed by the interplay of ferroelectric polarization, peculiarity of surface charge screening, and oxygen vacancies distribution. On this way, a detailed study of relative contribution of all above-mentioned factors to the resistive and FE properties of thin films opens a way for designing new materials with tailored functionality.

Toward this end, we study surface potential changes, dynamics of local FE properties, and oxygen vacancies migration accompanying the resistive switching effects in $SrTiO_3/La_{0.7}Sr_{0.3}MnO_3/BaTiO_3$ (STO/LSMO/BTO) heterostructures with thin FE films. In our previous studies, we experimentally observed that with increasing the thickness of epitaxial FE film in STO/LSMO/BTO structures the current through the structure associated with inelastic, trap-assisted tunnelling was slowly prevailing [13]. We attribute this result to the increase in the oxygen vacancies content in FE films with increasing their thickness. The shift from epitaxial state towards the polycrystalline structure is another way to increase the oxygen vacancy concentration in thin FE film due to the formation of crystallite boundaries. Thus, to elucidate the issues related to oxygen vacancies effect on resistive and ferroelectric properties, we opted for polycrystalline thin ferroelectric films. In order to analyse the nanoscale distribution of electric charges on the FE film surface, we combined Kelvin probe force microscopy (KPFM) and piezoresponse force microscopy (PFM) measurements in ambient conditions, with the resistive properties investigation by conductive atomic force microscopy (cAFM) in ambient conditions and scanning tunnelling microscopy (STM) under ultra-high vacuum (UHV).

## 2. Materials and methods

12-nm-thick polycrystalline BTO films were prepared on (001) LSMO/STO substrate by pulsed laser deposition (Fig. 1a,b). Gold top 10-15 nm thick electrodes were deposited on the BTO films by thermal evaporation sputtering. The electrodes were patterned using a stencil mask, which area ranged from 2 to 36 $\mu m^2$.

FE properties of fabricated heterostructures were measured by means of PFM, while for surface potential distribution noncontact KPFM regime was employed. Both KPFM and PFM measurements were performed under ambient condition using commercially available Pt-coated silicon tips and Veeco SPM instruments. During the PFM imaging, the bottom electrode was grounded, and an ac voltage was applied to the AFM tip. The poling procedure was done by applying dc voltage to the bottom electrode while keeping the grounded AFM tip in contact with the film surface.

In order to investigate the resistive properties of the fabricated heterostructures, I-V curves were measured using conductive AFM Pt-coated probes and tungsten STM tip by applying quasi-static voltage sweeps to the thick electrodes of the structure (in AFM regime) or bare film surface (in STM regime).

The tunnelling AFM and STM measurements were performed under UHV conditions at the resource centre “Physical Methods of Surface Investigation” (RC PMSI) of Research Park of Saint Petersburg State University. The pressure in the UHV chamber was below $3\times10^{-10}$ mbar. Before measurements, samples were heated up to 120° C and exposed for 24 hours in UHV to remove the adsorbed residual gas particles on a sample surface. The electrochemically etched STM tip was annealed in UHV to remove the oxide layer from the tip apex. The I-V curves were obtained in the tunnelling current range from 1 pA to 330 nA at room temperature using an Omicron VT AFM XA scanning probe microscope.

## 3. Results and discussion

The PFM image of the as-grown 12 nm thick BTO film (Fig. 1d-f) at ambient condition revealed the asymmetry of FE heterostructure, which may be partially attributed to the self-polarization effect [14]. The results of piezoresponse-voltage hysteresis loop measurements justify the existence of an internal bias field varying in the range of 0.1-0.4 V (Fig. 1c). The PFM images obtained after poling procedure showed that the spontaneous polarization had a preferred upward direction, as indicated by predominantly bright contrast of the area polarized with the positive (relatively to the BE) dc bias and areas outside of the polarized square on the film surface (Fig. 1e,f). In the absence of internal bias, positively charged oxygen vacancies should be compensated by negatively charged defects in the cation sublattice or electrons in accordance with the principle of electroneutrality [15]. The existence of internal bias should result in the internal charge redistribution in the film. For these reasons, we assume that in a pristine state of the BTO film the oxygen vacancies are mostly concentrated at the interface of the structure with the bottom electrode (Fig. 2d), combining negative screening charges and metallic charges of the bottom electrode. While at the film surface, the positive polarization charges are partially screened by chemisorbed water molecules in ambient conditions.

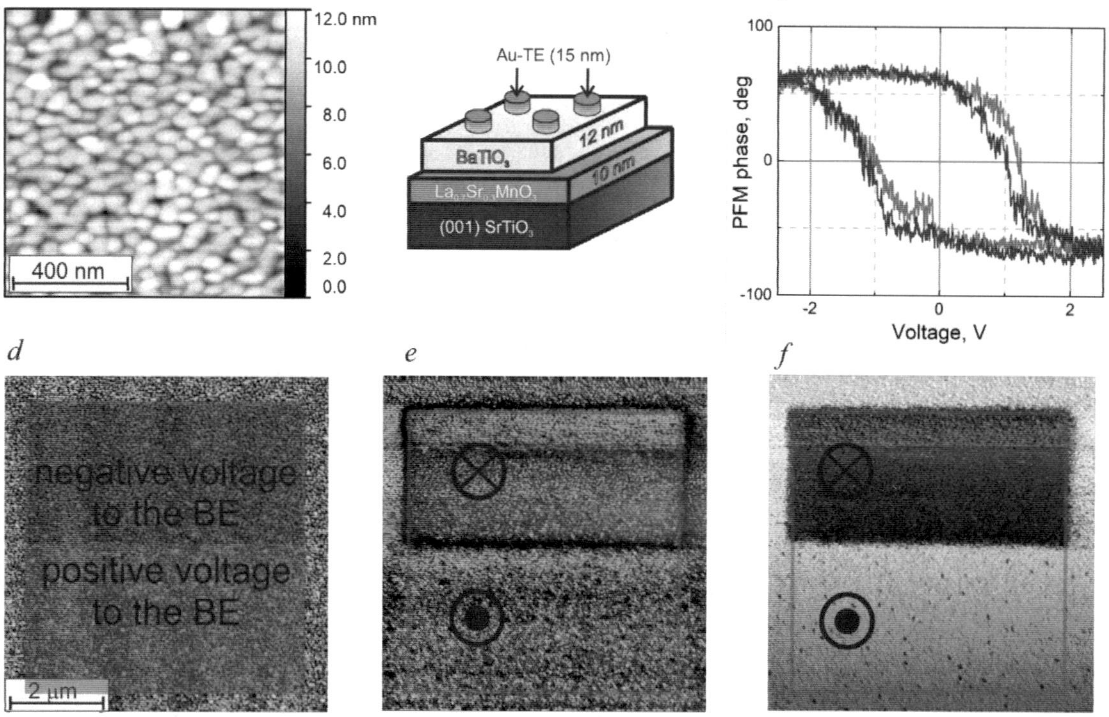


**Figure 1.** (a) AFM topography image of the 12 nm thick polycrystalline BTO film; (b) a sketch of a STO/LSMO/BTO structure; (c) local piezoresponse-voltage hysteresis loops of the 12 nm thick polycrystalline BTO film measured on bare film surface in spectroscopic PFM regime; (d) AFM topography image of BTO film, the area of a bias voltage application is indicated; (e) PFM amplitude and (f) PFM phase images of the poled area.

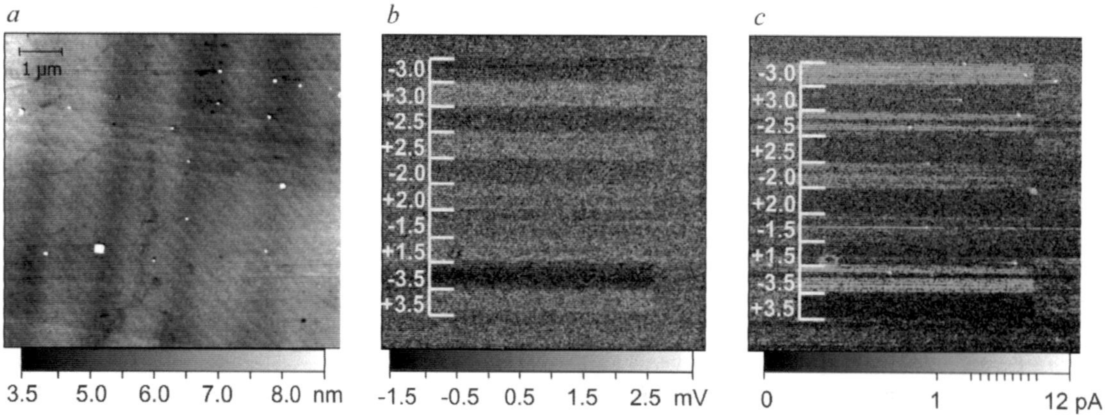


**Figure 2.** (a) AFM topography image for 3 nm thick epitaxial BTO film surface scanned with different applied biases from -3.5 to +3.5 V; (b) PFM signal and (c) current map of the same area (taken under 1.5 V applied between the tip and the bottom electrode).

During the poling procedure, the polarization reversal is accompanied by charge carrier injection in the FE films. Here, we consider a charge carrier injection mainly from the bottom electrode, as the barrier for this interface is significantly lower compared with the barrier "FE film - Pt-coated AFM tip" (Fig. 3f). The injected charged (electrons) and charges from the ambience influence the surface potential of the BTO films. Changes in surface potential could give a critical impact on the study of the local resistive properties of thin FE films by means of conductive AFM (for example, studies of a tunnel electroresistance of FE tunnel junctions). The long-range electrostatic contribution (due to the changes in the surface potential) to the AFM cantilever buckling [16] could induce changes in the image of local current distribution over the FE film surface measured with the conductive AFM probe and could be misunderstood as a change in resistive properties associated with FE polarization switching (Fig. 2).

In order to investigate the surface potential change during the polarization switching in our structures, we applied a series of dc voltage bias to the BTO film by scanning rectangular areas with different biases in the range of -6 - +6 V (indicated with the scale bar in Figure 4a). Subsequently, we acquired the potential distribution and piezoresponse from the same area on the film surface. The application of positive bias does not affect the polarization state (Fig. 4c) of the film, as evidenced from the PFM phase image, but lowers the surface potential (Fig. 4a). The appearance of the bright contrast on the PFM amplitude signal may be attributed to the electrostatic charge effect between the cantilever and the film surface due to changes in internal charges. We consider two possible sources of changes in internal charge concentration for this polarity of the applied bias: electron injection from the AFM tip and oxygen vacancies formation in the FE film. According to I-V curve measurements (Fig. 4f), the level of electron injection from the AFM tip is relatively low at the values of applied positive voltage below 6-7 V. Under these circumstances, we cannot eliminate the formation of oxygen vacancies at the values of positive

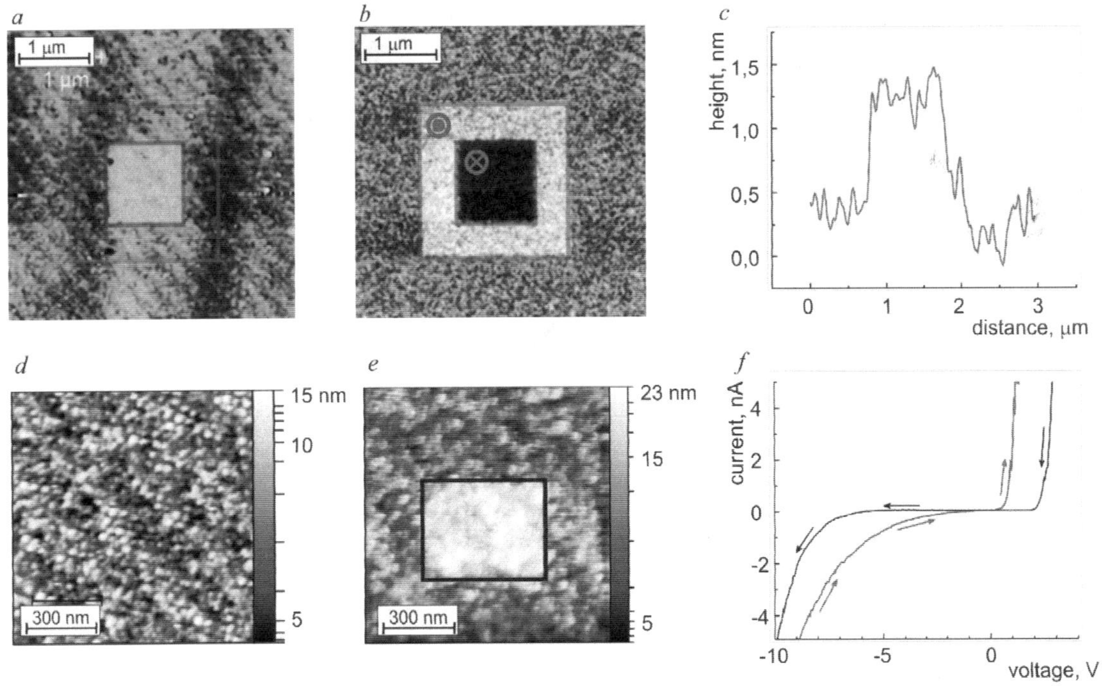


**Figure 3.** (a) AFM topography image of nanocrystalline 3 nm thick BTO film, the area of a bias voltage application is indicated; (b) PFM phase images of the poled area; (c) AFM topography line profile obtained from (a); (d) STM topography of nanocrystalline BTO film surface region (1x1 μm$^2$, 10 pA, 1.5 V); (e) STM topography of nanocrystalline BTO film surface region showing an expansion of the film volume after scanning with negative (relatively to the BE) voltage; (g) I-V characteristics of nanocrystalline BTO film (obtained from (d) measured with STM at T = 295 K under UHV conditions (voltage sweep is measured relatively to the STM tip).

voltages exceeding the coercive voltage of FE film. Thus, we relate the observed surface potential change mainly to extra charges from the ambience neutralizing the extra amount of oxygen vacancies in the FE film. The application of negative bias voltage to the film below the coercive voltage produces an increase in surface potential without disturbing both PFM amplitude and PFM phase signals. In this case, the low barrier of the structure with the bottom electrode (Fig. 4d) implies electron injection in the FE film. No change in corresponding PFM amplitude image suggests that at the given level of a charge carrier injection, charge compensation does not induce a significant electrostatic contribution to the AFM tip – sample interaction. The maximum change of the surface potential is observed from the negatively biased region (scanned with a voltage of -3 V) and coupled with a polarization reversal. Corresponding change in the electrostatic interaction, visualized in the PFM amplitude image, arose from the compensation of the negative polarization charges on the surface and electrons injected from the bottom electrode by positive charges from ambience. Eventually, as decreasing the negative bias to -6 V, the amplitude of the surface potential decreases gradually. As estimated from the I-V curves (Fig. 3f), starting from approximately -5 V there is a sharp increase in the current through the FE film, influencing the internal charge distribution and screening conditions of the polarization charges. Summarizing the experimental results, we assume that local surface potential reflects the changes in the surface charge screening conditions as response to complex interaction between electron injection, oxygen vacancy formation, and polarization reversal in thin FE films.

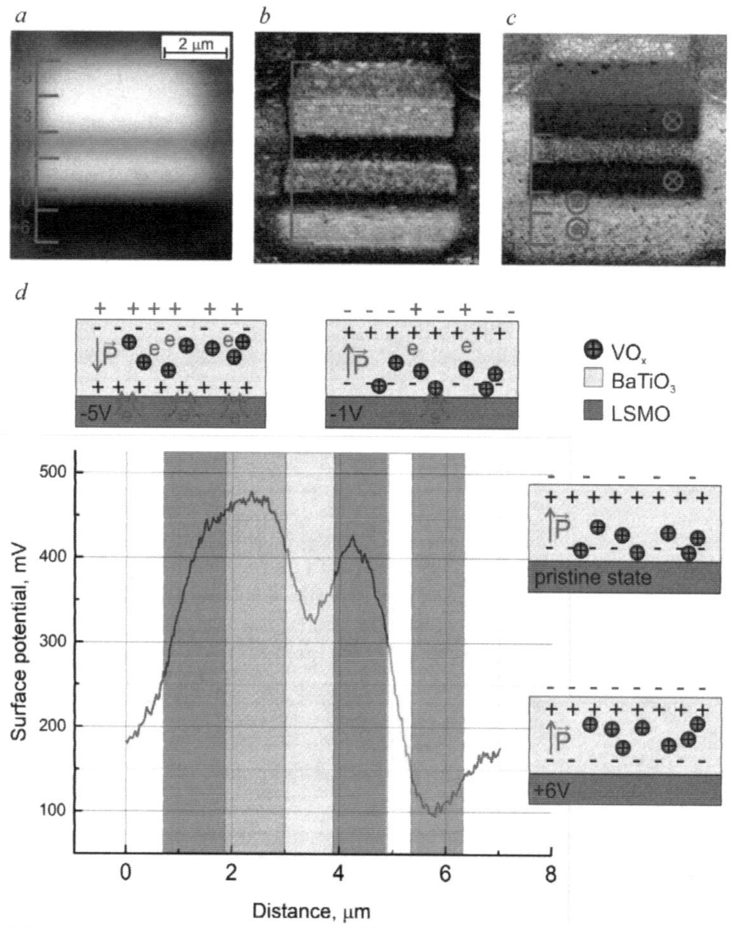


**Figure 4.** (a) KPFM surface potential distribution of the 12 nm thick polycrystalline BTO film surface scanned with different applied biases from -6 to +6 V; (b) PFM amplitude and (c) PFM phase images of the same area; (*d*) Surface potential line profile obtained from (a), insets show a scheme of an interplay between the polarization state, inner and surface charges distribution for different biases.

For further understanding of the role of oxygen vacancy formation in electrical properties of thin FE films, we study the correlation of the surface morphology with local electrical properties. Due to the fact that chemical expansion is accompanied by the presence of oxygen vacancies in perovskite-structure materials, an increase in the film volume as response to the voltage bias application could indicate oxygen vacancies formation in the film. To ascertain this effect, STO/LSMO/BTO structures with nanocrystalline BTO films with thicknesses in ranging from 3 to 7 nm (the description of the sample fabrication and characterization procedures can be found in [13]) were polarized by applying dc bias voltage applied between the conductive AFM tip and the bottom electrode of the structure at the ambient conditions. The corresponding polarization switching was visualized in the PFM mode. According to the results of PFM phase measurements (Fig. 3b), the positive voltage induces the upward polarization reversal, while the negative voltage – the downward one. Interestingly, the negatively biased regions exhibit an expansion of the film thickness (Fig. 3a). For 3 nm thick nanocrystalline BTO film, the increase in the thickness was equal to 1 nm (Fig. 3c). The same effect could be observed for the most nanocrystalline samples, when the applied voltage was comparable with the coercive voltage or exceeded it.

To eliminate the possible influence of screening charges from the ambience and electrostatic contribution to the "conductive AFM cantilever/FE film" contact on the results of the measurements, complementary studies were carried out under UHV condition in STM regime. Tunnel current as a control parameter for feedback loop in STM ensures a stable "STM probe/FE film" contact, in contrast with AFM feedback loop based on Van-der-Waals interactions. Moreover, STM has a better spatial resolution due to the fact that 90 % of tunnel current flows through the nearest to the sample atom of the STM tip. In case of dielectrics, the resistance of the tunnel barrier, between the film surface and the STM probe, is significantly lower than the resistance of the FE film, so, it could be considered as a series resistor between tip and the sample. Prior to the STM measurements, samples were placed in a UHV chamber for 24 hours to remove the chemisorbed water molecules. The results of STM topographical imaging of BTO film allow us to clearly distinguish nanocrystalline (Fig. 3d) surface structure with better spatial resolution as compared to the AFM measurements (Fig. 3a). Negatively biased regions of the BTO film surface demonstrate the vertical height increase while maintaining the lateral structure of the film (Fig. 3e). This behaviour could suggest that the crystal structure does not change at the volume expansion, being in accordance with the hypothesis about the formation of oxygen vacancies under bias voltage application accompanied by the chemical expansion in perovskite oxides. Unfortunately, the quantification of this effect is not simple and requires taking into account the dependence of the formation energy of oxygen vacancies on the potential landscape in thin FE film [11].

The resistive switching phenomena were observed both in ambient conditions and under UHV on the I-V curves measured on the STO/LSMO/BTO structures with polycrystalline BTO films. The existence of a pinched hysteresis effect is indicated on the experimental I-V curves (Fig. 3f). The change in slope of the I-V curves is associated with switching between different resistive states in BTO films. According to the results of our previous studies, we relate the appearance of the I-V hysteresis with the peculiarity of charge transport in the BTO films with thicknesses greater than the mean free pass of injected electrons [13]. In polycrystalline BTO films, we assume a contribution from inelastic, trap - assisted tunnelling in current through these structures (traps are presented by oxide defects). The questions on temporal stability of the observed resistive states and their changes after bias application go beyond a consideration of this paper. Here, we would just like to mention that pinched hysteresis I-V curves are observed for all studied nano- and polycrystalline BTO films.

## 4. Conclusions

In summary, we demonstrate that the local surface potential in thin FE films with resistive switching effects is influenced by the peculiarity of surface charge screening. Combined KPFM/PFM and PFM/cAFM measurements give evidence of strong impact of the long-range electrostatic interaction between AFM tip and FE surface on the results of local resistive property measurements by means of conductive AFM at the ambient conditions. Sometimes, the true current distribution over the FE film surface is completely veiled by electrostatics, driven by changes of surface charge screening conditions under polarization switching. Besides the chemisorbed water, electron injection in FE film together with oxygen vacancies formation could be considered as additional sources of screening charges, in ambient conditions.

The formation of oxygen vacancies in thin FE films under the bias voltage application is indirectly confirmed both by AFM measurements at the ambient conditions and the STM measurements under UHV. An increase in the thickness with retaining the lateral structure in the biased regions of the FE film is associated with a chemical expansion in perovskite oxide materials related to the presence of oxygen vacancies.

**Acknowledgements**

The research performed at the St. Petersburg Electrotechnical University "LETI" was funded by the grant of the Russian Ministry of Education and Science (Grant No. 16.2873.2017/4.6). The work carried out at the Christian-Albrechts-Universität zu Kiel was financially supported by the DAAD through their fellowship program under the contract number 8.9986.2017 (N.V. Andreeva). The authors are grateful to the resource centre "Physical Methods of Surface Investigation" of Research Park of Saint Petersburg State University for UHV SPM measurements performed at SPM module of the unique scientific facility "Nanolab".

**References**

[1] Li W, Ma J, Chen K, Su D and Zhu J S 2005 *Europhys. Lett.* **72** 131-6
[2] Erhart P, Träskelin P and Albe K 2013 *Phys. Rev. B* **88** 024107
[3] Ikegami S and Ueda I 1967 *J. Phys. Soc. Jpn.* **22** 725-34
[4] Kim Y-M, Morozovska A, Eliseev E, Oxley M P, Mishra R, Selbach S M, Grande T, Pantelides S T, Kalinin S V and Borisevich A Y 2014 *Nat. Mater.* **13** 1019-25
[5] Kim D J, Lu H, Ryu S, Bark C W, Eom C B, Tsymbal E Y and Gruverman A 2012 *Nano Lett.* **12** 5697-702
[6] Mao H J, Song C, Xiao L R, Gao S, Cui B, Peng J J, Li F and Pan F 2015 *Phys. Chem. Phys.* **17** 10146
[7] Li M, Zhou J, Jing X, Zeng M, Wu S, Gao J, Zhang Z, Gao X, Lu X, Liu J M and Alexe M 2015 *Adv. Electron. Mater.* **1** 1500069
[8] Hu W J and Wang Z, Yu W and Wu T 2016 *Nat. Commun.* **7** 10808
[9] Qin Q H, Äkäslompolo L, Tuomisto N, Yao L, Majumdar S, Vijayakumar J, Casiraghi A, Inkinen S, Chen B, Zugarramurdi A, Puska M and Dijken S V 2016 *Adv. Mater.* **28** 6852-9
[10] Jeen H, Choi W S, Biegalski M D, Folkman C M, Tung I-C, Fong D D, Freeland J W, Shin D, Ohta H, Chisholm M F and Lee H N 2013 *Nat. Mater.* **12** 1057-63
[11] Cazorla C 2017 *Phys. Rev. Appl.* **7** 044025
[12] Kim Y, Bae C, Ryu K, Ko H, Kim Y K, Hong S and Shin H 2009 *Appl. Phys. Lett.* **94** 032907
[13] Andreeva N, Petrov A, Petraru A, Petukhov A and Rybkin A 2019 *Mater. Res. Express* **6** 026427
[14] Kholkin A L, Brooks K G, Taylor D, Hiboux S and Setter N 1998 *Integr. Ferroelectr.* **22** 525-33
[15] Tuller H L and Bishop S R 2011 *Ann. Rev. Mater. Res.* **41** 369-98
[16] Hong S Nanoscale Phenomena in Ferroelectric Thin Films 2004 *Boston: Kluwer Academic Publishers*

IOP Conf. Series: Materials Science and Engineering **699** (2019) 012003 doi:10.1088/1757-899X/699/1/012003

# Phase composition investigation of titanium oxide nanostructures obtained by the local anodic oxidation

**V I Avilov, V A Smirnov, R V Tominov, N A Sharapov, N A Polupanov and O A Ageev**

Southern Federal University, Institute of Nanotechnologies, Electronics and Equipment Engineering, 344006 Taganrog, Russia

AvilovVI@sfedu.ru

**Abstract**. The paper presents the results of a phase composition study of titanium oxide nanostructures obtained by local anodic oxidation. A mathematical model was presented, which made it possible to calculate the distribution of oxide phases in the nanostructure volume. The results of the X-ray photoelectron spectroscopy (XPS) spectrum analysis are presented, which are consistent with theoretical calculations.

## 1. Introduction

Modern nanoelectronics development is closely related to the application of new methods for the formation of nanostructures [1-12]. An important task is the controlled formation of nanostructures with specified geometric and physical parameters. This problem is especially relevant for the formation of memristor structures in manufacturing RRAM resistive memory elements, since their characteristics are affected not only by the geometric dimensions, but also by the distribution of the phase composition in the oxide nanostructure (ONS) volume, which is the basis for the memristor manufacturing. One of the promising methods for the formation of forming-free memristor structures based on titanium oxide is local anodic oxidation, which allows the precise fabrication of ONS [13-23]. Nevertheless, existing models of the local anodic oxidation process can only predict the geometric parameters of the formed oxide nanoscale structures [24-26], while the oxide phase composition calculation is not presented in the literature. Therefore, this work aims at investigation of the phase composition distribution in the titanium ONS volume obtained by the local anodic oxidation nanolithography.

## 2. Theoretical model

ONS formation by the local anodic oxidation method is a multi-stage process including the oxygen ions generation in moist air due to the water molecules decomposition under the external electric field influence and oxygen ions mass transfer from the air to the oxide volume and to the metal-oxide boundary. Therefore, oxide volume increases during metal oxidation (Fig. 1). In addition, it is necessary to consider and calculate the metal oxide phase transitions rate by supplying an additional oxidizing agent to the oxide formed volume. For this, the Ti–O phase diagram was calculated, which showed that with an increase in the oxygen fraction, titanium sequentially oxidized to the TiO, $Ti_2O_3$, and $TiO_2$ phases (Fig. 2). Accordingly, when developing a mathematical model, it is necessary to consider the features of each of these processes separately and their influence on the process of ONS formation.

IOP Conf. Series: Materials Science and Engineering **699** (2019) 012003 doi:10.1088/1757-899X/699/1/012003

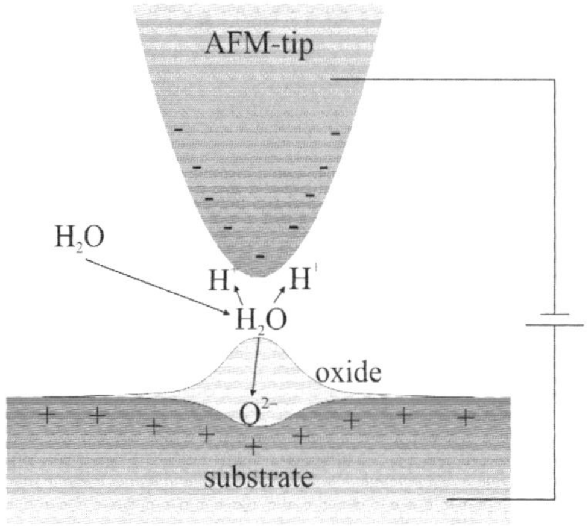


**Figure 1.** Oxygen ions generation and movement in the oxide volume.

Thus, to solve the problem, it is necessary to solve the system of differential equations:

$$\begin{cases} \nabla(\varepsilon\varepsilon_0\nabla\varphi) = -\rho(N), \\ \nabla(-\mu N\nabla\varphi + D\nabla N) = R(\varphi), \end{cases} \quad (1)$$

where φ and N are the electric potential and oxygen ions distribution in the oxide volume, εε0 is the dielectric constant of the medium, ρ(N) is the oxygen ions volume charge density, μ and D are the oxygen ions mobility and diffusion coefficient, and R(φ) is the rate of oxygen ions generation in the air.

A joint solution of the presented system of equations will allow obtaining the oxygen ions distribution in the oxide volume and the oxygen ions flow to the surface. In addition, thanks to the results obtained, it is possible to calculate the phase transition rate $Ti \rightarrow TiO \rightarrow Ti_2O_3 \rightarrow TiO_2$.

## 3. Experimental

To carry out experimental studies on a titanium film about 20 nm thick, formed by magnetron sputtering on a multifunctional Auto 500 installation (BOC Edwards, England), a titanium ONS with lateral dimensions of 300×300 μm$^2$ and a height of 4.5 nm was obtained using the probe nanolaboratory Ntegra

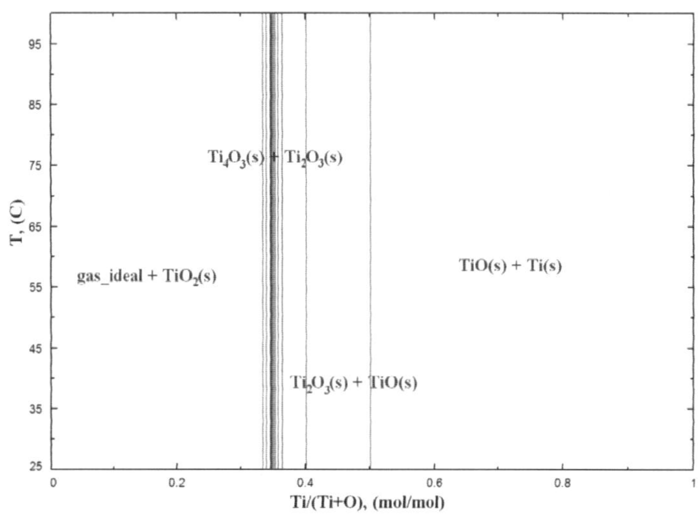


**Figure 2.** Phase diagram of Ti-O system.

(NT-MDT, Russia) by multiplication of 50×50 μm$^2$ area. The phase composition was studied by X-ray photoelectron spectroscopy (XPS) using a combined spectrometer for quantitative imaging and surface analysis of ESCALAB 250 (Thermo Scientific, USA) with monochromatizating of X-ray AlKα-line. For XPS analysis of the phase composition of the experimental sample in depth, ion profiling was used.

## 4. Results and discussion

Based on the developed model, the local anodic oxidation process of titanium film was numerically simulated using the MATLAB application package with the *Partial differential equation* module algorithms, which made it possible to solve nonlinear partial differential equations systems. For this, the electrochemical cell geometry was considered, including AFM probe, air medium, oxide, and metal film, for which a solution to a given differential equations system was found. Then, the distribution of the oxygen ions flow over the metal surface and the oxide growth rate were determined, and phase transitions in the oxide volume were calculated.

As a result, the distributions of the TiO, $Ti_2O_3$, and $TiO_2$ phases were obtained (Fig. 3), which showed that the TiO phase was formed near the oxide-metal interface, the $Ti_2O_3$ phase was in the oxide bulk, and the $TiO_2$ phase was present near the oxide surface. The results can be explained by the fact that in the local anodic oxidation process, oxygen ions diffuse through the oxide layer to the metal surface and, accordingly, the ion concentration decreases from the air to the metal. Thus, under oxygen deficiency conditions, the TiO phase is formed, and under increased oxygen concentration conditions, the oxide is transformed to $TiO_2$.

XPS analysis of the titanium oxide memristor nanostructure before and after ion etching showed the presence of Ti2p spectra of the core levels of elements (Fig. 4). The phase composition of the titanium ONS uppermost layer corresponds to titanium oxide $TiO_2$. Most likely, this degree of oxidation was obtained naturally under the influence of atmospheric oxygen. Deep profiling of ONS showed that the phase composition of Ti oxide contained three components. Energy of components 458.4, 456.6, and 454.8 eV corresponds to the intermediate oxidation states of $TiO_2$, $Ti_2O_3$, and TiO.

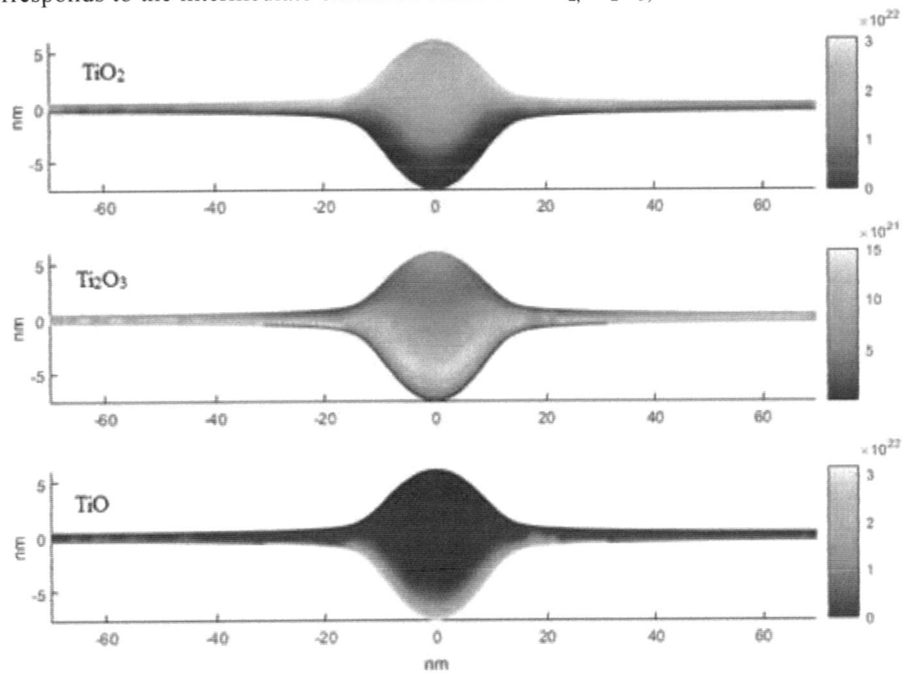


**Figure 3.** Titanium oxide phases concentration distribution in the O volume.

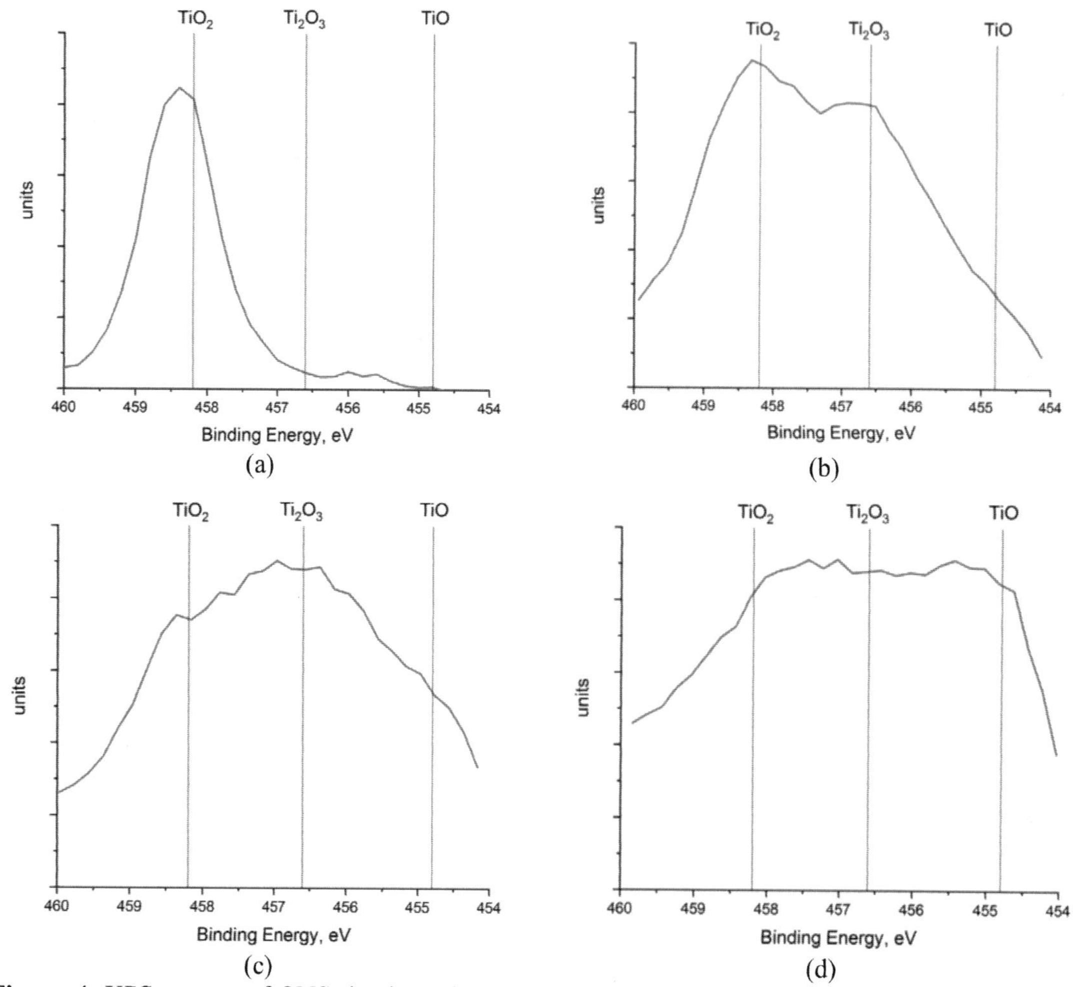


**Figure 4.** XPS spectra of ONS titanium obtained: (a) on the surface; (b) after etching time of 30 s; (c) after etching time of 60 s; (d) after etching time of 120 s.

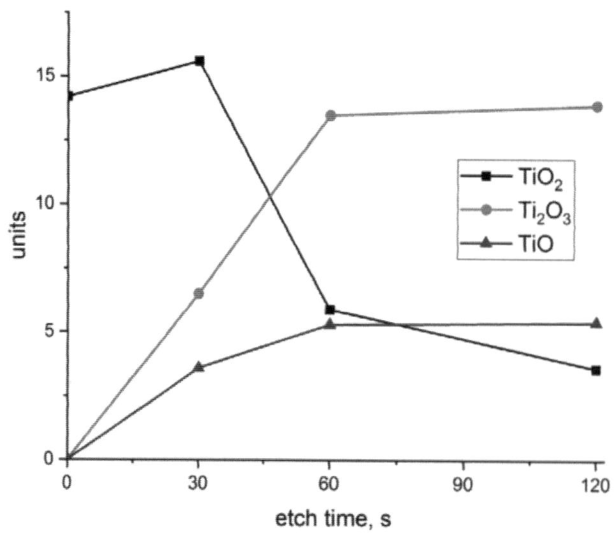


**Figure 5.** Ti oxide phases distribution content in depth.

From the data presented in Figure 4, it follows that with increasing depth of XPS analysis, the content of the $Ti_2O_3$ and TiO phases increases exponentially, while $TiO_2$ decreases according to the same exponential law (Fig. 5). So, with an etching time of 120 seconds, the fraction of the $TiO_2$ phase is 3.6 units, the $Ti_2O_3$ phase – 13.9 units, and TiO phases – 5.4 units.

The experimental results obtained are in good agreement with the data of mathematical modelling.

## 5. Conclusion

In this work, we developed a mathematical model that allowed simulating the formation of titanium oxide nanostructures obtained by local anodic oxidation, as well as calculating the distribution of titanium oxide phase composition. Theoretical results correlate with experimental data of the X-ray photoelectron spectroscopy spectrum of titanium oxide nanoscale structures obtained by the local anodic oxidation method.

The obtained results can be used in the development of technological processes for the fabrication of the RRAM element base based on titanium oxide nanostructures.

## Acknowledgements

This work was supported by RFBR according to the research projects № 19-29-03041 mk, № 18-37-00299 mol_a, and by Grant of the President of the Russian Federation No. MK-2721.2018.8.

## References

[1] Ha S D and Ramanathan S 2011 *J. Appl. Phys.* **110** 071101
[2] Chang T C, Chang K C, Tsai T M, Chu T J and Simon M S 2016 *Materials Today* **19** 254-64
[3] Ye C, Wu J, He G, Zhang J, Deng T, He P, and Wang H, 2016 *J. Mater. Sci. Technol.* **32** 1-11
[4] Zhu L, Zhou J, Guo Z and Sun Z 2015 *J. Mater.* **1** 285-95
[5] Omura Y and Kondo Y 2013 *J. Appl. Phys.* **114** 043712
[6] Nikonov D E and Young I A 2013 *Proc. IEEE* **101** 2498-533
[7] Strobel J, Hansen M, Dirkmann S, Neelisetty K K, Ziegler M, Haberfehlner G, Popescu R, Kothleitner G, Chakravadhanula V S K, Kübel C, Kohlstedt H, Mussenbrock T and Kienle L 2017 *J. Appl. Phys.* **121** 245307
[8] Klimin V S, Solodovnik M S, Lisitsyn S A, Rezvan A A and Balakirev S V *2018 J. Phys.: Conf. Ser.* **1124** 041024
[9] Ageev A O, Konoplev B G, Polyakov V V, Svetlichnyi A M and Smirnov V A 2007 *Russian Microelectronics* **36** 353
[10] Ageev O A, Smirnov V A, Solodovnik M S, Rukomoikin A V and Avilov V I 2012 *Semiconductors* **46** 1616-21
[11] Bousoulas P, Asenov P, Karageorgiou I, Sakellaropoulos D, Stathopoulos S and Tsoukalas D, 2016 *J. Appl. Phys.* **120** 154501
[12] Gusev E Yu and Avdeev S P 2018 *J. Phys.: Conf. Ser.* **1124** 022026
[13] Avilov V I, Ageev O A, Konoplev B G, Smirnov V A, Solodovnik M S and Tsukanova O G 2016 *Semiconductors* **50** 601-6
[14] Avilov V I, Kolomiytsev A S, Tominov R V, Alyabyeva N I and Bykova E M 2018 *J. Phys.: Conf. Ser.* **1124** 022019
[15] Bersuker G, Gilmer D C, Veksler D, Kirsch P, Vandelli L, Padovani A, Larcher L, McKenna K, Shluger A, Iglesias V, Porti M and Nafría M 2011 *J. Appl. Phys.* **110** 124518
[16] Do Y H, Kwak J S, Bae Y C, Jung K, Im H and Hong J P 2009 *Appl. Phys. Lett.* **95** 093507
[17] Szot K, Rogala M, Speier W, Klusek Z, Besmehn A and Waser R 2011 *Nanotechnology* **22** 254001
[18] Jeong D S, Schroeder H and Waser R 2009 *Phys. Rev. B - Condens. Matter Mater. Phys.* **79** 195317
[19] Pan F, Gao S, Chen C and Song C 2014 *Mater. Sci. Eng. R* **83** 1-59
[20] Avilov V I, Smirnov V A, Fedotov A A, Tominov R V, Sharapov N A and Polupanov N A 2018

*IOP Conf. Ser.: Mater. Sci. Eng.* **443** 012004

[21] Avilov V I, Polupanov N V, Tominov R V, Smirnov V A and Ageev O A 2017 *IOP Conf. Ser.: Mater. Sci. Eng.* **256** 012001

[22] Tsai J T H, Hsu C Y and Hsu C H 2015 *NANO* **10** 8

[23] Batko I and Batkova M 2016 *Materials Today: Proc.* **3** 803-9

[24] Avilov V I, Ageev O A, Blinov YF et al 2015 *Technical Physics* **60** 717-23

[25] Karen'kih O G, Avilov V I, Smirnov V A, Fedotov A A, Sharapov N A and Polupanov N A 2018 *IOP Conf. Ser.: Mater. Sci. Eng.* **443** 012013

[26] Orians A, Clemons C B, Golovaty D and Young G W 2006 *Surf. Sci.* **600** 3297-331

# Atomic force microscopy of titanium oxide nanostructures with forming-free resistive switching

**V I Avilov, V A Smirnov, R V Tominov, N A Sharapov, A A Avakyan, V V Polyakova and O A Ageev**

Southern Federal University, Institute of Nanotechnologies, Electronics and Equipment Engineering, 344006 Taganrog, Russia

AvilovVI@sfedu.ru

**Abstract**. The paper presents the results of study of the resistive switching effect in titanium oxide nanostructures obtained by local anodic oxidation. It was shown that the resulting structures exhibited a forming-free resistive switching effect. Analysis of the current-time characteristics made it possible to obtain a resistance ratio in the HRS and LRS states of about 70. The formed LRS regions persisted for 75 days.

## 1. Introduction

Modern computer technology based on standard microelectronics elements and methods of their formation has approached its technological limit and cannot be improved further [1-6]. This is because, firstly, the classical lithography methods approached their fundamental limitations, and, secondly, further microelectronics components miniaturization will lead to characteristics degradation due to the stray capacitance appearance, the breakdown voltages reduction, and the leakage currents appearance. Thus, an increase in the productivity and speed of modern electronics is associated with the development and investigation of perspective elemental base and nanolithography methods for formation of nanostructures [7-14]. One of the promising methods is the local anodic oxidation nanolithography, which allows preforming the oxide nanostructures precision formation with a thickness less than 5 nm. It can be used to form memristor structures that can switch between high (HRS) and (LRS) low resistance states [15-26]. The advantages of this method include obtaining non-uniform oxide phases of stoichiometric composition in the volume of titanium oxide nanostructures, which allows obtaining a forming-free memristor effect. To obtain such a result by other methods, it is necessary either to deposit sequential oxide layers with different compositions, or to carry out an additional annealing of memristor structures. Nevertheless, the issue of obtaining memristor structures based on titanium oxide is poorly studied in the literature. Therefore, for introducing into production, it is necessary to study the resistive switching patterns of memristor structures based on titanium oxide obtained by local anodic oxidation.

## 2. Experiment

Experimental studies were carried out on a 20 nm thick titanium film, obtained by magnetron sputtering on a dielectric $SiO_2$ substrate. A local anodic surface oxidation was preliminarily performed using a Solver P47 Pro scanning probe microscope. A titanium oxide nanostructure with lateral dimensions of 4×4 $\mu m^2$ and a thickness of 4.8 nm was formed (Fig. 1). Then, its current-voltage characteristics were measured, which showed that the obtained structure exhibited a memristor effect and switched between HRS and LRS (Fig. 2).

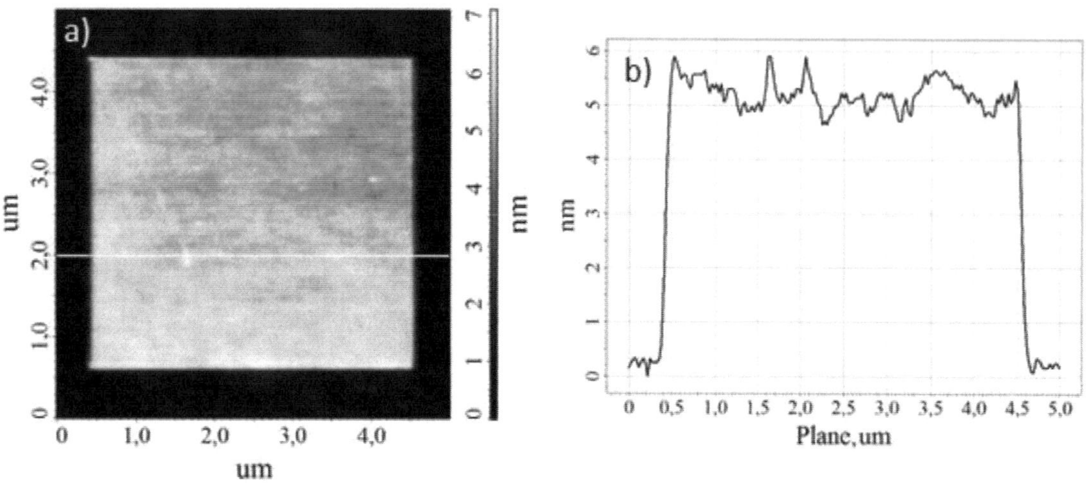


**Figure 1.** ONS formation results: (a) AFM-image; (b) profilogram along the line.

To study the current-time characteristics, a voltage pulses series with an amplitude of ± 1.5 V and a duration of 100 ms were applied to the applied structure.

Then, the duration of information storage in memristor structures was studied. For this, an oxide nanostructure with a lateral size of 5×5 μm$^2$ was formed in a similar way on the surface of such a structure using current SPM lithography by applying voltage pulses of -6 V to form LRS regions. Then, when applying voltage pulses of 1.5 V using the current AFM method, phase contrast AFM images of the formed regions were obtained 1 minute and 75 days after formation.

## 3. Results and discussion

Current-time characteristics analysis (Fig. 2) showed that in the initial state, upon application of positive voltage pulses, the current through the memristor structure was about 4.6 pA, which corresponded to the HRS state. When negative voltage pulses were applied, the current through the structure was of the order of 0.3 nA, which corresponds to the LRS state. The results obtained correspond to a memristor effect based on modulation of the width of potential barrier at the metal/oxide interface. The resistance

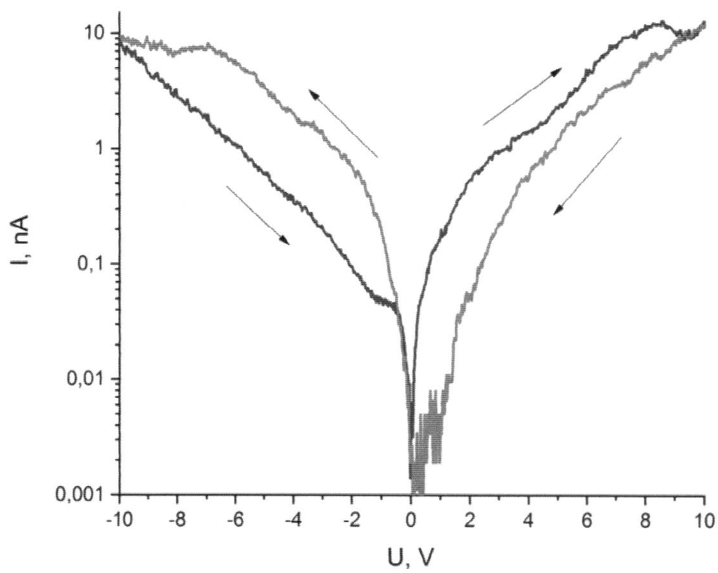


**Figure 2.** Titanium oxide nanostructure current-voltage characteristics.

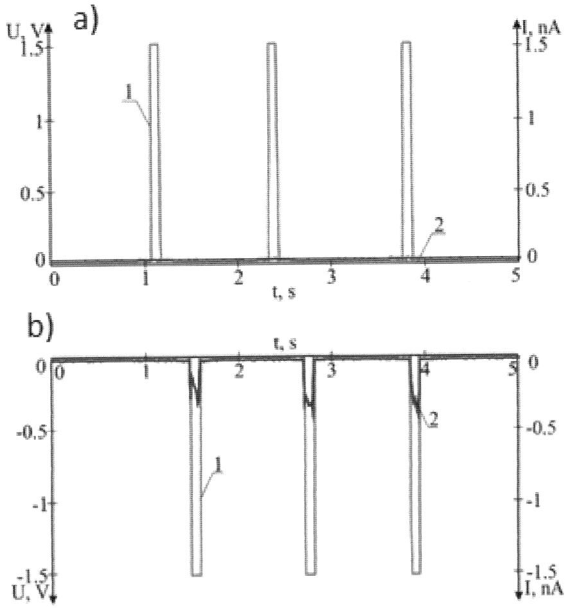


**Figure 3.** Current-time characteristics of a titanium oxide nanostructure obtained by the local anodic oxidation at different polarity of voltage pulses: (a) positive; (b) negative (1 – applied voltage impulses, 2 – corresponding current impulses).

ratio in the HRS and LRS was about 70. Then, the duration of information storage was studied. By the current AFM method, it was shown that as a result of current lithography, LRS regions were formed on the titanium oxide nanostructure surface (Fig. 3a). Repeated scanning of the structure after 75 days showed the presence of these areas (Fig. 3b).

## 4. Conclusion

It was shown that the local anodic oxidation method allowed forming oxide nanoscale titanium structures exhibiting a forming-free resistive switching effect. Such structures switched between the HRS and LRS states; the resistance ratio in the HRS and LRS states was about 70. The LRS regions formed by current lithography were preserved for a long time.

The obtained results can be used in the development of technological processes for manufacturing the elemental base of RRAM based on oxide nanosized titanium structures using probe nanotechnology.

## Acknowledgements

This work was supported by RFBR according to the research projects № 19-29-03041 mk, № 18-37-00299 mol_a, and by Grant of the President of the Russian Federation No. MK-2721.2018.8.

## References

[1] Kiazadeh A, Gomes H L, Rosa da Costa A M, Moreira J A, de Leeuw D M and Meskers S C J 2011 *Mater. Sci. Eng. B* **176** 1552-5
[2] Ha S D and Ramanathan S 2011 *J. Appl. Phys.* **110** 071101
[3] Hamdioui S, Taouil M, Du Nguyen H A, Haron A, Xie L and Bertels K 2016 *Int. Conf. Memristive Syst. MEMRISYS 2015* 9
[4] Haensch W, Nowak E J, Dennard R H, Solomon P M, Bryant A, Dokumaci O H, Kumar A, Wang X, Johnson J B and Fischetti M V 2006 *IBM J. Res. Dev.* **50** 339-61
[5] Nikonov D E and Young I A 2013 *Proc. IEEE* **101** 2498
[6] Shi L, Yi K, Ramanathan K, Zhao R, Ning N, Ding D and Chong T 2011 *Appl. Phys. A* **102** 865

[7] Ye C, Wu J, He G, Zhang J, Deng T, He P, and Wang H 2016 *J. Mater. Sci. Technol.* **32** 1-11
[8] Tominov R V, Zamburg E G, Khakhulin D A, Klimin V S, Smirnov V A, Chu Y H and Ageev O A 2017 *J. Phys.: Conf. Ser.* **917** 032023
[9] Tominov R V, Bespoludin V V, Klimin V S, Smirnov V A and Ageev O A 2017 *IOP Conf. Ser.: Mater. Sci. Eng.* **256** 012023
[10] Szot K, Rogala M, Speier W, Klusek Z, Besmehn A and Waser R 2011 *Nanotechnology* **22** 254001
[11] Ageev O A, Konoplev B G, Rubashkina M V, Rukomoikin A V, Smirnov V A and Solodovnik M S 2013 *Nanotechnologies in Russia* **8** 23
[12] Tominov R V, Smirnov V A, Avilov V I, Fedotov A A, Klimin V S and Chernenko N E 2018 *IOP Conf. Ser.: Mater. Sci. Eng.* **443** 012036
[13] Zhu L, Zhou J, Guo Z and Sun Z 2015 *J. Mater.* **1** 285-95
[14] Gusev E Yu, Jityaeva J Y and Bykov Al V 2017 *J. Phys.: Conf. Ser.* **929** 012055
[15] Kots I N, Klimin V S, Polyakova V V, Rezvan A A and Vakulov Z E *2018 J. Phys.: Conf. Ser.* **1124** 081035
[16] Solodovnik M S, Balakirev S V, Eremenko M M, Mikhaylin I A, Avilov V I, Lisitsyn S A and Ageev O A 2017 *J. Phys.: Conf. Ser.* **917** 032037
[17] Avilov V I, Ageev O A, Jityaev I L, Kolomiytsev A S and Smirnov V A 2016 *Proc.SPIE – Int. Soc. Opt. Eng.* **10224** 102240T
[18] Klimin V S, Solodovnik M S, Lisitsyn S A, Rezvan A A and Balakirev S V *2018 J. Phys.: Conf. Ser.* **1124** 041024
[19] Avilov V I, Ageev O A, Konoplev B G, Smirnov V A, Solodovnik M S and Tsukanova O G 2016 *Semiconductors* **50** 601-6
[20] Avilov V I, Smirnov V A, Fedotov A A, Tominov R V, Sharapov N A and Polupanov N A 2018 *IOP Conf. Ser.: Mat. Sci. Eng.* **443** 012004
[21] Chen A 2016 *Solid-State Electronics* **125** 25-38
[22] Bersuker G, Gilmer D C, Veksler D, Kirsch P, Vandelli L, Padovani A, Larcher L, McKenna K, Shluger A, Iglesias V, Porti M and Nafría M 2011 *J. Appl. Phys.* **110** 124518
[23] Strobel J, Hansen M, Dirkmann S, Neelisetty K K, Ziegler M, Haberfehlner G, Popescu R, Kothleitner G, Chakravadhanula V S K, Kübel C, Kohlstedt H, Mussenbrock T and Kienle L 2017 *J. Appl. Phys.* **121** 245307
[24] Omura Y and Kondo Y 2013 *J. Appl. Phys.* **114** 043712
[25] Jeong D S, Schroeder H and Waser R 2009 *Phys. Rev. B - Condens. Matter Mater. Phys.* **79** 195317
[26] Bousoulas P, Asenov P, Karageorgiou I, Sakellaropoulos D, Stathopoulos S and Tsoukalas D 2016 *J. Appl. Phys.* **120** 154501

IOP Conf. Series: Materials Science and Engineering **699** (2019) 012005 doi:10.1088/1757-899X/699/1/012005

# Microscopy investigation of conical and layered nanowires

**D A Cherkasov[1], D V Panov[1], I M Doludenko[1], V M Kanevskiy[1], A E Muslimov[1], D L Zagorskiy[1,2], D A Biziaev[3] and A A Bukharaev[3]**

[1] FSRC "Crystallography and Photonics" of RAS, 119333, Moscow, Russia
[2] Gubkin University of Oil and Gas, 119296, Moscow, Russia
[3] Kazan Institute of Physics and Technology named after E.K. Zavoisky, Federal Russia Research Centre "Kazan Research Centre of the Russian Academy of Sciences", 420008, Kazan, Russia

dzagorskiy@gmail.com

**Abstract.** This article discusses the use of scanning microscopy in the study of one-dimensional nanostructures – nanowires. The nanowires were produced by matrix synthesis. Atomic force microscopy (AFM) of conical nanowires was carried out, and layered nanowires were investigated by AFM and magnetic force microscopy (MFM) methods.

## 1. Introduction
Matrix synthesis is one of the promising methods for producing one-dimensional structures – nanowires (NWs). Track etched membrane (TM) is one of the most popular matrices [1, 2]. This method includes many advantages – the ability to adjust many parameters of both the matrix itself and the deposition conditions. Under certain conditions of etching of the TM, cone-shaped pores can be achieved [3]. At the certain electrolyte and deposition conditions it is possible to obtain a layered structure [4]. This work is devoted to the study of various structures using the methods of electron and probe microscopies.

This paper is devoted to production of cone-shaped nanostructures and their microscopy study, in particular, obtaining a nanocone with a small radius of curvature. Another part of the work is devoted to the production and study of layered NWs. Such NWs consist of alternating layers of magnetic and non-magnetic materials. Magnetic force microscopy (MFM) with application of an external magnetic field was used.

## 2. Experiment
Typically, TM is produced via etching of irradiated polyethylene terephthalate (PET) film in alkali. In normal mode, the rate of etching of the latent track is much higher than that of the undamaged polymer. As a result, a pore channel of almost cylindrical shape is formed during etching. However, when ethyl alcohol is added to the etcher, the etching rate of the latent track decreases, while that of the intact polymer increases. This makes it possible to obtain a cone-shaped cavity with a certain angle at the top by regulating etching parameters, such as the concentration of ethyl alcohol, etching time, temperature, etc. In this paper, to obtain conical-shaped pores, the irradiated film was etched in a two-normal solution of NaOH alkali with the addition of ethyl alcohol. The studied sample represented irradiated film of PET with the density of the pores $4*10^7$ $cm^{-2}$. Three concentrations of ethyl alcohol were used – 25%, 50%, 75%. In addition, etching was carried out at different temperatures – from 10°C to 60°C and durations – from 5 to 60 minutes. The films were etched on both sides. Based on the results of studies

by scanning electron microscopy (SEM) and scanning probe microscopy (SPM), the conditions for producing templates of a certain conical shape were determined.

For a contact layer creation, the surface of the matrices was metallized with a thin layer (about 30 nm) of copper using the method of thermal spraying. For the subsequent production of nanostructures-replicas of pore channels, the method of galvanic deposition of the required metal was used. Thus, the electrolyte of the following composition was used: $CuSO_4*5H_2O$ 200 g/l, $H_2SO_4$ 16 g/l. Moreover, in order to improve the quality of replicas, a brilliant additive CCN-74 in the amount of 5 g/l was added to this composition.

The prepared matrix with a pore diameter of 100 nm and a surface density of $1.1*10^8$ $cm^{-2}$ was used to obtain layered NWs. A single-bath method was chosen to obtain these structures. The following electrolyte composition was used: $MgSO_4{\bullet}7H_2O$ - 197 g/l; $CuSO_4{\bullet}5H_2O$ - 6.25 g/l; $H_3BO_3$ - 31.6 g/l. The deposition was carried out in a special galvanic cell, using the Ellins potentiostat as a source, which allowed to work both in the constant voltage mode and in the "pulse voltage" mode. The mode was selected in such a way that the layers were 100 and 250 nm thick. After the growth, NWs were released from the matrix, and then placed on a flat surface. The samples were covered with a thin layer of copper to prevent oxidation. MFM measurements were carried out using microscope Solver P47-PRO with Multi 75MG probes (BudgetSensor). During the experiment, an external magnetic field directed along the NW with an intensity of 16 mT was applied.

AFM measurements of conical NWs were carried out using microscope Solver P47. Electron microscopy of both conical and layered NWs was performed using a scanning electron microscope JCM-6000plus.

## 3. Results

### *3.1. Conical NWs*

Analysis of the results allows us to conclude that the angle at the top of the cone depends on the conditions for obtaining a conical pore. So, to increase the angle, you should increase the concentration of alcohol in the etching solution. The same effect is achieved by reducing the temperature of the etching process. Examples of cone structures with "large" and "small" angles at the top are shown in Figure . One of the problems of obtaining "sharp" cones is "smoothing" the angle at the top. Obviously, this is due to insufficient electrolyte filling of the top of the pores due to poor wettability. SEM and AFM images of replicas of cone-shaped pores are presented below (Figure ).

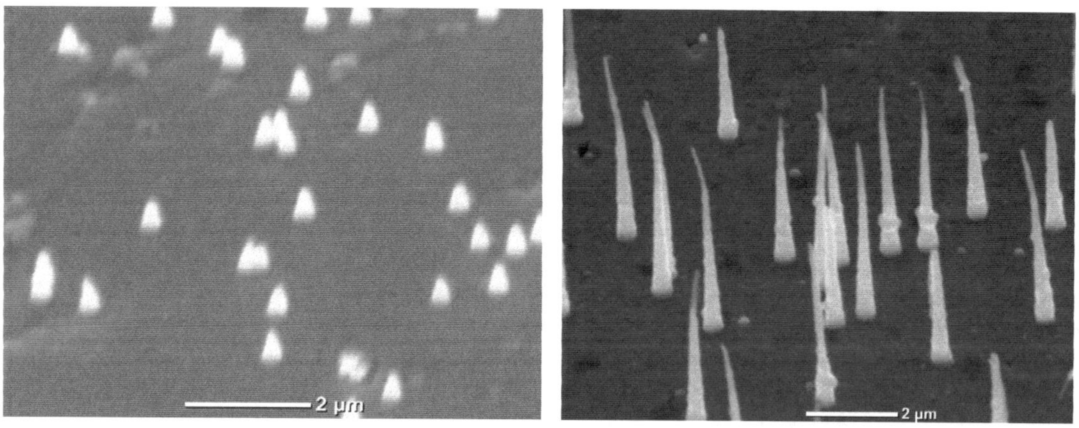


(a) (b)

**Figure 1.** SEM images of replicas of cone-shaped pores obtained in different conditions: (a) 75% concentration of alcohol, time of etching – 20 min, 20°C; (b) 50% concentration of alcohol, time of etching – 10 min, 40°C.

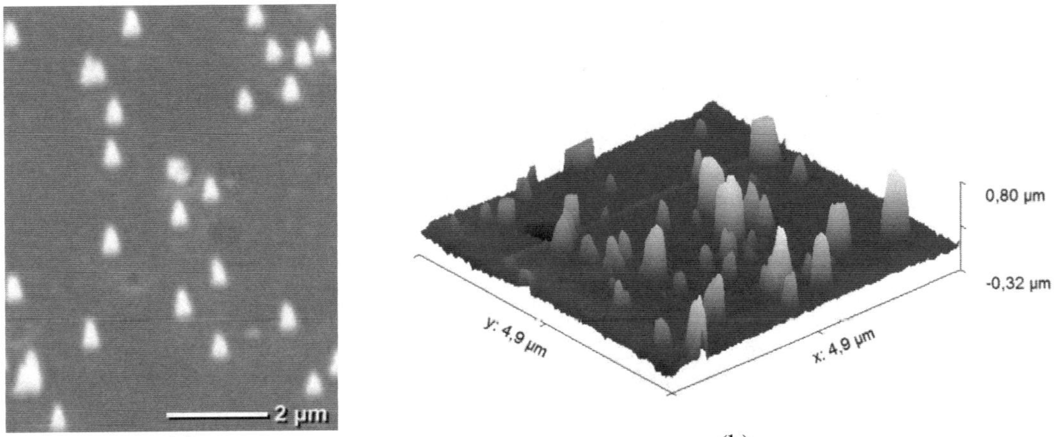


(a) (b)

**Figure 2.** (a) SEM and (b) AFM of conical nanostructures.

There can be seen a good match between SEM and AFM images. However, there is the rounded, smoothed angle at the top. To overcome this effect, several wetting agents have been tested. The best results were demonstrated by so-called bishopatallah supplements CCN-74.

It is shown that the use of this additive improves the adhesion of the electrolyte to the pore walls, so that NWs have a more elongated shape: the structures shown in Figure b are obtained using CCN-74.

### *3.2. Layered NWs*

A preliminary study was conducted using SEM (Fig. 3). It is shown that NWs completely fill the pores of the matrix, and their diameter corresponds to the pore diameter. TEM microscopy (previously performed in [5] for Ni/Cu nanowires) made it possible to determine the thickness of alternating layers (from 30 to 250 Nm). Elemental analysis showed that alternating layers consisted of pure copper and Nickel-copper alloy (up to 20% copper). To prepare the samples, an aqueous suspension of NWs was applied to the holder by the "irrigation" method with subsequent drying and then covered with a thin layer of metal (by thermal spraying). For MFM measurements, the one-pass technique was used in order to avoid the influence of neighbouring NWs on each other.

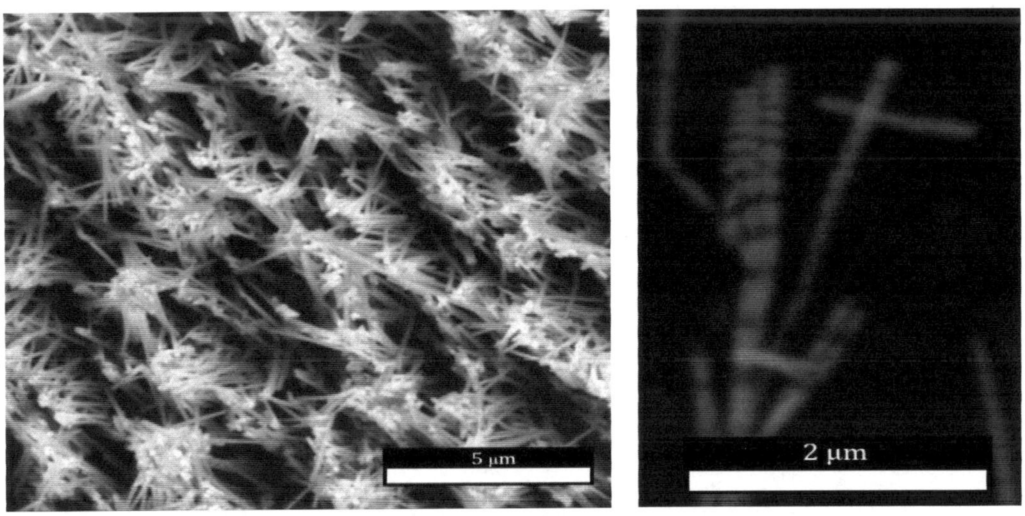


(a) (b)

**Figure 3.** (a) SEM and (b) TEM of layered NWs. Layers of Cu and Ni.

IOP Conf. Series: Materials Science and Engineering **699** (2019) 012005 doi:10.1088/1757-899X/699/1/012005

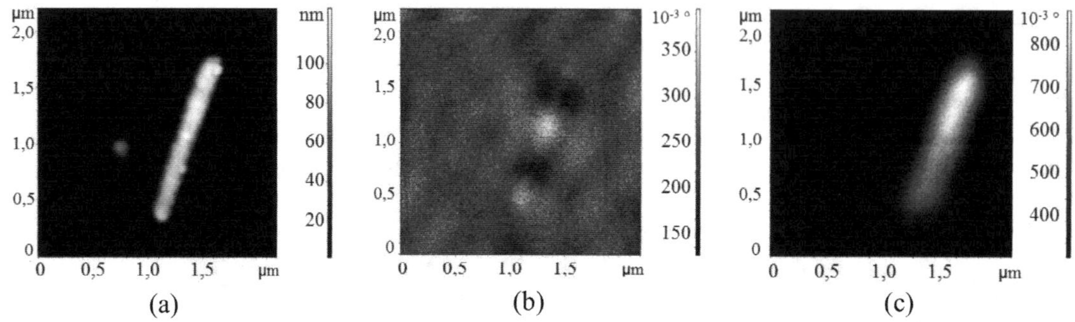


**Figure 4.** (a) AFM and (b, c) MFM images of layered Co/Cu nanowire. (b) without external magnetic field, (c) with external magnetic field of 16 mT. Period of layers is 500 nm.

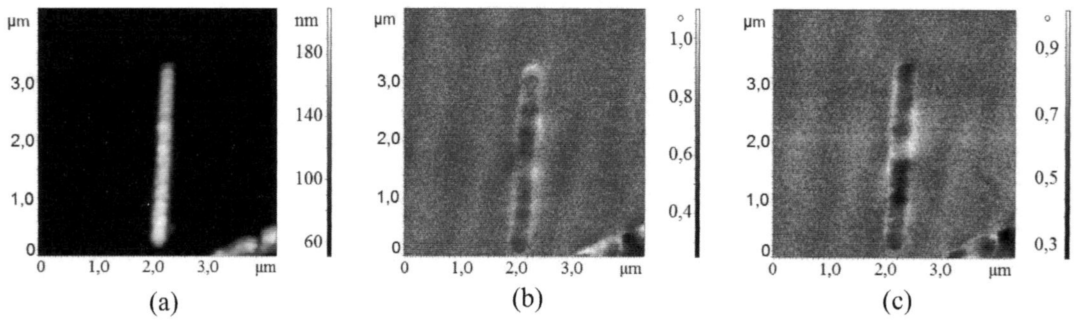


**Figure 5.** (a) AFM and (b, c) MFM images of layered Ni/Cu nanowire. (b) without external magnetic field, (c) with external magnetic field of 16 mT. Period of layers is 200 nm.

Figure 4a shows the AFM images of the topography of a single layer Co/Cu NW, Figure 4b shows the MFM image of the same NW, and Figure 4c shows the MFM image after an external magnetic field 16 mT was applied along the NW axis.

It is clearly seen that the NW has a layered structure with a period of about 500 nm, which corresponds to expectations, and it is also seen that the application of the magnetic field along the NW axis leads to the magnetization reversal of the nanowires.

Figure 5 represents the same AFM and MFM study for layered Ni/Cu NW. Period of layers is 200 nm. It can be seen also that external magnetic field affects magnetic state, but layers are not clearly visible.

## 4. Conclusion

In this work, arrays of conical and layered nanowires were obtained and the study by SEM and AFM was carried out.

It is shown that an increase in the concentration of alcohol and/or a decrease in temperature leads to an increase in the angle at the top of the cone.

An effect of smoothing peaks because of insufficient electrolyte wetting the top of the pores is revealed. Using brilliant additive CCN-74 reduces this effect.

It is shown that by alternating the growth potential in a two-component electrolyte, it is possible to obtain NWs with alternating layers. The correlation of topography and magnetic structure of layered NW is shown by AFM and MFM methods.

A method for fixing NW on a substrate is proposed and it is shown that the application of a small external magnetic field (16 mT) leads to the magnetization reversal of a single NW.

**Acknowledgements**

This work was carried out as the State Task of FNIC "Crystallography and Photonics" of RAS, and partially supported by Grant RFBR 16-29-11763 ofr-m (NWs synthesis). MFM investigation was done within the framework of state assignment of Zavoisky Physical-Technical Institute, FRC "Kazan Scientific Centre of RAS". Authors thank Prof. P. Yu. Apel (JINR, Dubna) for presenting the membranes.

**References**

[1] Martin C 1994 *Science* **266** 1961-6
[2] Lupu N (editor) 2010 Electrodeposited NWs and Their Applications *InTech* Rijeka Croatia
[3] Apel P, Ramirez P, Blonskaya I, Orelovitch O and Sartowska B 2014 Accurate characterization of single track-etched, conical nanopores *Phys. Chem. Chem. Phys.* **16** 15214-23
[4] Zhigalina O, Khmelenin D, Ivanov I, Zagorskiy D, Bedin S and Doludenko I 2018 Structure of Cu/Ni nanowires obtained by matrix synthesis *Cryst. Reports* **63** 480-4
[5] Zhigalina O, Khmelenin D, Ivanov I, Zagorskiy D, Bedin S and Doludenko I 2018 *Nanomaterials and nanostructures XXI Century* **9** 23

IOP Conf. Series: Materials Science and Engineering **699** (2019) 012006 doi:10.1088/1757-899X/699/1/012006

# Study of structure-property relationship in steels based on analysis of EBSD data

**O A Chikova[1,2], D S Chezganov[1], V V Yuzhakov[1] and N I Sinitsin[1]**

[1]Ural Federal University, 620002 Ekaterinburg, Russia
[2]Ural State Pedagogical University, 620017 Ekaterinburg, Russia

O.A.Chikova@urfu.ru

**Abstract.** In this work, we formulate novel data-driven assays for exploring the structure-property linkages for high-manganese austenitic wear-resistant steel 110G13L (Hadfield steel). Steel 110G13L has the following chemical composition, wt.%: C(0.95-1.50)-Mn(11.5-15.0). These assays are built on recent advances in high resolution quantification of material structure using correlations and principal analyses of electron backscatter diffraction (EBSD) data, as well as in the mechanical characterization using nanoindentation. These novel protocols are demonstrated on a steel 110G13L that exhibits various polycrystalline microstructures. A comparative analysis of EBSD data was carried out for samples of manganese steel 110G13L obtained by various methods. Analysis of the diffraction patterns of backscattered electrons allowed us to plot orientation maps, Schmid factor maps and distributions for austenite dendrites. Schmid factor maps are used to determine the degree of homogeneity of a possible deformation. The results of the measurement of hardness and Young's modulus for the austenite dendrites indicate the heterogeneity of the mechanical properties of the material in submicro-volumes due to lattice defects (dislocations) inside the crystallites.

## 1. Introduction

The structure-property linkages are leading to the development and deployment of advanced materials in emerging technologies [1]. In this work, we formulate novel data-driven assays for exploring the structure-property linkages for high-manganese austenitic wear-resistant steel 110G13L (Hadfield steel). Hadfield steel is known for its high hardness and wear resistance and is characterized by a content of 0.95-1.50 wt.% carbon and 11.5-15.0 wt.% manganese. In this paper, the results of a comparative electron backscatter diffraction (EBSD) study of crystalline structure of 110G13L steel ingots crystallized after heating to 1450 and 1630 °C are presented. Previously, the authors found that the heating of the 110G13L liquid steel to a temperature above 1630°C led to the destruction of microheterogeneity and change in conditions of metal crystallization [2]. Microheterogeneity of 110G13L liquid steel is understood as the presence of dispersed particles, which enriched by manganese, carbon and separated by interfacial surface from the melt. The relationship between destruction of the microheterogeneity state of the liquid steel 110G13L and changes in crystalline structure were investigated. These assays are built on recent advances in high resolution quantification of material structure using correlations and principal analyses of EBSD data, as well as in the mechanical characterization using nanoindentation [3-5].

The EBSD method allows revealing the structure of the studied material with subsequent quantitative processing of the images, obtaining the size and misorientation angle distribution of the structural element boundaries, estimating the proportion of low-angle and special grain boundaries of the material, and plotting a Schmid factor map and a misorientation histogram of the grains [3-5]. Schmid factor maps are used to determine the degree of homogeneity of a possible deformation. The effects of an inhomogeneous stress-strain state and the criteria for failure and wear of Hadfield steel based on an analysis of EBSD data are discussed in Refs. [6-9]. In particular, it was shown that twinning and dislocations play a key role in the strengthening of Hadfield steel [7]. An increase in grain size can increase the strength of austenitic steels with a high Mn content without compromising their ductility and toughness at low temperature [8]. A relationship between wear and crystallographic orientation of individual grains for high-manganese steel was established [10].

## 2. Experimental

The comparative analysis of EBSD data for samples of manganese steel 110G13L obtained by various methods (heating the melt to 1450 and up to 1630°C) was carried out. The ingots of steel 110G13L were crystallized at a rate of 0.2°C/s after heating the melt to 1450 and 1630°C. The chemical composition of Hadfield steel samples was determined using a SPECTROMAX (Spectro Analytical Instruments GmbH, Germany) emission spectrometer: C – 1.2%, Si – 0.5%, Mn – 12.6%, Cr – 1.2%, Ni – 0.2%, and Fe – the rest.

The analysis of the sample crystal structure was performed by EBSD using a scanning electron microscope (SEM) AURIGA CrossBeam workstation (Carl Zeiss, Germany) equipped with HKL Channel 5 EBSD analysis system (Oxford Instruments, UK). The EBSD data were acquired by Flamenco Acquisition software (Oxford Instruments, UK) using the SEM operated at the accelerating voltage of 20 kV and electron beam current of 5nA. The sample surface was prepared by diamond polishing down to 1 μm with a final polishing by colloidal silica SF1 (Logitech, UK). The areas with sizes of 21×21 $\mu m^2$ were scanned with a step size of 60 nm. The obtained data were processed and analysed by Tango software (Oxford Instruments, UK).

The mechanical properties (Young's modulus E, hardness H) were measured by nanoindentation method using NanoScan-4D (FSBI "TISNCM", Russia) following requirements of ISO 14577. The measurements were performed under continuous loading conditions with a linearly increasing load up to 50 mN at room temperature. The loading and unloading of the indenter, as well as the recording of the P-h diagram (applied load – depth of the indention) were carried out automatically. The hardness and Young's modulus of dendrites were determined as a result of 50 measurements for each of the regions. The size of the indenter footprint was measured at a maximum depth of indenter. The method of Oliver and Farah was used to process the results of mechanical material testing [11].

## 3. Results and discussion

The optical images of the microstructure of Hadfield steel ingots crystallized after heating the melt to 1450 and 1630°C are shown in Figure 1. The microstructure of the ingots is represented by austenite dendrites, the interdendrite space is filled with a ferrite-pearlite mixture with carbides and carbonitrides. The grain size and the distance between the secondary branches of dendrites differ slightly from one sample to another. Nonmetallic inclusions, such as sulfides (FeS+MnS), nitrides, oxides, and oxysulfides, are present both at the boundaries and grain body. The Mn heterogeneity of austenite dendrites was found for the ingot superheated to 1630°C before crystallization resulting in segregation upon solidification. During Electrical Discharge Machining (EDM) of Hadfield steel, a segregation layer with increased manganese content on the treated surface was observed earlier [12]. In addition, carbides were not found at the grain boundaries, which indicates a chemical composition alignment in dissolution. The decomposition of austenite at the boundaries of dendrites was not revealed for ingot superheated up to 1630°C as well. The porosity for the ingot overheated to 1630°C was greater than for the rest one.

The EBSD study of the crystal structure of austenite dendrites for ingots crystallized after heating the melt to 1450 and 1630°C are presented in the form of inverse pole figure (IPF) orientation maps, Schmid factor maps, and distributions (Fig. 2). EBSD analysis allows obtaining the distribution of the structural element (crystallite) boundaries according to the angle of misorientation at the IPF maps. The concept of grain in EBSD analysis is different from that in traditional optical microscopy. Two adjacent

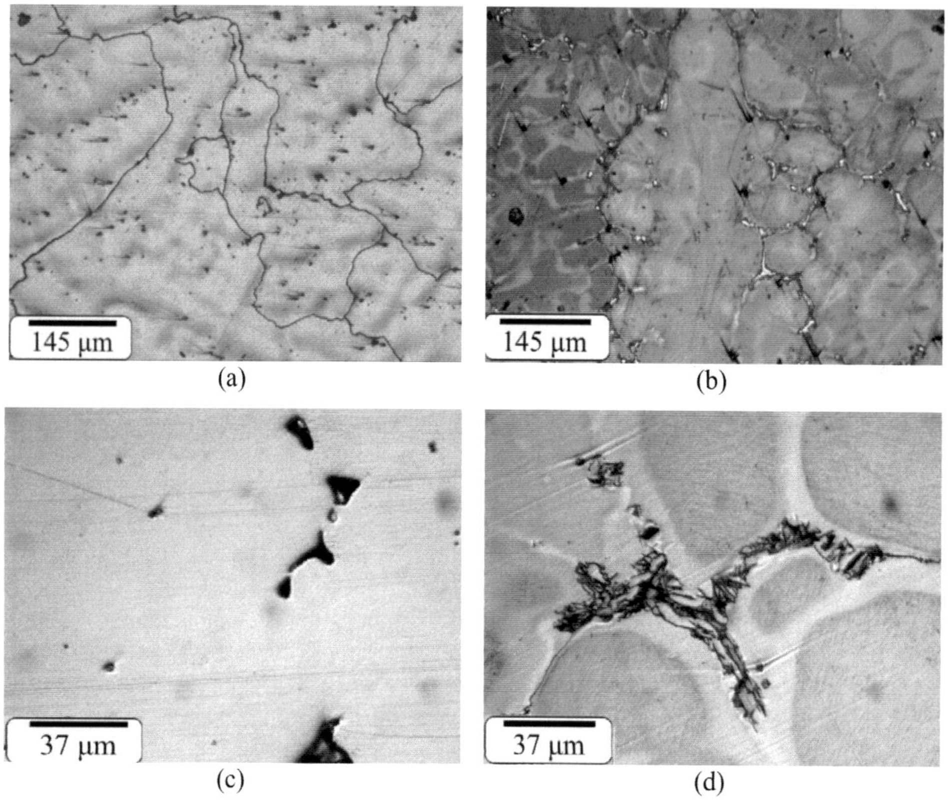


**Figure 1.** Microstructure of 110G13L steel ingots (optical microscopy): (a,c) crystallized after heating the melt to 1450°C; (b,d) crystallized after heating the melt to 1630°C.

points are considered to belong to the same crystallite, if the misorientation between them does not exceed the tolerance angle (subgrain structure) for all the studied samples. The Schmidt factor maps were plotted for a slip system typical for FCC crystals: (111) [110] with loading direction parallel to OZ axis. The distributions of the Schmid factor for samples made after heating of the melt to 1450 and 1630°C have one maximum, which indicates the homogeneity of its elastic characteristics (Fig. 2). The authors [13] investigated the mechanism of the formation of cast structures in Hadfield steel depending on changes in the cooling rate of the casting in two temperature ranges: crystallization (1200-1390°C) and separation of excess phases (560-790°C). A change in the cooling rate leads to decrease in the amount of excess phase (eutectic and carbides), change in their quantitative ratio and morphology. Such changes in the microstructure are reflected in changes in the magnitude of developing stresses. Thus, the features of the microstructure and crystalline structure characteristic of Hadfield steel ingots crystallized after the microheterogeneous state of the melt are destroyed similarly to the effects of changes in the cooling rate of the casting in the temperature range of the separation of excess phases.

It has been previously revealed that if almost all the boundaries are low-angle, the texture is stronger, than the Young's modulus and crystallite hardness is lower. The decrease in the modulus of elasticity due to high material texture significantly improves mechanical workability of the ingot [4].

The results of the measurement of mechanical properties are presented in Figure 3 and Table 1. For the austenite dendrites, there is no pronounced peak in the histograms corresponding to the average values of hardness and Young's modulus (Fig. 3). This type of E and H histograms indicates the heterogeneity of the mechanical properties of the material in submicro-volumes due to lattice defects (dislocations) inside the crystallites [4,14]. The interdendritic space has a complex structure (Fig. 1), which leads to heterogeneity of the mechanical properties of the material in submicro-volumes (Table 1). According to Ref. [14], the value of additional pressure, appeared at the interface between

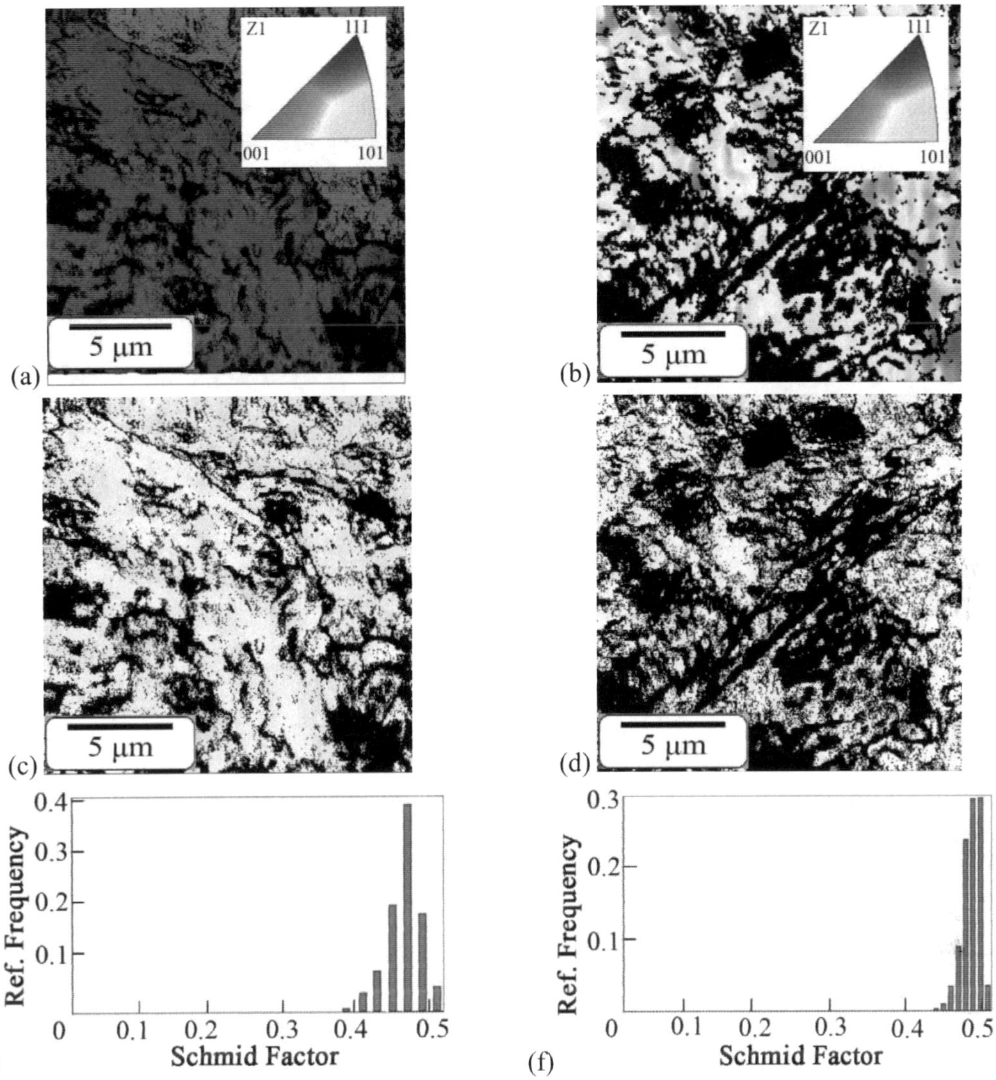


**Figure 2.** (a,b) IPF orientation maps; (c,d) Schmid factor maps; (e,f) Schmid factor distributions; (a,c,e) crystallized after heating the melt to 1450°C; (b,d,f) crystallized after heating the melt to 1630°C.

**Table 1.** Hardness and Young's modulus.

| Material | Melt heating (°C) | H (GPa) | E (GPa) |
|---|---|---|---|
| The austenite dendrites | 1630 | 4.7± 0.4 | 187 ± 6 |
| The interdendrite space | 1630 | 6.8 ± 0.7 | 101 ± 7 |
| The segregation layer with increased manganese content in austenite dendrites | 1630 | 11.6± 0.5 | 229±14 |
| The austenite dendrites | 1450 | 4.6± 0.2 | 174± 4 |

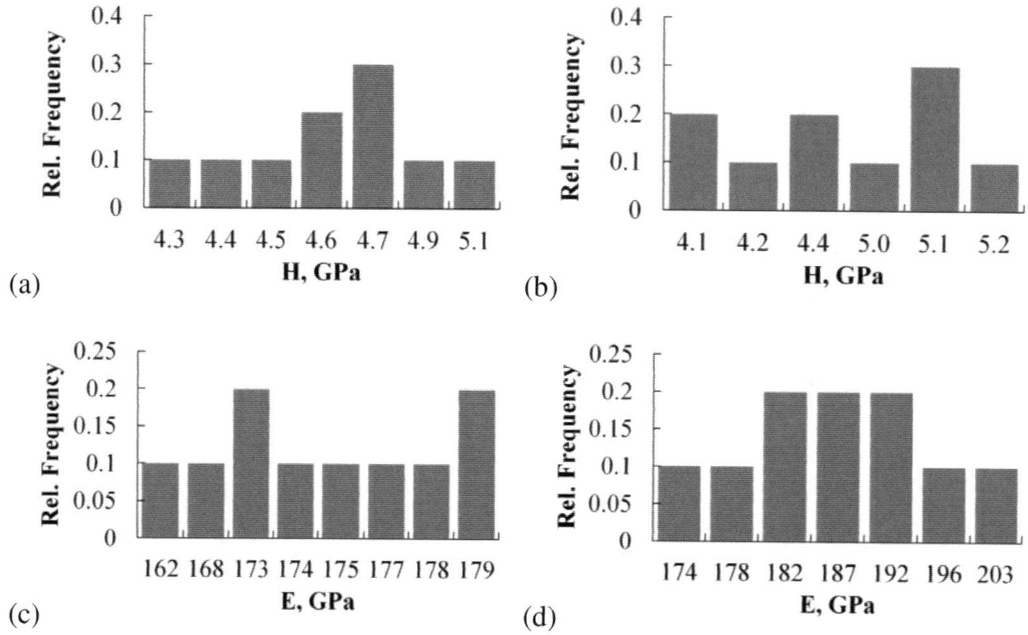


**Figure 3.** The distributions of (a,b) hardness H and (c,d) Young's modulus E of austenite dendrites for the ingots crystallized after heating the melts to (a,c) 1450°C and (b,d) 1630°C.

the segregation layer with increased manganese content and the body of austenite dendrites due to the difference in H and E, is significant and can cause the part to fail. Using the obtained values of the hardness and the Young's modulus of the body of austenite dendrites, the interdendritic space and the segregation layer with increased manganese content, the values of bonding strength and adhesion can be calculated [15].

## 4. Conclusion

A study was made of the relationship technology-structure-property based on analysis of EBSD data and nanoindentation for steel with a high content of manganese 110G13L (Hadfield steel). Steel 110G13L has the following chemical composition, wt.%: C(0.95-1.50)-Mn(11.5-15.0). A comparative analysis of EBSD and nanoindentation data was carried out for samples obtained by various methods. The samples crystallized at a rate of 0.2°C/s after heating the melt to 1450 and 1630°C. The results of the study of the structure of austenitic dendrites are presented using the basic analysis of EBSD data in the form of IPF maps, maps and histograms of Schmid factor. The results of measurements of mechanical properties in submicro-volumes by the nanoindentation method – hardness H (GPa) and Young's modulus E (GPa) - for austenitic dendrites, interdendritic space, segregation layer with a high content of manganese in austenitic dendrites are presented. EBSD data and nanoindentation data indicate the heterogeneity of the mechanical properties of the material in submicro-volumes due to lattice defects (dislocations) inside the crystallites.

**Acknowledgements**

The equipment of the Ural Centre for Shared Use "Modern nanotechnology" UrFU was used. The reported study was funded by RFBR (project No. 19-33-90198).

**References**

[1] Khosravani A, Cecen A and Kalidindi S R 2017 *Acta Mater.* **123** 55-69
[2] Chikova O A, Sinitsin N I and V'yukhin V V 2019 *Russ. J. Phys. Chem. A* **93** 1435-42

[3] Belonosov A V, Chikova O A, Yurovskikh V V and Chezganov D S 2013 *Russ. J. Nondestruct. Test.* **49** 196-205
[4] Chikova O A, Konstantinov A N, Shishkina E V and Chezganov D S 2013 *Russ. Metall.* **75** 35-44
[5] Chezganov D S, Chikova O A and Borovykh M A 2017 *Phys. Met. Metallogr.* **118** 857-63
[6] Harzallah R, Mouftiez A, Felder E, Hariri S and Maujean J-P 2010 *Wear* **269** 647-54
[7] Feng X, Zhang F, Zheng C, and Lü B 2013 *Sci. China Technol. Sci.* **56** 1151-4
[8] Wang X-J, Sun X-J, Song C, Chen H, Tong S, Han W and Pan F 2019 *Acta Metall. Sin. Engl.* **32** 746-54
[9] Machado P C, Pereira J I, Penagos J J, Yonamine T and Sinatora A 2017 *Wear* **376-7** 1064-73
[10] Li Y and Gu D 2014. *Mater. Des.* **63** 856-67
[11] Oliver W C and Pharr GM 2004 *J. Mat. Res.***19** 3-20
[12] Mouralova K, Benes L, Bednar J, Zahradnicek R, Prokes T, Matousek R, Hrabec P, Fiserova Z and Otoupalik J 2019 *J. Mech. Sci. Tech.* **33** 2371-86
[13] Gorlenko D, Vdovin K and Feoktistov N 2016 *China Foundry* **13** 433-42
[14] Chikova O A, Reznik P L and Ovsyannikov B V 2016 *Phys. Met. Metallogr.* **117** 1245-50
[15] Zhang C, Zhou H and Liu L 2014 *Acta Mater.***72** 239-51

IOP Conf. Series: Materials Science and Engineering **699** (2019) 012007 doi:10.1088/1757-899X/699/1/012007

# SEM investigation of the microstructure of cast CuNiAlCoFe equiatomic high entropy alloy

**O A Chikova[1,2], D S Chezganov[1], V S Tsepelev[1] and V Yu Ilyin[1]**

[1]Ural Federal University, 620002 Ekaterinburg, Russia
[2]Ural State Pedagogical University, 620017 Ekaterinburg, Russia

O.A.Chikova@urfu.ru

**Abstract.** Microheterogeneity and crystallization conditions of the equiatomic high-entropy CuNiAlCoFe alloy were studied by measuring the viscosity during heating/cooling, as well as visualizing the microstructure after solidification. The CuNiAlCoFe liquid alloy showed different temperature dependences of viscosity upon heating to 2070 K and subsequent cooling. Since the coinciding part of the temperature dependences was absent, complete destruction of the microheterogeneity of the melt after heating to 2070 K did not occur. Heating the melt to higher temperatures was not possible due to the evaporation of the metal. The melt viscosity during cooling is higher than that during heating; this is observed when microheterogeneity is dispersed. The CuNiAlCoFe alloy obtained as a result of the experiment at a crystallization rate of 10 K/s had a typical microstructure of cast dendrite. Dendrites consisted of a eutectic (α-Al $+CuAl_2+Cu_4NiAl_7$) and included primary crystals of $CoCu_2Al_7$ and $FeCu_2Al_7$. The interdendritic space contained small $Cu_4NiAl_7$ crystals.

## 1. Introduction

High entropy alloys (HEAs) CuNiAlCoFe have a combination of excellent magnetic and mechanical properties [1]. High-entropy alloys are an object of advanced extensive structural research; however, the properties of the liquid state and process features of liquid phase separation of these materials, which are relevant to their production, have been far less explored. In this work, the liquid phase separation in equiatomic alloys of CuNiAlCoFe has been qualitatively studied based on the theory of absolute reaction rate by measuring the viscosity during the heating/cooling process, as well as by imaging of microstructure after solidification.

The properties of highly entropic equiatomic alloys CuNiAlCoFe in the liquid state and the conditions for their crystallization were not studied previously. The alloys of AlCoCrCuFeNi and AlCoCuFe only were studied in a wide temperature range. It was found that a negative excess volume of the Al-containing liquid alloys correlated with a negative enthalpy of the mixture and the change of the excess volume was affected by two basic effects, namely, compression of the Al matrix and formation of compounds in the melt [2].

The separation of the liquid phase in the highly entropic CoCrCuFeNi alloy was studied. Liquid phase separation occurred upon supercooling equal to or exceeding 100 K [3]. The effect of the phase composition on the magnetic and mechanical properties of high-entropy alloys (HEA) $FeCoNi(CuAl)_x$ ($x = 0$-$1.2$, in molar ratios) was studied. The crystal structure of $FeCoNi(CuAl)_x$ HEAs transformed from single face-centred cubic (FCC) phase for $0 \leq x \leq 0.6$ to body-centred cubic (BCC) phase combined with minor FCC phase for $0.9 \leq x \leq 1.2$, whereas FCC plus BCC duplex phases were found in the range of $0.7 \leq x < 0.9$. Transmission electron microscopy images of $FeCoNi(CuAl)_{0.8}$ alloy showed large number of FCC Cu-rich nano-precipitates dispersed in the BCC matrix [1]. The alloys CuNiAlCoFe had a typical cast dendrite microstructure consisting of a dominated BCC and FCC solid solution [4].

## 2. Experimental

The samples CuNiAlCoFe alloy of equiatomic composition were obtained by vacuum arc melting at laboratory. High purity metals of aluminium (purity 99.9%), copper of the Mk00 brand (purity 99.98%) and carbonyl iron (purity 99.98%) were used as initial components.

The kinematic viscosity was measured by the oscillating cylinder method during heating and subsequent cooling. The measurements were carried out in the temperature range of liquidus up to 2070 K with the isothermal delay no less than 30 min at relatively small (50 K) step wise temperature changes. The temperature was maintained with precision of 1 K using a high-precision controller. The vibration parameters were measured by an optical method using a system of vibration photo recording. The BeO crucibles were used in all experiments. The experiments were carried out in a high-purity helium atmosphere at a pressure of $10^5$ Pa. The systematic error of measuring viscosity was 3% and including the random error was not greater than 1.5% at the confidence level of 95%.

Metallographic investigation of the structure of the alloys was carried out by conventional methods. The cooling rate during the crystallization of the samples for metallographic investigation was 10 K/s. The surface microstructure was studied by stereo optical microscopy using Neophot32 (Carl Zeiss, Germany) after electro-polishing. The optical microscopy study was carried out for the samples after etching (alcoholic solution of 3% $HNO_3$) and without etching. The microhardness testers PMT-3 at applied load of 2 N (measuring error 5%) were used for measuring the mechanical properties. The values of microhardness were averaged on the basis of ten measurements. A metallographic study was carried out on a scanning electron microscope (SEM). The surface chemical elemental distribution was measured by energy-dispersive X-ray spectroscopy (EDS) using SEM Merlin (Carl Zeiss, Germany) equipped with X-Max$^N$X-ray spectrometer (Oxford Instruments, UK). The data were acquired and analysed by Aztec software (Oxford Instruments, UK). The SEM was used for visualization of surface morphology and structural defects in secondary electron mode with the resolution down to 2 nm. The EDS was applied for revealing the elemental composition of defects and inclusions.

## 3. Results and discussion

Figure 1 presents the results of a viscosimetric study of CuNiAlCoFe liquid alloy with equiatomic composition. This melt demonstrates hysteresis-like viscosity temperature dependence for heating/cooling cycle. The coinciding parts of the temperature dependences of the viscosity, obtained by heating and cooling, are absent. This means that the transition of the melt to a state of true solution after heating up to 2070 K at complete destruction of the microheterogeneity did not occur. The viscosity changes occur nonmonotonically during heating. The significant increase in the viscosity during cooling at low temperatures (near liquidus) is shown. The increase of melt viscosity can explain the dispersing of microinhomogeneities at non-complete destruction of the microheterogeneity [5]. The microheterogeneity of CuNiAlCoFe liquid alloys is understood as the presence of dispersed particles, which are enriched in Fe, Co, Al, Ni, Cu and separated by an interfacial surface from the rest of the melt.

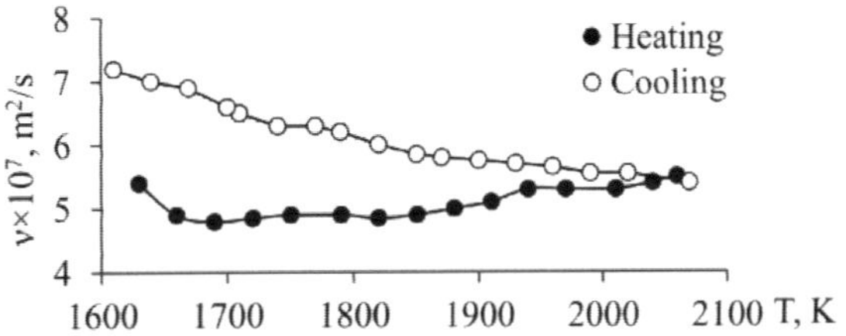


**Figure 1.** The temperature dependences of the kinematic viscosity of the CuNiAlCoFe melts.

IOP Conf. Series: Materials Science and Engineering **699** (2019) 012007 doi:10.1088/1757-899X/699/1/012007

The temperature dependency of viscosity for cooling is monotonic and obeys an Arrhenius law [5]:

$$\nu = A\exp(E/_{RT}) \tag{1}$$

Our attention was focused on the changing characteristics of the viscous flow of the CuNiAlCoFe melt: the activation energy of viscous flow E and the entropy factor A in the Arrhenius equation.

According to the theory of absolute reaction rate [6], melt viscosity can be described by the equation:

$$\nu = \frac{hN_A}{\mu}\exp(\Delta G^{\neq}/_{RT}) = \frac{hN_A}{\mu}\exp(-\frac{\Delta S^{\neq}}{R})\exp(\frac{\Delta H^{\neq}}{RT}) \tag{2}$$,

where $h$ - Planck's constant, $N_A$ - Avogadro constant, $\Delta G^{\neq}$ - the free activation energy of viscous flow, $\mu$ - the molar mass, $\Delta H^{\neq}$ - the enthalpy of viscous flow activation, $\Delta S^{\neq}$ - the entropy of viscous flow activation, $R$ - the universal gas constant and $T$ - temperature (K). The Entropy factor in the Arrhenius equation depends on the entropy of the viscous flow $\Delta S^{\neq}$:

$$A = \frac{hN_A}{\mu}\exp(-\frac{\Delta S^{\neq}}{R}) \tag{3}$$.

We calculated the activation energy E = 27.29 $kJ \cdot mol^{-1}K^{-1}$ and the entropy of activation $\Delta S^{\#} = -21.48 J \cdot K^{-1}$ of viscous flow upon cooling.

We carried out the metallographic study of the microstructure of the ingots obtained after the viscosity measurement. Images of microstructure of samples are shown in Figures 2 and 3.

The microstructure of ingots is homogeneous over the cross section and consists of dendrites and the interdendritic layers of multi-phase composition. The multiple phases in the dendrites and the interdendritic space were distinguished by optical microscopy after chemical etching. The different of phases show sharply a differ in microhardness HV (MPa). According to data on microhardness, we conclude about complex phase composition of dendrites and interdendritic space. For example, microhardness from the periphery to the centre of the dendrite changes the value: 2834-3050 MPa. In the interdendritic space, inclusions having microhardness of 3511, 2187, and 2932 MPa were found.

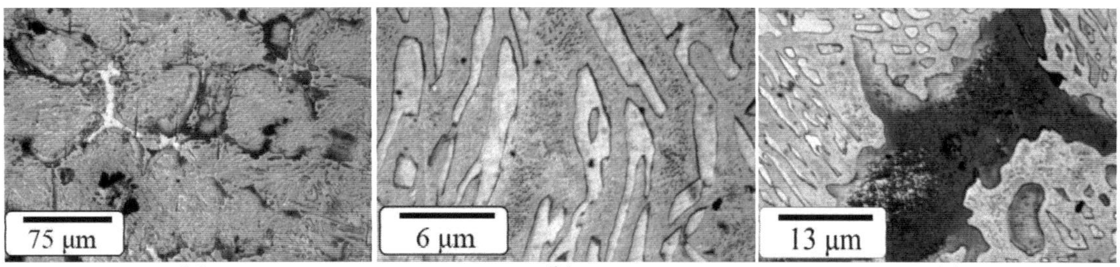

(a) (b) (c)

**Figure 2.** The microstructure of CuNiAlCoFe alloy ingots (optical microscopy): (a) dendrites, (b) intrinsic microstructure of dendrites, (c) interdendritic space.

**Table 1.** Elemental composition of CuNiAlCoFe alloy at the points marked at Figure 3.

| | C | Al | Cu | Fe | Co | Ni |
|---|---|---|---|---|---|---|
| Spectrum 1, wt.% | 4.68 | 6.78 | 33.73 | 17.73 | 17.84 | 19.24 |
| Spectrum 2, wt.% | 4.60 | 12.15 | 22.85 | 16.86 | 20.30 | 23.24 |

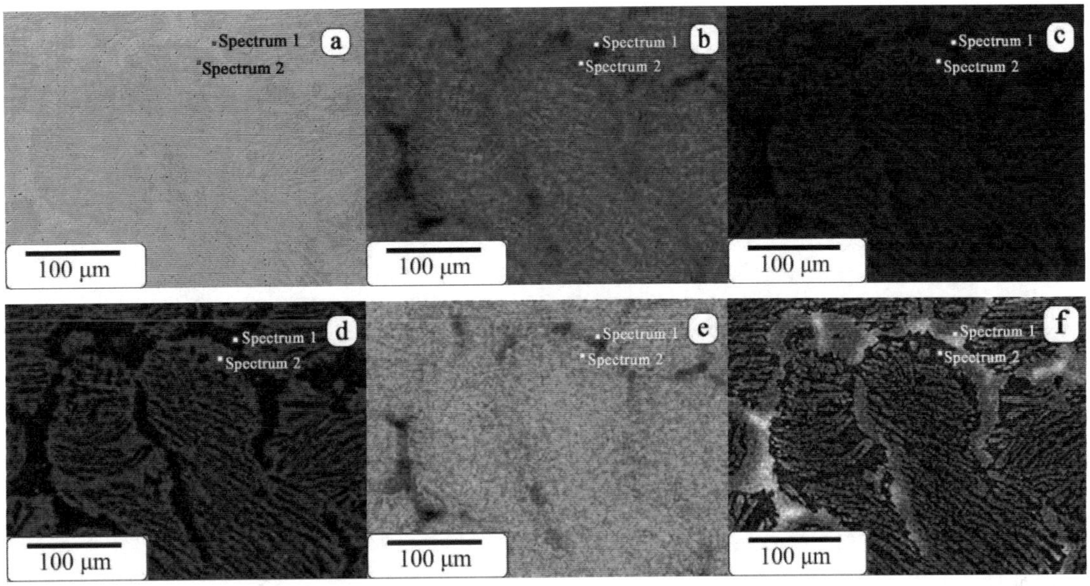


**Figure 3.** (a) SEM image of surface microstructure of CuNiAlCoFe alloy ingots. Elemental distributions: (b) Fe, (c) Ni, (d) Co, (e) Al, (f) Cu.

The dendrites consist of ternary eutectic (α-Al + $CuAl_2$ + $Cu_4NiAl_7$) and large primary crystals of $CoCu_2Al_7$ and $FeCu_2Al_7$ isomorphic compounds [1,3,4,7-11]. The interdendritic space contains small primary crystals of $Cu_4NiAl_7$.A significant interdendritic porosity was found. Ab-initio molecular dynamics (AIMD) simulations were used to determine crystal structures of phases at different temperatures in CuNiAlCoFe equiatomic composition [12]. The AIMD results showed a possible coexistence of FCC and BCC phase at the room temperature. The phase diagrams of CuNiAlCoFe system was calculated using a modified thermodynamic approach based on Calculation of Phase Diagrams (CALPHAD) and Muggianu's methods [8-9]. The characterization utilizing SEM, X-ray diffraction, and EBSD confirmed that crystal structures and composition of phases calculated by AIMD and phase diagram simulations.

## 4. Conclusion

High entropy CuNiAlCoFe (HEA) alloys have a combination of superior magnetic and mechanical properties. Microheterogeneity and crystallization conditions of the equiatomic high-entropy CuNiAlCoFe alloy were qualitatively studied by measuring viscosity based on the theory of absolute reaction rates, as well as visualization of the microstructure after solidification.

The equiatomic liquid alloy CuNiAlCoFe exhibits a hysteresis-like temperature dependence of viscosity upon heating to 2070 K and cooling. The coincident parts of the temperature dependences of viscosity, which are obtained by heating and cooling, are absent. This means that the melt did not undergo the transition to the true solution state after heating to 2070 K, and microheterogeneity remained. An increase in the melt viscosity during cooling can explain the process of microheterogeneity dispersion. A significant increase in viscosity at low temperatures (near liquidus) during cooling is characteristic.

The CuNiAlCoFe alloys demonstrated a typical cast dendrite microstructure consisting of dominated body-centred-cubic and the face-centred-cubic solid solutions. The dendrites consist of eutectic (α-Al + $CuAl_2$+ $Cu_4NiAl_7$) and include the primary crystals of $CoCu_2Al_7$ and $FeCu_2Al_7$. The interdendritic space contains small primary crystals of $Cu_4NiAl_7$. A significant interdendritic porosity was found.

**Acknowledgements**
The equipment of the Ural Centre for Shared Use "Modern nanotechnology" UrFU was used. Authors are grateful for the support of experimental works by Act 211 Government Russian Federation, contract № 02.A03.21.0006.

**References**

[1] Zhang Q, Xu H, Tan XH, Hou XL, Wu S W, Tan G S and Yu L Y 2017 *J. Alloys Compd.* **693** 1061-7
[2] Plevachuk Y, Brillo J and Yakymovych A 2018 *Metall. Mater. Trans. A* **49** 6544-52
[3] Guo T, Li J, Wang J, Wang Y, Kou H and Niu S 2017 *Intermetallics* **86** 110-5
[4] Zhuang Y X, Liu W J, Chen Z Y, Xue H D and He J C 2012 *Mat. Sci. Eng. A* **556** 395-9
[5] Chikova O A, Sinitsin N I and V'yukhin V V 2019 *Russ. J. Phys. Chem. A* **93** 1435-42
[6] Chikova O A, Tsepelev V S and Moskovskikh O P 2017 *Russ. J. Phys. Chem. A* **91** 979-83
[7] Raghavan V 2006 *J. Phase Equilib. Diff.* **56** 389-91
[8] Kundin J, Wang P, Emmerich H and Schmid-Fetzer R 2014 *Eur. Phys. J. Spec. Top.* **223** 567-90
[9] Wang C-H, Chen S-W, Chang C-H and Wu J-C 2003 *Metall. Mater. Trans. A* **34** 199-209
[10] Yadav T P, Shaz M A, Tiwari R S and Srivastava O N 2003 *Zeitschrift fur Kristallographie* **218** 12-6
[11] Yadav T P, Mukhopadhyay N K, Tiwari R S and Srivastava O N 2007 *Mat. Sci. Eng. A* **448-451** 1052-6
[12] Kivy M, Beyramali Z, Asle M and Lekakh S 2017 *Mater. Des.* **127** 224-32

# Domain structure of CoNi microparticles under mechanical stress

**N I Nurgazizov, A P Chuklanov, D A Biziyaev, A A Bukharaev and V V Chirkov**

Zavoisky Physical-Technical Institute, FRC Kazan Scientific Centre of RAS, 420029 Kazan, Russia

achuklanov@kfti.knc.ru

**Abstract**. In this work, the results of the study of the domain structure of planar CoNi microparticles are presented. The microparticles represented squares with a side of 7 μm and height of 0.03 μm. It was shown that in the non-stressed state the particles had the regular four-domain state. The mechanical stress (compression) of the particles led to a change in domain size, which depended on the degree of compression, the angle between the direction of magnetization, and the side of the mechanical stress. Under the mechanical stress, the change in the coercive force and the residual magnetization of the particles was observed.

## 1. Introduction

The magnetic properties of a solid may change under mechanical stress. This effect is called "the magnetoelastic effect" (or the Villary-effect). Recently, this phenomenon has been studied in connection with the possibility of its application for the magnetization reversal of microparticles. The mechanical stress can significantly reduce the energy required to magnetization reversal of a single-domain or multi-domain microparticle [1-5]. Straintronics is separated as special area of magnetoelastic effect that studies this phenomenon for micro and nano-objects [3].

The materials with crystalline anisotropy being close to zero and high magnetostriction coefficient (usually the alloys of ferromagnetic metals) are interesting for studying the magnetic properties under mechanical stress. The effect of crystalline anisotropy on the magnetic properties can be reduced almost to zero in polycrystalline materials. This is due to the random orientation of the crystallographic planes in each grain of a polycrystal.

In this paper, the effect of mechanical stresses on the magnetic structure of CoNi microparticles was studied. The magnetic force microscopy and microscopy based on magneto-optical Kerr effect (MOKE) were used for this study. The composition of alloy (Co18%, Ni82%) was chosen in order to get a sufficiently large value of the magnetostriction coefficient and to exclude the influence of crystalline anisotropy on the magnetic properties. Therefore, the anisotropy of the studied microparticles was induced by mechanical stress and shape.

## 2. Sample preparation

The planar CoNi microparticles of size 7.5×7.5×0.03 $\mu m^3$ has been studied. An array of these microparticles was formed on the thin glass substrate. The lateral size of substrate was 18×3 $mm^2$ and the thickness was 0.15 mm. The microparticles were fabricated by electron beam evaporation technique using the ultrahigh vacuum setup "Multiprobe P" (Omicron). The deposition on the substrate was performed through the metal grid tightly pressed to the surface. The uniformly spaced square holes of

IOP Conf. Series: Materials Science and Engineering **699** (2019) 012008 doi:10.1088/1757-899X/699/1/012008

the grid were located on the substrate in the way that one of the sides of the resulting particle was parallel to the axis of compression of the particle and the long axis of the substrate. The area with the particles had a diameter about 2.5 mm. The size of this area was limited by the grid.

Before deposition, some part of the substrates was elastically bent by placing into a special holder. After deposition, the sample was removed from the holder, this caused substrate straightening and the compression of particles on the surface. A similar technique was used to induce stresses in thin films [6,7] and particles [5]. Due to uniform bending of the substrate, the particles were also compressed uniformly throughout the sample. This made it possible to study samples by MOKE, when a signal is collected from a certain array of particles and it is important that all the particles under study had the same physical properties. In our case, the size of region under study was $10^{-2}$ mm$^2$.

The P47-Pro and Solver HV setups (NT–MDT) in magnetic force microscope (MFM) mode and atomic force microscope (AFM) mode were used. Both setups allow making MFM-measurements of the samples in an external magnetic field directed in the sample plane. For research purposes, the commercial Multi75M-G cantilevers (BudgetSensor) were used. The MFM-measurements were performed by single-pass technique, when the probe moved during scanning at a constant height above the sample surface. In this case, the influence of magnetic field of MFM probe on the magnetization distribution in microparticles is significantly reduced. This allows one to determine more precisely the value of external magnetic field corresponding to a change in the particle magnetic structure.

## 3. Magneto-optical Kerr effect (MOKE)

The setup using the Kerr-effect based on modified LEFT-3M-1 ellipsometer was used to register hysteresis loops. The setup allows measuring the dependence of the polarization rotation of the light reflected from the sample surface on the external magnetic field. In addition, it is possible to make in-plane rotation (with resolution of 4.5°) of the sample and measure in different sample points with spatial resolution about 0.1 mm. The initial sample orientation was approximately parallel to the long axis of the substrate and one of the sides of the square particle. The initial orientation of the sample was close to the direction of compression under mechanical stress. Each hysteresis loop is the result of averaging of 200 measurements. The hysteresis loops in the direction of easy axis and hard axis are presented in Figure 1a and Figure 1b, respectively. The angular dependences of the relative remnant magnetization (Fig. 1c) of the particles were calculated by analysis of the recorded hysteresis loops.

It is seen at the angular dependences that in the absence of mechanical stresses the microparticles have two easy axes (EA) oriented along the square diagonals and two hard axes (HA) of magnetization oriented along the sides of the square. The corrective force along EA is about 14 Oe and 21 Oe along HA. The anisotropy field is about 60 Oe along EA and HA. The compression of the particles leads to the formation of the one EA in the direction of compression and one HA rotated 90 degrees relative to EA. The coercive force along EA increased to 24 Oe, the anisotropy filed decreased to 40 Oe. Anisotropy field along HA is above 85 Oe and coercive force is 6.7 Oe.

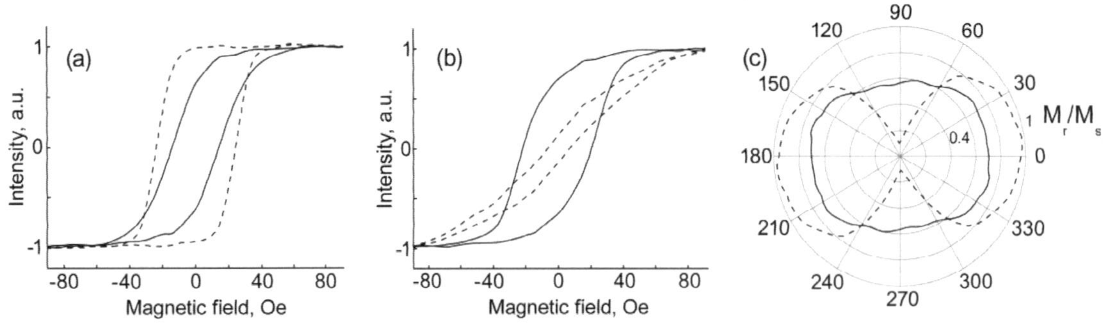


**Figure 1.** Hysteresis loops obtained by the MOKE for non-stressed (solid lines) and stressed (dashed lines) CoNi particles with the direction of the external magnetic field along: (a) easy and (b) hard axes. (c) The angular dependences of the normalized remnant magnetization.

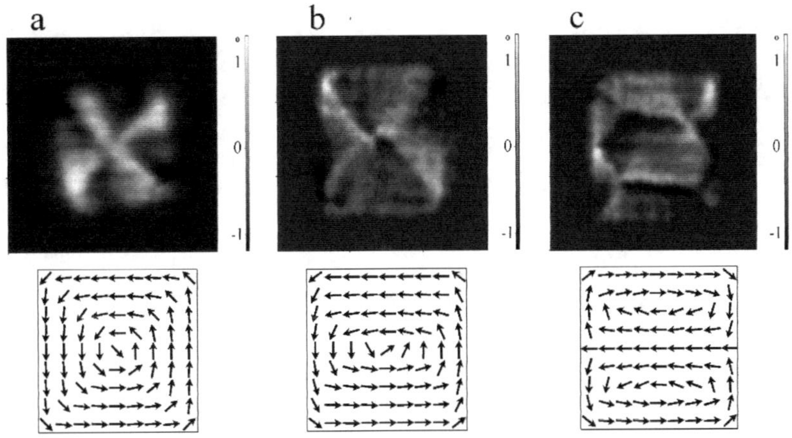


**Figure 2.** MFM images (up) and corresponding distribution of the magnetization (down) (a) for unstressed CoNi particle in four-domain state, for compressed particle: (b)in four-domain state and (c) in seven- domain state.

Thus, based on the analysis of hysteresis loops, it can be concluded that particle compression leads to an increase in the coercive force in about 1.7 times and decrease in anisotropy field on 30% along the compression axis. This fact indicates a significant change in the magnetic properties of particles under the mechanical stress.

## 4. Magnetic properties of CoNi particles studied by MFM

MFM-image of a CoNi particle without mechanical stress is shown in Figure 2a. According to the MFM-image, the particle is in a four-domain state with equal domain size. The direction of magnetization in each domain is parallel to the nearest square side. In order to establish the magnetization distribution in a particle, the following steps were performed. First, the magnetization distribution in the particle was calculated by means of the Object Oriented Micro Magnetic Framework (OOMMF) [8] using the sizes and shape of the particle obtained by AFM (Fig. 3a). Second, a virtual MFM-image was calculated on the base of the magnetization distribution by the algorithm described in [9] and if the experimental MFM-image coincided with the virtual one, it was concluded that the calculated magnetization distribution was correct.

If the uniaxial mechanical compression was applied to the microparticles, the sizes of two domains with magnetization direction parallel to the compression axis were increased (Fig. 2b). A characteristic domain wall in the shape of a bridge is formed between these two domains. The length of the bridge depends on the magnitude of the mechanical stress in the particle. In addition to the four-domain state, some particles were in a seven-domain state (Fig. 2c). The energy of the magnetic subsystem in the seven-domain state is higher than that in the four-domain state according to the OOMMF calculations; therefore, it is energetically unfavourable. It can be assumed that the seven-domain state of a microparticle is attributed to the presence of local defects, which is not taken into account in the OOMMF calculation.

MFM studies of the magnetization reversal in CoNi particles were carried out in an external magnetic field ranged from +160 Oe to -160 Oe applied with the step of 5 Oe in-plane along two directions (EA and HA). A uniform magnetization with the direction parallel to the magnetic field was observed for the particles without mechanical stresses along HA in the range from +160 Oe to +60 Oe. This distribution is displayed as black and white stripes at the opposite ends of the MFM-image (Fig. 3b). In the field range from +55 Oe to -55 Oe, the microparticles were in the multidomain state (Fig. 3c,d) and the magnetic contrast at the edges of the particles switched when the direction (sign) of the external magnetic field changed. The breaks on these images are related to the influence of the magnetic field of MFM probe. It is impossible to eliminate this effect, because the magnitude of MFM-probe field is

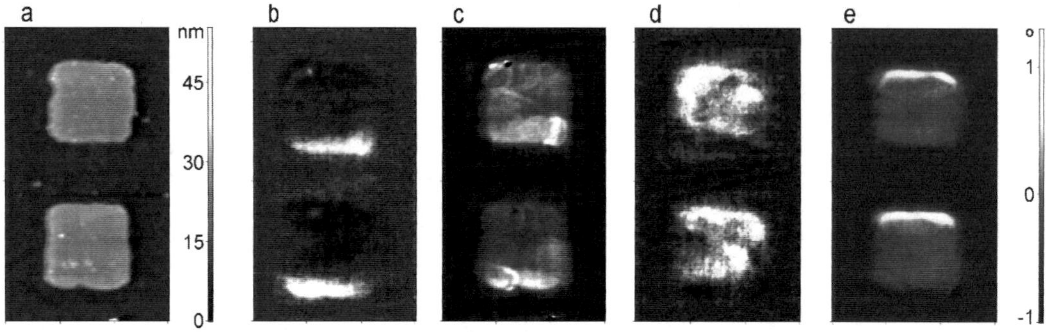


**Figure 3.** (a) AFM image and (b-e) MFM images of two unstressed CoNi particles in external magnetic field: (b) +70 Oe, (c) +20 Oe, (d) -20 Oe, (e) -70 Oe. Magnetic field is directed along the long side of the scan. Scan size 21×12 μm$^2$.

comparable with external filed. The particles were uniformly magnetized again with the magnetization direction opposite to the initial in the field range from -60 Oe to -160 Oe (Fig. 3e). The registered field value (taking into account the field step during measurements) is comparable with the value of the anisotropy field (60 Oe), which is observed for this direction in MOKE. If the external field was applied along EA (coinciding with the diagonal of the particle), the uniform magnetization remained in the field range from +160 Oe to +45 Oe. In the fields from +40 Oe to -40 Oe the multidomain state was observed. In the fields from -45 Oe to -160 Oe, the particles became uniformly magnetized in the direction opposite to the original one.

For particles under uniaxial mechanical stress, the magnetic field was also applied in two directions: parallel (EA) and perpendicular (HA) to the axis of stress, which coincided with one of the sides of the particle. In this case, the MFM images of the particles were similar to the images obtained for uncompressed particles with the field parallel to the particle side (Fig. 3). Only the values of external magnetic fields, at which magnetization transform occurred, were changed. In a magnetic field directed along the compression axis (i.e. along EA), the particles retained uniform magnetization in the range from +160 Oe to +10 Oe. A multi-domain structure was observed in the fields from +5 Oe to -30 Oe. In the fields from -35 Oe to -160 Oe, the particles became uniformly magnetized in the opposite direction. For a magnetic field directed perpendicular to the compression direction (along HA), the particles retained uniform magnetization in the range from +160 Oe to +95 Oe. A multi-domain structure was observed in field range from +90 Oe to -90 Oe. In the fields from -95 Oe to -160 Oe, the particles became uniformly magnetized in the opposite direction.

The MFM data allow making analysis of the changes in the anisotropy field of a particle under mechanical compression in the selected direction. Here, the anisotropy field means the value of the external magnetic field, at which the particle becomes uniformly magnetized in the direction of this field. In the direction of compression (along one of particle sides), an EA is formed and the anisotropy field decreases from 60 Oe to 35 Oe. Along the other side of the particle and perpendicular to EA, the HA is formed and the anisotropy field increases from 60 Oe to 95 Oe. A significant change in the particle anisotropy field under the mechanical stress is due to the high value of the magnetostriction coefficient of CoNi alloy (-25×10$^{-6}$), and the directions of EA and HA are caused by its negative sign. The values of the anisotropy field obtained by MFM measurements are in good agreement with the data obtained by MOKE (anisotropy field decreases from 60 Oe to 40 Oe along EA and increases from 60 Oe to more than 85 Oe along HA when mechanical stressed).

## 5. Conclusion

According to the experimental data, in the absence of mechanical stress, CoNi particles have a four-domain structure with domains of the equal size. Each particle has two axes of easy magnetization (parallel to the diagonals of the particle) and two axes of hard magnetization (parallel to the side of the particle) due to the shape anisotropy. The mechanical compression of the particles leads to a change in their domain structure. The two axes are formed: one easy axis is parallel to the direction of compression of the particle, and one hard axis is perpendicular to the direction of compression of the particle. The coercive force of the compressed particle along the EA increases about 1.7 times, and the anisotropy field decreases about 2 times. Thus, using materials with high magnetostriction coefficient for producing particles allows varying their magnetic properties over a wide range by applying mechanical stress.

**Acknowledgements**
The research by MOKE method was partially supported by "Russian Foundation for Basic Research" (RFBR No. 17-08-00915). The studying by MFM method was partially supported by RFBR No. 18-02-00204.

**References**
[1] Bur A, Wu T, Hockel J, Hsu C-J, Kim H K D, Chung T-K, Wong K, Wang K L and Carman G P 2011 *J. Appl. Phys.* **109** 123903
[2] Morozov A I 2014 *Phys. Solid State* **56** 865-72
[3] Bukharaev A A, Zvezdin A K, Pyatakov A P and Fetisov Yu K 2018 *Phys.-Usp.* **61** 1175-212
[4] Finizio S, Foerster M, Buzzi M, Kruger B, Jourdan M, Vaz C A F, Hockel J, Miyawaki T, Tkach A, Valencia S, Kronast F, Carman G P, Nolting F and Klaui M 2014 *Phys. Rev. Appl.* **1** 021001
[5] Bizyaev D A, Bukharaev A A, Churkov V V, Chuklanov A P, Khanipov T A and Nurgazizov N I 2018 *J. Phys.: Conf. Ser.* **1068** 012004
[6] Belyaev B A, Izotov A V 2007 *Phys. Solid State* **49** 1731-9
[7] Cao D, Wang Z, Pan L, Feng H, Cheng X, Zhu Z, Wang J, Liu Q and Han G 2016 *Appl. Phys.* A **122** 938
[8] Donahue M J and Porter D G OOMMF URL:http://math.nist.gov/oommf
[9] Ovchinnikov D V and Bukharaev A A 2001 *Tech. Phys.* **46** 1014-9

# Dielectric characterization of erythrocytes by electrostatic force microscopy

**N A Davletkildeev[1,2], D V Sokolov[1], E Yu Mosur[1,2] and A A Lopandina[2]**

[1]Omsk Scientific Centre, Siberian Branch, Russian Academy of Sciences, Omsk, 644024 Russia

[2]Dostoevsky Omsk State University, Omsk, 644077 Russia

classicsub-zero@mail.ru

**Abstract**. The method of determination of the dielectric constant of individual erythrocytes, based on a quantitative analysis of the profiles of their images obtained by electrostatic force microscopy, is presented. A model of electrostatic interaction between the atomic force microscope probe and the biological cell is proposed. The obtained values of the permittivity of individual erythrocytes agree well with known literature data.

## 1. Introduction

The human cells have unique biophysical and biochemical properties that ensure interaction with the surrounding physiological environment to realize the specific functions. The study of mechanical, electrical, and optical parameters of cells is important for clinical diagnostics.

The electrical properties of cells are investigated generally by measuring their response on the external electric field [1], which depends on the shape, size, internal structure, and conductivity of the cell, as well as susceptibility of the cellular components. It should be noted that the response registration is usually performed on large cell populations that leads to averaging the measured parameters. In this case, the properties of individual cells are not taken into evaluation that leads to limited measurement accuracy. The study of individual cells allows determining the heterogeneity of their characteristics within cells population and measuring the samples containing various cell types without their separation [2]. Using the electrostatic force microscopy (EFM) to determine the dielectric response of individual bacterial cells with high precision and reproducibility was demonstrated in [3]. The method of determining the dielectric constant of individual hemoglobin molecules based on the modelling of EFM profiles was presented in [4]. This technique can be adapted to determine the dielectric constant of individual erythrocytes, also called red blood cells (RBC) that perform a fundamental physiological function of living organisms – transport of respiratory gasses. The purpose of this work is application of EFM method for dielectric characterization of individual RBC.

## 2. Materials and methods

RBC were separated by centrifugation of whole blood, and then washed three times with isotonic solution. In the next step, RBC were fixed of a 2%-paraformaldehyde solution for 2 h, then transferred to deionized water to achieve concentration equal to $10^8$ cells/ml. Aqueous suspension of RBC (15 μl) was deposited on a freshly cleaved surface of highly oriented pyrolytic graphite (HOPG) and incubated for 30 min. Then the samples were washed in deionized water to remove non-adsorbed cells and air-dried at room temperature.

IOP Conf. Series: Materials Science and Engineering **699** (2019) 012009 doi:10.1088/1757-899X/699/1/012009

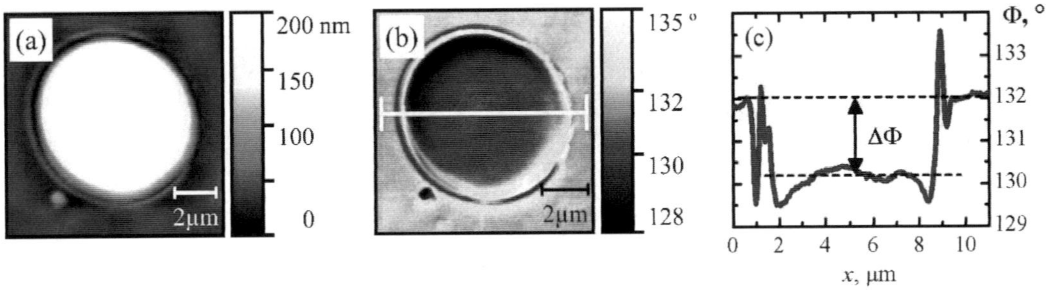


**Figure 1.** 2D-images of the individual RBC on HOPG. (a) First and (b) second passes (h = 100 nm, + 3 V), and (c) the corresponding profile of the EFM-image.

The HOPG surface with immobilized RBC was scanned by atomic force microscope MFP-3D SA (Asylum Research) in EFM-mode at various applied voltages. The used conductive cantilevers of the ETALON HA_FM series (NT-MDT SI, Russia) with Pt coating have the following parameters: $k$~3.4 N/m, $Q$~234, $R$~35 nm.

## 3. Results and discussion

Figure 1a shows an example of topography and corresponding EFM-contrast (Fig. 1b) for the individual RBC on HOPG surface at + 3 V at the probe. From topography profiles analysis, the shape of the cells was determined as an almost symmetrical disk with average height of about 80 nm. Figure 1c presents the phase shift ($\Delta\Phi$) between RBC and HOPG surface on EFM-profile.

Figure 2 demonstrates the EFM measurements for quantitative analysis of experimental data. RBC modelled as a disc with the height $H$ and dielectric constant $\varepsilon$ is located on the HOPG. The probe tip is assumed to be a hemisphere with the curvature radius $R$ that is at a fixed height $h$ above the sample.

The dependence of the phase shift tangent on the square of the applied voltage for studied samples is linear. This indicates that the contrast of EFM-images for RBC is determined only by the probe-sample capacitive coupling.

The EFM-profiles modelling was performed according to the method described in [4]. To calculate the phase shift between RBC and the probe, the second derivative of the probe-RBC-HOPG capacitance was calculated using the equation:

$$\frac{\partial^2 C}{\partial z^2} = 4\varepsilon_0 \int_0^R \int_{y_1}^{y_2} \left[ h + R - \sqrt{R^2 - x^2 - y^2} + \frac{H}{\varepsilon} \right]^{-3} dxdy \qquad (1)$$

where $\varepsilon_0$ – vacuum permittivity, $\varepsilon$ – permittivity of individual RBC, $y_1 = -\sqrt{R^2 - x^2}$ and $y_2 = +\sqrt{R^2 - x^2}$ – the limits of integration along the $y$-axis.

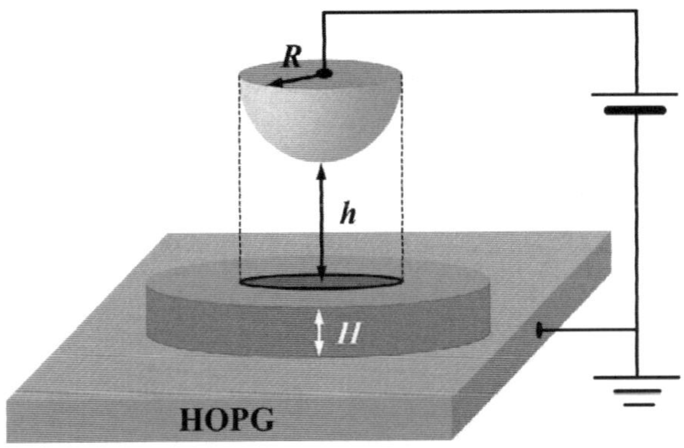


**Figure 2.** Scheme of EFM experiment: probe tip is placed in the centre above the RBC.

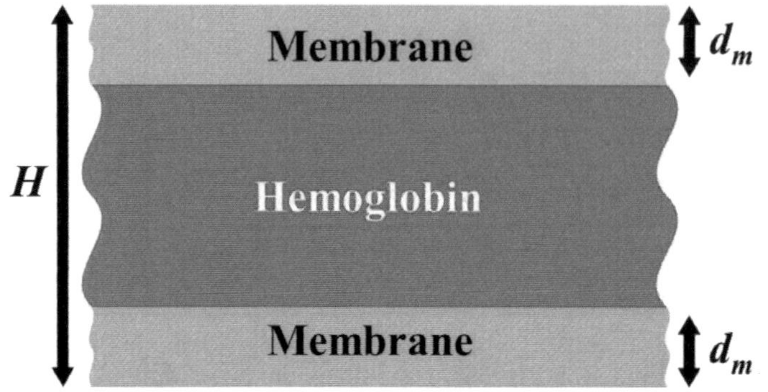


**Figure 3.** RBC membrane model.

By fitting the model phase shift to the experimental one, the mean value and the permittivity range of individual RBC were determined to be equal to 2.6 ± 0.8. Since the volume of the RBC is filled with proteins, which, according to the literature, have dielectric permittivity values of 2-4 [3], the obtained value of $\varepsilon$ is adequate.

For comparison, RBC permittivity ε was estimated using an effective medium approximation for multilayer dielectric systems [5]. The model of cross profile of RBC has a three-layer structure: a cytoplasmic layer (cytoskeleton filled with hemoglobin), which is located between two membrane layers consisting from lipids (Fig. 3). The fraction of integral membrane proteins is approximately 0.5; however, their dielectric constant is not known; therefore, they were not taken into account in the permittivity calculations.

According to effective medium model, the effective permittivity of two-component medium, in this case, an RBC is written as:

$$\frac{1}{\varepsilon_{RBC}} = \frac{f_{Hb}}{\varepsilon_{Hb}} + 2\frac{f_m}{\varepsilon_m} \tag{2}$$

where $\varepsilon_{RBC}$, $\varepsilon_{Hb}$, $\varepsilon_m$ – permittivity of RBC, hemoglobin and membrane, $f_{Hb}$ and $f_m$ – volume fractions of hemoglobin and membrane, respectively.

The measured values of the RBC height from the topography images and the known thickness of the membrane layer ($d_m \approx 8$ nm [6]) were used for estimation of the volume fractions. As a result, the calculated data were $f_{Hb} \approx 0.8$ and $f_m \approx 0.1$, respectively. The necessary permittivity was taken from the literature: $\varepsilon_{Hb} \approx 2.9$ [4], $\varepsilon_m \approx 2.0$ (permittivity of the membrane was taken for lipids [3]). As a result, permittivity of RBC determined by the relationship (2) was equal to 2.7. This value agrees well with the mean value for permittivity of RBC obtained by EFM. Calculations by effective medium approximation for RBC show that the cytoplasmic membrane provides the lower permittivity for the RBC (equal to 2.7) compared with the permittivity of its main component – hemoglobin (equal to 2.9).

## 4. Conclusion

The technique described in [4] is adapted to determine the permittivity of individual erythrocytes taking into account the features of electrostatic interaction between an atomic force microscope probe and a biological cell on the substrate surface. This method allows determining the values of the permittivity of individual erythrocytes, which is in good agreement with the known literature data and corresponds to the value obtained by the effective medium approximation. The mentioned technique can be applied to study the dielectric characteristics of other individual cells.

## Acknowledgements

The work was carried out according to the state task of the Omsk Scientific Centre SB RAS (project registration number AAAA-A17-117041210227-8).

**References**

[1] Du E, Ha S, Diez-Silva M, Dao M, Suresh S and Chandrakasan A P 2013 *Lab Chip* **13** 3903-9
[2] Brown M and Wittwer C 2000 *Clin. Chem.* **46**:8(B) 1221-9
[3] Esteban-Ferrer D, Edwards M A, Fumagalli L, Jurez A and Gomila G 2014 *ACS nano* **8** 9843-9
[4] Davletkildeev N A, Sokolov D V, Mosur E Yu, Lopandina A A and Bolotov V V 2019 *Tech. Phys. Lett.* **45** 981
[5] Fujiwara H 2007 *Spectroscopic ellipsometry: principles and applications* (Chichester: John Wiley & Sons Ltd) p 369
[6] Lombard J 2014 *Biol. Direct* **9** 32

# Gas sensing properties of individual composite nanostructures $TiO_{2-x}$/MWCNT and $SnO_x$/MWCNT measured by scanning probe microscopy

**N A Davletkildeev[1,2], D V Sokolov[1], E Yu Mosur[1,2] and V V Bolotov[1,2]**

[1]Omsk Scientific Centre, Siberian Branch, Russian Academy of Sciences, Omsk, 644024 Russia
[2]Dostoevsky Omsk State University, Omsk, 644077 Russia

classicsub-zero@mail.ru

**Abstract**. The gas sensitivity properties of individual composite nanostructures based on multi-walled carbon nanotubes (MWCNT), functionalized by argon ions and coated with tin and titanium oxides, were investigated by the methods of scanning probe microscopy. The Fermi level shift and the change in longitudinal conductance of nanostructure upon absorption of ammonia and nitrogen dioxide are determined. It is shown that the change in conductance of the composite nanostructures correlates with the Fermi level shift under gases exposure as a result of charge carrier concentration changing in the metal oxide surface layers upon adsorption of gas molecules.

## 1. Introduction

Composite materials based on multi-walled carbon nanotubes (MWCNT) and metal oxides are promising materials for various applications in nanoelectronics [1], in particular, for gas-sensing elements [2]. The trend of miniaturization of electronic devices and the fundamental replacement of massive layers by nanoscale elements attracts a great interest in the study of electron transport processes in individual nanostructures. For study of these processes, the methods development for measuring the electrical parameters of individual nanostructures is needed. Scanning probe microscopy is an effective method to study the electrical properties of single nanostructures, for example, conductive atomic force microscopy (C-AFM) and electrostatic force microscopy (EFM) [3-5].

In this work, the electrical conductivity of individual MWCNTs and metal oxides/MWCNT nanocomposites was determined by the C-AFM method. The work function of MWCNTs and metal oxides/MWCNT nanocomposites was measured using EFM, and then the Fermi level shifts in these nanostructures were found. The gas-sensing properties of the nanostructures were studied by the change in conductivity and the Fermi level shift under the exposure to a reducing and an oxidizing gas.

## 2. Materials and methods

MWCNTs synthesized by chemical vapour deposition were studied. According to X-ray photoelectron spectroscopy data, the nitrogen concentration in nanotube walls is 3.7 at%. The functionalization of as-grown MWCNTs was carried out by argon ions irradiation with energy 5 keV and dose $10^{16}$ $cm^{-2}$. The composite nanostructures of metal oxide/MWCNTs were fabricated by deposition of non-stoichiometric titanium dioxide ($TiO_{2-x}$) and tin oxide ($SnO_x$) on the surface of functionalized MWCNTs using magnetron sputtering [6,7].

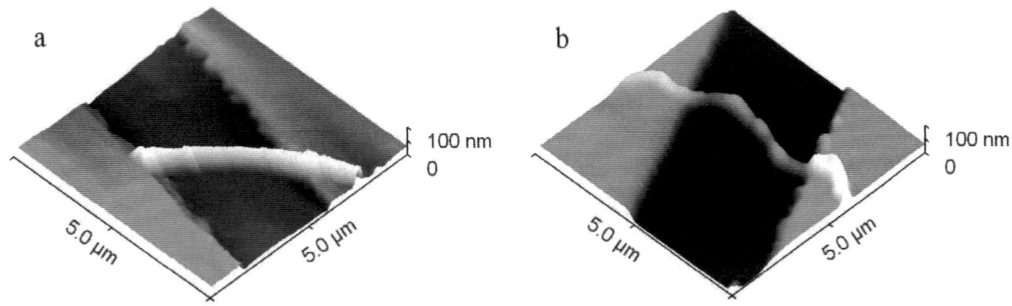


**Figure 1.** AFM images of individual composite nanostructures located on two adjacent gold microelectrodes: (a) $TiO_{2-x}$/MWCNT and (b) $SnO_x$/MWCNT.

Individual MWCNTs, $TiO_{2-x}$/MWCNTs, and $SnO_x$/MWCNTs were deposited on gold microelectrode array from suspensions by spin coating. The suspensions of nanotubes and nanostructures were obtained by ultrasonic dispersion of their layers in dichloromethane for 20 min. C-AFM and EFM measurements were performed using an atomic force microscope MFP-3D SA (Asylum Research) with conductive cantilevers HA_FM/Pt (NT-MDT) in dry nitrogen (RH ~ 2%) at room temperature. Gas sensing properties were investigated on exposure to a reducing gas (ammonia) and an oxidizing gas (nitrogen dioxide) with concentrations of 1000 ppm.

## 3. Results and discussion

Optical and atomic-force microscopes were used to detect the individual MWCNTs and composite nanostructures located on two adjacent electrodes (Fig. 1).

The work function of individual nanoobjects, placed between the electrodes on $SiO_2$ surface, was determined by the technique described in [8]. The mean values of the Fermi level shift for obtained samples upon adsorption of ammonia and nitrogen dioxide were calculated from the values of the work function (Fig. 2).

The work function of MWCNTs functionalized by argon ions averaged 4.86 eV, which is 0.36 eV more than non-functionalized MWCNTs (4.5 eV). By comparing the work function values for highly oriented pyrolytic graphite as calibration sample and MWCNTs, it has been found that the Fermi energy in functionalized MWCNTs is just below the Dirac point by 0.13 eV. This fact indicates that the argon ions form acceptor-like defects in nanotube walls and holes become the major charge carriers in functionalized MWCNTs. During the adsorption of the reducing gas $NH_3$ known as the electron donor [9], the Fermi level shift in MWCNTs increases slightly due to electron-hole recombination. In the case of adsorption of the oxidizing gas $NO_2$, which is an electron acceptor [10], the Fermi level shift decreases significantly caused by the increase of hole concentration in MWCNT walls.

The Fermi level shift in $TiO_{2-x}$/MWCNTs is similar to functionalized MWCNTs, but in contrast to MWCNTs, the titanium oxide formed by magnetron sputtering is n-type semiconductor and major charge carriers therein are electrons. Adsorption of a reducing gas leads to the increase in the electron concentration on the $TiO_{2-x}$/MWCNT surface and the Fermi level shifts upward. During the adsorption of an oxidizing gas, the electron concentration in $TiO_{2-x}$/MWCNT decreases, band-bending at surface increases, and the Fermi level shifts deeper in the energy gap (Fig. 2).

In the $SnO_x$/MWCNTs nanostructure the Fermi level shift during the gas adsorption is the opposite (Fig. 2). Tin oxides are a wide-gap semiconductor with band-bending near the $SnO_x$ surface due to surfaces states. This is responsible for formation of space charge region in tin oxide. It has been previously shown that $SnO_x$ obtained by magnetron sputtering is an amorphous oxide with defect structure and high density of dangling bonds on the surface [6]. As a result, a high density of acceptor-type surface states exists in the band gap of the material, which leads to the Fermi level pinning deeper in band gap and strong band bending at the $SnO_x$ surface.

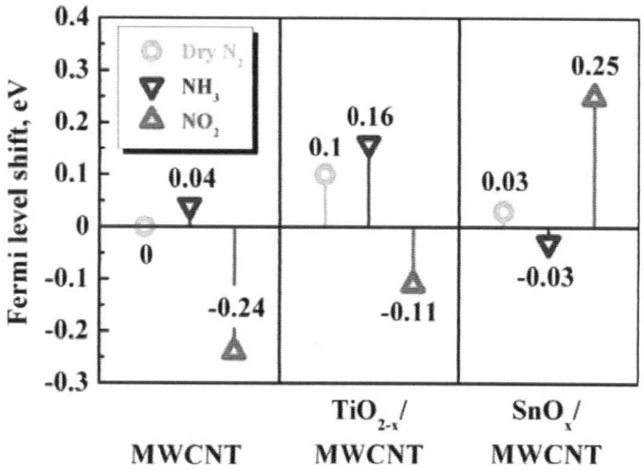


**Figure 2.** The mean value of the Fermi level shift relative to its position in MWCNT for individual MWCNTs, $TiO_{2-x}$/MWCNTs and $SnO_x$/MWCNTs in dry nitrogen, ammonia, and nitrogen dioxide.

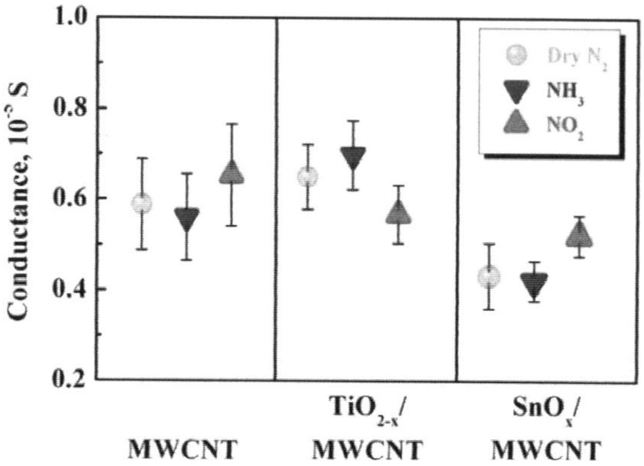


**Figure 3.** The mean value and the range of the conductivity for individual MWCNTs, $TiO_{2-x}$/MWCNTs and $SnO_x$/MWCNTs in dry nitrogen, ammonia, and nitrogen dioxide.

After adsorption of the oxidizing gas, electrons from deep acceptor levels transfer to adsorbate energy levels. The positive charge of surface defect states, which occurs after electrons leave, can partially compensate the surface negative charge thus leading to an increase in the Fermi level shift upwards (Fig. 2). These processes can be considered for explanation of the observed effect for $SnO_x$/MWCNTs nanostructure.

The mean value and range of the conductivity of individual MWCNTs and nanostructures upon exposure to gases were determined by C-AFM measurements (Fig. 3). The electrical conductivity of individual $TiO_{2-x}$/MWCNTs and $SnO_x$/MWCNTs is of the same order of magnitude as that of individual nanotubes. This fact indicates the essential contribution of the MWCNT walls to the conductivity of the composite nanostructures. The conductivity change during the gas exposure is caused by change in the Fermi level shift for individual MWCNTs, $TiO_{2-x}$/MWCNTs, and $SnO_x$/MWCNTs nanostructures.

A decrease and increase in the mean value of the conductivity for MWCNTs functionalized by argon ions correlate with the Fermi level shift due to variation of the hole concentration in the MWCNT walls under gases exposition.

A similar conductivity change upon gas adsorption was observed also for $TiO_{2-x}$/MWCNTs nanostructures. However, $TiO_{2-x}$, in contrast to nanotubes, is n-type semiconductor with electrons as the main charge carriers. Thus, during the ammonia adsorption, the Fermi level shifts upward due to an increase of the electron concentration leading to an increase in electrical conductivity of $TiO_{2-x}$/MWCNTs. Conversely, the Fermi level shifts downward upon nitrogen dioxide adsorption that leads to a decrease in the conductivity of $TiO_{2-x}$/MWCNTs nanostructures (Fig. 3).

For $SnO_x$/MWCNTs nanostructures, a decrease and increase in the mean value of the conductivity upon exposure to gases also correlate with a decrease and increase in the Fermi level shift as a result of a change in the charge carrier concentration in the surface layers of the metal oxide for composite nanostructures.

## 4. Conclusion

The Fermi level shift and the longitudinal conductivity of individual nanotubes and $TiO_{2-x}$/MWCNTs and $SnO_x$/MWCNTs nanostructures upon adsorption of oxidizing and reducing gas molecules have been determined with high resolution using electrostatic force microscopy and conductive atomic force microscopy. The conductivity of the individual $TiO_{2-x}$/MWCNTs and $SnO_x$/MWCNTs is very close to that of individual MWCNT indicating the essential contribution of the nanotube walls in the total conductivity of the functionalized MWCNT coated with tin and titanium oxides. It is shown that the change of the conductivity is consistent with the Fermi level shift caused by change of the free charge carrier concentration in the surface layers of the composite nanostructures upon gas exposure.

**Acknowledgements**
The authors are grateful to Sten'kin Yu.A. for synthesis of MWCNTs, Povoroznyuk S.N. for argon ion beam irradiation of MWCNTs, and Kang V.E. for synthesis of composites layers.

The work was carried out according to the state task of the Omsk Scientific Centre SB RAS (project registration number AAAA-A17-117041210227-8).

**References**
[1] Mallakpour S and Khadem E 2016 *Chem. Eng. J.* **302** 344-67
[2] Espinosa E H, Ionescu R, Chambon B, Bedis G, Sotter E, Bittencourt C, Felten A, Pireaux J-J, Correig X and Llobet E 2007 *Sensor Actuat. B: Chem.* **127** 137-42
[3] Wang S, Cheng G, Cheng K, Jiang X and Du Z 2011 *Nanoscale Res. Lett.* **6** 541
[4] Baldacchini C and Cannistraro S 2007 *Appl. Phys. Lett.* **91** 122103
[5] Melin T, Zdrojek M and Brunel D 2010 *Scanning Probe Microscopy in Nanoscience and Nanotechnology*. (Heidelberg: Springer-Verlag Berlin Heidelberg) p 89
[6] Nesov S N, Bolotov V V, Korusenko P M, Povoroznyuk S N and Vilkov O Y 2016 *Phys. Solid State* **58** 997-1003
[7] Bolotov V V, Kan V E, Knyazev E V, Davletkildeev N A, Nesov S N, Ponomareva I V, Sokolov D V and Korusenko P M 2017 *AIP Conf. Proc.* **1876** 020063
[8] Davletkildeev N A, Stetsko D V, Bolotov V V, Stenkin Y A, Korusenko P M and Nesov S N 2015 *Mater. Lett.* **161** 534-7
[9] Nguyen L Q, Phan Ph Q, Duong H N, Nguyen Ch D and Nguyen L H 2013 *Sensors* **13** 1754-62
[10] Naje A N, Ibraheem R R and Ibrahim F T 2016 *Photonic Sensors* **6** 153-7

IOP Conf. Series: Materials Science and Engineering **699** (2019) 012011 doi:10.1088/1757-899X/699/1/012011

# The effect of optical cooling of the SPM probe in the optomechanical resonator

**M S Dunaevskiy[1], P A Alekseev[1], V A Sharov[1] and A Baranov[2]**

[1]Ioffe Institute, Saint-Petersburg 194021, Russia
[2]Institute of Electronics and Systems, UMR 5214 CNRS – University of Montpellier, Montpellier, France

Mike.Dunaeffsky@mail.ioffe.ru

**Abstract.** Optomechanical resonators formed by the surface of an emitting semiconductor laser and the reflective surface of the probe of scanning probe microscope (SPM) were studied. The use of scanning probe microscopy techniques allowed studying the properties of such optomechanical resonators in details. The effect of optical cooling of an SPM probe in an optomechanical resonator was investigated by measuring the spectra of thermal vibrations of an SPM probe. Minimal achievable temperatures of SPM probe cooling were experimentally established.

## 1. Introduction

The optomechanical resonator is a connected system consisting of an optical resonator and a mechanical resonator. The simplest example of an optomechanical resonator is a Fabry-Perot resonator, in which one mirror is stationary, and the second one is attached to a spring and can oscillate. However, there is a huge number of other various designs of optomechanical resonators. Optomechanical resonators have been studied since the 1970s, one of the first ideas in this area was proposed by Braginskii [1]. In the last decade, studies of processes in optomechanical resonators were actively carried out [2,3]. Ongoing studies indicate the possibility of substantial optical cooling of microscopic mirrors in optomechanical resonators down to the ground state [4,5]. This opens up a number of new research opportunities in the field of precision measurements of ultra-small forces [6], quantum computing [7], as well as in other areas. The ability to detect ultra-small forces is extremely important, for example, in scanning probe microscopy (SPM) measurements. It is worth noting the work [8] proposing a method for creating a special SPM-probe, which is connected to a high-Q optical resonator and forms an optomechanical resonator, which allows: (i) increasing the sensitivity of motion detection to values ~ 1 fm/(Hz)$^{1/2}$ and (ii) tuning the stiffness coefficient of the SPM-probe in a wide range from 0.01 N/m up to 290 N/m. It is worth noting that such an optomechanical SPM-probe is quite difficult to manufacture, extremely expensive, and, therefore, cannot be widely used in SPM research.

In this paper, we study the optomechanical resonators formed by the surface of an emitting semiconductor laser and the reflective surface of SPM-probe. The use of SPM techniques allows studying the properties of such optomechanical resonators in details. In this paper, we study the effect of optical cooling of the SPM probe. The optical cooling effect can be used to increase significantly the sensitivity of the SPM during the detection of small forces.

IOP Conf. Series: Materials Science and Engineering **699** (2019) 012011 doi:10.1088/1757-899X/699/1/012011

## 2. Experimental details

Recently, the authors of this work have discovered the effect of a shift in the resonant frequency $\Delta\omega$ of an atomic force microscope (AFM) probe under illumination [9]. This effect is related with the light absorption in the probe. Based on this effect, a new probe method was developed to measure light-induced frequency shifts (LIFS) for the near-field mapping of the intensity distribution J(x,y,z) of the radiation of semiconductor laser structures [9]. In addition, the authors discovered the interference effect between the light wave emerging from a laser and the wave reflected from the SPM probe [10]. In this case, the laser surface and the reflecting surface of the SPM-probe form an optomechanical resonator. As a result, when the SPM-probe was placed at a distance (from the laser surface) equal to an integer number of half-waves, multiple amplification of the detected signal occurred. This allows one to use this effect for (i) increasing sensitivity when detecting light emission, (ii) performing LIFS-spectroscopy experiments, and (iii) performing optical cooling experiments.

It is worth noting that the SPM-probe is usually inclined at an angle of 15 degrees to the surface under study, which does not allow creating a high-Q resonator. To improve the quality factor, we placed the SPM-probe surface parallel to the light-emitting surface of the semiconductor laser (Fig. 1). Such a plane-parallel optical resonator, in which one mirror is stationary and the second one can oscillate (the SPM-cantilever is a mechanical oscillator), is a classic example of an optomechanical resonator. A feature of such optomechanical resonators is that when light interacts with the SPM-probe, an additional force arises, which can either suppress the thermal vibrations of the SPM-probe (optical cooling), or vice versa, significantly increase them (optical heating). In this work, for the experimental detection of the effect of optical cooling, we will use the standard technique for measuring the spectrum of thermal vibrations of the SPM-probe.

The SPM measurements were carried out using Ntegra-Aura (NT MDT, Russia) device. Silicon PPP-FMAu (Nanosensors) and silicon nitride ORC-8 (Bruker) probes were used to study the optical cooling effect. We used semiconductor stripe laser emitting on 1.05 μm (threshold current $I_{th}$=30 mA) as a part of the optomechanical resonator. This laser is based on the AlGaAs/InGaAs/GaAs heterostructure grown by metalorganic chemical vapour deposition.

## 3. Results and discussion

Figure 2 shows the measured spectrum of thermal vibrations of the silicon PPP-FMAu probe (in the dark) and the result of its fitting by the Lorentzian function. In further analysis, we will give only the smoothed spectra of thermal vibrations that are already free from noise. Figure 3 shows the measured spectra of thermal vibrations of the silicon PPP-FMAu probe (i) in the dark and (ii) when the laser is turned on and pumped with a current above the laser generation threshold. It can be seen that at certain probe-surface distances, increase in the power of the laser radiation leads to a multiple attenuation of the thermal noise amplitude, which corresponds to a multiple decrease in temperature.

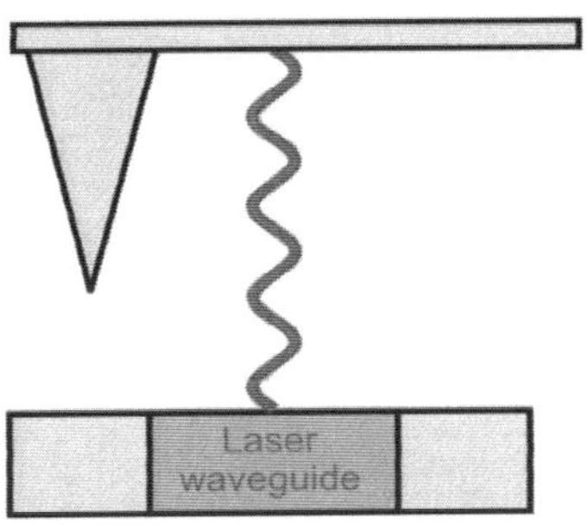


**Figure 1.** The scheme of the optomechanical resonator "semiconductor laser - SPM probe".

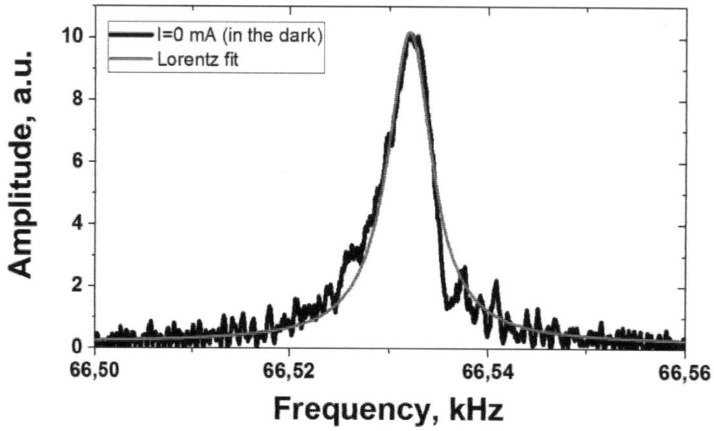


**Figure 2.** The spectrum of thermal vibrations of the silicon PPP-FMAu probe (black curve) and its fitting (red curve) by the Lorentzian function.

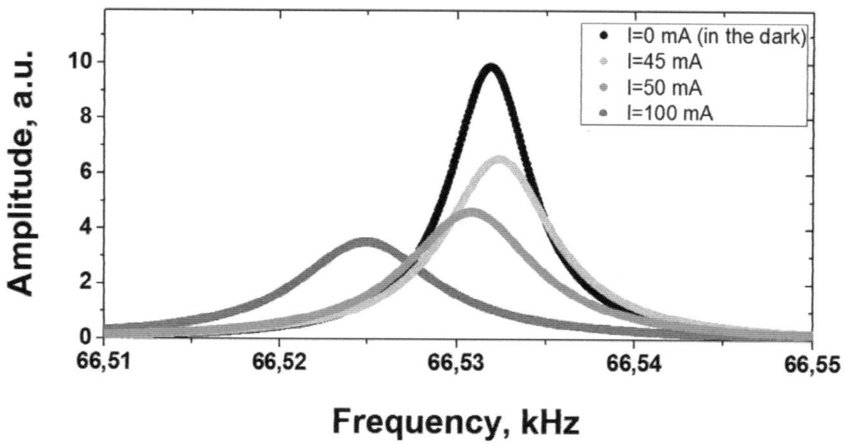


**Figure 3.** The fitted spectra of thermal vibrations of the silicon PPP-FMAu probe in the dark and with increase of light intensity in the optomechanical resonator.

It should be noted that with a further increase in the power of the light entering into the optomechanical resonator, the temperature increase was observed. That is a certain limiting value of the achieved cooling temperature ($T_{min}$). In our experiments, it was possible (i) to cool silicon PPP-FMAu probes down to 100 K and (ii) to cool silicon nitride probes down to 80 K. It can be assumed that heating is associated with an increase in light absorption in the probe. In order to decrease $T_{min}$ during optical cooling, probes with a bottom reflective coating can be used. This will reduce absorption in the probe and allow the probe to be cooled down to $T_{min}$.

## 4. Conclusion

The formation of optomechanical "probe-laser" resonators using silicon and silicon nitride probes was investigated using scanning probe microscopy methods. The possibilities of optical cooling of probes in an optomechanical resonator were investigated. It is shown that the minimum achievable optical cooling temperatures are 80-100 K. To improve the optical cooling process, it is necessary to deposit a bottom reflective coating to the probes thus decreasing absorption in the probes.

IOP Conf. Series: Materials Science and Engineering **699** (2019) 012011 doi:10.1088/1757-899X/699/1/012011

**Acknowledgements**
M.S.D. acknowledges support by RFBR grant (research project No 17-52-16029). P.A.A. and V.A.S. acknowledge support by Goszadanie project.

**References**
[1] Braginskii V B, Manukin A B and Tihonov M Yu 1970 *Zh. Eksp. Teor. Fiz.* **58** 1549-52
[2] Kippenberg T J and Vahala K J 2008 *Science* **321** 1172-6
[3] Groblacher S, Hammerer K, Vanner M R and Aspelmeyer M 2009 *Nature* **460** 724-7
[4] Schliesser A, Rivière R, Anetsberger G, Arcizet O and Kippenberg T J 2008 *Nat. Phys.* **4** 415-9
[5] Chan J, Alegre T P M, Safavi-Naeini A H, Hill J T, Krause A, Gröblacher S, Aspelmeyer M and Painter O 2011 *Nature* **478** 89-92
[6] LIGO Collaboration 2011 *Nat. Phys.* **7** 962-5
[7] Stannigel K, Rabl P, Sørensen A S, Zoller P and Lukin M D 2010 *Phys. Rev. Lett.* **105** 220501
[8] Liu Y, Miao H, Aksyuk V and Srinivasan K 2012 *Opt. Express* **20** 18268
[9] Dunaevskiy M S, Alekseev P A, Baranov A N, Monakhov A M, Teissier R, Arinero R, Girard P and Titkov A 2013 *Appl. Phys. Lett.* **103** 053120
[10] Dunaevskiy M, Dontsov A, Alekseev P, Monakhov A M, Baranov A, Girard P, Arinero R, Teissier R and Titkov A 2015 *Appl. Phys. Lett.* **106** 171105

IOP Conf. Series: Materials Science and Engineering **699** (2019) 012012 doi:10.1088/1757-899X/699/1/012012

# Investigation of resistive switching by non-stationary signals in $ZrO_2$(Y) films by atomic force microscopy

**D O Filatov[1], M N Koriazhkina[1,3], D A Antonov[1], I N Antonov[2], D A Liskin[1,3], M A Ryabova[1,3] and O N Gorshkov[1]**

[1]Research and Education Centre for Physics of Solid State Nanostructures, Lobachevskii State University of Nizhnii Novgorod, Nizhnii Novgorod, 603950, Russia
[2]Research Institute for Physics and Technology, Lobachevskii State University of Nizhnii Novgorod, Nizhnii Novgorod, 603950, Russia
[3]Department of Physics, Lobachevskii State University of Nizhnii Novgorod, Nizhnii Novgorod, 603950, Russia

dmitry_filatov@inbox.ru

**Abstract.** Resistive switching in the yttria stabilized zirconia films on the conductive substrates has been studied using Conductive Atomic Force Microscopy. Switching was performed by triangle voltage pulses with superimposed high-frequency sinusoidal signal applied between the probe and the sample. The performance of the resistive switching was characterized quantitatively by the ratio of the electric current flowing through the contact of the probe to the dielectric film surface (together constituting a nanometre-sized virtual memristor) in the low resistance state and in the high resistance one (ON/OFF ratio). The increase in this ratio when applying the sinusoidal signal as compared to the switching by the triangle pulses has been observed. Also, the long-time scale stability of the mean values of the probe current in both states was found to improve when applying the sinusoidal signal. The effect of the high frequency sinusoidal signal on the performance of the virtual memristor was attributed to the resonant activation of the oxygen ion migration via the oxygen vacancies by the alternating external electric field.

## 1. Introduction

In recent years, the investigations of resistive switching (RS) have attracted much attention [1]. The effect of RS consists in the bistable (or multistable) switching of the resistance of a thin (from several nanometres to several tens of nanometres in thickness) dielectric films sandwiched between two conductive electrodes under the external voltage applied between the electrodes [2]. The electronic devices based on the RS effect are called *memristors* [3]. At present, the understanding of the RS mechanism in the metal oxides is based on the concept of forming the conductive filaments composed of the oxygen vacancies between the conductive electrodes in the electric field applied between the electrodes [4]. The switching of a memristor from the low resistance state (LRS) to the high resistance state (HRS) is achieved by the rapture of the filament by a voltage pulse of an appropriate polarity applied between the electrodes (the RESET process). The filament can be restored again by another voltage pulse of the opposite polarity that results in the switching from the HRS back to LRS (the SET process). The processes described above constitute the physical mechanism of so-called *bipolar* RS.

The memristors are considered to be promising for various applications, for instance, in the next generation of non-volatile computer memory (Resistive Random Access Memory, ReRAM) [5], in the neuromorphic computers [6], etc. However, at present the application of the memristors is limited by an insufficient stability of the memristor parameters, which is a fundamental property of the RS [7]. The origin of the RS instability is the stochastic nature of the rapture and restoring of the filaments in the functional dielectric layer. A very limited (countable) number of oxygen vacancies are involved into the RS [8]. So far, memristor appears to be a typical mesoscopic system, where the fluctuations of any physical parameter may be much larger than its averaged value.

Traditional approaches to the improvement of the RS stability include the appropriate choice of the functional insulator and electrode materials, the engineering of the grain boundaries in the functional insulator, the application of electric field concentrators inside the insulator, etc. [9–11]. The alternative approaches include the circuit solutions, the application of adaptive switching protocols, etc. [12].

In the present study, the local RS in the yttria stabilized zirconia (YSZ) films was investigated by Conductive Atomic Force Microscopy (CAFM) [13] using the triangular switching voltage pulses with superimposed sinusoidal high-frequency (HF) signal.

## 2. Experiment

The YSZ films of ≈4 nm in thickness (the molar fraction of the stabilizing oxide $Y_2O_3$ in the film material was ≈0.12) were deposited by HF magnetron sputtering at 300 °C using Torr International® 2G1-1G2-EB4-TH1 vacuum setup. The Si(001) substrates with pre-deposited $SiO_2$ film (≈500 nm thick), Ti adhesion layer, and TiN conductive one (25 nm in thickness each) were used.

The RS in the YSZ films was investigated using Omicron® UHV AFM/STM LF1 in ultra-high vacuum (UHV) at room temperature in the contact mode. The base residual gas pressure inside the AFM/STM chamber was ~$10^{-10}$ Torr. The NT-MDT® NSG-11 DCP™ diamond-like film coated AFM probes were used. The schematic representation of the experimental setup is shown in Figure 1. The bias voltage $V_g$ applied between the CAFM probe and the TiN conductive sublayer was supplied by the digital-to-analogue converter (DAC) of NT-MDT® Solver Pro™ AFM controller used as an external computer-controller programmable voltage source.

The endurance diagrams for the probe current values $I_t$ in the LRS and HRS $I_{ON}$ and $I_{OFF}$, respectively during multiple cyclic write/erase operations were recorded in the experiments. The switching/measurement protocol is shown in Figure 2. After the triangular switching voltage pulses with an amplitudes $V_{SET}$ = 5 V and $V_{RESET}$ = –6 V and the pulse durations $T_{SET}$, $T_{RESET}$ = 1 ÷ 5 s, the values of $I_{ON}$ and $I_{OFF}$, respectively, were recorded at $V_g$ set to the read voltage $V_{READ}$ = 3 V several times $N_{READ}$ = 10 ÷ 20 and averaged. The HF sinusoidal signal with the amplitude $A$ = 0.2 V and frequency $f$ = 6.5 kHz (matching the frequency of the $O^{2-}$ ion jumps to the nearest neighbouring oxygen vacancies in YSZ at 300 K [14]) was superimposed onto the triangle switching pulses.

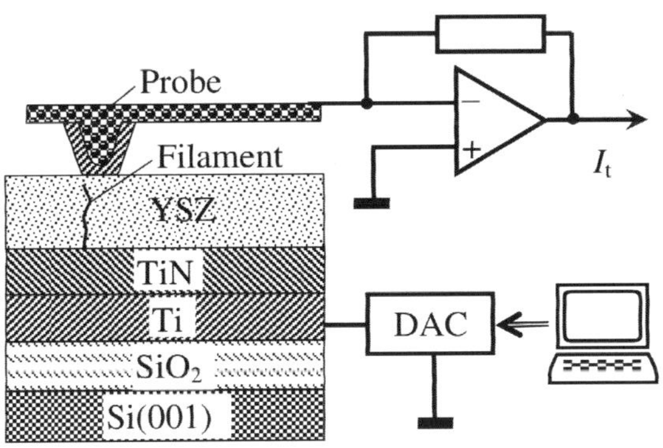


**Figure 1.** Schematic representation of the experimental setup for studying the local RS in thin YSZ film on a conductive substrate by CAFM. $I_t$ – probe current signal; DAC – digital-to-analogue converter.

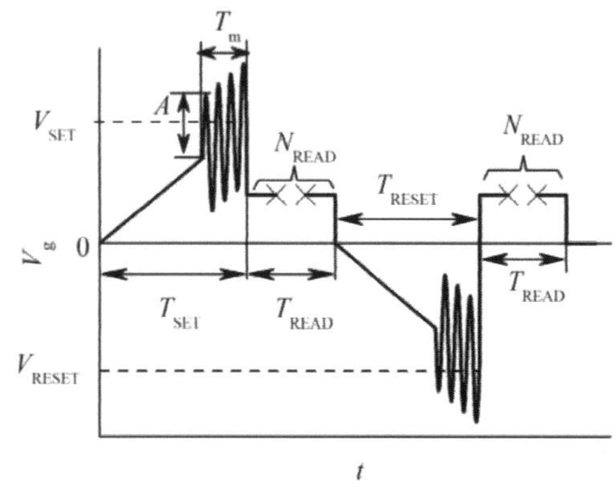


**Figure 2.** The protocol (the dependence of $V_g$ on the time $t$) for measuring the endurance diagrams to study the effect of the superposition of the HF sinusoidal signal onto the triangle switching pulses on the RS parameters.

## 3. Results and discussion

A pronounced hysteresis typical for the bipolar RS was observed in the cyclic current-voltage ($I$–$V$) curves of the CAFM probe contact to the YSZ film (together constituting a nanometre-sized virtual memristor) (Fig. 3). The decrease of the area within the hysteresis loops with number of switching cycles $N$ was the manifestation of the virtual memristor degradation during recording of the cyclic $I$–$V$ curves. Accordingly, the $I_{ON}$ values approached $I_{OFF}$ with increasing $N$ (Fig. 4). The adding of the HF sinusoidal signal to the triangular switching pulses resulted in an increase of the local RS durability: the decrease of $I_{ON}$ began at larger values of $N$ than in the case of switching by the triangle pulses without the HF sinusoidal signal. The observed effect of enhancing the RS durability was attributed to the resonant activation of the $O^{2-}$ ion migration via the oxygen vacancies under the external alternating electric field, which stimulates the elementary jumps of the $O^{2-}$ ions and, this way, favours the formation and rupture of the filaments under the slowly varying external electric field during the SET and RESET processes, respectively. The effect of resonant activation belongs to a wide class of phenomena inherent to the stochastic multistable systems (along with stochastic resonance, noise-induced stabilization, etc.) and was manifested in many fields of physics and chemistry [15].

The characteristic frequency of the $O^{2-}$ ion jumps onto the nearest neighbouring oxygen vacancy $f_i$ can be estimated from the ion jump activation energy $E_a$ using the following equation [16]:

$$f_i \sim f_0 \exp\left(-\frac{E_a}{kT}\right), \quad (1)$$

where $f_0 \sim 10^{13}$ Hz is typical lattice vibration frequency, $k$ is Boltzmann constant, and $T$ is the temperature. Assuming $E_a = 0.55$ eV for YSZ [14], one gets $f_i \approx 6.5$ kHz at 300 K. Based on this estimate, the frequency of the HF sinusoidal signal used in the experiment was selected.

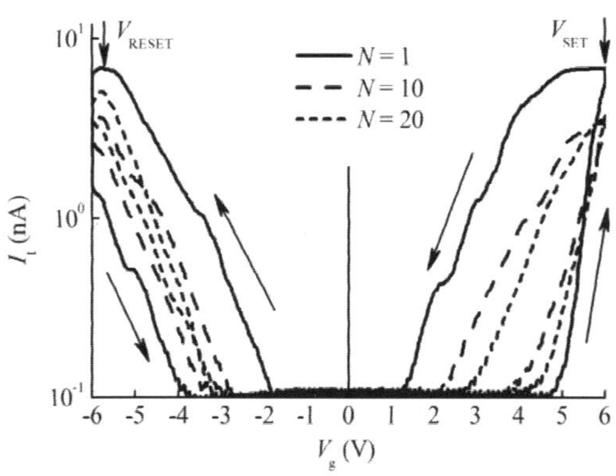


**Figure 3.** Cyclic $I$–$V$ curves of the CAFM probe contact to the YSZ film surface (virtual memristor). $N$ – the number of switching cycles.

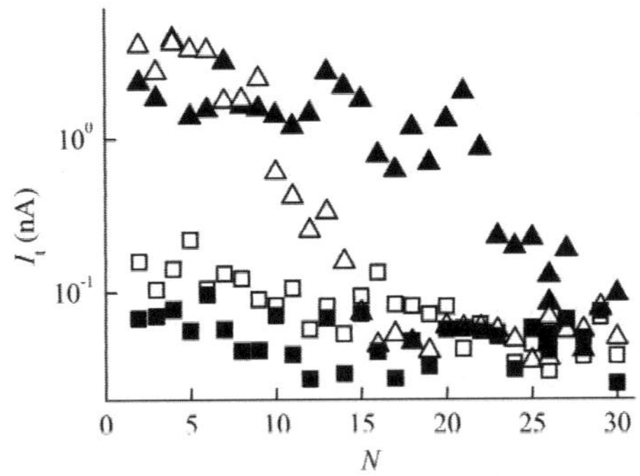


**Figure 4.** Endurance diagrams for the AFM probe contact to the YSZ film (virtual memristor): ▲,△ – $I_{ON}$, ■,□ – $I_{OFF}$, ■,▲ – with HF sinusoidal signal ($A$ = 0.2 V, $f$ = 6.5 kHz) on the triangle switching pulses; □,△ – without the sinusoidal signal.

## 4. Conclusion

The present study demonstrates a beneficiary impact of additional high frequency sinusoidal signal on the triangle switching voltage pulses on the stability of the resistive switching in the contact of a CAFM probe to the YSZ film on a conductive substrate. The observed effect was attributed to the resonant activation of the oxygen ion migration via the oxygen vacancies in YSZ in external alternating electric field oscillating with corresponding frequency. The obtained results open the prospects for the development of the innovative switching protocols applicable to the next-generation RRAM devices to improve the durability of these ones.

**Acknowledgements**
The present study was supported by Russian Foundation for Basic Research (Project # 18-42-520059). The CAFM investigations were carried out using the shared research facilities of Research and Educational Centre for Physics of Solid State Nanostructures at Lobachevskii State University of Nizhnii Novgorod.

## References

[1] Ielmini D and Waser R 2016 *Resistive Switching: From Fundamentals of Nanoionic Redox Processes to Memristive Device Applications* (Saarbruken: Wiley-VCH)
[2] Waser R and Aono M 2007 *Nature Mater.* **6** 833-40
[3] Strukov D B, Snider G S, Stewart D R and Williams R S 2008 *Nature* **453** 80-3
[4] Riess I 2017 *J. Electroceram.* **39** 61-72
[5] Ouyang J 2016 *Emerging Resistive Switching Memories* (Berlin-Heidelberg: Springer)
[6] Huang A, Zhang X, Li R and Chi Yu 2018 Memristor Neural Network Design *Memristor and Memristive Neural Networks* ed A James (Rijeka: IntechOpen) chapter 12 pp 249-81
[7] Yi W, Savel'ev S E, Medeiros-Ribeiro G, Miao F, Zhang M-X, Yang J J, Bratkovsky A M and Williams R S 2016 *Nature Commun.* **7** 11142
[8] Parreira P, Paterson G W, McVitie S and MacLaren D A 2016 *J. Phys. D: Appl. Phys.* **49** 095111
[9] Sun Y, Song C, Yin J, Chen X, Wan Q, Zeng F and Pan F 2017 *ACS Appl. Mater. Interf.* **9** 34064
[10] Trapatseli M, Cortese S, Serb A, Khiat A and Prodromakis T 2017 *J. Appl. Phys.* **121** 184505
[11] Wu W, Wu H, Gao B, Deng N and Qian H 2018 *J. Appl. Phys.* **124** 152108
[12] Alibart F, Gao L, Hoskins B D and Strukov D B 2012 *Nanotechnology* **23** 075201
[13] Lanza M 2014 *Materials* **7** 2155-82
[14] Filatov D, Tikhov S, Gorshkov O, Antonov I, Koryazhkina M and Morozov A 2018 *Adv. Cond. Mat. Phys.* 2028491
[15] Gammaitoni L, Hanggi P, Jung P and Marchesoni F 1998 *Rev. Mod. Phys.* **70** 223-87
[16] Yakimov A V, Filatov D O, Gorshkov O N, Antonov D A, Liskin D A, Antonov I N, Belyakov A V, Klyuev A V, Carollo A and Spagnolo B 2019 *Appl. Phys. Lett.* **114** 253506

IOP Conf. Series: Materials Science and Engineering **699** (2019) 012013 doi:10.1088/1757-899X/699/1/012013

# Microscopic studies of disordered carbon-rich inclusions in ultra-high pressure glasses from impactites of the Kara astrobleme

**Ye A Golubev[1], T G Shumilova[1,2], S I Isaenko[1] and A A Utkin[3]**

[1]Institute of Geology of Komi SC of Uralian Branch of RAS, Syktyvkar, 167982 Russia
[2]Hawaii Institute of Geophysics and Planetology, University of Hawaii at Manoa, Honolulu, HI, 96822, USA
[3]Department of Radiophysics, Syktyvkar State University, Syktyvkar, 167000 Russia

golubev@geo.komisc.ru; shumilova@geo.komisc.ru

**Abstract.** The paper presents the results of study of the carbon-containing phase discovered in the impact glass of the Kara astrobleme. We used the following research methods: optical, scanning electron, and atomic force microscopies, as well as Raman spectroscopy. This phase represents carbon-containing inclusions up to several tens of micrometres in size with an amorphous diamond-like structure. At the nanoscale, the studied phase is characterized mainly by a homogeneous structure.

## 1. Introduction

Diamond-like carbon (DLC) is a general term for various forms of amorphous carbon materials demonstrating the unique properties of natural diamond [1]. DLC is a mixture of $sp^2$ and $sp^3$ hybridized carbon atoms. Generally, this material is a diamond-like $sp^3$-matrix with $sp^2$-clusters.

Amorphous diamond-like carbon films and nanocomposites based on them are promising new materials, which determines the relevance of their research [2,3]. The amorphous diamond-like state of carbon is metastable. Therefore, despite the large number of works performed recently on amorphous diamond-like carbon materials, there is a need for further research to determine their properties, structure, and stability limits.

Impact glasses are present among natural objects as a product of impact (shock) transformation of rocks of the Earth's crust [4–8]. Ultra-high-pressure high-temperatures (UHPHT) glasses are particularly interesting, because their formation is caused by the highest pressures (several tens or hundreds of GPa) at temperatures above 3300 K [9-11]. We expect to obtain interesting information from detailed studies of the structure and physical properties of melt impact silicate glasses of the vein type, which have been recently discovered in the impactites of the Kara astrobleme (Pay-Khoy, Russia) [12] formed about 70 Ma. These glasses form ribbon-like subparallel vein bodies, secant suevites, and are characterized by multilevel liquation structures with relic coesite [12]. Detailed results of studying the structure of this impactite by spectroscopic methods showed [13, 14] that it generally consisted of amorphous feldspar material and contained small drops of coesite-containing quartz glass with a low degree of polymerization of the silicate carcass. At that, the spectroscopy indicated the presence of an amorphous carbon phase in these impactites [13]. The discovery and study of the carbon particles with an amorphous structure in natural impact glasses is of great interest.

This paper reports on the discovery of an amorphous carbon matter in UHPHT glasses of the Kara astrobleme and on the results of the study of this matter by atomic force and scanning electron microscopies.

## 2. Experimental

For the studies, we used natural UHPHT vein-bodied impact glasses from the Kara astrobleme (Pay-Khoy, Russia). The material was sampled by T.G. Shumilova in 2015 at the right bank of the Kara River.

We studied the polished surface of the sample. It was covered with a gold layer with a thickness about 0.5 nm by magnetron sputtering, and then elemental mapping was performed. Then the gold layer was removed together with the upper layer of the surface to the depth about 0.5 μm by a grinding machine using aluminium oxide powder, and atomic force microscopy (AFM) studies of the selected objects were carried out. The objects on the surface of the sample after AFM study were additionally diagnosed by energy-dispersive X-ray spectroscopy (EDS) mapping and local elemental analysis.

Initially, the samples have been observed with optical microscopy (microscope POLAM-312 and Nikon Eclipse E400 Pol equipped with a camera DCMOS14000KPA). Raman spectroscopy tests were provided by a spectrometer LabRam HR800 instrument (Horiba, JobinYvon) equipped by a He-Ne laser (632.8 nm, 2 mW), an optical microscope Olympus BX41, and a Si-based CCD detector. Spectra were obtained in the 100–4000 $cm^{-1}$. Zero position, spectra background correction, and individual bands deconvolution have been provided with a curve-fitting procedure from the LabSpec 5.39 software.

The chemical composition has been analysed using a scanning electron microscope (SEM) Tescan MIRA3 equipped by an X-ray energy-dispersive spectrometer.

The topography has been examined with AFM in a tapping and phase-contrast modes using an Integra Prima (NT-MDT, Russia) by super sharp silicon cantilevers SSS-NCH (Nanosensors) with the resonant frequency 330 kHz, the radius at the tip 2-4 nm, and the stiffness constant about 35 N/m. The images have been recorded with resolution of 512 × 512 pixels at a scan frequency 0.9-1 Hz. The sample surfaces were grounded by a silver paste to avoid the influence of static electricity.

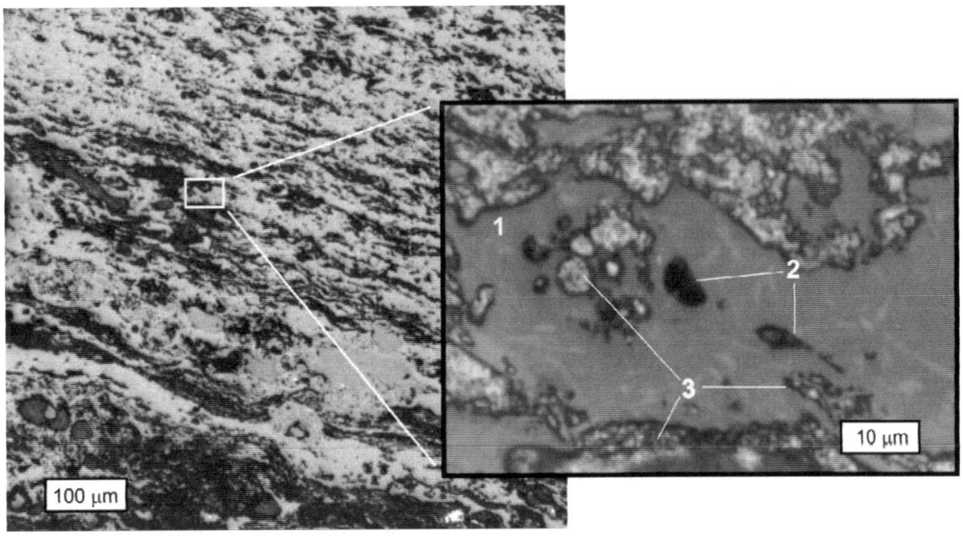


**Figure 1.** General view of the microstructure of the impact Kara glass: optical image in reflected light. 1 – aluminosilicate glass; 2 – inclusions of carbonaceous matter; 3 – separate inclusions of mixed-layer aluminosilicates.

## 3. Results

Vein UHPHT glass has a fluid microstructure with an aluminosilicate matrix and silicate inclusions of hundreds of micrometres in size [15] (Fig. 1). In addition, numerous rounded inclusions of mixed-layer aluminosilicates are presented. They are elongated and often curved according to the matrix texture.

The aluminosilicate glass has a variable composition with a significant presence of iron, calcium, and magnesium (Table 1). Silicate inclusions are chemically homogeneous shapeless areas without contrasting boundaries with the surrounding feldspar glass [15]. The carbon inclusions are presented primarily in the silicate regions. These inclusions have a vitreous luster, rounded shapes, in a longitudinal section similar to beans (Fig. 1). The sizes of the inclusions are from 1 to 10 microns, and thickness is below 2 microns (Fig. 2). The composition of these inclusions cannot be determined correctly by EDS method because of their small thickness. The electron beam penetrates through the carbon inclusion, and the final analysis includes elements being a part of the surrounding matter. First of all, it is silicon, oxygen, and aluminium. Therefore, the carbon content measured in the inclusions using EDS cannot be considered reliable; it likely exceeds this value significantly.

**Table 1.** The content of rock-forming oxides in basic phases of impact glass of the Kara astrobleme (in at.%).

| Phases | $SiO_2$ | MgO | $Al_2O_3$ | $Na_2O$ | $K_2O$ | CaO | $TiO_2$ | $Fe_2O_3$ | C |
|---|---|---|---|---|---|---|---|---|---|
| Alumino-silicate matrix | 61–78 | 2–4 | 12–25 | 0.3–2 | 1.75–2.5 | 1.5–2.5 | 0.1–0.3 | 2.5–5 | n.d. |
| Silicate inclusions | 90–93 | 0.5–1 | 0.5–1.5 | 0.5–1 | 0.5 | n.d. | 0.1 | 0.5 | n.d. |
| Carbon-rich inclusions | 41–43* | 3–4* | 6.5–7* | 0.75* | 0.75* | 1* | 0.3–0.6* | 2–3* | 36–39 |

n.d. – not detected, * – probably from the host glass matrix.

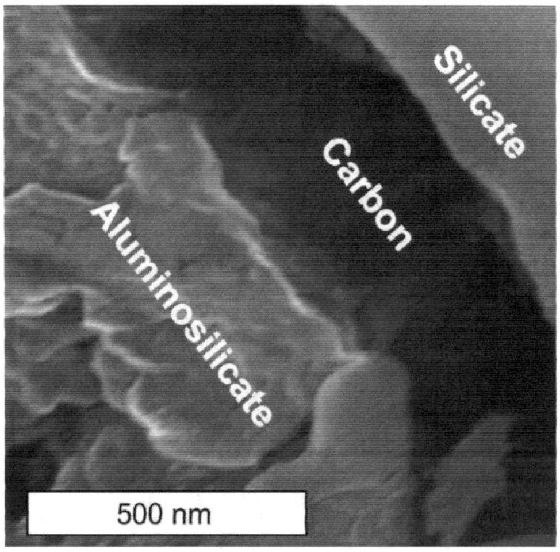


**Figure 2.** SEM image of a cut of amorphous carbon grain.

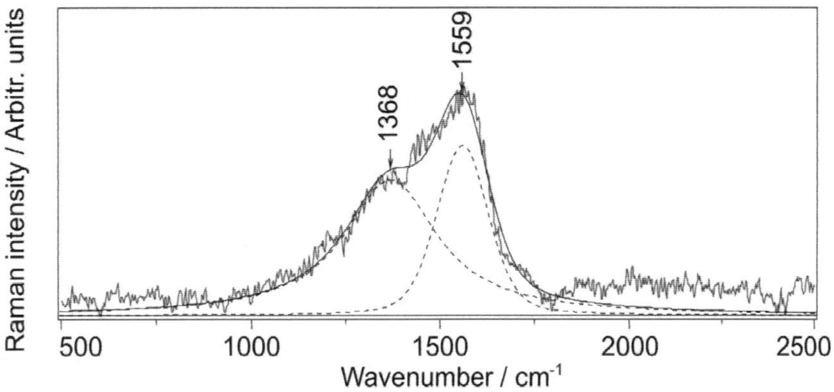


**Figure 3.** Raman spectrum of carbon inclusion in the aluminosilicate glass matrix. Typical Raman spectrum of the carbon inclusion and its decomposition into two Gaussian components with maxima near 1368 and 1559 $cm^{-1}$ and FWHM values 300 and 156 $cm^{-1}$, respectively. The solid curve is a sum of two Gaussian components, dashed lines show individual components.

The Raman spectrum of carbon inclusions (Fig. 3) has two wide intersecting bands that correspond to position to D and G bands. The maximum of G-band lies in the range from 1558 to 1568 $cm^{-1}$, the maximum of D-band varies over a wider range from 1360 to 1380 $cm^{-1}$. FWHM values are from 150 to 170 $cm^{-1}$ for G-band and about 300 $cm^{-1}$ for D-band. The ratio of D and G bands maxima values is about 0.7 and the ratio of the integrals ranges from 1.8 to 1.9. A wide band of water in 3200-3500 $cm^{-1}$, which is present in silicate and aluminosilicate glasses [15], is absent in the carbon matter.

Scanning electron and atomic force microscopies show the predominant morphological homogeneity of the carbon inclusion. SEM images allowed revealing that the cleavage surface is broken by sinuous isolated fractures into partially isolated rounded blocks 50-200 nm in size with serrated edges. According to AFM data, the amorphous carbon inclusions are mainly homogeneous and composed of films up to 500 nm long and 50-150 nm wide (Fig. 4).

## 4. Discussion

Raman spectroscopy is the best method to obtain the bonding structure of DLC materials in details. In the Raman spectra of amorphous carbon with mixed $sp^2$–$sp^3$ hybridization, even a small content of $sp^2$-

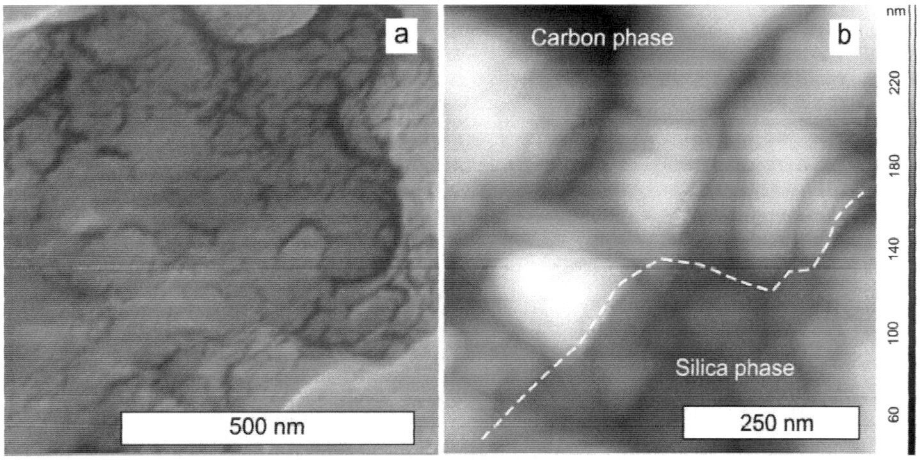


**Figure 4.** Cleavage surface of carbon inclusion: (a) SEM image; (b) AFM image.

bound carbon atoms determines their predominant contribution to the spectra, overlapping the component associated with $sp^3$-hybridized carbon atoms [1]. This phenomenon is well known and is related to the fact that in the visible range the capture cross-section for $sp^2$-bound carbon atoms is much larger, since π-state has a lower energy and is much more polarized than π-state, which gives ten times difference in the spectrum intensity [1]. Therefore, even if the proportion of $sp^2$ bond in the carbon matter is 10-15%, it will dominate in the intensity of the bands in the Raman spectrum.

The shift of G band from the standard graphite 1580 $cm^{-1}$ to 1560 $cm^{-1}$ is probably conditioned by the presence of $sp^3$ bond [1]. The abnormally broadened reflections D and G show a strong distortion of interplanar spacings. This shift, a large band width, and a small ratio of band intensities confirm a significant disorder of the bonds due to increasing $sp^3$-bond formation.

The atomic force and scanning electron microscopies have shown that amorphous diamond-like carbon of geological origin has a more homogeneous structure at the nanoscale, in contrast to natural and synthetic glasses and glassy carbon [16].

Pressure-temperature conditions for the formation of diamond-like films in the laboratory are significantly softer than the impact conditions (pressure less than 200 kPa and temperature less than 2000 K). Diamond-like carbon is formed after rapid cooling of scattered carbon atoms and annealing on relatively cold surfaces at high energy. Nevertheless, there is evidence of the formation of amorphous diamond-like carbon from glassy carbon at high (up to 100 GPa) pressures [18] and amorphous carbon glass under “stellar” (7000-13000 K, 40 GPa) conditions [19]. Therefore, the formation of amorphous diamond-like carbon upon impact of a meteorite into rocks is quite probable. Organic objects (trees, coal, bitumen of different degrees of transformation) could act as precursors.

## 5. Conclusion

Unusual natural carbon matter was found in UHPHT impact glasses of Kara astrobleme using EDS. It presents the isolated inclusions up to several tens of micrometres in size. Raman spectra allowed comparing structure of studied matter to amorphous diamond-like carbon. The atomic force and scanning electron microscopies were used to study the nanoheterogeneous structure of the amorphous carbon matter. Microscopy showed a predominantly homogeneous layered structure of carbonaceous matter at the nanoscale. However, it looks like it is not the pure carbon, but rather a carbon-containing composite. Additional high-resolution study is required to clarify this point.

## Acknowledgements

The study was supported by Russian Science Foundation (project 17-17-01080). The studies were conducted with the help of scientific equipment at the Centre for collective use “Geonauka” at the Institute of Geology of Komi SC of Uralian Branch of RAS supported by NIR # AAAA-A17-117121270036-7 and at the Syktyvkar State University (Syktyvkar, Russia).

## References

[1] Robertson J 2002 Diamond-like amorphous carbon *Mater. Sci. Eng. R* **37** 129-281
[2] McKenzie D R, Muller D and Pailthorpe B A 1991 Compressive-stress-induced formation of thin-film tetrahedral amorphous carbon *Phys. Rev. Lett.* **67** 773-6
[3] Grill A 1999 Diamond-like carbon: state of the art *Diam. Relat. Mater.* **8** 428-34
[4] Impact craters at the turn of the Mesozoic and Cenozoic. Sat Art. / Ed. Masaytis V L, L.: Nauka, 1990. 191 p
[5] Lyutoev V P and Lysyuk A Yu 2016 The structure and texture of silica of impactites of the Karsky astrobleme *Vestnik of the IG of the Komi Science Center* **9** 24-32
[6] French B M 1998 Traces of catastrophe: A handbook of shock-metamorphic effects in terrestrial meteorite impact structures. LPI Contribution No. 954. Lunar and Planetary Institute, Houston, Texas. 120 p
[7] Melosh H J 1989 Impact cratering – A geological process. Oxford Univ. Press, New York
[8] Robertson P B and Grieve R A F 1977 Shock attenuation at terrestrial impact structures. In: Roddy

D J, Pepin P O, Merill R B. (eds) Impact and explosion cratering. Pergamon Press, New York, 687-702

[9] Langenhorst F and Deutsch A 1994 Shock experiments on pre-heated α-and β- quartz: I. Optical and density data *Earth Planet. Sc. Lett.* **125** 407-20

[10] Schmitt R T 2000 Shock experiments with the H6 chondrite Kernouvé: pressure calibration of microscopic shock effects *Meteorit. Planet. Sci.* **35** 545-60

[11] Stöffler D 1984 Glasses formed by hypervelocity impact *J. Non-Cryst. Solids* **67** 465-502

[12] Shumilova T G, Isaenko S I, Makeev B A, Zubov A A, Shanina S N, Tropnikov Ye M and Askhabov A A 2018 Ultrahigh-pressure liquation of an impact melt. Part 1 *Dokl. Earth Sci.* **480** 595-8

[13] Shumilova T G, Lutoev V P, Isaenko S I, Kovalchuk N S, Makeev B A, Lysiuk A Yu, Zubov A A and Ernstson K 2018 Spectroscopic features of ultrahigh-pressure impact glasses of the Kara astrobleme *Sci. Rep.* **8** 6923

[14] Shumilova T G, Zubov A A, Isaenko S I, Karateev I A and Vasiliev A L 2019 Mysteriouslong-living ultrahigh pressure or secondary impact crisis? *Geoscience Frontiers* (in press)

[15] Golubev Y A, Shumilova T G, Isaenko S I, Radaev V A, Utkin A A, Makeev B A and Ernstson K 2019 Microscopic studies of ultra-high pressure glasses from impactites of the Kara astrobleme *J. Non-Cryst. Solids* (in press)

[16] Golubev Y A, Shumilova T G, Isaenko S I, Utkin A A, Makeev B A, Suvorova E I and Ernstson K, 2018 Nano-heterogeneity of natural impact silica-rich glasses according to atomic force microscopy and spectroscopy data *J. Non-Cryst. Solids* **500** 388-400

[17] Golubev Y A, Shumilova T G and Isaenko S I 2018 Structure of natural impact glasses on microscopic data *IOP Conf. Ser.: Mater. Sci. Eng.* **443** 0120082018

[18] Lin Y, Zhang L, Mao H K, Chow P, Xiao Y, Baldini M, Shu J and Mao W L 2011 Amorphous diamond: a high-pressure superhard carbon allotrope *Phys. Rev. Lett.* **107** 175504

[19] Shumilova T G, Tkachev S N, Isaenko S I, Shevchuk S S, Rappenglück M A and Kazakov V A, 2016 A "diamond- like star" in the lab. Diamond-like glass *Carbon* **100** 703-9

# Creation of thin superconducting MoCN thin film by cathode sputtering technique as a basic material for functional cryogenic nanoelements

**B V Goncharov[1], B A Gurovich[1], M M Dementyeva[1], K E Prikhodko[1,2], E D Olshansky[1] and L V Kutuzov[1]**

[1]National Research Centre «Kurchatov Institute», 123182 Moscow, Russia
[2]National Research Nuclear University (MEPhI), 115409 Moscow, Russia

goncharov_bv@nrcki.ru

**Abstract**. Thin MoCN films were synthesized by a method of cathode sputtering of mosaic Mo-C targets by nitrogen ions at temperature of 800°C on substrates of the oxidized silicon. The temperature dependences of electrical resistance of the films in the range of 4.2–300°K were measured. Volt-ampere characteristics of the films and critical current density vs film thickness were obtained at a temperature of 4.2°K. The microstructure and film-depth chemical composition profiles of thin films were investigated by analytical techniques of the transmission electron microscopy (HRTEM, EELS) on a cross-section samples made by the focused ion beam (FIB). It was established that MoCN thin film at a thickness of 8 nm contained MoC and $Mo_5N_6$ layers. The high density of critical current in these films indicated prospects of their application in the field of cryogenic nanotechnologies.

## 1. Introduction

At present, studies of thin superconducting films as a basis for nanoscale electronic devices such as superconducting single-photon detectors (SSPD), hot electron bolometers (HEB), and other nano-devices are of great interest in connection with high values of critical current densities (3-6 MA/cm$^2$) and relatively high superconducting transition temperatures (12-14°K). Nowadays, different types of thin film superconductive materials, for instance NbN, are used to build up SSPD and HEB devices. However, with the rapid development of cryoelectronics, one needs to search for new materials.

The literature contains data on a possibility of MoCN thin films deposition by a cathode sputtering technique, in which X-ray diffraction peaks from phases of molybdenum carbide and molybdenum nitride were found [1]. In this regard, thin films of MoCN were found to be a prospective material for the formation of functional cryogenic nanodevices, because carbides and nitrides of molybdenum were showing superconducting properties with a critical temperature of superconducting transition of about 10-14°K [2].

## 2. Experimental

Samples of molybdenum carbon and nitrogen (MoCN) thin films were prepared by cathode sputtering of mosaic Mo and C target on oxidized (~0.15°μ amorphous $SiO_2$) monocrystalline silicon substrate. The temperature of a substrate was 800°C. The thickness of the initial films was in the range of 4-12°nm.

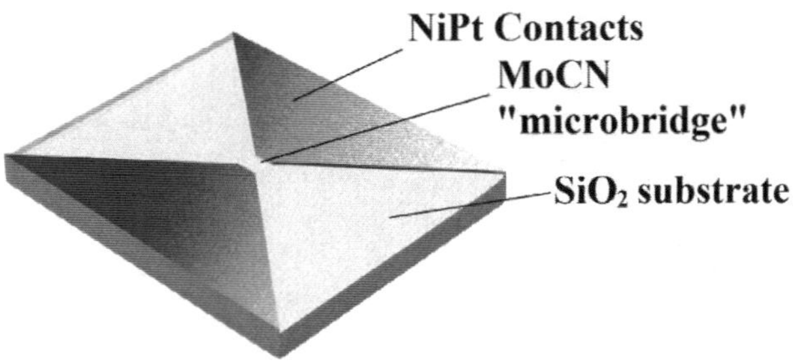


**Figure 1.** Scheme of the sample for electrical measurements with 20x20°μm² MoCN microbridge.

The phase and chemical composition of the thin MoCN film were studied using the "Titan 80-300 ST" transmission electron microscope, operated at 200°kV, equipped with a GIF-2001 electron energy loss spectrometer (EELS). The EEL spectra were obtained with an energy dispersion of 0.5°eV/ch, convergence angle α was 10°mrad and collection angle β was 14.85°mrad. A cross-section samples MoCN/$SiO_2$/Si were prepared by Focusing Ion Beam (FIB "Helios Nanolab 650") technique.

Quantitative EELS analysis of relative atomic concentrations of A and B elements was carried out using equation:

$$\frac{N_A}{N_B} = \frac{I_A(\beta,\Delta)}{I_B(\beta,\Delta)} \cdot \frac{\sigma_B(\beta,\Delta)}{\sigma_A(\beta,\Delta)}, \quad (1)$$

where $I_A$, $I_B$ – integrated intensities of the EELS peaks after background subtracting, and $\sigma_A$ and $\sigma_B$ – ionization cross sections [3]. Determination of the phase composition in the initial samples was performed by the Fourier transform diffraction pattern obtained from the corresponding HRTEM image similar to the method used in Ref. [4].

Formation of structures for electrical measurements was performed by photolithography and plasma-chemical etching, in a manner similar to that described in Ref. [5].

For electrical measurement, the microbridges with the sizes of 20x20 μm² were used (Fig. 1).

Samples were cooled by placing them in a liquid helium. Measurements of temperature dependences of electrical resistance were performed in a temperature interval from 4.2 to 300 K at direct measuring current 100°μA. Pseudo "4-wire" measuring method was used. I(V) characteristic were measured at temperature 4.2 K using a Keithley-4200SCS parameter analyser. Critical current densities were calculated using equation:

$$j_c = \frac{I_c}{S}, \quad (2)$$

where $j_c$ – critical current density, $I_c$ – direct critical current, $S$ – sample cross-sectional area.

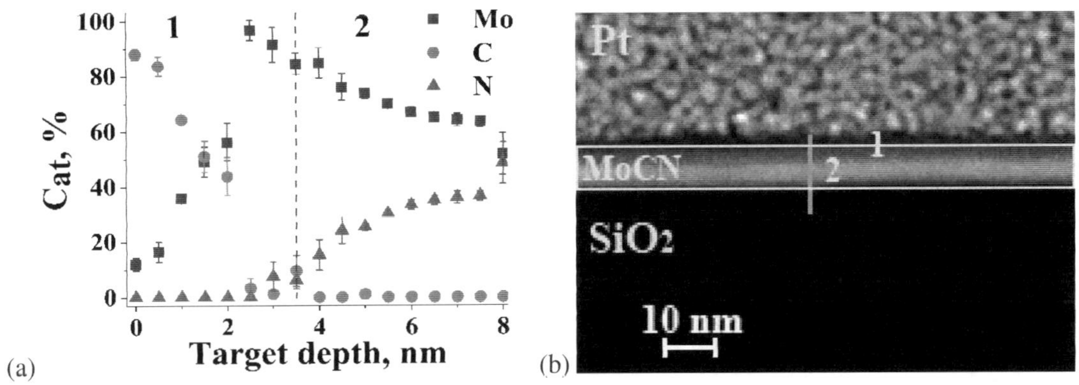


**Figure 2.** (a) Elements depth distribution profiles of the MoCN thin film in initial state, calculated according to the EELS data, (b) HAADF STEM image.

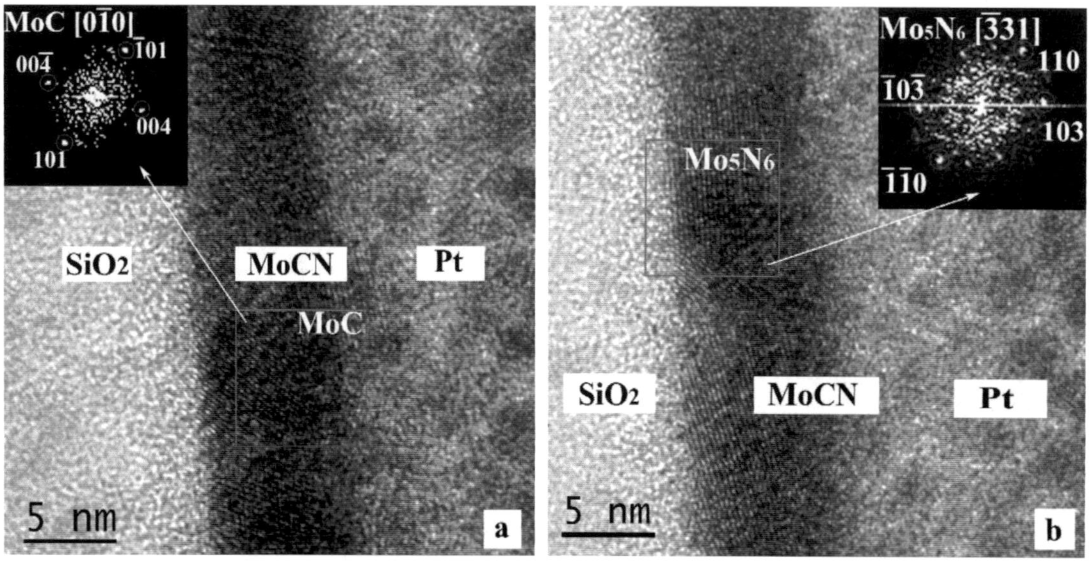


**Figure 3.** HRTEM image of MoCN thin film cross section sample, diffraction (insert): (a) MoC phase (layer 1) and (b) $Mo_5N_6$ phase (layer 2).

## 3. Results and discussion

Structural studies showed that MoCN film was multilayer. Figure 1a shows depth profile distribution of Mo, C, and N atomic concentrations, calculated according to the EELS data. The total film thickness was 8 nm. In this case, MoCN films were bilayer and consisted of a layer of nitrogen atoms (region 1 in Fig. 2a) and a layer of Mo and N atoms only (region 2 in Fig. 2b). In obedience to the EELS studies results, the thickness of the upper layer (layer 1) was 3.5°nm and that of the lower one (layer 2) being close to the $SiO_2$/Si substrate was 4.5°nm. The corresponding contrast was observed on STEM dark-field image of the Pt/MoCN/$SiO_2$/Si cross-section structure. The MoCN thin film layers are labelled 1 and 2, similarly to Figure 2a.

Quantitative analysis data were completely consistent with the results of phase composition studies. Phase analysing data are shown in Figure 3. It was established that phase composition of individual grains corresponded to the P63/mmc hexagonal MoC phase with cell parameters: a=b=2.932 Å, c=10.97 Å and the P63/m hexagonal $Mo_5N_6$ phase with cell parameters a=b=4.8924 Å, c=11.0643Å.

Figure 4 shows R(T) dependence for MoCN film, and it is typical for a superconducting thin film. The value of superconducting transition temperature $T_c$ gets to a presumable interval 7-11°K. One can see a small additional down step of R at a temperature of approximately 2°K below $T_c$ for all

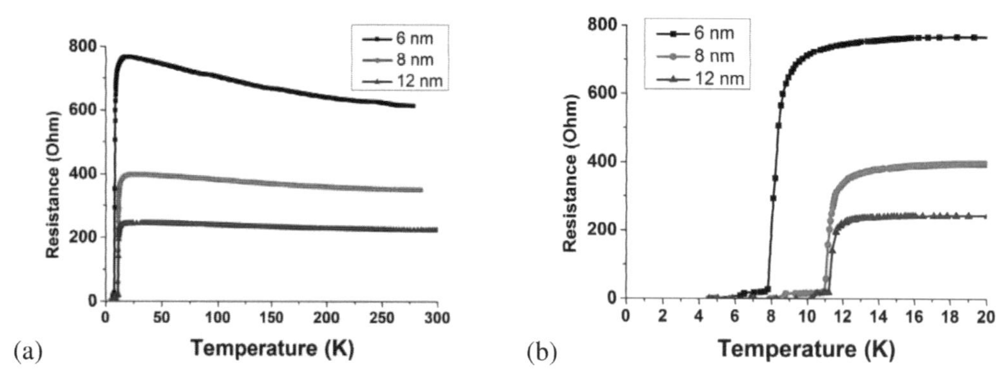


**Figure 4.** Temperature dependence of electrical resistance for MoCN films of different thickness for (a) 4-300°K, (b) 4-20°K.

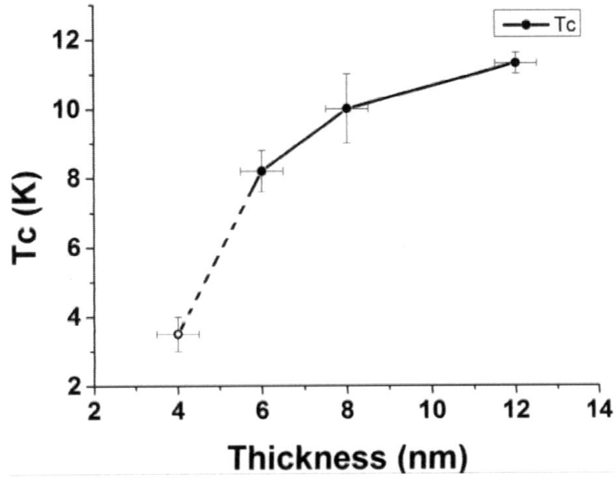


**Figure 5.** Dependence of superconducting transition temperature on film thickness.

characteristic curves (Fig. 4). It was apparently due to the superconducting transition of a MoCN film under the measuring contacts (MoCN-Ni/Pt) as a result of proximity effect of superconducting suppression by the normal metal at the interface.

The dependence of critical temperature $T_c$ for the various thicknesses of MoCN films on the film thickness is presented in Figure 5. The value of $T_c$ for film thickness 4 nm in Figure 5 was evaluated by the trend of R(T) to fall down just above the transition temperature, because the full superconducting transformation did not appear at this thickness due to the 4.2 K liquid helium temperature limitation.

It is clear that at total MoCN film thickness of 4 nm and lower there is no superconductivity at the temperature of liquid helium.

Figure 6a shows the V(I) characteristic for MoCN film thickness of 12 nm (microbridge 20x20 $\mu m^2$). Figure 6b shows the dependence of critical current density on the total MoCN film thicknessб which is comparable with corresponding value for NbN films [6].

## 4. Conclusion

Thin MoCN films were synthesized by the cathode sputtering method of mosaic Mo-C targets nitrogen ions at a temperature 800°C on substrates of the oxidized silicon. TEM studies showed the two-layer structure of MoCN film at a total thickness of 8 nm. The inside layer close to a substrate (~4.5 nm) consists of molybdenum nitride ($Mo_5N_6$) grains, while the external layer (~3.5 nm) contains grains mainly of molybdenum carbide (MoC).

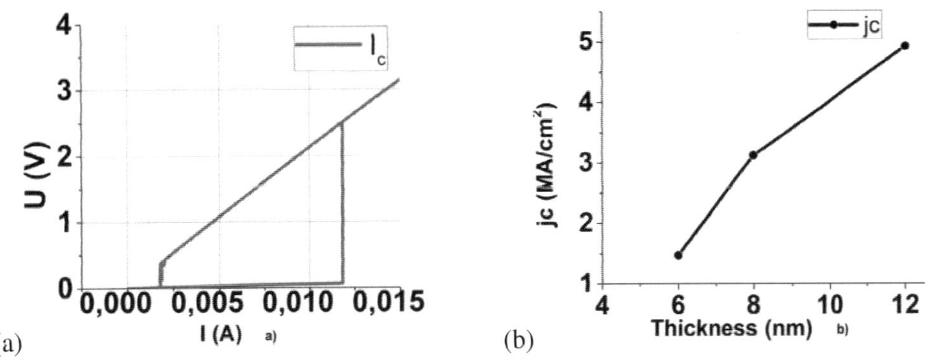


(a) (b)

**Figure 6.** (a) I(V) for MoCN film thickness of 12 nm (microbridge 20x20 $\mu m^2$) with NiPt contacts, (b) critical current density calculated from V(I) for different film thicknesses.

The measured dependences of electrical resistance of films on the temperature in the range 4.2–300°K showed a typical trend for thin superconducting films. The received volt-ampere characteristics at a temperature of 4.2°K allowed calculating the values of critical current density for the films and also its dependence on the total thickness of MoCN. The high density of critical current at a total thickness ~8 nm indicates wide range of application for a thin MoCN films in the field of cryogenic nanotechnologies.

It was shown that MoCN thin films characterized by a high superconducting transition temperature in range of about 8-11°K (for different thickness) and high values of critical current density in range of ~1.4-5x$10^6$ A/$cm^2$.

**Acknowledgements**
The work was supported by the NRC "Kurchatov Institute" (№1359).

**References**

[1] Gilewicz A, Jedrzejewski R, Kochmanska A E and Warcholinski B 2015 *Thin Solid Films* **577** 94-6
[2] Vendenberg J M and Matthias B T 1974 *Mater. Res. Bull.* **9** 1085-90
[3] Williams D and Carter A 2009 *Transmission Electron microscopy* New York: Springer
[4] Gurovich B, Prikhodko K and Dement'eva M 2016 *IOP Conf. Ser.: Mater. Sci. Eng.* **130** 012046
[5] Gurovich B A, Prikhodko K E, Tarkhov M A, Domantovsky A G, Komarov D A, Goncharov B V and Kuleshova E A 2015 *Micro and Nanosystems* **7** 172-9
[6] Gurovich B A, Tarkhov M A, Prikhodko K E et al. 2014 *Nanotechnologies in Russia* **9** 386-90

IOP Conf. Series: Materials Science and Engineering **699** (2019) 012015 doi:10.1088/1757-899X/699/1/012015

# Charged domain walls in lithium tantalate with compositional gradients produced by partial VTE process

**E D Greshnyakov, B I Lisjikh, V I Pryakhina, M S Nebogatikov and V Ya Shur**

School of Natural Sciences and Mathematics, Ural Federal University,
620002 Ekaterinburg, Russia

viktoria.pryahina@urfu.ru

**Abstract**. The morphology of a single charged domain wall, appeared under the action of composition gradients produced by partial VTE procedure by Cherenkov-type second harmonic generation microscopy, was observed in detail. The width of the charged domain wall was estimated as 70 μm. Non-through and through narrow domains, grown from the charged domain wall, were revealed. The maximum length of non-through domains with submicron diameter was about 100 μm. The growth of narrow domains from the charged domain wall was demonstrated and attributed to the action of pyroelectric field. The widening of domains occurred after achieving the polar surface.

## 1. Introduction

Formation of the initial (as-grown, virgin) domain structure during zero field cooling from the temperature above the transition point in ferroelectric crystals depends on many factors, one of which is a spatial nonuniformity of the material composition [1-5]. Crystal inhomogeneities, such as point defects [6], growth layers [1-3], etc., influence the polarization reversal process due to variations of physical properties and lattice imperfections.

Ferroelectric phase transition in uniaxial crystal without external electric field results in formation of random domain structure, which is determined by spatial distribution of the sign of local built-in fields produced by various defects or concentration gradients [3,4]. It was shown that growth striations oriented perpendicular to the polar axis in lead germanate ($Pb_5Ge_3O_{11}$, PGO) crystals resulted in the formation of charged domain walls [8,9] localized in the points, where the gradient changed its sign [3]. We have shown recently that the single charged domain wall was formed during phase transition in lithium tantalate ($LiTaO_3$, LT) with artificially produced composition gradient [10,11].

In our research, we used LT single crystal as a model uniaxial ferroelectric, popular for piezoelectric, electro-optic, pyroelectric, and nonlinear optical applications. LT crystals exist in a wide range of compositions characterized by Li concentration ($c_{Li}$) ranged from 48.5 to 50 mol.%. Polarization reversal threshold fields [12-14], domain shape [7], and Curie temperature [15,16] depend essentially on composition. The vapour transport equilibration (VTE) procedure, which is mostly used for increasing $c_{Li}$ from congruent CLT (48.5 mol.%) to stoichiometric SLT (50 mol.%) composition [12,13,16], allowed us to create spatially inhomogeneous distribution of $c_{Li}$ by decreasing duration of the procedure. As a result, the LT plates, being stoichiometric at the surfaces and congruent in the bulk, with composition gradient along the polar axis were created [10,11].

As-grown domain structure, appeared due to phase transition during cooling after termination of VTE procedure, depends on the compositional gradient. In the certain range of VTE procedure durations, the single charged domain wall appeared in the sample centre [10,11]. With advantages of Cherenkov-type second harmonic generation microscopy (CSHG) for imaging domains [17,18] in the crystal bulk, it was possible to obtain the set of domain wall positions at different depths and 3D image of as-grown domain structure.

In this paper, we present the first detail observation of the morphology of single charged domain wall appeared under the action of composition gradients produced by partial VTE procedure.

## 2. Experimental

### *2.1. Samples preparation by VTE process*

Half-millimetre-thick *Z*-cut plates of CLT with single-domain initial state were used. During VTE procedure, CLT plate was placed in a crucible with Li-rich powder (60 mol.% $Li_2CO_3$ + 40 mol.% $Ta_2O_5$) and was annealed at 1100°C during 40-80 h in high-temperature furnace LHT 01/17 (Nabertherm, Germany). Heating and cooling rates were 1 °C/min.

### *2.2. Composition measurements*

Confocal Raman spectroscopy by means of Alpha 300 AR (WiTec, Germany) was used for a nondestructive measurement of $c_L$ spatial distribution along the polar axis. Since Raman spectrum in LT depends on Li concentration [19, 20], it is possible to evaluate the composition by measuring the full width at half maximum (FWHM) of the 143 cm$^{-1}$ line corresponding to E(TO1) mode. The local $c_{Li}$ value was estimated by $c_{Li}$ = 52.04 – 0.3535 FWHM, received from calibrating by homogeneous CLT and SLT plates [11].

### *2.3. Domain imaging*

Optical microscopy (Olympus BX-61, Japan) was used for imaging of the domain structure, revealed by selective etching in HF during about 5 min at room temperature at polar surfaces and Y cross-section. Y cross-section was obtained by cutting VTE-plate and polishing the cut to the optical grade.

Cherenkov-type second harmonic generation (CSHG) was used for non-destructive domain imaging in the bulk with spatial resolution about 500 nm. The measurement setup was realized on the base of Ntegra Spectra (NT-MDT, Russia) using Yb fibre laser (1064 nm, 40 mW).

## 3. Results and discussion

LT plates with nonuniform spatial distributions of $c_{Li}$ along polar axis were produced by changing the VTE annealing time. Annealing allowed producing LT plates with stoichiometric composition at the polar surfaces and near-congruent in the centre of LT plate. Change of gradient sign ($dc_{Li}/dz$) in the centre resulted in formation of the single charged domain wall.

The created charged domain wall had a complex morphology with elongated narrow bumps (Fig. 1). Some of bumps reached the polar surfaces. It is clear that 2D cross-section did not allow revealing precise characteristics of 3D domain structure.

CSHG imaging of the domain structure in the bulk allowed to reveal domains between the charged domain wall in the crystal bulk and polar surfaces (Fig. 2). These domains formed as bumps at the charged domain wall and grew in polar direction. Two types of domains have been separated: (1) non-through domains with submicron diameter located in the vicinity of charged domain wall and (2) through domains with diameter below 5 μm reaching the polar surface.

We extracted the area of the reversed polarization from CSHG domain images (Fig. 3). The width of the charged domain wall of the irregular wavy shape was about 70 μm. The total area of submicron non-through domains was negligible compared with through domains, which diameter enlarged after achieving polar surface (completing direct growth). The maximum depth of non-through domains was about 100 μm (Fig. 4). The irregular wavy shape of the charged domain wall can be attributed to a result of the counter motion of the phase boundaries during cooling through the transition point [10, 11]. The narrow domains appeared during subsequent cooling under the action of the pyroelectric field [10, 11].

IOP Conf. Series: Materials Science and Engineering **699** (2019) 012015 doi:10.1088/1757-899X/699/1/012015

**Figure 1.** Optical image of domain structure at Y cross-section revealed by selective etching.

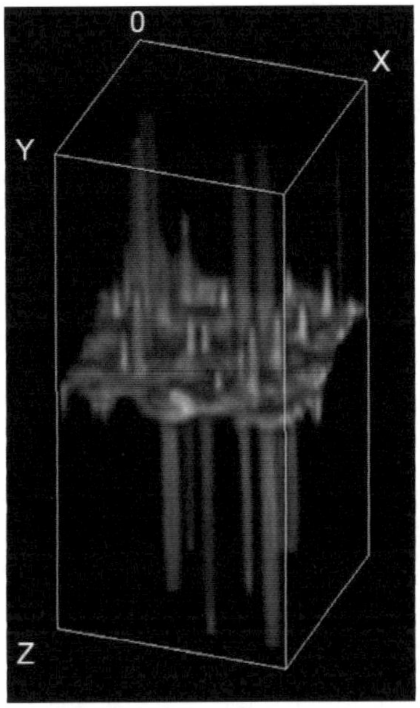


**Figure 2.** 3D reconstruction of domain structure in the bulk by CSHG imaging. Sizes of volume block XYZ 40×40×500 μm.

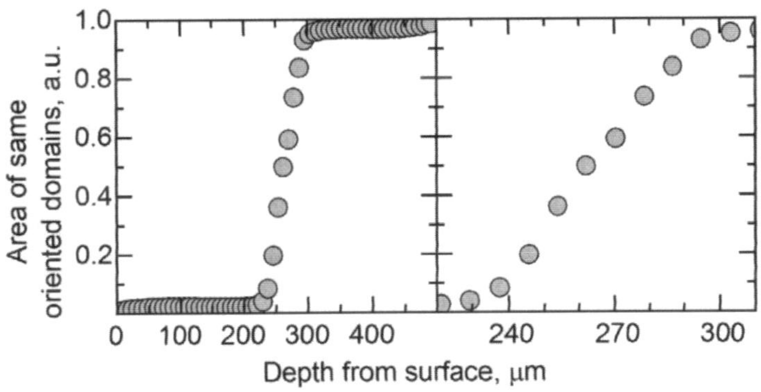


**Figure 3.** (a) The dependence on the crystal depth of the fraction of the total area of domains with the same orientation. (b) Enlarged region of the charged domain wall.

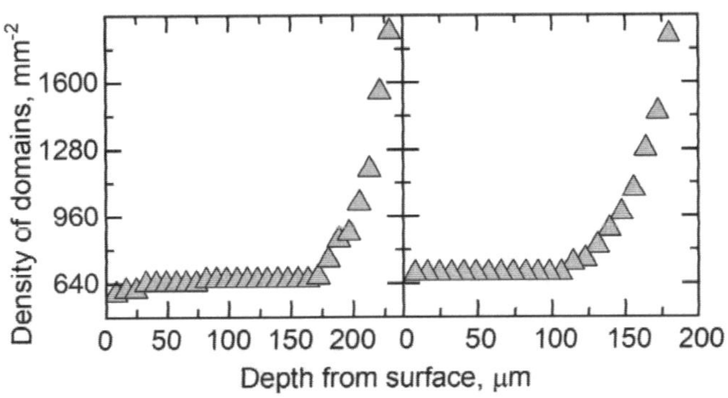


**Figure 4.** The dependence on the crystal depth of the density of domains appeared between charged domain wall and both polar surfaces.

## 4. Conclusion

The detail study of the morphology of single charged domain wall appeared under the action of composition gradients produced by partial VTE procedure was realized by Cherenkov-type second harmonic generation microscopy. The width of the charged domain wall was estimated as 70 μm and the maximum length of non-through domains – as 100 μm. The growth of narrow domains from charged domain wall was demonstrated and attributed to the action of the pyroelectric field during cooling. The widening of domains occurred after achieving of polar surfaces.

**Acknowledgements**

The equipment of Ural Centre for Shared Use "Modern Nanotechnology" Ural Federal University was used. The research was made possible by Russian Science Foundation (project № 19-12-00210).

## References

[1] Shur V Ya, Gruverman A L, Letuchev V V, Rumyantsev E L and Subbotin A L 1989 Domain structure of lead germanate *Ferroelectrics* **98** 29-49

[2] Shishkin E I, Shur V Ya, Schlaphof F and Eng L M Observation and manipulation of the as-grown maze domain structure in lead germanate by scanning force microscopy *Appl. Phys. Lett.* 2006 **88** 252902

[3] Shur V and Rumyantsev E 1998 Arising and evolution of the domain structure in ferroics *J. Kor. Phys. Soc.* **32** S727-32

[4] Kugel V D and Rosenman G 1993 Domain inversion in heat-treated $LiNbO_3$ crystals *Appl. Phys. Lett.* **62** 2902-4

[5] Åhlfeldt H 1994 Single-domain layers formed in heat-treated $LiTaO_3$ *Appl. Phys. Lett.* **64** 3213-5

[6] Gopalan V, Dierolf V and Scrymgeour D A 2007 Defect-domain wall interactions in trigonal ferroelectrics *Annu. Rev. Mater. Res.* **37** 449-89

[7] Shur V Ya, Nikolaeva E V, Shishkin E I, Chernykh A P, Terabe K, Kitamura K, Ito H and Nakamura K 2002 Domain shape in congruent and stoichiometric lithium tantalate *Ferroelectrics* **269** 195-200

[8] Shur V Y, Rumyantsev E L, Nikolaeva E V. and Shishkin E I 2000 Formation and evolution of charged domain walls in congruent lithium niobate *Appl. Phys. Lett.* **77** 3636-8

[9] Eliseev E A, Morozovska A N, Svechnikov G S, Gopalan V and Shur V Y 2011 Static conductivity of charged domain walls in uniaxial ferroelectric semiconductors *Phys. Rev. B* **83** 235313

[10] Pryakhina V I, Greshnyakov E D, Lisjikh B I, Akhmatkhanov A R, Alikin D O, Shur V Y and Bartasyte A 2018 As-grown domain structure in lithium tantalate with spatially nonuniform composition *Ferroelectrics* **525** 47-53

[11] Pryakhina V I, Greshnyakov E D, Lisjikh B I, Nebogatikov M S and Shur V Ya 2019 Influence of composition gradients on heat induced initial domain structure in lithium tantalate *Ferroelectrics* **542** 13-20

[12] Hum D S, Route R K, Miller G D, Kondilenko V, Alexandrovski A, Huang J, Urbanek K, Byer R L and Fejer M M 2007 Optical properties and ferroelectric engineering of vapor-transport-equilibrated, near-stoichiometric lithium tantalate for frequency conversion *J. Appl. Phys.* **101** 093108

[13] Tian L, Gopalan V and Galambos L 2004 Domain reversal in stoichiometric $LiTaO_3$ prepared by vapor transport equilibration *Appl. Phys. Lett.* **85** 4445

[14] Kitamura K, Furukawa Y, Niwa K, Gopalan V and Mitchell T E 1998 Crystal growth and low coercive field 180° domain switching characteristics of stoichiometric $LiTaO_3$ *Appl. Phys. Lett.* **73** 3073-5

[15] Wöhlecke M, Corradi G and Betzler K 1996 Optical methods to characterise the composition and homogeneity of lithium niobate single crystals *Appl. Phys. B Laser Opt.* **63** 323-30

[16] Bordui P F, Norwood R G, Bird C D and Carella J T 1995 Stoichiometry issues in single-crystal lithium tantalate *J. Appl. Phys.* **78** 4647-50

[17] Sheng Y, Best A, Butt H-J, Krolikowski W, Arie A and Koynov K 2010 Three-dimensional ferroelectric domain visualization by Čerenkov-type second harmonic generation *Opt. Express* **18** 16539

[18] Kämpfe T, Reichenbach P, Schröder M, Haußmann A, Eng L M, Woike T and Soergel E 2014 Optical three-dimensional profiling of charged domain walls in ferroelectrics by Cherenkov second-harmonic generation *Phys. Rev. B* **89** 035314

[19] Kostritskii S M, Aillerie M, Bourson P and Kip D 2009 Raman spectroscopy study of compositional inhomogeneity in lithium tantalate crystals *Appl. Phys. B* **95** 125-30

[20] Shi L, Kong Y, Yan W, Liu H, Li X, Xie X, Zhao D, Sun L, Xu J, Sun J, Chen S, Zhang L, Huang Z, Liu S, Zhang W and Zhang G 2005 The composition dependence and new assignment of the Raman spectrum in lithium tantalate *Solid State Commun.* **135** 251-6

IOP Conf. Series: Materials Science and Engineering **699** (2019) 012016 doi:10.1088/1757-899X/699/1/012016

# Integrated cryogenic resistors formed from NbN thin film under ion beam irradiation

**B A Gurovich[1], B V Goncharov[1], M M Dementyeva[1], K E Prikhodko[1,2], L V Kutuzov[1], D A Komarov[1] and A G Domantovsky[1]**

[1]National Research Centre «Kurchatov Institute», 123182 Moscow, Russia
[2]National Research Nuclear University (MEPhI), 115409 Moscow, Russia

goncharov_bv@nrcki.ru

**Abstract**. Niobium nitride ultrathin films were synthesized by cathode sputtering technique. The cryogenic resistors built in the superconducting nanowire were formed using the radiation technology developed in National Research Centre "Kurchatov Institute". This technique was based on the transformation of NbN superconducting film to a normal state (at 4.2K) under the composite ion beam irradiation through the resistive mask. The formation of the cryogenic resistors in the broad range of values was demonstrated.

## 1. Introduction

The traditional "lift-off" method of creation of the resistive elements integrated into the superconducting line from Ti-Au has a number of disadvantages: rather big sizes of elements, a large number of technological operations, and also essential difficulties of such method when forming elements of the nanometre sizes.

In National Research Centre "Kurchatov Institute" the new method of functional nanoelements creation under ion beam irradiation was developed for different applications [1,2].

The essence of a method consists of composite ion beam irradiation of thin NbN films. The irradiation leads to a controlled change in the chemical composition and a modification of the niobium nitride electrical properties. The change in the properties occurs due to the replacement of part of the nitrogen atoms by oxygen atoms. The advantages of using this approach are: a small number of required technological operations and a possibility of formation of small integrated elements (with sizes <100°nm). Moreover, this method allows forming different types of passive functional elements (resistors and capacities) by varying radiation dose, beam composition, and the geometry sizes of irradiated region.

## 2. Materials and experimental methods

Thin NbN films (5°nm), obtained by cathode sputtering on $SiO_2$ substrates, were used to create the cryogenic resistors. Formation of structures for electrical measurements was performed by electron lithography and plasma-chemical etching [3]. NbN films were modified by the composite ion beam irradiation of $H^+$ and $OH^+$ ions [3,4]. Ions were extracted from RF plasma discharge by applying a square pulse alternating voltage bias to the holder with samples. The value of the negative voltage part of square bias was in the range from 0.1 to 4°keV to get more uniform distribution of oxygen ions to the depth of the film. To minimize the charging effect during irradiation, the electrons were extracted from the plasma discharge. Every high voltage square negative pulse was followed by the low voltage square

IOP Conf. Series: Materials Science and Engineering **699** (2019) 012016 doi:10.1088/1757-899X/699/1/012016

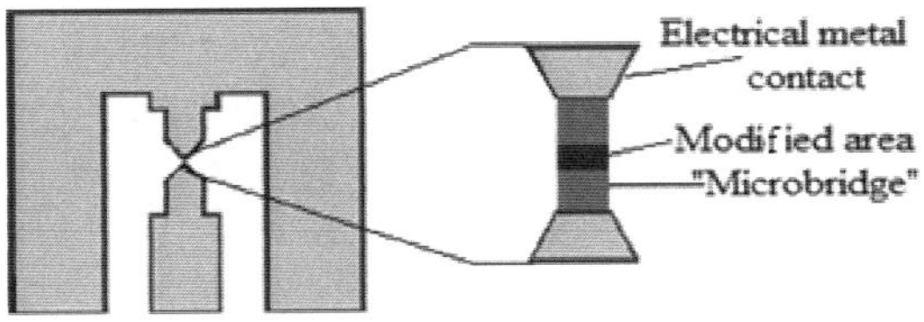


(a) (b)

**Figure 1**. Scheme of the functional nanoelement – "resistor" (top view): (a) the scheme of the chip with macroscopic electric contacts; (b) the scheme of a functional nanoelement.

positive pulse (+100°V). The typical frequency of high voltage bias was about 100°kHz. The ratio of the concentration of $OH^+$ ions to protons in the supplied gas mixture was controlled by measuring the partial pressures of the components (water vapour and hydrogen, respectively), and in our case it was $C = 1.2\cdot 10^{-3}$ [2,3]. Composite irradiation damage rate (for nitrogen) was calculated by means of the SRIM [5], and in our case was $2.25\cdot 10^{-2}$°dpa/s.

The samples of "micro-bridge" type, obtained by a photolithography method with metal contacts, were used to study a dose dependence of sheet resistance at room temperature and at a temperature of liquid helium. These samples were used for analysis of the temperature dependences of electrical resistance. The geometry sizes of "micro-bridges" were 20x20°$\mu m^2$ (Fig. 1).

Measurements of temperature dependences of resistance of the irradiated samples were performed in a temperature interval from 4.2 to 300°K at direct measuring current 100°nA.

We have created cryogenic passive nanoelements (resistors) using thin (5 nm) superconducting films of NbN. Normal state of the film for forming cryogenic resistors was created due to modification of NbN atomic composition under composite ion beam irradiation. Irradiated elements showed resistive properties at working liquid helium temperature (4.2°K) due to transition of an initial superconducting NbN film to a new state, which was characterized by metallic conductivity.

The irradiation conditions to modify the superconducting NbN to metallic NbNO were found in Ref. [6]. The analysis of the ion irradiation dose dependencies [6] (Fig. 2) showed that at radiation doses <1.8°d.p.a. the material kept its superconducting properties at temperatures over 4.2°K, and at doses higher than 8°d.p.a. electrical properties began approaching dielectric ones. Thus, the optimum range of doses for a composite ion beam irradiation was about 1.8-8°d.p.a. on nitrogen. Using the data on dose dependences of films [6] of NbN irradiated with the composite ion beam irradiation, we formed the resistive elements (resistors) integrated into the superconducting nanowire. Advantages of this approach were: a possibility to form cryogenic resistors with the broad range of values (by varying only 2 parameters: a dose of radiation and geometry of the irradiated area), and a possibility to produce

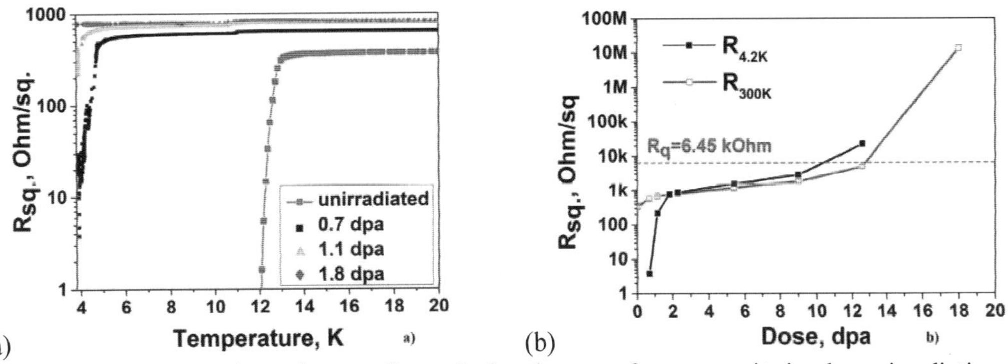


(a) (b)

**Figure 2.** Temperature dependences of samples' resistance after composite ion beam irradiation to various doses of radiation (a) dependence of $R_{300K}$ and $R_{4.2K}$ on a radiation dose (b) [6].

IOP Conf. Series: Materials Science and Engineering **699** (2019) 012016 doi:10.1088/1757-899X/699/1/012016

integrated resistors for active devices (the cryogenic switches, SNSPD, PNDs, etc.). Resistive element can be integrated into the superconducting current line during one technological process with active devices. The final resistance R will be defined by the expression:

$$R = \frac{R_{sq}L}{W}, \quad (1)$$

where R – final value of resistance, $R_{sq}$ – sheet resistance of NbN after modification, L – length, and W – width of irradiated area.

## 3. Results and discussion

Using the described method and expression (1), resistors of different topology and values were formed. Scanning electron microscopy images of different types of geometry are shown in Figure 3.

To get high resistance values, a large number of squares was required. The topology "meander" (Fig. 3b) was used for this purpose. It was also possible to create different types of connecting elements to the scheme: in series (Fig. 3a) and in parallel (Fig. 3c).

Resistors integrated into the superconductive current line were made using the above mentioned radiation technique. The formed values of resistor were in the range of 2°Ohm-6°MOhm (Fig.°4). The dose of irradiation was ~2°d.p.a. (on nitrogen) and a corresponding sheet resistance was 850 Ohm/sq. Resistors were measured at a temperature of liquid helium by a "four-contact" method. Figure 4 shows that when a dose of irradiation is close to the lower bound of optimum range (1.8-8°d.p.a. on nitrogen), the resistors below 100°Ohms have a wide scatter of values. It was due to the difficulties to form open areas in the mask with a very small ratio of L/W ≈ 0.01. At the same time, during the irradiation process, the edges of the open mask areas changed their shape and dimensions. As a result, the final irradiated region of the material had a different geometry than the initial one.

The lack of stability of mask shape and size could be overcome by the selection of optimum parameters of radiation, using more stable to composite ion beam irradiation protective mask material, and also by optimization of the topology of elements. Work in these directions will be continued.

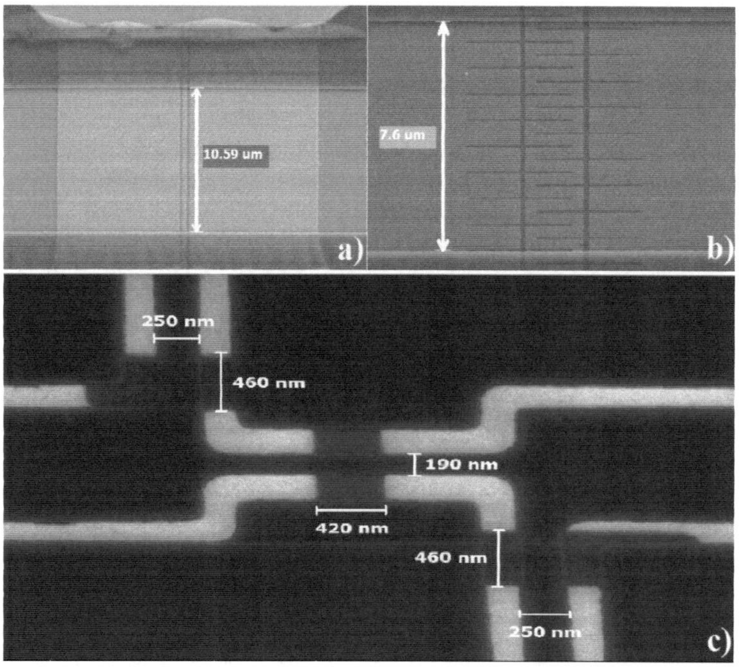


**Figure 3.** The resistors of different topologies created using the SDA method: (a) built in current line by "straight line," (b) "meander," (c) three built-in superconducting current line resistors of different face values.

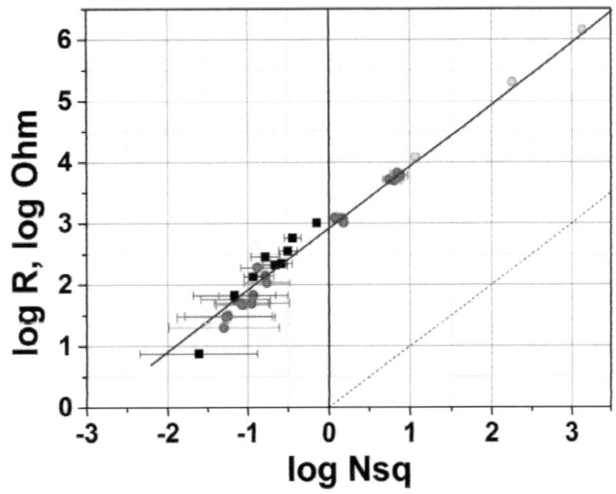


**Figure 4.** Dependence of the experimentally formed resistive elements on a number of squares (for a ~2°d.p.a. on nitrogen dose of irradiation and 850°Ohm/sq sheet resistance).

## 4. Conclusion

The cryogenic resistors, integrated into the superconducting nanowire, of different values (2°Ohm-6°MOhm) were made. It was shown that the composite ion beam irradiation was a new technique to form a nanoscale functional resistive elements in NbN thin films. The main advantage of this approach was that resistors integrated into the superconducting nanowire gave a chance to avoid additional sputtering on "lift-off" technology.

Integrated resistors can be used when forming nano-dimensional cryogenic devices of different functions – for example, cryogenic logical switches or cryogenic counters of single photons and in other cryoelectronic devices.

## Acknowledgements

The work was supported by the NRC "Kurchatov Institute" (№1359).

## References

[1] Gurovich B A and Prikhodko K E 2009 *Physics-Uspekhi* **52** 165-78
[2] Gurovich B A, Prikhodko K E, Kuleshova E A, et al 2013 *JETP* **116** 916-27
[3] Gurovich B A, Tarkhov M A, Prikhodko K E, et al 2014 *Russ. Nanotechnologies* **9** 386-90
[4] Gurovich B A, Prikhodko K E, Tarkhov M A, et al 2015 *Russ. Nanotechnologies* **10** 530-6
[5] Ziegler J F, Ziegler M D and Biersack J P 2010 *SRIM - The stopping and range of ions in matter* Nucl. Instruments Methods Phys. Res. Sect. B Beam Interact. with Mater. Atoms. Elsevier B.V. **268** 1818-23
[6] Gurovich B A, Prikhodko K E, Tarkhov M A, Domantovsky A G, Komarov D A, Goncharov B V and Kuleshova E A 2015 *Micro and Nanosystems* **7** 172-9

# Electron beam processing of 6H-SiC substrate to obtain graphene-like carbon films

**E Yu Gusev[1,2], V G Dudko[3], M V Avramenko[4], M P Karmanov[1], S P Avdeev[1,2], A S Kolomiytsev[1] and O A Ageev[1,2]**

[1]Southern Federal University, Institute of Nanotechnologies, Electronics and Equipment Engineering, Taganrog, 347922, Russia
[2]Southern Federal University, Research and Education Centre "Nanotechnologies", Taganrog, 347922, Russia
[3]Teksent Ltd, 347900, Taganrog, Russia
[4]Southern Federal University, Faculty of Physics, Rostov-on-Don, 344090, Russia

eyugusev@sfedu.ru

**Abstract.** We report on growth of graphene-like carbon films on 6H-SiC {0001} substrate by electron-beam. The processing was carried out on a specialized electron beam system with the Pierce electron gun. The D, G, and 2D peaks as well as D/G (0.2-0.9) and 2D/G (0.7-0.9) ratios are detected on processed samples by Raman spectroscopy. The prominent bands D, G, and 2D are located at 1350, 1584, and 2707 $cm^{-1}$, respectively. Atomic force microscopy showed that the average roughness lies in the range from 5 to 30 nm, and ten point height – from 40 to 200 nm. The results demonstrate that the electron-beam technique is appropriate to form graphene-like structures directly on 6H-SiC substrates and could be used for electronic device fabrication.

## 1. Introduction

The ever-growing development of technologies brings demands for discoveries of new physical and chemical phenomena, design and fabrication of multifunctional materials, and manufacturing of next-generation electronic devices [1-5]. Over the past few decades, carbon materials have been attracting a growing research interest for variety of applications [5-8]. Graphene has tremendous perspectives due to its unique properties including high electron mobility with low electronic scattering that make it a very promising candidate for next generation nanoscale electronics and photonics devices [1,9-12]. Therefore, achieving quality synthesis of graphene, graphene-like carbon films, and related materials is in demand.

Several outstanding advancements in graphene fabrication technology have been explored including research grade mechanical exfoliation of graphite [2], chemical vapour deposition [13], thermal decomposition of SiC [11], and chemical methods [14]. Among the various methods of its obtaining the most interesting is the synthesis directly on silicon carbide substrates [5], e.g. thermal decomposition [9,10,15]. However, this high-temperature treatment prevents commercial use of SiC substrates for large-scale production of graphene-based electronic devices.

Recently, various approaches using irradiation technique or particle beam (laser, ion and electron beams) were extensively paid attention to due to high precision, efficiency, mature control method, and fast speed [5,16-18]. A method for converting the area of 6H-SiC substrates into homogeneous epitaxial graphene by axial electron beam irradiation is proposed [5,16]. 6H-SiC has a hexagonal structure and

IOP Conf. Series: Materials Science and Engineering **699** (2019) 012017 doi:10.1088/1757-899X/699/1/012017

represents an ideal pattern for graphene growth. One of the main advantages over other methods is SiC decomposition, i.e. graphene layers are formed on a semiconductor substrate, so no transfer is required to process the device. In localized interaction, scattering, excitation, and ionization between the primary electrons and the SiC surface lead to SiC bond rupture, and the excess electron energy is dissipated as heat, which leads to selective sublimation of the Si, contributing to the formation of epitaxial graphene. Using I-shaped electron beams [17,18] instead of axial ones to produce graphene can have obvious advantages, such as suitability for large-scale graphene fabrication.

## 2. Experimental

Semi-insulating 6H-SiC wafer was used as a substrate. The substrates were cleaned in acetone and isopropyl alcohol prior to placing into the chamber. Then electron beam processing (EBP) was carried out on a specialized electron beam vacuum system with the Pierce electron gun [17]. The chamber was evacuated to the pressure of 0,3-0,5 mTorr. The emission current was fixed for and finely tuned between 50 and 180 mA. Since the beam energy and irradiation time are precisely controllable, the corresponding temperature ranged from 1680 to 2000 K [18,19]. The as-grown layers and initial substrate surfaces on the reference sample were characterized by Raman spectroscopy and atomic force microscopy.

## 3. Results and discussion

A series of {0001} 6H-SiC samples were electron beam processed and nanostructured carbon films were available on the sample surfaces. Raman spectra of some samples are shown in Figure 1.

The D, G, and 2D peaks as well as D/G (0.2-0.9) and 2D/G (0.7-0.9) ratios are detected by Raman spectroscopy. The prominent bands D, G, and 2D are located at 1350, 1584, and 2707 $cm^{-1}$, respectively. The full-width-at-half-maximum (FWHM) of 2D peaks is in the range of 61-81 $cm^{-1}$. It is widely reported that the FWHM of the 2D band for the tri- and tetra-layers graphene grown on SiC is ~ 60 and 85 $cm^{-1}$, respectively, whereas for the single and bilayer epitaxial graphene, it is found to be 30 and 45 $cm^{-1}$, respectively [20-22]. The dominant G band intensity indicates good quality of graphene. The results of Raman spectroscopy of all samples are presented in Table 1.

The surface morphology of the 6H-SiC samples processed by electron beam was effectively examined by AFM with semi-contact mode (Fig. 2). These images show several types of morphology. The average roughness lies in the range from 5 to 30 nm, and ten point height – from 40 to 200 nm (Fig. 2a and 2b, respectively). Figure 2c demonstrates terraces of 300-1100 nm in width with step height of 150-750 nm.

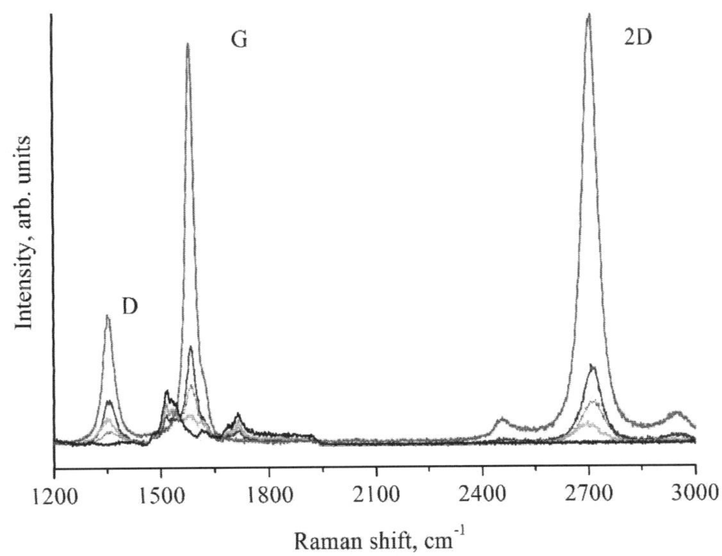


**Figure 1.** Raman spectra of reference (black line) and electron beam processed (coloured lines) 6H-SiC samples.

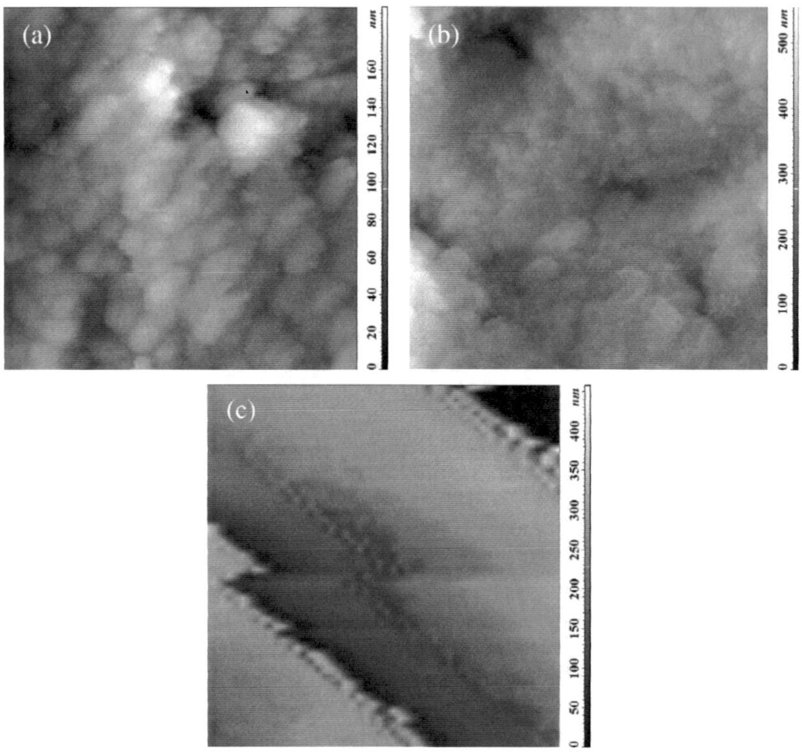


**Figure 2.** AFM images of electron beam processed 6H-SiC samples. Scan area: (a), (b) 3×3 μm$^2$, (c) 1×1 μm$^2$.

**Table 1.** The results of Raman spectroscopy.

| Sample | FWHM D (cm$^{-1}$) | FWHM G (cm$^{-1}$) | FWHM 2D (cm$^{-1}$) | D/G | 2D/G |
|---|---|---|---|---|---|
| 1 | 48 | 72 | 81 | 0.7 | 0.9 |
| 2 | 11 | 56 | 76 | 0.2 | 0.7 |
| 3 | 22 | 50 | 62 | 0.4 | 0.8 |
| 4 | 15 | 43 | 61 | 0.4 | 0.7 |

Comparison of the obtained results with known Raman spectra and atomic force microscopy data [9,10,16] allows us to assume that nanostructured carbon films, from monolayer to multilayer graphene and turbostratic graphite with various degrees of defects, are formed under the conditions of electron beam processing.

## 4. Conclusion

Distinct nanostructured carbon films have been synthesized on {0001} 6H-SiC by electron beam processing. The observed FWHM of the 2D band at 60 and 85 cm$^{-1}$ on Raman spectra, as well as the average roughness in the range from 5 to 30 nm obtained by AFM, allows us to assume that nanostructured carbon films, from monolayer to multilayer graphene and turbostratic graphite with various degrees of defects, are formed under the conditions of electron beam processing. It is expected that further optimization of processing conditions will allow producing the required number of layers and quality of graphene.

The results could be used to optimize the processing conditions in order to obtain single-layer and double-layer graphene films with good electrical properties. The preparation of graphene on silicon carbide substrates by electron beam processing should be a promising technique in carbon electronic device fabrication.

**Acknowledgements**

This work was financially supported by Southern Federal University (grant VnGr-07/2017-02: obtaining graphene-like carbon films on SiC by electron beam processing, Raman spectroscopy studies) and grant of the President of the Russian Federation for the state support of young scientists – candidates of sciences (project MK-1811.2019.8: atomic force microscopy studies). The results were obtained using the equipment of the Research and Education Centre "Nanotechnologies" of Southern Federal University.

**References**

[1] Lee H C, Lie W W, Chai S P, Mohamed A R, Lai C W, Khe C S, Voon C H and Hashim U 2016 *Procedia Chem.* **19** 916-21
[2] Novoselov K S, Geim A K, Morozov S V, Jiang D, Zhang Y, Dubonos S V, Grigorieva I V and Firsov A A 2004 *Science* **306** 666-9
[3] Avilov V I, Ageev O A, Konoplev B G, Smirnov V A, Solodovnik M S and Tsukanova O G 2016 *Semiconductors* **50** 601-6
[4] Avilov V I Smirnov V A, Fedotov A A, Tominov R V, Sharapov N A and Polupanov N A 2018 *IOP Conf. Ser.: Mater. Sci. Eng.* **443** 012004
[5] Mishra N, Boeckl J, Motta N and Iacopi F 2016 *Phys. Status Solidi* A **213** 2277-89
[6] Il'ina M, Il'in O, Blinov Y, Konshin A, Konoplev B and Ageev O 2018 *Materials* **11** 638
[7] Klimin V S, Rezvan A A and Ageev O A 2018 *J. Phys.: Conf. Ser.* **1124** 071020
[8] Klimin V S, Rezvan A A and Ageev O A 2018 *J. Phys.: Conf. Ser.* **1124** 022035
[9] Lebedev A A, Kotousova I S, Lavrent'ev A V, Lebedev S P, Dement'ev P A, Petrov V N, Smirnov A N and Titkov A N 2010 *Phys. Solid State* **52** 855-61
[10] Konakova R V, Kolomys O F, Okhrimenko O B, Strelchuk V V, Volkov E Yu, Grigoriev M N, Svetlichnyi A M and Spiridonov O B 2013 *Semiconductors* **47** 812-4
[11] Emtsev K V, Bostwick A, Horn K, Jobst J, Kellogg G L, Ley L, McChesney J L, Ohta T, Reshanov S A, Rohrl J, Rotenberg E, Schmid A K, Waldmann D and Seyller T 2009 *Nat. Mater.* **8** 203-7
[12] Jityaev I L, Svetlichnyi A M, Kolomiytsev A S, Volkov E Yu, Polyakova V V and Ageev O A 2017 *IOP Conf. Ser.: Mater. Sci. Eng.* **256** 012021
[13] Li X, Cai W, An J, Kim S, Nah J, Yang D, Piner R, Velamkanni A, Jung I, Tutuc E, Banerjee S K, Colombo L and Ruoff R S 2009 *Science* **324** 1312-4
[14] Park S and Ruoff R S 2009 *Nat. Nanotech.* **4** 217-24
[15] Jityaev I L, Ageev O A, Svetlichnyi A M, Kolomiytsev A S and Spiridonov O B 2016 *J. Phys.: Conf. Ser.* **741** 012011
[16] Kim J, Park M, Shin H K, Choia J, Panta B, Saud P S, An T, Chae S-H and Kim H-Y 2015 *Mater. Lett.* **149** 15-8
[17] Avdeev S P, Kravchenko A A, Gusev E Yu and Petrov S N 2007 *Proc. SPIE* **6636** 66360K
[18] Gusev E Yu and Avdeev S P 2018 *J. Phys.: Conf. Ser.* **1124** 022026
[19] Gusev E Yu, Avdeev S P and Ageev O A 2018 *J. Surf. Invest.* **3** 256-60
[20] Krauss B, Lohmann T, Chae D H, Haluska M, Klitzing K and Smet J H 2009 *Phys. Rev. B* **79** 165428
[21] Das A, Pisana S, Chakraborty B, Piscanec S, Saha S K, Waghmare U V, Novoselov K S, Krishnamurthy H R, Geim A K, Ferrari A C and Sood A K 2008 *Nat. Nanotech.* **3** 210-5
[22] Reina A, Jia X, Ho J, Nezich D, Son H, Bulovic V, Dresselhaus M and Kong J 2009 *Nano Lett.* **9** 30-5

IOP Conf. Series: Materials Science and Engineering **699** (2019) 012018 doi:10.1088/1757-899X/699/1/012018

# Styrene-terminated polyphenylene oxide and polybutadiene low dielectric reactive blend

**W Hu[1], J Shen[1], J Zhou[1], W Jin[2] and W Chen[1]**

[1]State Key Laboratory of Advanced Technology for Materials Synthesis and Processing, School of Materials Science and Engineering, Wuhan University of Technology, Wuhan 430070, P. R. China
[2]State Key Laboratory of Silicate Materials for Architectures, Wuhan University of Technology, Wuhan 430070, P. R. China

zhoujing@whut.edu.cn

**Abstract**. Styrene-terminated polyphenylene oxide (SPPO) and low polar thermosetting polybutadiene (PB) resin were employed to obtain a curing system with excellent electrical properties and enhanced thermal properties. By terminal styrenisation, SPPO is cured with PB, forming a continuous phase in the blend. The SPPO/PB blends have excellent solvent resistance and thermal stability, while the good dielectric performance is retained. Especially, when the mass ratio of PB is 0.5, the curing system has a high $T_g$ of 226.5°C and excellent dielectric properties with Dk = 2.46 and Df = 0.0022 at 10 GHz, which is suitable for high frequency electrical applications.

## 1. Introduction

The rapid development of electronic information requires access to Copper-Clad Laminates (CCLs) with high thermal stability, low dielectric constant (Dk), and loss factor (Df) [1-3]. Thermoset polymers, such as phenolic and epoxy resins, have been widely used as base materials for CCLs. However, their dielectric properties are not able to meet the requirements of high frequency application [4]. PTFE resin is a popular polymeric matrix for high frequency CCLs [5-7] because of its outstanding dielectric properties. However, its disadvantages are also glaring, such as low glass transition temperature, high coefficient of thermal expansion (CTE), difficult processability, poor rigidity, and high cost. Therefore, new materials need to be developed to meet the requirements for high frequency CCLs applications.

Polyphenylene oxide (PPO) resin is known for its low Df, which is the second after PTFE resin in common polymeric matrices of CCLs. Besides, PPO has additional merits, such as high glass transition temperature among the thermoplastic polymers, low CTE, outstanding mechanical properties, and good resistance against acids and alkalis. Therefore, PPO resin is considered as a potential candidate for the matrix of high frequency CCLs. However, PPO is easily corroded by solvents such as halogenated hydrocarbons or aromatic hydrocarbons, used in the clean processing for CCLs. Moreover, the progress of lead-free welding process requires higher thermal stability for high performance CCLs. In order to make PPO meeting these requirements, numerous efforts have been made [8-10]. A general approach is to blend with other thermosetting resins, especially epoxy resins [11], which cross-linked structures can restrain the movement of PPO molecular chains to make PPO insoluble in organic solvents and improve the temperature of its glass transition or degradation. Unfortunately, the polar hydroxy groups, introduced via the curing of the epoxy resins, drastically increase the dielectric constant and loss factor of polymeric dielectrics. Hence, it is significant to develop a facile and effective method to improve the solvent resistance and thermal properties of PPO with its electrical properties maintaining.

PB resin has a very weak polarity and cross-linkable groups, making it an effective curing agent to blend with PPO. Thus, it is expected to make polymer matrix for the high-performance CCLs by modifying PPO with PB. However, the structural differences between the two resins lead to poor compatibility of the PPO/PB blend system, especially for the high molecular weight PPO [12,13]. Furthermore, the molecular chain of PPO is only interspersed in the cross-linked network of PB, resulting in bad structural stability of the blend. The compatibility and phase behaviour for the blend of PPO with other resins also have been investigated by several research groups [14-17]. Weng et al. [16,17] have found that the molecular weight of PPO influences its compatibility with other resins and the decrease in PPO molecular weight leads to the improvement of its compatibility and processability.

Previously, compatibilisers were widely used to promote blending of incompatible polymers through reducing the interfacial energy between different phases. It is reported that copolymeric compatibilisers were introduced into the PPO/PB blends to construct the blends with good compatibility [18]. These non-reactive compatibilisers only make the blend components finer and enhance dispersion uniformity upon blending, that is, they act only as a general emulsifier. Compared with the utilization of compatibilisers, reactive blending by chemical functionalization of polymer chains is a more effective way for preparing polymer with better compatibility and structural stability with incompatible raw materials. The reactive functional groups can react to form cross-linked structure at the interface, which can reduce the interfacial tension and stabilize the morphology of blends [19]. Grafting the functional groups such as allyl and methacrylate to PPO's backbones or chain ends was also reported [20,21]. However, in these works, the synthesis process of allylated PPO is too complicated, while the telechelic PPO oligomers with methacrylate end group contain polar ester bonds, which are harmful to the dielectric properties. Compared with above functional groups, introduction of styrene, which structure is similar to the main chain structure of PPO, will retain the high symmetry of PPO molecular chain and its dielectric properties. Therefore, it would be an ideal modification group of PPO for high-performance CCLs. However, as far as we know, the blend of styrene-terminated PPO and PB has rarely been explored and its curing reaction mechanism is still far from being well understood. Moreover, the effects of terminal modification of PPO on the curing behaviour and phase morphology of PPO/PB blends have been scarcely investigated.

In this study, the styrene-terminated low-molecular-weight PPO was synthesized and reactively blended with low polar thermosetting PB resin. PPO curing system with good compatibility and structural stability was obtained. The curing behaviour, phase morphology, solvent resistance, thermal and dielectric properties of cured SPPO/PB blends were investigated.

## 2. Experimental

### *2.1. Materials*

PPO was purchased from SABIC Innovative Plastics Co., Ltd. PB with Mn of 4000 g/mol was purchased from Toyo Soda Manufacturing Co., Ltd. The terminating agent 4-Chloromethylstyrene (≥ 95.0%) was purchased from Saen Chemical Technology Co., Ltd. The phase transfer catalyst Tetrabutylammonium bromide (TBAB) (99.0%), Sodium hydroxide, and Dicumyl peroxide initiator were purchased from Sinopharm Chemical Reagent Co., Ltd. All reagents were used without further purification.

### *2.2. Synthesis of styrene-terminated PPO*

Figure 1 illustrates the synthetic procedure of styrene-terminated PPO (SPPO). PPO was dissolved in a mixture of toluene and isopropanol. TBAB and 4-Chloromethylstyrene were added and the reaction mixture was heated to 80°C under mechanical stirring. Then 50% aqueous NaOH solution was added dropwise over a period of 30 min and the mixture was kept at 80°C for 5 h. After the reaction, the mixture was poured into a mixed solvent containing methanol and deionized water. The filtrated product was dried at 80°C for 24 h under vacuum.

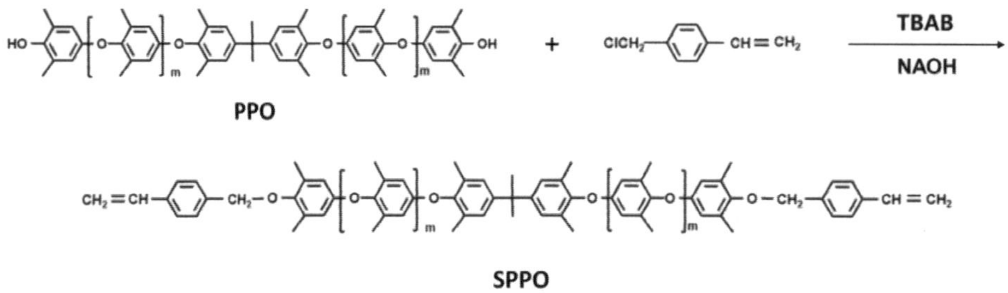


**Figure 1.** Synthetic route of SPPO.

### 2.3. *Preparation and blends of SPPO*

SPPO was thermally curing with PB, and the PPO/PB blends were also prepared for the investigation of terminal modification effects on the curing behaviour and phase morphology of PPO/PB blends. The chemical structures of PPO, SPPO, PB, and curing agent DCP are illustrated in Figure 2. SPPO (or PPO), PB, and DCP with the predetermined mass fraction (Table 1) were dissolved in toluene. The solvent was removed under ventilating cabinet at room temperature for 48 h and under vacuum at 40°C for 24 h. Then the blends were heated in a quartz tube furnace at 200°C for 30 min and then at 220°C for 10 min. The blends were named as SPPO/PB-*x* or PPO/PB-*x*, where *x* corresponds to mass fraction of PB.

**Table 1.** Composition of the blends.

| Sample | Mass fraction (%) | | |
|---|---|---|---|
| | PPO/SPPO | PB | DCP |
| SPPO | 100 | 0 | |
| SPPO/PB-0.1 | 90 | 10 | |
| SPPO/PB-0.3 | 70 | 30 | |
| SPPO/PB-0.5 | 50 | 50 | 3[a] |
| PPO/PB-0.1 | 90 | 10 | |
| PPO/PB-0.3 | 70 | 30 | |
| PPO/PB-0.5 | 50 | 50 | |

[a] Relative to the total mass of the resin.

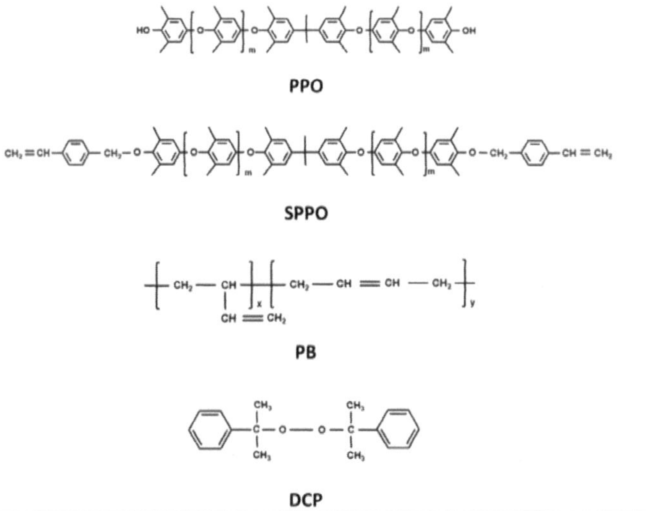


**Figure 2.** Chemical structures of PPO, SPPO, PB, and DCP.

### *2.4. Characterization*

$^1$H-NMR spectra were recorded on a 300 MHz Bruker AVANCE II nuclear magnetic resonance spectrometer (Bruker Biospin Corp., Switzerland) using CDCl3 as the solvent. Gel permeation chromatography (GPC) was performed by a Hitachi L2400 (Hitachi, NY, USA) in tetrahydrofuran (THF), using polystyrene as a standard. Fourier transform infrared (FTIR) spectroscopic analysis was carried out on a NICOLET 6700 spectrometer (Thermo Electron Corp., USA) with KBr pellets at room temperature. Differential scanning calorimetric (DSC) analysis was recorded on a TA-DSC2500 instrument, the sample was heated from 30 to 350°C at a heating rate of 10°C/min. The peak areas in the DSC curing curves were taken as the curing enthalpy. The phase morphology of blends was examined with a JSM-IT300 scanning electron microscope (JEOL Corp., Japan). The cryofracture surface of blends was obtained by quenching fracture in liquid nitrogen. Thermogravimetric analysis (TGA) was performed with a Discovery TGA (Waters Corp., USA) at a heating rate of 10 °C/min in an atmosphere of nitrogen. Dynamic mechanical analysis (DMA) was performed using a PE-DMA8000 instrument (Perkin-Elmer Corp., USA). The tan $\delta$ and storage modulus were determined under the bending mode at a frequency of 1 Hz and a heating rate of 5°C/min. Dielectric properties were performed with an Agilent 8722ET network analyser (Agilent, Santa Clara, CA, USA) at 10 GHz at room temperature.

## 3. Results and discussion

### *3.1. Synthesis and characterization of SPPO*

SPPO was prepared from the electrophilic substitution of PPO and 4-chloromethylstyrene (Fig. 1). PPO in the organic phase interacts with sodium hydroxide in the aqueous phase at the interface to generate aromatic oxygen anion, which forms ion pair with TBAB cation and transfers to the organic phase for alkyl reaction with 4-chloromethylstyrene. Figure 3 shows the $^1$H-NMR spectra of PPO and SPPO. The signals of phenolic OH at 4.0–4.5 ppm [22, 23] are observed in PPO. However, new signals of the vinyl group at 5.0-6.0 ppm [24] appear for SPPO, whereas no phenolic OH signals are observed, suggesting the successful modification of PPO chain terminals to styrene through the completion of the alkyl reaction. Furthermore, as shown in Table 2, the molecular weight of PPO increases and its distribution remains narrow after the terminal modification, indicating the successful transformation of chain terminals.

### *3.2. Curing behaviour*

Styrene-terminated PPO and PB can be cured with peroxides. In this work, we used high-vinyl polybutadiene with the 1,2-structure content more than 70%, which can provide a large amount of reactive vinyl groups and improve the cross-linking density during curing [25]. The reaction of vinyl modified PPO and PB is generally considered as addition polymerization, which follows the mechanism of free radical crosslinking [26-28]. Figure 4 shows the curing procedure of SPPO/PB blend based on the widely proposed reaction mechanism. At first, the homolytic cleavage of the curing agent DCP occurs and forms two alkoxy radicals. The subsequent addition reaction of the alkoxy radicals to the carbon-carbon double bond takes place, and then two adjacent polymer chain radicals combine. These elementary reactions repeat successively to give the thermosetting SPPO/PB blend.

The actual curing process of SPPO/PB blend was investigated by FTIR analysis. All the blends show similar curing behaviours, thus SPPO/PB-0.5 is taken as an example to depict the IR spectrum, TGA and DSC curves. Figure 5 shows the FTIR spectra of SPPO/PB blend before and after curing. As shown, the characteristic absorption peaks at 1640 and 910 cm$^{-1}$ corresponding to the vinyl group stretching and

**Table 2.** Molecular weights of PPO and SPPO.

| Sample | $M_n$ | $M_w$ | PD |
|---|---|---|---|
| PPO | 3186 | 5271 | 1.65 |
| SPPO | 3359 | 5848 | 1.74 |

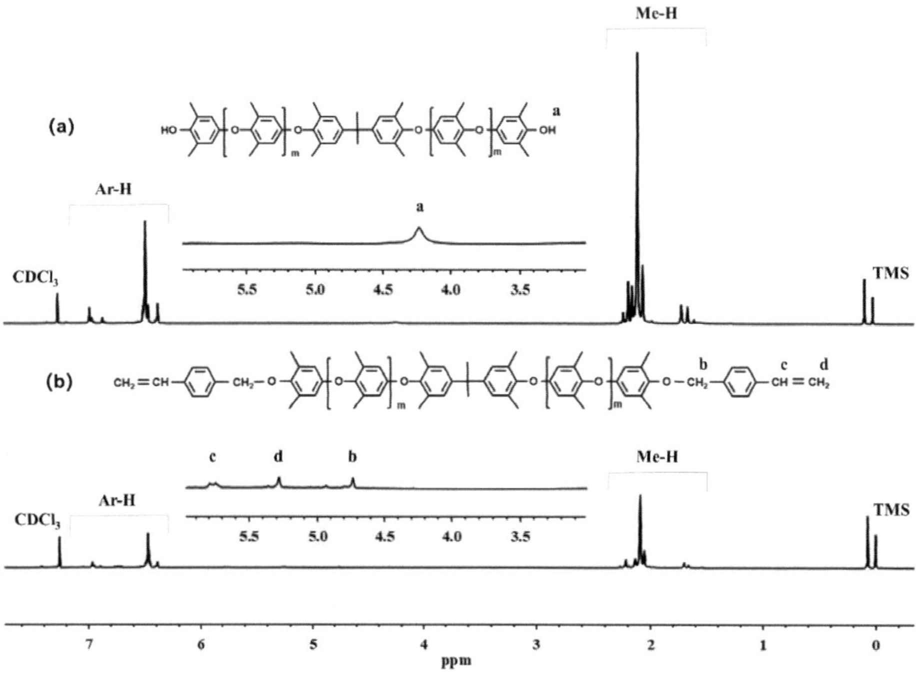


**Figure 3.** $^1$H-NMR spectra of (a) PPO and (b) SPPO in CDCl$_3$.

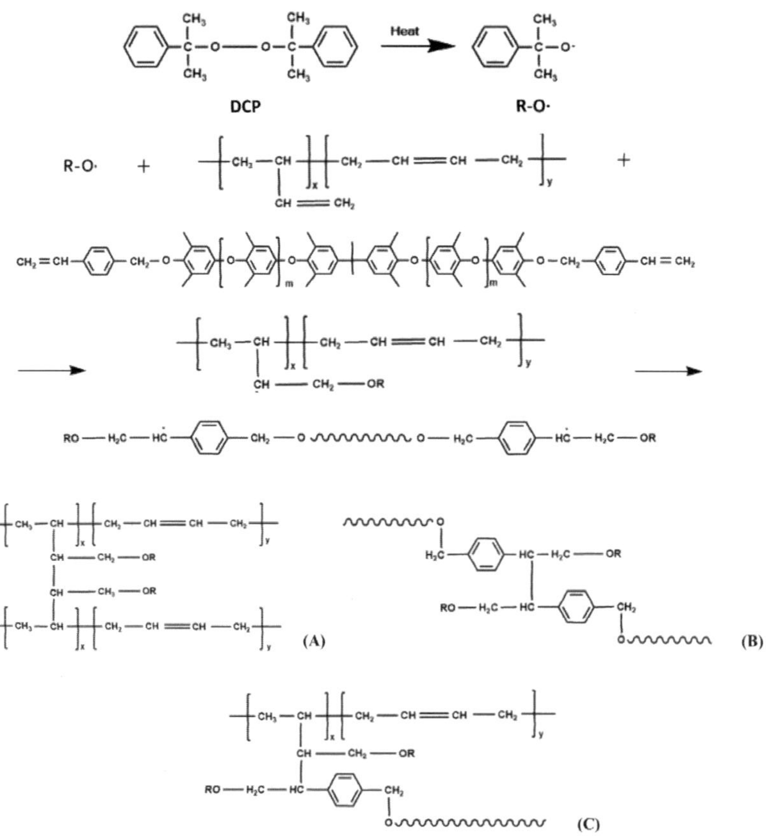


**Figure 4.** Supposed procedure for the synthesis of SPPO/PB blend.

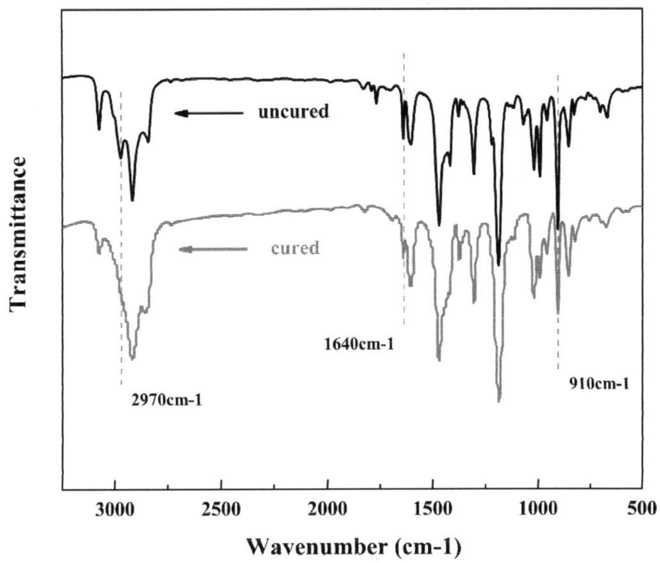


**Figure 5.** FT-IR spectra of SPPO/PB blend before and after curing.

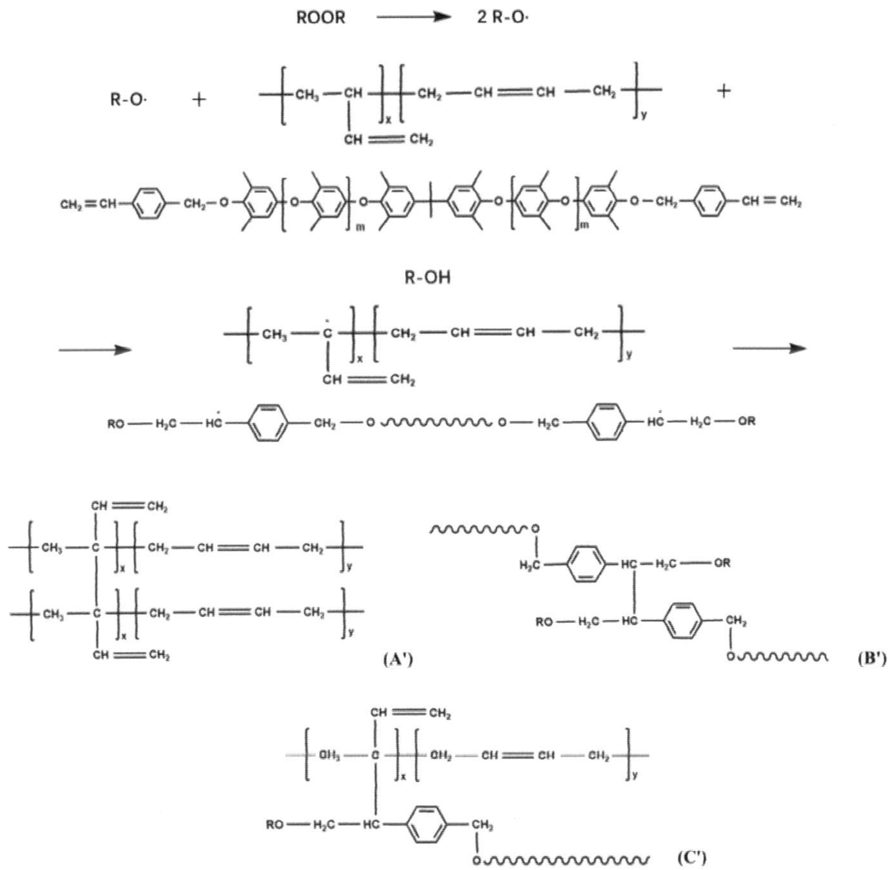


**Figure 6.** Corrected procedure for the synthesis of SPPO/PB blend.

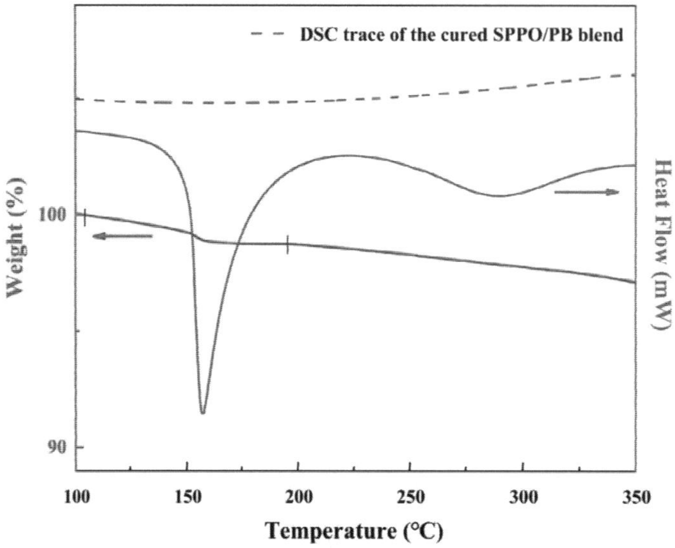


**Figure 7.** TGA and DSC curves of SPPO/PB blend before (solid line) and after (dotted line) curing.

distorting [24] still exist after curing, while absorption peak at 2970 $cm^{-1}$ attributed to the stretching vibration of the methine [29] disappears. The α-H, which is bonded to the carbon-carbon double bond of pendant vinyl group in the polybutadiene, belongs to methine hydrogen, and its activation energy required for the substitution reaction with the radical is very small [30]. Thus, the transfer of methine hydrogen easily occurs from polybutadiene to the alkoxy radicals generated by DCP split during curing (Fig. 6).

To further examine the curing process of SPPO/PB blend, DSC and TGA analysis was employed to determine the thermal behaviour of the blend. As shown in the Figure 7, the SPPO/PB blend with 3 wt% of DCP starts to form cross-linked structures under the temperature of ~120°C, and experiences an unobvious weight loss process in the range 120-200°C. This can be ascribed to the formation of small molecules (2-phenylisopropanol) through the substitution reaction between the alkoxy radical and methine hydrogen. When the temperature is elevated to 220°C, a wide exothermic peak appears, indicating that the cross-linking reaction is further carried out. This could be explained by the complicated curing reactions in the SPPO/PB curing system. The alkoxy radicals not only capture the methine hydrogen of PB, but also add to the terminal styrene groups of SPPO, which leads to the appearance of multiple curing peaks. In addition, the DSC trace of the cured SPPO/PB blend exhibits no obvious exothermic peak, reflecting the curing reaction of SPPO/PB is completed.

The effects of terminal modification of PPO on the curing process of the blend were studied via DSC. Figure 8 shows the DSC curves of unterminated PPO/PB blends and styrene-terminated PPO/PB blends recorded during the heating process. All curves show one major exothermal curing peak in the temperature range of 120-220°C. The major peak temperature ($T_p$) and curing enthalpy ($\Delta H$) were evaluated in Figures 9a,b, respectively. As seen in Figures 8 and 9a, SPPO/PB blend presents a lower $T_p$ than the corresponding PPO/PB blend under the same blend composition, suggesting a reduced curing temperature. For the SPPO/PB blends, the tail-end styrene group of the SPPO reacts with the vinyl groups of PB to extend the network chain. Thus, the DSC curves appear as relatively sharp peaks and the curing peaks shift to lower temperature in contrast to the unterminated PPO/PB blends. This result indicates the faster curing rate of the SPPO/PB blends. Moreover, obvious shoulder peaks appear at high temperature for the SPPO/PB blends, which could be attributed to the self-crosslinking reaction of SPPO. As shown in Figure 9b, due to introducing reactive styrene groups in PPO, the $\Delta H$ of the blending of SPPO/PB blend, which is proportional to the degree of curing, is much higher than that of PPO/PB blend with the same composition. In addition, the $T_p$ of blends decreases with an increase in the fraction

IOP Conf. Series: Materials Science and Engineering **699** (2019) 012018 doi:10.1088/1757-899X/699/1/012018

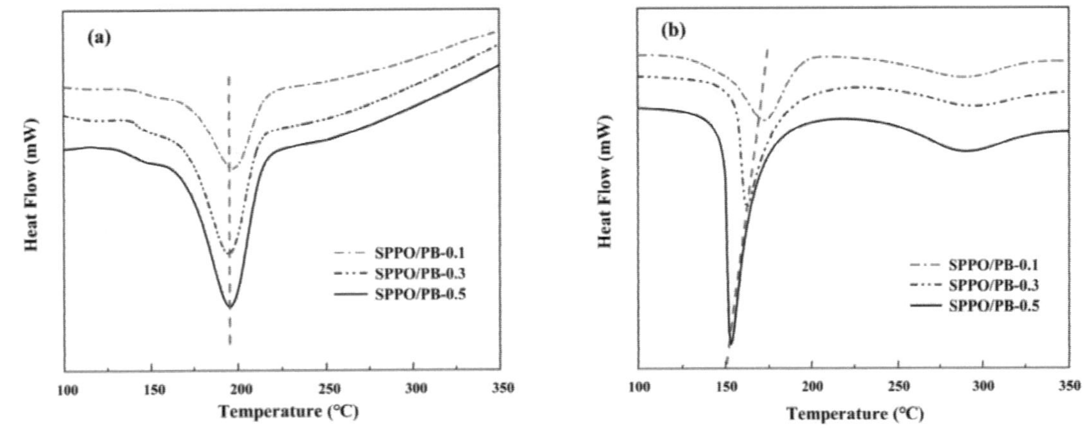


**Figure 8.** DSC curves of (a) PPO/PB and (b) SPPO/PB blends with different compositions.

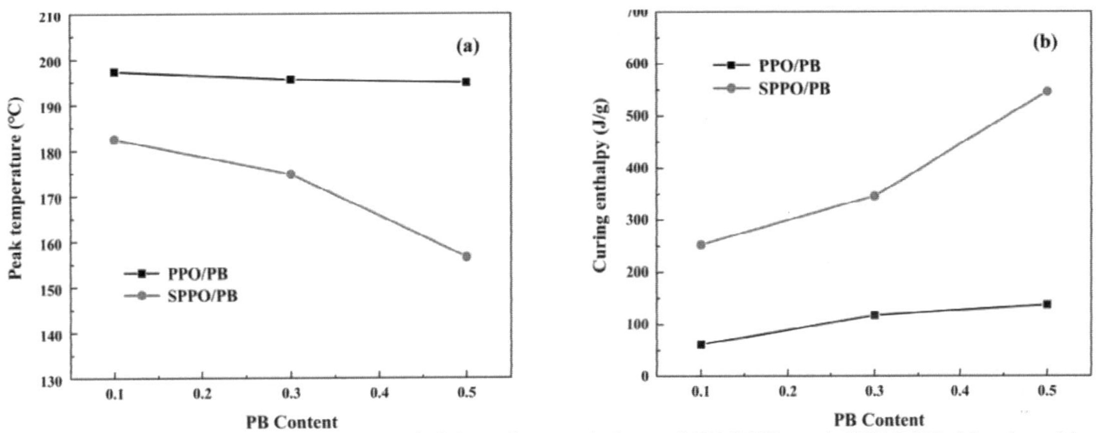


**Figure 9.** (a) Peak temperature and (b) curing enthalpy of PPO/PB and SPPO/PB blends with different compositions.

of PB and the opposite one appears to the ΔH change, which can be more clearly observed in the SPPO/PB blends. High-vinyl polybutadiene contains many active sites for crosslinking, the higher PB content contributes to promote the reaction between the blend components and increase the crosslinking density of curing system.

### *3.3. Phase morphology*

The effects of terminal modification of PPO on phase morphology of the blends were investigated via SEM. Figure 10 shows the SEM images of cryofracture surfaces of PPO/PB and SPPO/PB blends. For the PPO/PB blends, although the low-molecular-weight PPO can partly improve the compatibility of blends, the obvious granular structure and phase interface are observed with different compositions due to the difference in molecular structure [18]. Comparatively, the SPPO/PB blends give rise to the novel morphological features due to the reaction between the blend components in curing. As seen in the SEM images, PB phase is dispersed throughout the SPPO matrix as the irregular-shaped islands, while the phase interface is fuzzy at a very low PB concentration (10%). The reaction between the styrene-terminated PPO and PB is difficult due to low amount of reactive groups, thus the transfer of PB phase is restrained by the cross-linked structure of SPPO. As the PB content increases, no obvious phase interface and an evenly continuous morphology is formed on the whole. The unique morphology of SPPO/PB blends can be ascribed to the reaction between the blend components, PB can be involved in the cross-linked structure of SPPO matrix and a continuous phase is ultimately formed at a high PB content.

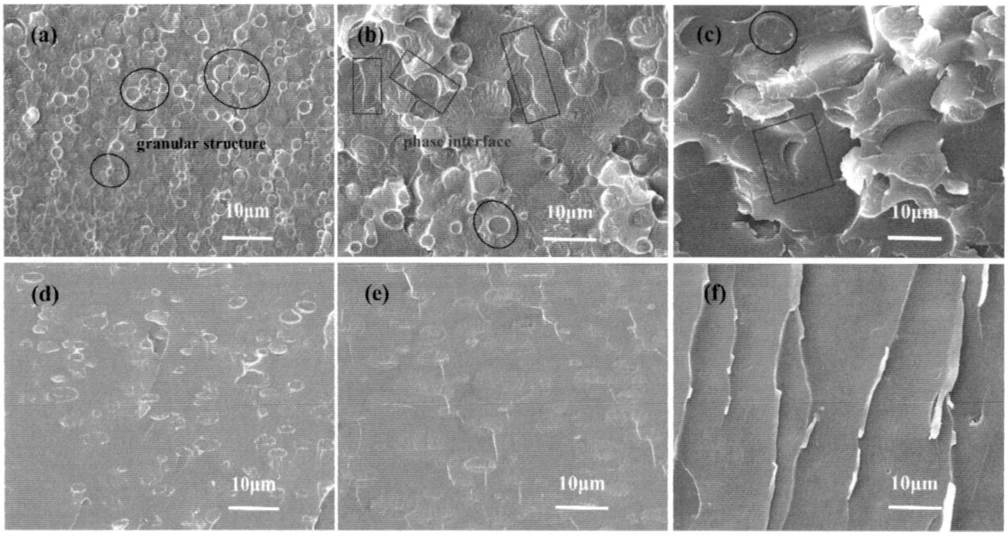


**Figure 10.** SEM images of PPO/PB blends with (a) 0.1, (b) 0.3, and (c) 0.5 PB fractions and SPPO/PB blends with (d) 0.1, (e) 0.3, and (f) 0.5 PB fractions.

### *3.4. Solvent resistance*

Table 3 presents the solvent resistance of PPO/PB blends with different compositions before and after PPO modified. All the cured blends show good resistance against acids and alkalis. The cured SPPO/PB blends are insoluble in the common organic solvents, in contrast to the blends of unterminated PPO and PB that show bad resistance against these organic solvents due to their poor structural stability. This indicates that the solvent resistance of PPO/PB is improved, which is a result of the formation of cross-linked networks between the blend components by introducing styrene groups to the PPO chain ends. The good solvent resistance makes the SPPO/PB blend ready for cleaning process in the copper clad laminate industry.

### *3.5. Thermal properties*

The thermal properties of cured SPPO/PB blends were investigated by TGA and DMA. Figure 11a shows the TGA curves of original SPPO and its reactive blends with PB. Both SPPO and PB have a good thermal stability, because they bear the high-density benzene ring structure and cross-linked structure, respectively. As depicted in Figure 11b, SPPO exhibits 5-wt% loss temperature (Td5%) of 365.8°C, higher than that of PB (346.2°C). The Td5% values of SPPO/PB blends with 0.1, 0.3, and 0.5 PB fractions are 421.5, 432.2, and 415.8°C, respectively, which are higher than that of the original SPPO. First, it increases and then decreases as the content of PB increases. At a relatively low PB content, the increase in Td5% of SPPO/PB blends with an increase in the PB content is due to the increased degree of curing under this condition, proved by the increases in the weight residues of SPPO/PB blends. However, when the PB content is increased to 50%, since the thermal stability of PB is inferior to that of SPPO, the Td5% values of SPPO/PB blend decreased at high PB content.

**Table 3.** Solvent resistance of PPO/PB and SPPO/PB blends with different compositions.

| Sample | Toluene | THF | Chloroform | Acid | Alkali |
|---|---|---|---|---|---|
| PPO/PB-0.1 | Soluble | Soluble | Soluble | Insoluble | Insoluble |
| PPO/PB-0.3 | Soluble | Soluble | Soluble | Insoluble | Insoluble |
| PPO/PB-0.5 | Soluble | Soluble | Soluble | Insoluble | Insoluble |
| SPPO/PB-0.1 | Insoluble | Insoluble | Insoluble | Insoluble | Insoluble |
| SPPO/PB-0.3 | Insoluble | Insoluble | Insoluble | Insoluble | Insoluble |
| SPPO/PB-0.5 | Insoluble | Insoluble | Insoluble | Insoluble | Insoluble |

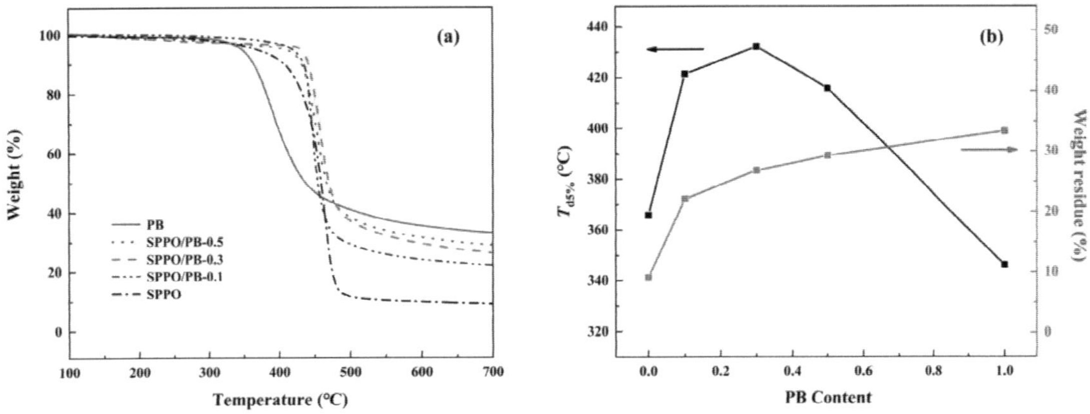


**Figure 11.** (a) TGA curves and (b) temperatures with 5wt% weight loss and weight residues of SPPO/PB blends with different PB contents.

It is notable that high performance PPO should have high $T_g$, which can be identified by the tan $\delta$ peak of DMA curves. Figure 12a shows the tan $\delta$ values versus temperature of SPPO/PB blends as a function of PB content, with the $T_g$ of all reactive blends increased, and the $T_g$ of original PB and SPPO occurring at 144.8 and 171.7°C, respectively. The copolymerization between SPPO and PB restricts dynamic motion, resulting in the great increase in $T_g$. Meanwhile, the higher PB content promoted the reactivity of tail-end styrene groups and led to less spacing between the crosslink networks, also resulting in a higher $T_g$ as well. Remarkably, the cured SPPO/PB blends with 0.3 and 0.5 PB fractions exhibit relatively high $T_g$ (214.6 and 226.5°C) compared with other PPO, demonstrating that enhancing thermal properties of PPO are archived successfully in this work. In addition, the DMA curve of SPPO/PB blend has two peaks at a very low PB concentration (10%), while increasing the PB content results in the appearance of relatively sharp single peaks in DMA curves, indicating that the reactive blends have a relatively uniform phase structure. These are consistent with the results observed by SEM.

Based on rubber elasticity theory, the crosslinking density of cured resins can be estimated from rubbery plateau modulus [31,32]. The temperature dependent storage modulus of cured SPPO/PB blends with different PB contents is shown in Figure 12b. The increase of the storage modulus suggests that the crosslinking density increases with increasing PB content. This can be explained by the functionality of SPPO and high reactivity of PB, which leads to a more densely crosslinked network.

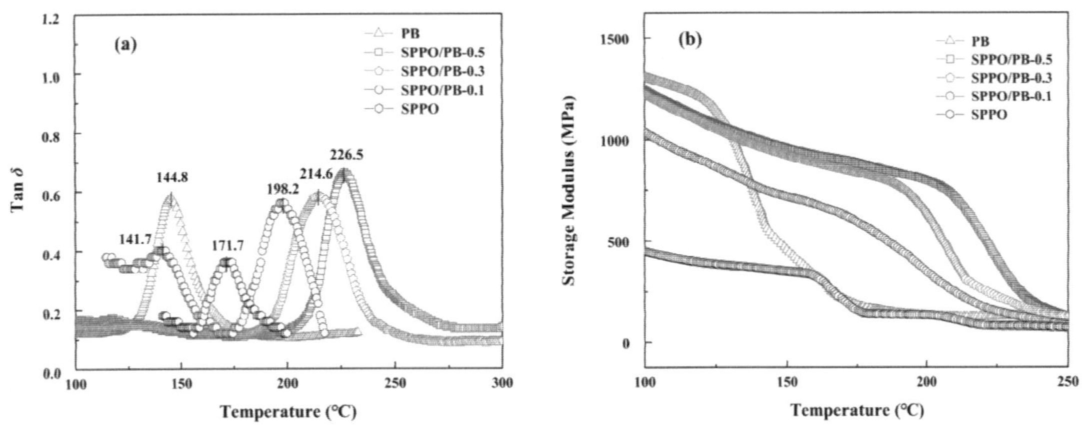


**Figure 12.** DMA curves of temperature dependent (a) tan δ and (b) storage modulus of SPPO/PB blends with different PB contents.

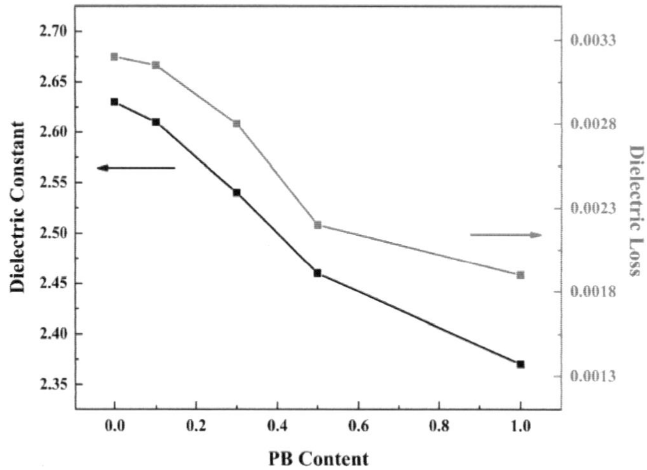


**Figure 13.** Dielectric properties of SPPO/PB blends as a function of PB content.

### *3.6. Dielectric properties*

Figure 13 shows the dielectric constant (Dk) and loss factor (Df) of the blends with different compositions. As seen in this figure, all the cured blends have low Dk and Df, and those are lower than original SPPO. The Dk and Df values generally decrease with increasing the PB content, which is due to the fact that PB has a very low polarity. However, the Df values decrease more slowly with respect to the Dk, when the PB fraction increases from zero to 0.3, which is considered to be attributed to the multiphase morphology of the blends. In the case of a multiphase material, the dielectric properties are mainly dominated by the polarity of each phase, but the dielectric loss is dominated not only by composition, but also by phase morphology [33]. The compatibility between SPPO and PB is poor at a low PB concentration, and the appearance of the second phase increases the quantity of the accumulated charge due to an additional contribution of the polarization at phase interfaces [19]. As the PB content further increases, the compatibility of the resin system is improved, leading to the Df of the blend decreasing significantly.

## 4. Conclusion

In this study, the terminally styrene-modified PPO is cured with PB to form SPPO/PB blends successfully. The results show that higher PB content promotes the reactivity of tail-end styrene groups and leads to a higher degree of curing. The terminal styrenisation of PPO enhances the compatibility of SPPO and PB, and thus continuous phase is formed in the blend at high PB contents. The cross-linked network enhances organic solvent-resistance of the blends, and the high density of the benzene ring structure and cross-linked structure provide excellent thermal properties with good dielectric performance retained. Especially, when the mass ratio of PB in the blend is 0.5, the curing system has a high $T_g$ of 226.5 °C, and simultaneously extremely low Dk (2.46) and Df (0.0022) at 10 GHz, indicating that the reactive blend of styrene-terminated PPO can meet the requirements of matrix resins for the fabrication of printed circuit board in electronic industry.

**Acknowledgements**

This work was financially supported by the National Natural Science Foundation of China (No. 51572205), the Equipment Pre-Research Joint Fund of EDD and MOE (No. 6141A02022262), and the Fundamental Research Funds for the Central Universities (WUT: 2018III019).

**References**

[1] Zhuo D, Gu A, Liang G, Hu J and Yuan L 2010 *J. Mater. Sci.* **46** 1571-80
[2] Zhang B, Ye F, Liu S, Liu Q and Gao Y 2015 *J. Mater. Sci.* **26** 7823-8

[3] Kim Y, Lim Y W, Kim Y H and Bae B S 2016 *ACS Appl. Mater. Interfaces* **8** 8335-40
[4] Gu X, Zhang Z, Yuan L, Liang G and Gu A 2016 *Chem. Eng. J.* **298** 214-24
[5] Murali K P, Rajesh S, Jacob K S, Prakash O, Kulkarni A R and Ratheesh R 2010 *J. Mater. Sci.: Mater. Electron.* **21** 192-8
[6] Wu K T, Yuan Y, Zhang S R, Yan X Y and Cui Y R, 2013 *J. Polym. Res.* **20** 223
[7] Pan C, Kou K, Jia Q, Zhang Y, Wang Y, Wu G and Feng A 2016 *J. Mater. Sci.: Mater. Electron.* **27** 11909-16
[8] Su C T, Lin K Y, Lee T J and Liang M 2010 *Eur. Polym. J.* **46** 1488-97
[9] Habaue S, Iwai S, Kubo H, Nagura K, Watanabe T, Muraki Y and Tsutsui Y 2014 *React. Funct. Polym.* **83** 49-53
[10] Yang M C, Higashihara T, Su H W, Ueda M and Chen W C 2016 *J. Polym. Sci. Part A: Polym. Chem.* **54** 3218-23
[11] Wu S, Tung N, Lin T and Shyu S 2000 *Polym. Int.* **49** 1452-7
[12] Liang G, Meng J and Zhao L 2003 *Polym. Int.* **52** 966-72
[13] Zhu W and Weil E D 1998 *J. Appl. Polym. Sci.* **67** 1405-14
[14] Merfeld G D, Yeager G W, Chao H S and Singh N 2003 *Polymer* **44** 4981-92
[15] Soulé E R, Fine T, Borrajo J, Williams R and Pascault J P 2006 *J. Appl. Polym. Sci.* **100** 1742-7
[16] Weng L, Zhang Y, Zhang X, Liu L and Zhang H 2017 *Polym. Compos.* **39** 2334-45
[17] Weng L, Zhang Y, Zhang X, Liu L and Zhang H 2018 *J. Mater. Sci.: Mater. Electron.* **29** 2831-40
[18] Wang H B, Wang Y X, Huang C and Chen S J 2008 *China Plastics Industry* **36** 47-50
[19] Mo J, Xia L, Pan P, Shentu B and Weng Z 2013 *J. Appl. Polym. Sci.* **127** 4879-88
[20] Wang Y, Cheng S, Li W, Huang C, Li F and Shi J 2007 *Polym. Bull.* **59** 391-402
[21] Wu Y, Gui L, Wang H, Shentu B and Weng Z 2013 *Ind. Eng. Chem. Res.* **52** 17351-9
[22] Chen C H, Lee K W, Lin C H and Juang T Y 2018 *Polymers* **10** 411
[23] Hwang H J, Hsu S W and Wang C S 2008 *J. Appl. Polym. Sci.* **110** 1880-90
[24] Fukuhara T, Shibasaki Y, Ando S and Ueda M 2004 *Polymer* **45** 843-7
[25] Yuan X, Wang J, Wang C, Gao S, Guo S and Zhang Y 2016 *J. Appl. Polym. Sci.* **133** 43280
[26] Zeng X P 2012 China Patent No.102807658A
[27] Chen M H and Liu S F 2013 China Patent No.104031377A
[28] Jung D H, Kwon J D, Kim M H and Hong D W 2017 United State Patent No. 20170342264A1
[29] Yang Q J 1997 *Modern rubber technology* (Beijing: Chemical Industry Press)
[30] Li X L, Shao Y G, Xiao Y M, Hua J and Chen X G 2017 *Rubber Sci. Technol.* **15** 18-24
[31] Flory P J 1953 *Principles of Polymer Chemistry* (New York: Cornell University Press)
[32] Lv J, Meng Y, He L, Qiu T, Li X and Wang H 2013 *J. Appl. Polym. Sci.* **128** 907-14
[33] Gao R, Gu A, Liang G, Dai S and Yuan L 2011 *J. Appl. Polym. Sci.* **121** 1675-84

IOP Conf. Series: Materials Science and Engineering **699** (2019) 012019 doi:10.1088/1757-899X/699/1/012019

# Development and research carbon nanotube-based resistive gas sensor

**O I Il'in, M V Il'ina, N N Rudyk, A V Guryanov and A A Fedotov**

Southern Federal University, Institute of Nanotechnologies, Electronics and Electronic Equipment Engineering, 347922 Taganrog, Russia

nnrudyk@sfedu.ru

**Abstract**. Experimental studies of the processes of formation of catalytic centres (CC) and carbon nanotubes (CNTs) at ITO contacts have been carried out. The regularities of the influence of the annealing temperature on the geometric dimensions of CC have been established. An array of interwoven CNTs with a highly developed surface has been grown. A model of a gas sensor with a sensitive element based on a CNT network has been created. The reaction time and reaction of the sensor, its sensitivity to $N_2$, $O_2$, and Ar have been experimentally investigated. It has been shown that the sensor has a maximum sensitivity of 17.2% to $N_2$, 16.3% to Ar, and 18.7% to $O_2$ in the range of gas concentrations from 30 to 70 ppm. It has been shown that gas detection is possible at room temperature, despite a rather long reaction and reduction time. In this case, an almost complete restoration of the sensitive element of the initial resistance has occurred without additional heating.

## 1. Introduction

In a sophisticated man-made situation and terrorist threat danger, one of the problems is gas sensors for environmental monitoring. The main parameters of gas sensors are sensitivity, selectivity, response and recovery times, dimensions, and power consumption. Traditional materials used in resistive sensors (SnO, ZnO, et al.) have good sensitivity, but do not show sufficient selectivity. The long reaction and recovery times, high energy consumption associated with the necessity of heating the sensitive layer [1,2] are the disadvantages of these sensory materials. Therefore, there is a regular research of new materials, which can improve the sensors characteristics. Graphene and carbon nanotubes (CNT) are ones of the most promising materials both for electronic chips [3-5] and gas sensors, which are designed to eliminate the disadvantages of sensors available on the market [6,7].

The aim of this work is the development of the CNT-based resistive gas sensor and experimental study of its sensitivity parameters.

## 2. Experimental

A model of a gas sensor with a sensitive layer was created on the basis of an array of interwoven multiwalled carbon nanotubes (MWCNT). To create a model, interdigit electrodes based on an ITO film 100 nm thick were formed on a glass substrate. ITO was deposited using the Auto500 unit (BOC Edwards, UK) through a mask. A 10 nm thick Ni film acting as a catalytic layer of CNT growth was deposited over the ITO layer through a mask. After that, the catalytic film was annealed to form catalytic centres (CC) of CNT growth.

The effect of heating temperature on the formation of CC for CNT growth was studied in the range of 700-800°C. Investigation of the CC parameters of the experimental samples was carried out using a probe nanolaboratory Ntegra (NT-MDT, Russia) (Fig. 1). Geometric parameters of Ni CCs were determined using special software Image Analysis.

Then, at a temperature of 750°C, a network of disordered CNTs was grown in the NANOFAB PECVD module (Fig. 2). CNTs were grown in an atmosphere of acetylene and ammonia.

The gas sensor was placed in a gas measuring stand, where its response to nitrogen, argon, and oxygen was checked. The measuring stand consisted of a vacuum chamber, a controlled gas supply system, and a pumping system. The choice and study of the sensor's reaction to the presence of these gases is associated with their potential danger during accumulation in confined spaces.

Measurement of resistance was carried out using a universal ohmmeter B7-78/1, at a temperature in the measuring chamber of 300 K. The sensitivity of the device was calculated by the formula:

$$S_k = \frac{R_{gas}-R_{air}}{R_{air}} \cdot 100\% \qquad (1)$$

where $R_{gas}$ – sensor resistance in the presence of gas, $R_{air}$ – sensor resistance before gas supply.

The response time was defined as a time interval of 90% of the time required for the release ($R_{gas}$ - $R_{air}$) to a stationary level after injecting a certain concentration of the analysed gas. Sensor recovery time was defined as the time interval, during which the sensor resistance recovered after injection of the analysed gas to 10% of the maximum value $0.1(R_{gas}-R_{air})+R_{air}$.

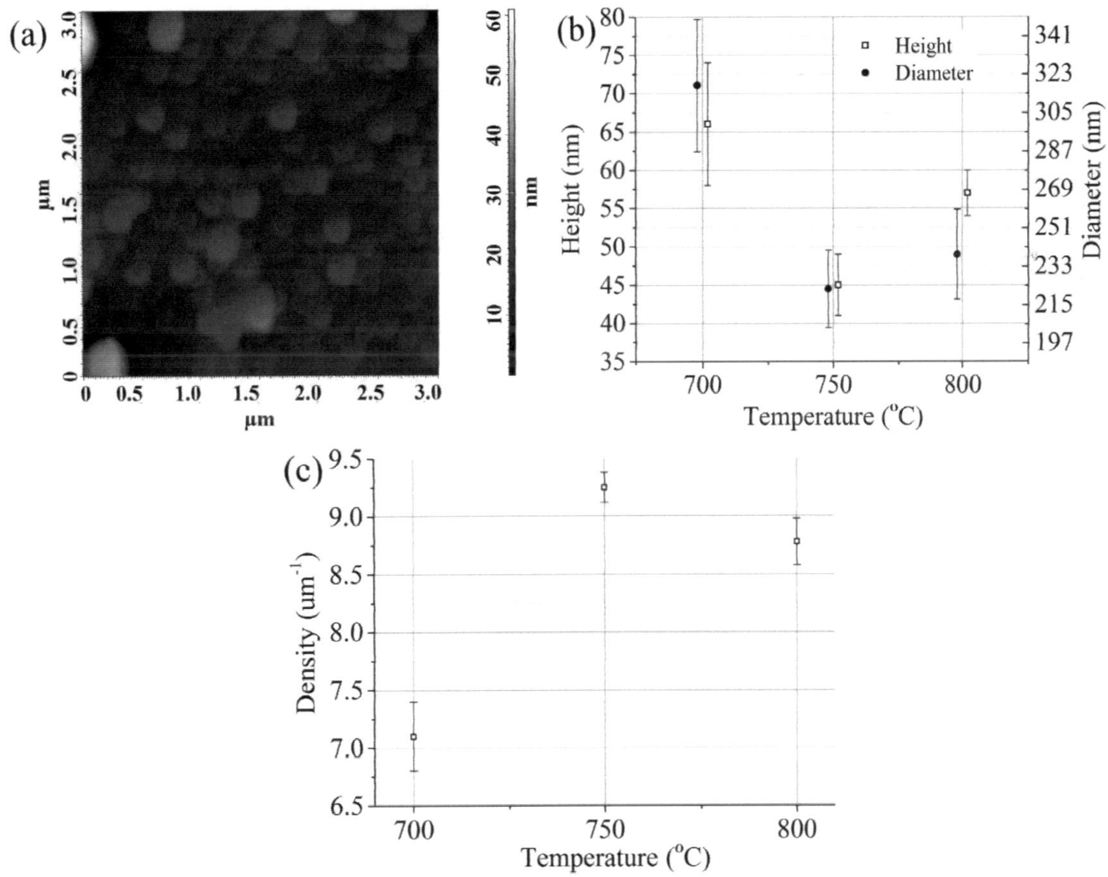


**Figure 1.** Dependence of CC geometric parameters on processing temperature: (a) example of AFM image of CC; (b) dependence of CC geometric parameters on processing temperature; (c) dependence of CC surface density on processing temperature.

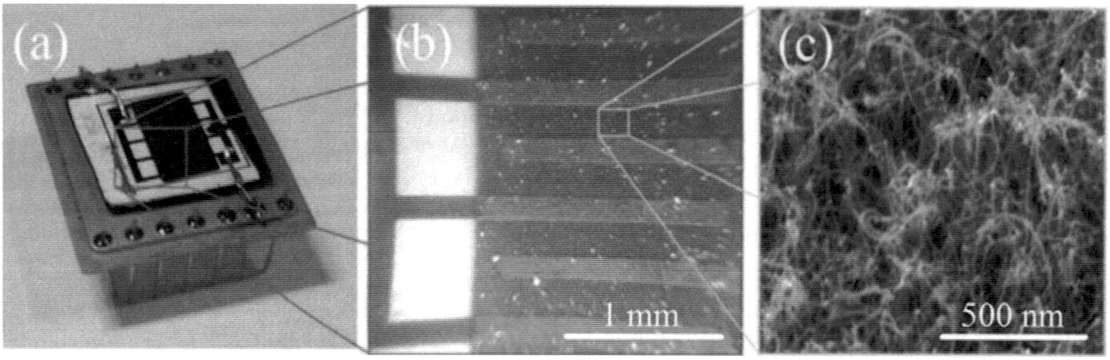


**Figure 2**. The model of a gas sensor: (a) side view; (b) sensitive layer with interdigital electrodes; (c) SEM-image of CNT film.

## 3. Results and discussion

The geometric dimensions (height and diameter) and the surface density of CC are the main factors affecting the geometric parameters of the grown CNTs. The study of the size of the CC at the annealing stage made it possible to optimize the modes of their formation. The results of the AFM study of the CC showed that the annealing temperature had a significant effect on the geometric parameters of the CC. Analysis of the obtained AFM images made it possible to determine the average diameter, height, and surface density of the formed CCs (Fig. 1). It can be seen that with increasing annealing temperature from 700 to 750°C, both the height and diameter of the CC decrease, which indicates the occurrence of evaporation and sublimation of Ni from the substrate surface. This leads to a decrease in the average geometric dimensions of the CC, and also reduces their dispersion in the array. With a further increase in temperature, the processes of atomic diffusion over the surface are manifested, which is reflected in the combination of small droplets into larger ones and, correspondingly, an increase in the size of formed CCs. It is established (Fig. 1) that the temperature of 750°C allows the formation of the smallest CC (223 ± 20 nm) with the maximum density (9.2 $\mu m^{-2}$).

Thus, for the subsequent formation of the sensitive element of the model (Fig. 2a,b) of a CNT-based gas sensor, a temperature of 750°C was chosen. To ensure maximum sensitivity of the sorption gas sensor, an array of interwoven MWCNTs with a high surface development was grown (Fig. 2c). The formation of interwoven CNTs is due to the fact that the sensitivity of the sensor is proportional to the surface area of the CNTs. Thus, nanotubes can adsorb more gas molecules onto themselves.

The initial resistance of the structure was found to be 516 kΩ. Figure 3 shows the sensor's response time to different gases.

Table 1 summarizes the results of measuring the change in sensor resistance, converted to gas sensitivity according to formula (1), as well as the response and recovery time for each gas.

**Table 1.** Measured characteristics of gas sensor model.

| *Parameters* | | *Value* | | |
|---|---|---|---|---|
| | | $N_2$ | *Ar* | $O_2$ |
| Gas sensitivity, ppm | 10 | - | (5 ± 1.3) % | - |
| | 30 | (8.7 ± 2.1) % | (9.7 ± 3.2) % | (10.4 ± 2.2) % |
| | 50 | (9.4 ± 1.4) % | (16.3 ± 4.1) % | (16.6 ± 2.5) % |
| | 70 | (17.2 ± 3.6) % | - | (18.7 ± 3.3) % |
| Response time, s | | 153 ± 12 | 186 ± 15 | 141 ± 7 |
| Recovery time, s | | 280 ± 19 | 130 ± 11 | 190 ± 8 |

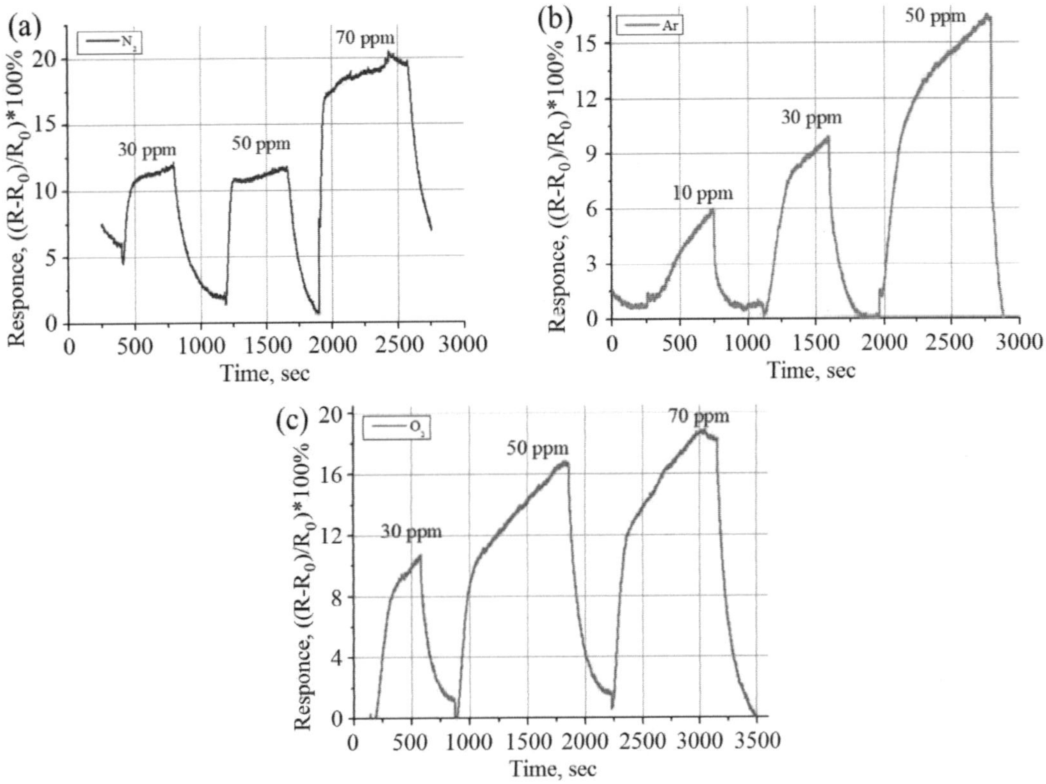


**Figure 3**. Sensor's response in the presence of gases with different concentrations: (a) nitrogen; (b) argon; (c) oxygen.

Thus, the maximum measured sensitivity to oxygen, argon, and nitrogen was 18.7%, 16.3%, and 17.2%, respectively, with a response time of 141 s, 186 s, and 153 s, respectively. The onset of exposure to all three gases causes a fairly fast response of the sensor (on average 160 sec), which is a good result for a sensitive layer operating at room temperature. The fastest recovery time of 130 s was shown after exposure to argon. After exposure to oxygen, the sensor was restored for 60 s longer, which might be due to the high chemical activity of oxygen and a large mass of adsorbed molecules. The long recovery time of the sensor after exposure to nitrogen may be associated with the formation of C-N bonds and partial nitrogen chemisorption. Also, the argon atom is 2.5 times heavier than oxygen one, which also complicates the desorption process. However, the patterns that affect the recovery time of the gas-sensitive layer require additional research. Moreover, in all experiments, the restoration of the sensor model occurred without heating, at room temperature. From the obtained results, we can conclude that the developed gas sensor sensitive element has a high sensitivity and speed, as well as does not need to be heated for degassing.

## 4. Conclusion

A model of a sorption gas sensor with a CNT-based sensitive element was developed, and its performance was measured. It was shown that the sensor had a maximum sensitivity of 17.2% to $N_2$, 16.3% to Ar, and 18.7% to $O_2$ in the range of gas concentrations from 30 to 70 ppm. It was also found that even at room temperature, despite a rather long reaction and recovery times, an almost complete restoration of the sensitive element of the initial resistance occurred.

Obtained results can be used in the development of promising sensor devices based on carbon nanotubes, in particular portable wearable devices as well as multisensory systems, which will expand the range of sensitivity and increase the selectivity of devices.

**Acknowledgements**
This work is supported by the Russian Science Foundation under grant № 18-79-00176. The results were obtained using the equipment of the Research and Educational Centre and the Centre for Collective Use "Nanotechnology" of the Southern Federal University.

**References**

[1] Rigoni F, Tognolini S, Borghetti P, Drera G, Pagliara S, Goldoni A and Sangaletti L 2013 Enhancing the sensitivity of chemiresistor gas sensors based on pristine carbon nanotubes to detect low-ppb ammonia concentrations in the environment *Analyst.* **138** 7392-9
[2] Joshi N, Hayasaka T, Liu Y, Liu H, Oliveira O N and Lin L 2018 A review on chemiresistive room temperature gas sensors based on metal oxide nanostructures, graphene and 2D transition metal dichalcogenides *Microchim. Acta* **185** 213
[3] Gusev E Y and Avdeev S P 2018 Preparation of flat surfaces on silicon carbide substrate using electron beam processing *J. Phys. Conf. Ser.* **1124** 022026
[4] Il'ina M V, Il'in O I, Blinov Y F, Smirnov V A, Kolomiytsev A S, Fedotov A A, Konoplev B G and Ageev O A 2017 Memristive switching mechanism of vertically aligned carbon nanotubes *Carbon N. Y.* **123** 514-24
[5] Il'ina M, Il'in O, Blinov Y, Konshin A, Konoplev B and Ageev O 2018 Piezoelectric response of multi-walled carbon nanotubes *Materials* **11** 638
[6] Chen K, Gao W, Emaminejad S, Kiriya D, Ota H, Nyein H Y Y, Takei K and Javey A 2016 Printed carbon nanotube electronics and sensor systems *Adv. Mater.* **28** 4397-414
[7] Dai H, Feng N, Li J, Zhang J and Li W 2019 Chemiresistive humidity sensor based on chitosan/zinc oxide/single-walled carbon nanotube composite film *Sensors Actuators B Chem.* **283** 786-92

IOP Conf. Series: Materials Science and Engineering **699** (2019) 012020 doi:10.1088/1757-899X/699/1/012020

# In situ supported $VO_x$ on carbon nanotubes for the low-temperature selective catalytic reduction of NO with $NH_3$

**W Jin[1], B Q Jiao[1], W Chen[2], M Y Li[1], J Zhao[1] and X W Zhang[1]**

[1]State Key Laboratory of Silicate Materials for Architectures, Wuhan University of Technology, 430070 Wuhan, China
[2]School of Materials Science and Engineering, Wuhan University of Technology, 430070 Wuhan, China

chenw@whut.edu.cn

**Abstract**. The vanadium oxide/carbon nanotube ($VO_x$/CNTs) composite catalyst was prepared by in-situ growth of $VO_x$ nanoparticles on CNTs with a cetylpyridinium chloride (CPC) assisted reflux route. The in-situ prepared catalysts exhibit better $NH_3$-SCR activity in a broader temperature window at low temperature than those prepared by impregnation or a mechanically mixed method. The structural characterizations show that vanadium oxide nanoparticles have a good dispersion on the CNTs surface, with much lower valence vanadium species and chemisorbed oxygen species. The physics-chemistry properties indicate that there is a strong interaction between the $VO_x$ and CNTs, and the catalysts present a larger amount of stronger acid. The abovementioned reasons have eventually led to the enhancement of $NH_3$-SCR activity.

## 1. Introduction

$NO_x$ emitted from stationary source plants can cause problems, such as photochemical smog, acid rain, and ozone hole, which harm the environment and human health [1]. Due to high denitration efficiency, and good selectivity, selective catalytic reduction (SCR) has become the mainstream technology for removing $NO_x$ nowadays [2]. As a typical commercial $NH_3$-SCR catalyst, $V_2O_5$-$WO_3$($MoO_3$)/$TiO_2$ catalysts are widely used due to their high $NO_x$ conversion and $SO_2$ tolerance. However, there are still some inevitable disadvantages, such as the narrow operation temperature window (300-400°C) and especially the poor low-temperature catalytic activity [3]. Therefore, much focus has been placed on low temperature SCR Catalyst.

Carbonous materials have rich physical and chemical properties, which lend to their high porosity and good surface chemistry. In line with this, carbon-based materials are suitable for use in most catalytic process applications [4]. Studies show that the introduction of carbon materials with large specific surface area (such as activated carbon, graphene and carbon nanotubes) can improve the NO conversion of the catalyst at low temperature [5-10]. Nowadays, carbon nanotubes (CNTs) are recognized as an attractive catalyst support for SCR catalysts due to their electronic properties and unique nanostructure [11,12]. CNT-supported metal oxide catalysts are well-known in the adsorption field and present interesting properties for the denitrification of $NO_x$ species. Wang et al. [13] reported high NO conversion (≈90%) at a relatively low reaction temperature of 200 to 300°C on the Mn-Ce/CNT catalyst prepared using the incipient-wetness method. Bai et al. [14] investigated the activity of $V_2O_5$/CNT catalysts at a low SCR reaction temperature. They found that $V_2O_5$/CNT catalysts had a higher catalytic activity compared to that of $V_2O_5$ particles supported on AC ($V_2O_5$/AC) at 200°C reaction temperature. In addition, they also found that the conversion of $NO_x$ momentously increased and became almost linear at a high level of SCR reaction by $V_2O_5$/CNT (0.1-1 wt %) catalysts in the presence of $SO_2$. Therefore, due to the unique promoting effect of $SO_2$, the $V_2O_5$/CNT catalyst is justified as one of the most promising catalytic materials for SCR at low temperature.

As the core of the $NH_3$-SCR technology, the catalyst needs to work under high temperature, high air flow rate, and high dust environment. Abovementioned catalysts, prepared by the traditional impregnation method or mechanical mixing method, are prone to inactivation by aggregations or fall away from the carrier during using, due to the weaker bonding strength between active component and the carrier during using, which hinders its practical application [15].

Herein, in order to enhance the bonding strength between the active component and the supports, we use a facile method for the fabrication of $VO_x$ nanoparticles in-suit grown on the surface of CNTs with cetylpyridinium chloride (CPC) assisted reflux method. In the fabrication process, CPC is a kind of cationic surfactant, bonded with CNTs by π-π conjugate. As a result, the surface charges of CNTs were shifted from negative to positive after mixing with CPC, which served as centres for the anchoring and dispersion of metavanadate anions. Subsequently, the in-situ growth of uniform $VO_x$ nanoparticles on CNTs was achieved viå a heat treatment process. The vanadium oxide/carbon nanotube ($VO_x$/CNTs) composite catalyst was investigated for its catalytic activities in low temperature SCR reactions. The effect of the weight ratio of CNTs and CPC was also studied.

## 2. Experimental

### *2.1. Catalyst preparation*

The employed multiwall CNTs were purchased from Qinhuangdao Tai Chi Ring nano products Co. Ltd (China). Cetylpyridinium chloride was purchased from Aladdin. All other chemicals were purchased from Sinopharm Chemical Reagent Co., Ltd. (China). All the reagents had an analytical grade and were used without further purification.

The employed multiwall CNTs were refluxed and oxidized with concentrated $HNO_3$ (65-68 wt%) under stirring for 6 h to remove carbon nanoparticles and metal species and introduce oxygenated surface groups. The treated CNTs were then washed fully with deionized water and ethanol, and dried at 80°C for 12 h. The purified CNTs was denoted as P-CNTs.

In-situ deposition of $VO_x$ nanoparticles on the surface of CNTs was prepared by CPC reflux route with different weight rations 1:1, 2:1, and 3:1 of CPC: CNTs, which were labelled as $V_2O_5$/CNTs-1, $V_2O_5$/CNTs-2, and $V_2O_5$/CNTs-3, respectively. Typically, 0.36 g purified CNTs were dispersed with CPC in 80 mL deionized water upon ultrasonic vibration for 1 h to get a stable suspension. Subsequently, 0.27 g ammonium metavanadate ($NH_4VO_3$) was added to 50 mL deionized water and stirred with heating at 80°C for 30 min to form a clear transparent solution. Then the two solutions were mixed and stirred for 30 min to get a uniform solution. The final mixture was refluxed under continuous stirring at 80°C for 12 h. After that, the suspension was filtered and washed fully with distilled water and ethanol, followed by desiccation at 80°C for 12 h. Finally, the sample was calcined at 400°C for 5h in $N_2$ atmosphere, and then calcined at 300°C for 30 min in air to make vanadium oxide crystallize.

For comparison, the catalysts were also prepared by impregnation (denoted as $V_2O_5$/CNTs-I) and mechanically mixed method (denoted as $V_2O_5$/CNTs-M). $V_2O_5$/CNTs-I catalyst was prepared by dissolving ammonium metavanadate into deionized water and then stirring for 12 h at 80°C. Subsequently, the sample was dried at 80°C for 12 h and calcined at 400°C for 5 h in $N_2$ atmosphere followed by calcination at 300°C for 30 min in air. The $V_2O_5$/CNTs-M catalyst was also prepared by fully grinding the mixture of the purified CNTs and ammonium metavanadate, and then calcined at 400°C for 5 h in $N_2$ atmosphere followed by calcination at 300°C for 30 min in air.

### *2.2. Characterization methods*

X-ray diffraction (XRD) patterns of the samples were obtained using an X-ray diffractometer (Pert-Pro, PANalytical, Netherlands) with Cu $K_\alpha$ radiation. The microstructure and morphology of the catalysts were observed using a field emission transmission microscope (JEM-2100F, Japan Electronics Co. Ltd). Specific surface areas and pore volumes were measured on a fully automated Quantachrome Nova 2200e surface area analyser using nitrogen adsorption at 77 K. The Brunauer-Emmett-Teller (BET) equation was used for the determination of the specific pore volume. The samples were characterized by XRD using a Rigaku D/MAX2500V system with Cu $K_\alpha$ radiation. The operating voltage was 40 KV, the operating current was 40 mA, and the scanning speed was 5°min$^{-1}$. X-ray Photoelectron Spectroscopy measurements (XPS, ESCALAB 250Xi, Thermo Fisher, USA) were performed using Al $K_\alpha$ radiation as the excitation source. The emission angle between the photoelectron beam and the sample surface was 45°, and the calibration of the binding energy for the electron spectrometer was made using the maximum adventitious C 1s signal at 284.6 eV with the solution of the full width at half maximum (FWHM) being 0.8 eV.

$NH_3$ temperature programmed desorption ($NH_3$-TPD) experiments were carried out on a PX200 apparatus (Tianjin Pengxiang Technology Co. Ltd.) with a thermal conductivity detector (TCD). Before $NH_3$ adsorption, the samples (0.15 g) were pre-treated under He flow (50 mL min$^{-1}$) at 400°C (ramp 10°C min$^{-1}$) for 60 min. $NH_3$ adsorption was done at 100°C using 10 vol% $NH_3$ in He with flow rate of 50 mL min$^{-1}$ for 60 min and then flushed with He during 1 h to remove physisorbed $NH_3$. Finally, the ammonia was desorbed in helium flow of 30 mL min$^{-1}$ from 100 to 600°C using a heating rate of 10°C min$^{-1}$. $H_2$ temperature programmed reduction ($H_2$-TPR) was carried out on the same PX200 apparatus (Tianjin Pengxiang Technology Co. Ltd.) with a TCD. Prior to $H_2$-TPR measurements, the catalyst (0.1 g) was pre-treated under Ar flow of 50 mL min$^{-1}$ rate at 400°C (ramp 10°C min$^{-1}$) for 60 min. After being cooled down to 50°C in the same atmosphere, the sample was exposed to a flow containing 10 vol% $H_2$ in Ar (50 mL min$^{-1}$) for 1 h, and then heated between 50 and 800°C with a heating rate of 10°C min$^{-1}$.

### *2.3. Catalyst performance tests*

The SCR activity tests were conducted in a vertical fixed-bed quartz reactor with an internal diameter of 25 mm and heated by a tube-type electric furnace. The simulated flue gas consisted of 500 ppm NO, 500 ppm $NH_3$, 3 vol% $O_2$, and the balance gas was mixed with Ar. A total flow rate of 1000 mL/min was maintained in all the tests. The amount of catalyst was 2 g, the reaction space velocity was 30,000 h$^{-1}$, and the reaction temperature ranged from 100 to 300°C. The gas concentrations of NO were measured online by using a Laoying 3026 infrared flue gas comprehensive analyser. NO conversion was calculated according to the following expression:

$$\text{NO conversion (\%)}=\frac{[NO]_{in}-[NO]_{out}}{[NO]_{in}}\times 100\% \qquad (1)$$

where the $[NO]_{in}$ and $[NO]_{out}$ indicated the inlet and outlet concentration.

## 3. Results and discussion

### *3.1. Catalytic activity*

Figure 1 shows the SCR activities of NO with $NH_3$ over all catalysts. It is clearly seen that NO conversion over all the catalysts increased with temperature rising from 100 to 300°C except P-CNTs. For P-CNTs, NO conversion slightly increases less than 150°C. Once exceeding the temperature, the NO conversion starts to decline gradually. As for $V_2O_5$/CNTs-I and $V_2O_5$/CNTs-M, the maximum efficiencies of denitration are 85% and 72% at 300°C, respectively. However, their temperature window is relatively narrow. Compared with these catalysts, those prepared by in-situ growth show better NO conversion and more wide operating temperature window. The $NO_x$ efficiencies of in-situ growth catalysts are more than 70%, when the operation temperature reaches 175°C. The results show that the operation temperature of catalysts prepared by in-situ growth has obviously dropped. Meanwhile, the $NO_x$ efficiencies slightly increase first and then decrease with increasing the weight ratio of CPC and CNTs. When the weight ratio of CPC and CNTs is 2:1, the $V_2O_5$/CNTs-2 catalyst exhibits the best low

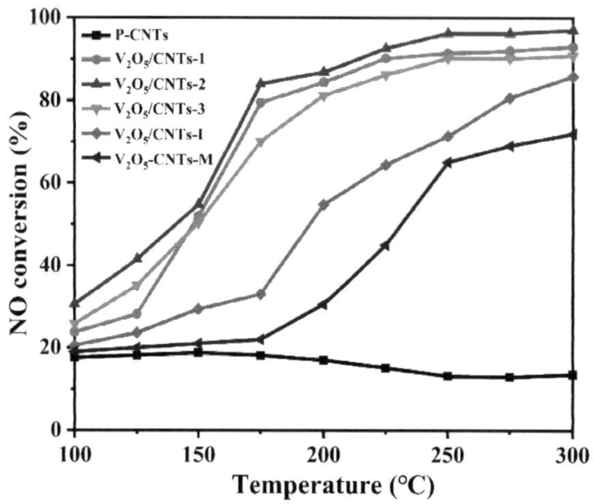


**Figure 1.** $NH_3$-SCR performance of the catalysts. Reaction conditions: [NO] = [$NH_3$] = 500 ppm, [$O_2$] = 3 vol%, $N_2$ balance, and GHSV = 30000 $h^{-1}$.

temperature catalytic activity. The highest NO conversion of $V_2O_5$/CNTs-2 is up to 97% at 250°C, which is much higher than that of $V_2O_5$/CNTs-I and $V_2O_5$/CNTs-M. Meanwhile, the operation temperature needed for reaching a maximum NO conversion is significantly lower than that of $V_2O_5$/CNTs-I and $V_2O_5$/CNTs-M. The results suggest that in-situ growth $V_2O_5$/CNTs catalysts are promising catalytic materials for low-temperature SCR of NO with $NH_3$.

### *3.2. Characteristics of the catalysts*

Figure 2 shows the XRD patterns of all catalysts. It can be observed that the catalysts prepared by in-situ deposition exhibited similar XRD patterns. The broad diffraction peaks at 26.2° and 44.3° are attributable to graphite of CNTs. Besides that, there is no obvious diffraction signal related to the vanadium species, which could be attributed to the highly dispersed or amorphous vanadium oxide on the surface of CNTs. In contrast, for the $V_2O_5$/CNTs-I and $V_2O_5$/CNTs-M catalysts, some sharp diffraction peaks around 15.4°, 20.3°, 21.7°, 25.3°, and 31° are observed, ascribed to the characteristics diffraction peaks of $V_2O_5$ (PDF#01-085-0601). This indicates that the vanadium species of the two catalysts may be agglomerated over the surface of CNTs.

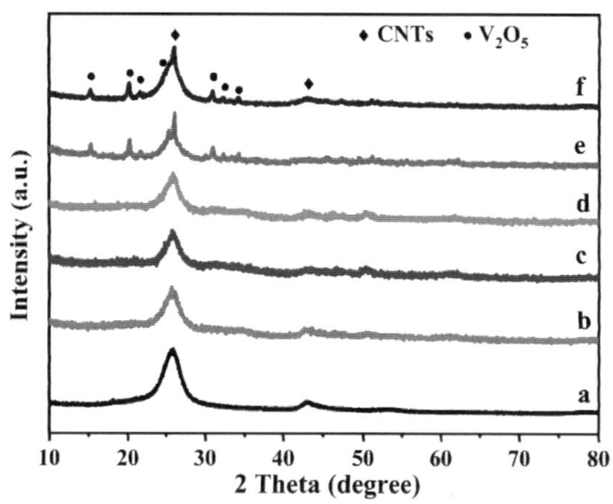


**Figure 2.** XRD patterns of (a) P-CNTs, (b) $V_2O_5$/CNTs-1, (c) $V_2O_5$/CNTs-2, (d) $V_2O_5$/CNTs-3, (e) $V_2O_5$/CNTs-I, and (f) $V_2O_5$/CNTs-M.

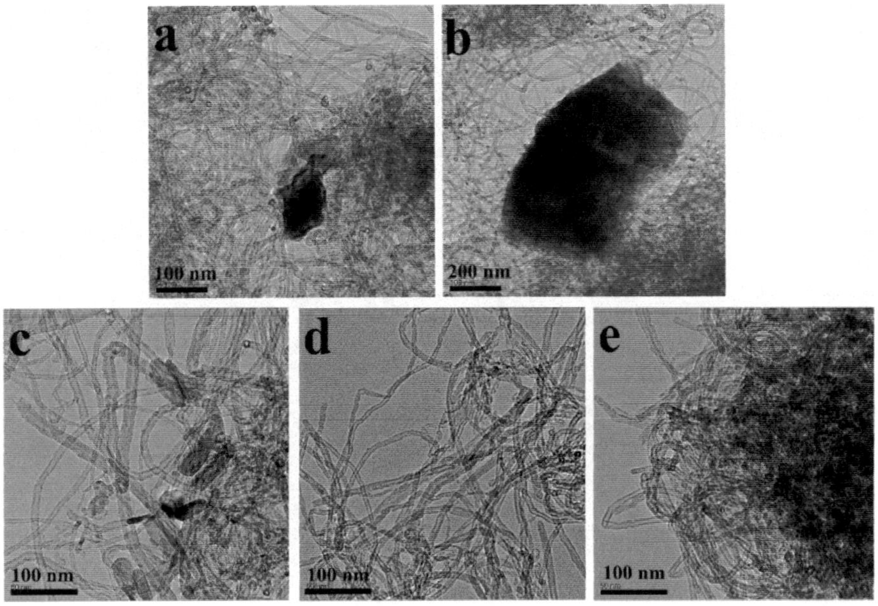


**Figure 3.** TEM images of (a) $V_2O_5$/CNTs-1, (b) $V_2O_5$/CNTs-2, (c) $V_2O_5$/CNTs-3, (d) $V_2O_5$/CNTs-I, and (e) $V_2O_5$/CNTs-M.

Figure 3 shows the TEM images of the catalysts prepared by different methods. $V_2O_5$/CNTs-2 catalyst shows the best dispersion compared to $V_2O_5$/CNTs-1 and $V_2O_5$/CNTs-3 samples. But no obvious nanoparticles anchored on the surface of CNTs can be observed. This is due to the high dispersion of $VO_x$ on the CNTs surface. Besides that, the block $V_2O_5$ was found in Figures 3d and 3e, which indicates that the active components in samples $V_2O_5$/CNTs-I and $V_2O_5$/CNTs-M are agglomerated severely. In addition, the agglomeration of $V_2O_5$/CNTs-M catalyst is the most serious. Good dispersion could provide more active sites, and thus improve the $NH_3$-SCR activities of the catalysts. EDS mapping of $V_2O_5$/CNTs-2 catalyst was used to further confirm the dispersion state of $VO_x$ supported on the surface of CNTs. As shown in Figure 4, the atomic ordinal contrast of the high angle annular dark field image of the sample indicates that $VO_x$ is successfully loaded on the surface of the CNTs. EDS-mapping indicates that V and O elements are evenly distributed on the surface of CNTs, which proves that vanadium oxide is highly dispersed on the surface of the CNTs.

The $N_2$ adsorption-desorption isotherms and pore size distributions of the catalysts are exhibited in Figure 5. As shown in Figure 5a, all the samples show the typical IV isotherms with H3 hysteresis loops [16]. Figure 5b shows the pore size distributions of six catalysts. It reveals that all catalysts have similar pore structure. As shown in Table 1, P-CNTs possess the largest surface area. After the vanadium oxide is supported, the specific surface area of the composite catalysts is decreased, because vanadium oxide

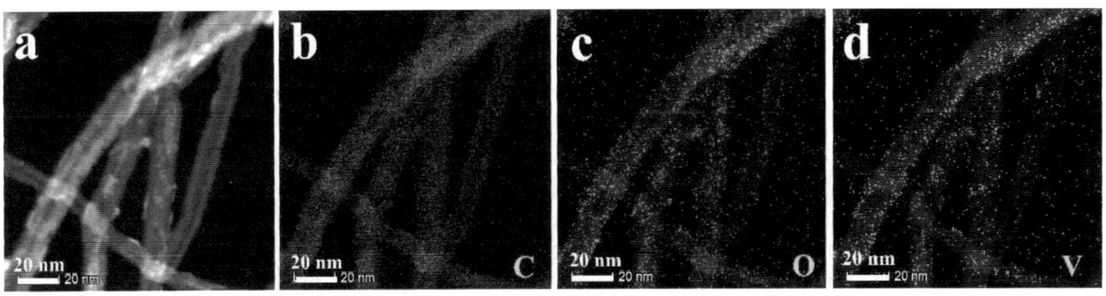


**Figure 4.** EDS mapping of the $V_2O_5$/CNTs-2 catalyst.

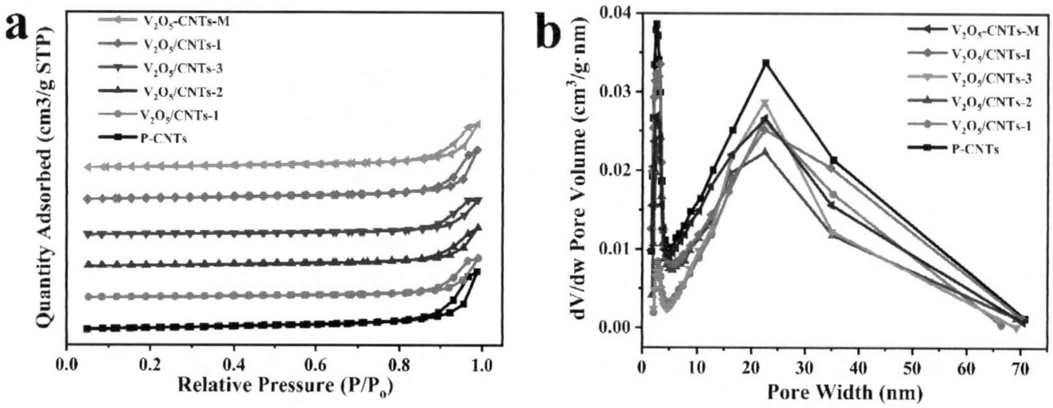


**Figure 5.** $N_2$ adsorption-desorption isotherms (a) and pore size distributions (b) of the catalysts prepared by different methods.

**Table 1.** Surface area and pore characterization of catalysts.

| Catalysts | BET surface area ($m^2$ $g^{-1}$) | Pore volume ($cm^3$ $g^{-1}$) | Pore size (nm) |
|---|---|---|---|
| P-CNTs | 295.2 | 1.442 | 19.54 |
| $V_2O_5$/CNTs-1 | 146.7 | 0.975 | 26.60 |
| $V_2O_5$/CNTs-2 | 188.6 | 0.969 | 20.54 |
| $V_2O_5$/CNTs-3 | 138.3 | 0.857 | 24.78 |
| $V_2O_5$/CNTs-I | 252.4 | 1.236 | 19.59 |
| $V_2O_5$-CNTs-M | 239.4 | 1.111 | 18.56 |

blocks a part of the pores of the CNTs. In addition, the catalysts prepared by in-situ growth possess a smaller surface area than those prepared by impregnation and mechanically mixed method. This indicates that in catalyst $V_2O_5$/CNTs-2, $VO_x$ more closely contacts with CNTs, thereby blocking more pores. In addition, compared with $V_2O_5$/CNTs-1 and $V_2O_5$/CNTs-3, $V_2O_5$/CNTs-2 catalyst has the largest specific surface area due to the best dispersion, which is beneficial to the SCR reaction.

The XPS measurements were carried out to study the valence state of elements on the surface of catalysts. The obtained XPS spectra of V2p and O1s are shown in Figure 6. As shown in Figure 6a, the V2p peaks of $V_2O_5$/CNTs-2 could be deconvoluted to three peaks: the peaks at 517.3, 515.8, and 514.2 eV represented $V^{5+}$, $V^{4+}$, and $V^{3+}$, respectively. However, the V2p peaks of $V_2O_5$/CNTs-I and $V_2O_5$/CNTs-M could only be deconvoluted to two peaks, which were attributed to $V^{5+}$ (517.3 eV for $V_2O_5$/CNTs-I, 517.2 eV for $V_2O_5$/CNTs-M) and $V^{4+}$ (515.9 eV for $V_2O_5$/CNTs-I, 516.0 eV for $V_2O_5$/CNTs-M). As shown in Table 2, the $V^{4+} + V^{3+}/ V^{n+}$ (n = 3, 4, and 5) of the three catalysts were 44.9, 25.2, and 20.1%, respectively. According to the literature, vanadium oxides with low valence states can easily adsorb oxygen to generate active oxygen species during SCR reaction, which is important to the low temperature SCR activity [17-20]. Figure 6b shows O 1s XPS spectra of the catalysts prepared by different methods. The peak at 530 eV was assigned to the lattice oxygen species ($O_\beta$) and that at 532 eV was assigned to the chemisorbed oxygen species ($O_\alpha$). The $O_\alpha/(O_\alpha+O_\beta)$ ratios of three catalysts were calculated and are shown in Table 2. It is interesting that the $O_\alpha/(O_\alpha+ O_\beta)$ ratio of $V_2O_5$/CNTs-2 (40.9%) is higher than $V_2O_5$/CNTs-I (35.1%) and $V_2O_5$/CNTs-M (23.1%). It is reported that the chemisorbed oxygen species are more active than the lattice ones, which is attributed to a higher mobility of chemisorbed oxygen. In addition, the chemisorbed oxygen species could easily oxide NO to $NO_2$, which could accelerate the $NH_3$-SCR reaction due to the occurrence of the 'fast SCR' [3].

The acid sites on the surface of the catalyst are closely related to the adsorption and activation of $NH_3$. The $NH_3$-TPD was performed to investigate the surface acid amount and strength of the catalysts.

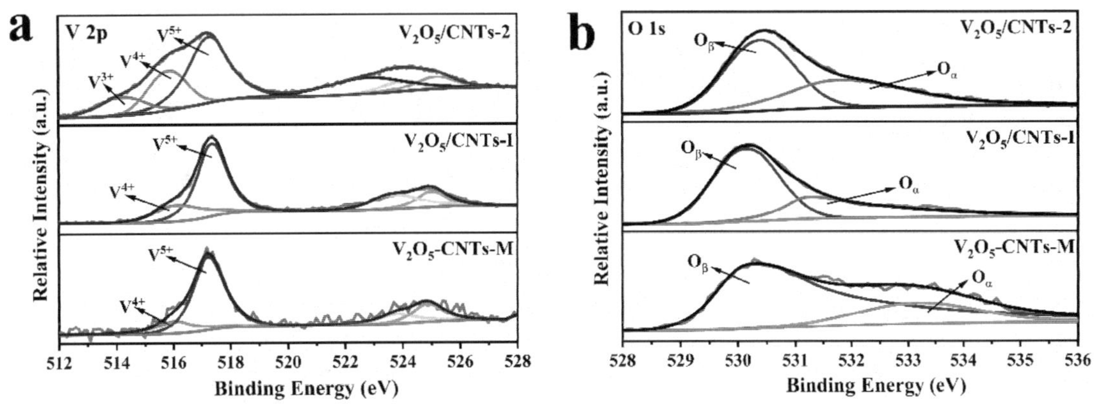


**Figure 6.** XPS spectra for (a) V2p and (b) O1s of the catalyst samples.

**Table 2.** Some results of XPS, $H_2$-TPR, and $NH_3$-TPD.

| Catalysts | $V^{4+}$+$V^{3+}$/$V^{n+}$ n=3, 4, and 5 | $O_\alpha$/( $O_\alpha$+ $O_\beta$) | Peaks area ($H_2$-TPR) | Peaks area ($NH_3$-TPD) |
|---|---|---|---|---|
| $V_2O_5$/CNTs-2 | 44.9% | 40.9% | 11.55 | 12.65 |
| $V_2O_5$/CNTs-I | 25.2% | 35.1% | 8.96 | 2.38 |
| $V_2O_5$-CNTs-M | 20.1% | 23.1% | 6.55 | 1.45 |

As shown in Figure 7a, there are two desorption peaks of $V_2O_5$/CNTs-M catalyst. The weak peak centred at about 167°C is mainly caused by desorption of ammonia bounded to weak Brønsted acid sites and physically adsorbed ammonia molecules. Another peak centred at 438°C is attributed to desorbed ammonia from strong Lewis acid sites. However, a new peak cantered around 358°C is appeared in the $NH_3$-TPD profile of the $V_2O_5$/CNTs-I sample, which is corresponded to desorbed ammonia from medium strong acid sites [21]. As for $V_2O_5$/CNTs-2 catalyst, in addition to the broad peak centred at 167°C, there are two strong peaks centred at 470 and 520°C, which can be attributed to desorbed ammonia from strong acid sites. The higher the temperature corresponding to the desorption peak, the stronger the acid site. So the acid sites of $V_2O_5$/CNTs-2 catalyst are the strongest among all the catalysts.

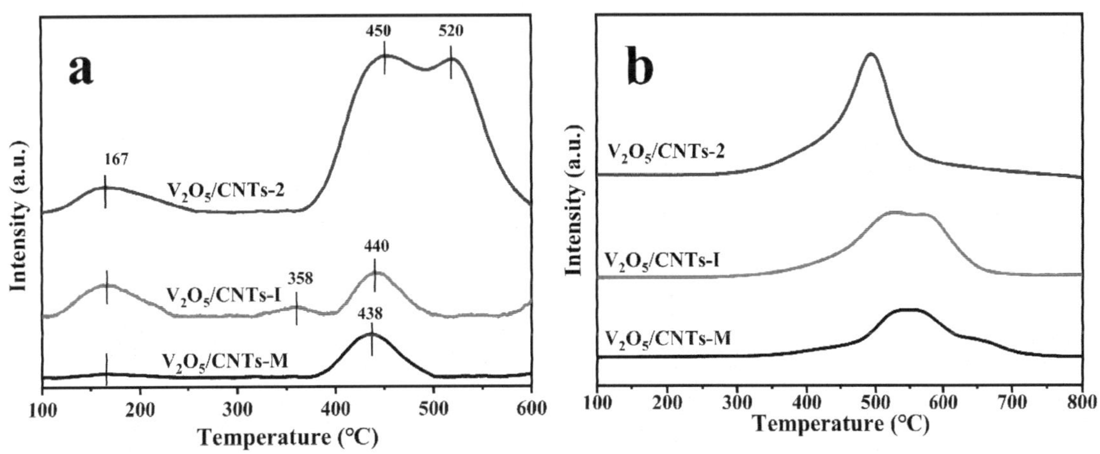


**Figure 7.** $NH_3$-TPD (a) and $H_2$-TPR (b) profiles of the catalyst samples.

In addition, the area of the peaks represents the amount of adsorbed $NH_3$, which is also equivalent to the number of acid sites on the catalyst surface. Therefore, Table 2 allows concluding that $V_2O_5$/CNTs-2 catalyst possesses most acid sites. The results show that $V_2O_5$/CNTs-2 catalyst exhibits a stronger ability to adsorb $NH_3$ than $V_2O_5$/CNTs-I and $V_2O_5$/CNTs-M, which may result in a higher SCR activity.

The redox properties of different catalyst samples were investigated by $H_2$-TPR experiments. It is commonly recognized that the redox capacity of the catalyst plays a crucial role in the SCR cycle. As shown in Figure 7b, the redox peaks in the range of 400 - 600°C are assigned to the reduction of the $VO_x$ from $V^{5+}$ to $V^{3+}$. The redox peaks of $V_2O_5$/CNTs-2 are at about 419 and 497°C, while the peaks of $V_2O_5$/CNTs-I and $V_2O_5$/CNTs-M are at 512, 582 and 549, 657°C, respectively. The area of the reduction peak is positively correlated with the consumption of $H_2$. As shown in Table 2, the area of the reduction peak for $V_2O_5$/CNTs-2 is larger than that of the other two samples. The above results suggest that the $V_2O_5$/CNTs-2 possesses stronger redox properties than $V_2O_5$/CNTs-I and $V_2O_5$/CNTs-M catalysts due to interaction between the vanadium oxide and CNTs, which is associated with higher SCR activity.

## 4. Conclusion

A series of catalysts of in situ supported $VO_x$ on CNTs were successfully synthesized by a CPC assisted reflux route, and their performance on the $NH_3$-SCR reaction was evaluated. All the catalysts prepared by in situ growth display better $NH_3$-SCR activity and more extensive operating-temperature window than the those prepared by impregnation and mechanically mixed method. In addition, the catalyst $V_2O_5$/CNTs-2 shows the highest NO conversion of 97% at 250°C. Characterization revealed that $V_2O_5$/CNTs-2 catalyst displays best dispersion of the active component on the surface of CNTs. Besides that, the $V_2O_5$/CNTs-2 catalyst possesses more vanadium oxides with lower valence states ($V^{4+}$ and $V^{3+}$) and chemisorbed oxygen species, which could promote $NH_3$-SCR reaction. $NH_3$-TPD result indicates that the $V_2O_5$/CNTs-2 catalyst possesses a stronger ability to adsorb $NH_3$, and the $H_2$-TPR result shows that it has stronger redox properties, which is critical in the SCR cycle.

## References

[1] Meng D M, Zhan W C, Guo Y, Guo Y L, Wang L and Lu G Z 2015 A highly effective catalyst of Sm-$MnO_x$ for the $NH_3$-SCR of $NO_x$ at low temperature: Promotional role of Sm and its catalytic performance *ACS Catal.* **5** 5973-83
[2] Hu G, Yang J, Tian Y M, Kong B W, Liu Q C, Ren S, Li J and Kong M 2018 Effect of Ce doping on the resistance of Na over $V_2O_5$-$WO_3$/$TiO_2$ SCR catalysts *Mater. Res. Bull.* **104** 112-8
[3] Zhang D S, Zhang L, Shi L Y, Fang C, Li H R, Gao R H, Huang L and Zhang J P 2013 In situ supported $MnO_x$-$CeO_x$ on carbon nanotubes for the low-temperature selective catalytic reduction of NO with $NH_3$ *Nanoscale* **5** 1127-36
[4] Muradov N, Smith F and T-Raissi A 2005 Catalytic activity of carbons for methane decomposition reaction *Catal. Today* **102-3** 225-33
[5] Wang M X, Liu H N, Huang Z H and Kang F Y 2014 Activated carbon fibers loaded with $MnO_2$ for removing NO at room temperature *Chem. Eng. J.* **256** 101-6
[6] Sousa J P S, Pereira M F R and Figueiredo J L 2013 Modified activated carbon as catalyst for NO oxidation *Fuel Process. Technol.* **106** 727-33
[7] Guo Z Y, Wang M X, Huang Z H and Kang F Y 2015 Preparation of graphene/carbon hybrid nanofibers and their performance for NO oxidation *Carbon* **87** 282-91
[8] You X, Sheng Z and Yu D 2017 Influence of Mn/Ce ratio on the physicochemical properties and catalytic performance of graphene supported $MnO_x$-$CeO_2$ oxides for $NH_3$-SCR at low temperature *Appl. Surf. Sci.* **423** 845-54
[9] Lu X, Song C and Chang C C 2014 Manganese oxides supported on $TiO_2$-graphene nanocomposite catalysts for selective catalytic reduction of $NO_x$ with $NH_3$ at low temperature *Ind. Eng. Chem. Res.* **53** 11601-10
[10] Wang X, Zheng Y and Lin J 2013 Highly dispersed Mn-Ce mixed oxides supported on carbon nanotubes for low-temperature NO reduction with $NH_3$ *Catal. Commun.* **37** 96-9

[11] Santillan-Jimenez E, Miljković-Kocić V, Crocker M and Wilson K 2011 Carbon nanotube-supported metal catalysts for NOx reduction using hydrocarbon reductants. Part 1: Catalyst preparation, characterization and $NO_x$ reduction characteristics *Appl. Catal. B: Environ.* **102** 1-8

[12] Santillan-Jimenez E, Crocker M, Bueno-López A and Salinas-Martínez de Lecea C 2011 Carbon nanotube-supported metal catalysts for $NO_x$ reduction using hydrocarbon reductants: gas switching and adsorption studies *Ind. Eng. Chem. Res.* **50** 7191-200

[13] Wang X, Zheng Y, Xu Z, Liu Y and Wang X L 2014 Low-temperature NO reduction with $NH_3$ over Mn-$CeO_x$/CNT catalysts prepared by a liquid-phase method *Catal. Sci. Technol.* **4** 1738-41

[14] Bai S, Jiang S, Li H and Guan Y 2015Carbon nanotubes loaded with vanadium oxide for reduction NO with $NH_3$ at low temperature *Chin. J. Chem. Eng.* **23** 516-9

[15] Zhang X L, Wu Q, Diao Q C, Wang J W, Xiao K S, Yang B J and Wu X P 2019 Performance study for $NH_3$-SCR at low temperature based on different methods of Mnx/SEP catalyst *Chem. Eng. J.* **370** 364-71

[16] Sing K S W, Everett D H, Haul R A W, Mascou L, Pierotti R A, Rouquerol J and Siemieniewska T 1985 Reporting physisorption data for gas/sold systems with special reference to the determination of surface area and porosity *Pure & Appl. Chem.* **57** 603-19

[17] Chen H F, Xia Y, Fang R Y, Huang H, Gan Y P, Liang C, Zhang J, Zhang W K and Liu X S 2018 The effects of tungsten and hydrothermal aging in promoting $NH_3$-SCR activity on $V_2O_5$/$WO_3$-$TiO_2$ catalysts *Appl. Surf. Sci.* **459** 639-46

[18] Aguilar-Romero M, Camposeco R, Castillo S, Marín J V, Rodríguez-González L Z and Mejía-Centeno García-Serrano I 2017 Acidity, surface species, and catalytic activity study on $V_2O_5$-$WO_3$/$TiO_2$ nanotube catalysts for selective NO reduction by $NH_3$ *Fuel.* **178** 123-33

[19] Li F K, Shen B X, Tian L H, Li G L and He C 2016 Enhancement of SCR activity and mechanism stability on cordierite supported $V_2O_5$-$WO_3$/$TiO_2$ catalyst by substrate acid pretreatment and addition of silica *Powder Technol.* **297** 384-91

[20] Zhang S L and Zhang Q 2013 Promotional effect of $WO_3$ on $O_2^-$ over $V_2O_5$/$TiO_2$ catalyst for selective catalytic reduction of NO with $NH_3$ *J. Mol. Catal. A: Chem.* **373** 108-13

[21] Arfaoui J, Ghorbel A, Petitto C and Delahay G 2018 Novel $V_2O_5$-$CeO_2$-$TiO_2$-$SO_4^{2-}$ nanostructured aerogel catalyst for the low tempe rature selective catalytic reduction of NO by $NH_3$ in excess $O_2$ *Appl. Catal. B: Environ.* **224** 264-75

IOP Conf. Series: Materials Science and Engineering **699** (2019) 012021 doi:10.1088/1757-899X/699/1/012021

# Dependence of pyroelectric response on inter-electrode capacitance for integrated-optical circuits utilizing x-cut $LiNbO_3$ chips

**S M Kostritskii[1], A V Yatsenko[2], Yu N Korkishko[1] and V A Fedorov[1]**

[1]Optolink Ltd, Zelenograd, Moscow, Russia
[2]Crimean Federal University, Simferopol, Russia

skostritskii@optolink.ru

**Abstract**. Experimental study of the pyroelectric effect has been made for multi-function integrated-optical circuits (MIOC) utilizing x-cut $LiNbO_3$ chips. It has been experimentally established that pyroelectric voltage is proportional to the temperature scanning rate, and a voltage magnitude depends significantly on capacitances of chip parts between the MIOC electrodes. The pyroelectric effect is considered by us to be an important source of the thermal instability of MIOC. The model of the pyroelectric contribution to the MIOC thermal instability is used to suggest the new methods reducing this instability.

## 1. Introduction

In the past few decades, the closed-loop fibre optic gyroscope (FOG) has gradually become the mainstream sensor for angular velocity measurement [1-3]. In the discussed all-digital closed-loop FOG products, the multifunction integrated-optical circuits (MIOC) serves as a beam splitter/combiner (important part of a Sagnac interferometer), light polarizer, and reciprocal phase modulator [4-7]. It is fabricated on lithium niobate ($LiNbO_3$) wafers [1], using conventional techniques to form large numbers of identical components at a low cost and with high reliability. It has been established that the MIOC half-wave voltage depends on temperature [5]. This dependence has been regarded as an important source of the MIOC modulation phase error, which has a significant influence on temperature stability of FOG [5-7]. The compensation model based on the temperature dependence of the equivalent capacitance, considering the simple linear relation between the modulation phase error and temperature, was proposed previously [6, 7]. However, the related compensation scheme can only slightly reduce the normalized value of phase error [7]. For suppressing the temperature dependence, a new more complicated model needs to be developed. Subsequently, the pyroelectric effect [8-10] is considered by us to be an additional source of the thermal instability of MIOC and the appropriate model has been proposed in our paper. $LiNbO_3$ is the pyroelectric crystal, and quite large electric field of the order of a few kV/cm/°C can be generated. This pyroelectric field has only a z-axis component $E_3$. The same electric field component is used as external electrical loading in the $LiNbO_3$-based electro-optic modulators; thus, a change in device temperature can drastically affect the modulator performance [8-10], as some extra contribution in MIOC modulation phase error should be expected. The pyroelectric field causes also the migration of the compensating charges in a near-surface layer and their redistribution between the electrodes. This field can change the effective electric field in the

IOP Conf. Series: Materials Science and Engineering **699** (2019) 012021 doi:10.1088/1757-899X/699/1/012021

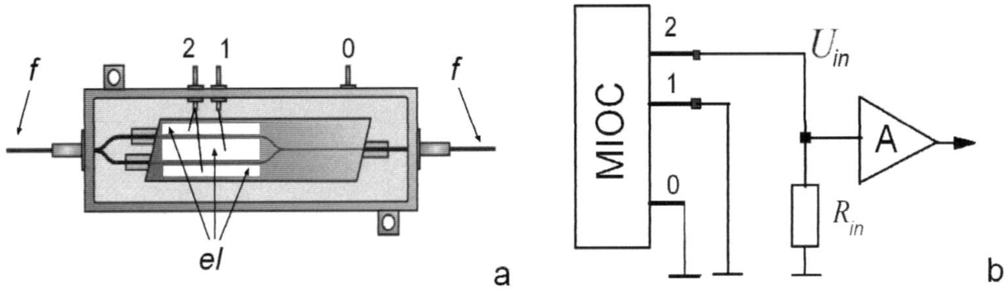


**Figure 1.** (a) The intrinsic structure of MIOC: Pins 1 and 2 are connected by Au-wires with Au/Cr-electrodes of the electro-optic modulator on a $LiNbO_3$ chip, *el* are Au/Cr-electrodes, *f* are input and output optical fibres; (b) Scheme of measurement of pyroelectric response of MIOC: $R_m$ is measuring resistor at electrometer amplifier input; A is electrometric amplifier.

waveguide. During the heating, the pyroelectric field causes decrease in effective field, in contrast to the increase at the cooling down [8].

## 2. Experimental

The experimental study of the pyroelectric response has been made for the hermetically-sealed fibre-pigtailed x-cut $LiNbO_3$ MIOC used in fibre-optical gyroscopes [11]. The schematic sketch of the MIOC is given in Figure 1a. Also, the pyroelectric response has been studied for an unpigtailed integrated-optical chip (IOC) fabricated on a x-cut $LiNbO_3$ plate. Each IOC contains the coplanar Au/Cr electrodes consisting of the two electro-optic phase modulators of MIOC on main –x surface. The IOCs studied were subjected to the additional hard cleaning and drying treatments. Besides, several IOCs studied have an additional metallization: the two extra In-Ga electrodes were deposited on +z and –z surfaces of chip's lateral edges and these extra electrodes were grounded or connected together with Au-wire. According to the previous finding [9], such a conductive coating must decrease the pyroelectric response, that is undesirable phenomenon at practical application of IOC [8-10].

The pyroelectric response was measured by the method described in the previous paper [12], i.e. the pyroelectric voltage $U_{in}$ was directly recorded while changing the temperature $T$ of the sample in the range from 295 to 380 K. The range of temperature scanning rate $V = dT/dt$ covered the interval from $\leq 0.1$ to 2.5 K/min. A load (measuring) resistor $R_{in}$ was wired in parallel to the electrometric amplifier input, and $U_{in}$ appeared on an MIOC pin (or IOC electrode) was amplified by factor of 5 and measured with a H307/2 two-coordinate potentiometer, connected to the output of amplifier, while a stable zero position of the signal is carefully maintained (Fig. 1b). The signals that are superimposed by higher frequency noise or other interference can be smoothed using a low pass filter with selected time constants (1-3 s). Assuming that for appropriate evaluation of $U_{in}$ the amplifier load resistance $R_{in}$ should be much smaller than the crystal resistance, the measurements were performed at the following $R_{in}$ values: 1 MΩ, 91 MΩ, and 1.06 GΩ. The pyroelectric response after amplification was measured as voltage drop at this resistor. The signals that are superimposed by higher frequency noise or other interference can be smoothed using a low pass filter with selected time constants (1-3 s). Besides, the measurements without any filtration and with high temporal resolution (about 1 μs) were performed with an AKIP 72205A digital oscilloscope.

## 3. Results

The slowly changed voltage $U_{in}$ appearing on the MIOC pins #1 and #2 is found to be monotonously increasing with the temperature scanning rate $V$ (Fig. 2). Therefore, according to the Ref. [12], it has been assigned by us to the pyroelectric response of the MIOC. Sharp pulses in the electrometer output are related to electric discharges between the lateral ±z edges of MIOC and between electrodes within

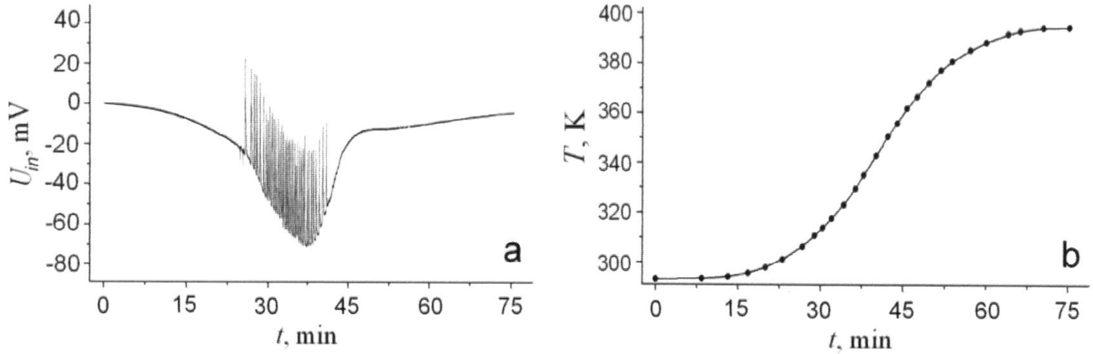


**Figure 2.** (a) Pyroelectric voltage $U_{in}$ and (b) temperature $T$ as functions of the scanning time $t$ during heating of MIOC. Pulses in the electrometer output are related to electric discharges. $R_{in}$ = 1.06 GΩ.

inter-electrode gaps. These numerous electrical discharges are detected by electrometer at connection to the MIOC pin #1 or #2 (Fig. 1a). These discharges are established to be the other specific feature of the MIOC pyroelectric response in addition to the slowly changed voltage $U_{in}$. The data shown in Figure 2 were obtained with the low-pass filter. Hence, a pulse amplitude is limited by transmission band of the low-pass filter (1 Hz) and, also, by the response speed of a recording system of our electrometer set-up. However, the real amplitude of these pulses is much larger than is shown in Figure 2. For example, our measurements with the aid of the digital oscilloscope AKIP without a low pass filter show that the real amplitude of the short (≈ 0.08 ÷ 0.11 ms) pulses is about several Volts that are larger by two orders of magnitude than the amplitude recorded with the electrometer (Fig. 2).

The pyroelectric response of MIOC is explained by the previous finding that x-cut $LiNbO_3$-based IOC would generate voltages across the 3 mm width of the chip of above 150 volts per one degree centigrade [9]. The field that produces the voltage differences across the surface of the chip is caused by any change in the bulk temperature of the chip. Our measurements of the pyroelectric response of the single IOCs have demonstrated the results similar to one shown in Figure 2 and, thus, it confirms the previous finding [9-10] on the origin of MIOC pyroelectric response.

The +z and -z faces of several IOCs were coated with a conductive coating. A conductive path couples the +z and -z faces to prevent a charge differential from developing between the +z and -z faces due to pyroelectric effect at any change in temperature of the chip, as unscreened bound charge build-up occurs on *z* surfaces of the lateral edges [9]. However, the direct measurements of the pyroelectric response of the IOCs with conductive coating demonstrate that this method allows only marked decrease in the slow-varying component of $U_{in}$ and the pulses amplitude, but a full suppression of the pyroelectric response was not observed in contrast to the previously reported findings [9, 10]. It may be explained by the fact that the electrodes create strong gradients in the local surface potential near the electrodes when unscreened pyroelectrically induced surface charges are present [8]. The resulting differential electric field strength gives rise to the thermal instability, which is in five orders of magnitude larger than the calculated instability due to the pyroelectric effect alone in the case of z-cut $LiNbO_3$ modulators [8]. The effect of thermal instability was found to be weaker on x-cut $LiNbO_3$ substrates than on z-cut substrates, although the pyroelectric field has the same magnitude in both cases and, thus, it cannot be neglected in application for the phase sensitive devices. Therefore, we consider a contribution of the so-called electrostatic component [8] for qualitative explanation of the pyroelectric response of x-cut $LiNbO_3$ IOC. The capacitive loading of modulator electrodes makes the potential of these electrodes change much less than the potential of the rest of the x-cut surface, leading to a large potential difference between the IOC modulator electrodes, i.e. to appearance of the electrostatic component. This transient field has a pyroelectric origin and its strength will be limited by leakage and breakdown processes [8, 9]. Thus, the strength of this electric field must depend on the inter-electrode capacitances of $LiNbO_3$-based IOC. To verify this assumption clearly, we study an IOC preform, which is much wider than a single IOC and, therefore, it is composed of different capacitances (Fig. 3).

A simple circuit composed of capacitances is chosen to model how the charge distribution, and thus the electric field, may depend on the electrode connection topology. The scheme given in Figure 3 divides the IOC capacitance into different contributions. Direct measurement of the capacitances of different parts of IOC chip gave the following results: $C_{11} \cong 5.1$ pF, $C_{22} \cong 2.6$ pF, $C_{21} \cong 4.1$ pF, and $C^*_{21} \cong 1.9$ pF. $C_{21}$ and $C^*_{21}$ are capacitances measured, when a second modulator electrode is ungrounded and grounded, respectively. The ratio of the capacitances for the different-length inter-electrode gaps scales roughly with the gaps length ratio. It should be noted that a simple evaluation of $C^*_{21}$ and $C_{21}$ capacitances ratio cannot be made directly, because the shape of the field lines should be substantially different in these cases.

The pyroelectric crystal (in conjunction with the electrodes) acts as a capacitor. When the crystal experiences a change in temperature, it frees electric charges that appear on the extra electrodes deposited on the polar surfaces of lateral edges of IOC chip. The pyroelectrics are oxides, and they have very high resistances. Thus, these accumulated charges cannot leak through the material. Thus, under homogeneous change in temperature, the pyroelectric crystal acts as a charge generating capacitor [13]. It means that $U_{in}$ is determined by dynamics of an excess charge on the extra electrode on polar surface and the modulator electrode connected to measuring circuit with $C_{in}$. It is important to note that our channel-waveguide device is fundamentally different from the bulk-optics $LiNbO_3$ devices. There is inhomogeneity caused by the presence of waveguide within a certain part of the inter-electrode gap. In our device, the ratio of the width of the gap and channel is 2.4. This inhomogeneity appears when narrow channel waveguides are formed through proton exchange reaction, as the capacitance and resistance within the proton-exchanged areas are different from ones in the rest crystal.

It was experimentally established [12], that pyroelectric voltage $U_{in}$ is proportional to the temperature scanning rate $V$. Our measurement of the pyroelectric response the IOC chips demonstrates that a magnitude of $U_{in}$ at any given value of $V$ depends significantly on the electrodes connection topology (Figs. 3 and 4). Thus, the pyroelectric voltage $U_{in}$ depends on capacitance $C_{ij}$ of a chip part between the modulator electrode, which is connected to measuring circuit, and ungrounded extra electrode deposited on the right edge of IOC (Fig. 3).

To estimate quantitatively $U_{in}$, we consider that during the temperature increase at a rate $V$ the change in the charge $q$ at an effective capacitor $C_{ij}$ in time is given by the difference of the charge transferred through the measuring circuit and the charge released due to the pyroelectric effect:

$$dq/dt \sim \gamma \times V \quad (1)$$

where $\gamma$ is pyroelectric coefficient. According to our measurements, a $LiNbO_3$ wafer (supplied by YCC Ltd) used for IOC preform fabrication is characterized by $\gamma \approx 100$ $\mu C/m^2K$ [14].

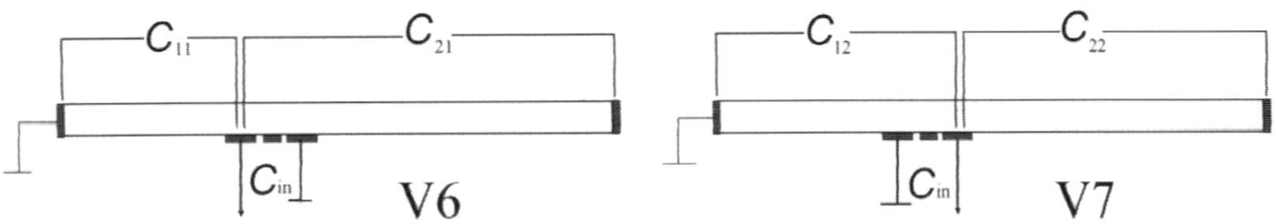


**Figure 3.** Schemes of the different topologies for electrical connection of the IOC preform. $C_{in}$ is input integral capacitance of modulator and amplifier for measuring circuit, $C_{ij}$ are capacitances of different parts of IOC preform. One modulator electrode is connected to measuring circuit with $C_{in}$.

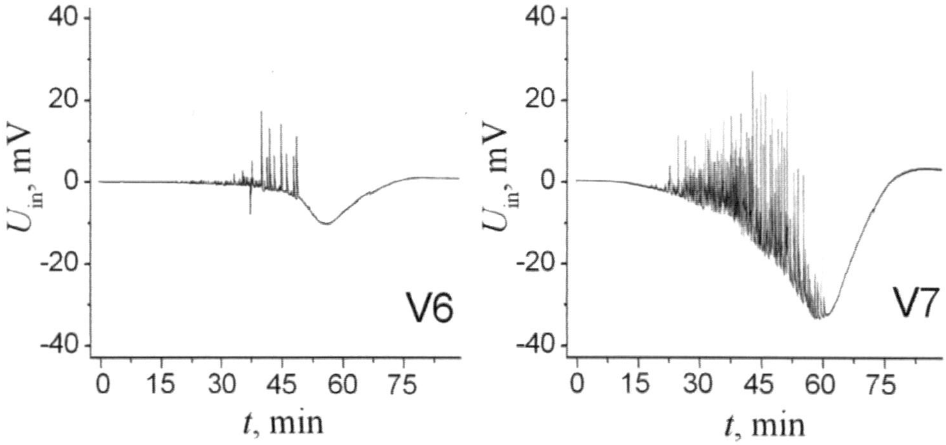


**Figure 4.** Pyroelectric voltage $U_{in}$ measured for the two different electrodes connection topologies (Fig. 3) as functions of scanning time $t$ during heat-up of IOC. Rate $V = dT/dt$ of temperature change was gradually increased from $V \leq 0.1$ K/min (at $t \leq 5$ min) to the maximum value of 1.5 K/min at $53 \leq t \leq 62$ min, after $V$ decreased and $V = 0$ at $t \geq 84$ min. $R_{in} = 1.06$ GΩ.

The voltage $U_{in}$ across $C_{in}$ equals to the voltage across the corresponding capacitor of the IOC chip:

$$U_{in} \sim R_{in} C_{21}^{*} \frac{dq}{dt} \text{ and } U_{in} \sim R_{in} C_{22} \frac{dq}{dt} \quad (2)$$

for V6 and V7 electrodes connection topologies (Fig. 3), respectively.

In fact, the experimental results obtained for $U_{in}$ (Fig. 4) are consistent with our simple model of the IOC as the charge generating capacitor divided into the different contributions $C_{ij}$, according to the electrodes topology. Besides, these results show that a further suppression of undesirable pyroelectric response is possible by decrease in appropriate contributions in IOC capacitance, which may be achieved by increase in the inter-electrode gap or/and IOC chip width. Moreover, some additional suppression would be possible by optimization of electrodes configuration [13, 15-17].

## 4. Conclusion

The model of the pyroelectric contribution to the MIOC thermal instability has been developed and used to suggest the methods reducing this instability. The effective inter-electrode capacitance could be decreased to the point, where no significant screening of the pyroelectrically induced surface polarization charge occurs, thus greatly reducing the pyroelectric voltage and, hence, the thermal instability of the MIOC phase error. The other materials-oriented approach from our model is to increase the surface or bulk conductivity of the $LiNbO_3$ to the point, where pyroelectrically induced fields are screened out as they are generated by slow thermal variations of the environment (at the expense of losing sensitivity to the low-frequency applied fields). A semi-insulating layer is sufficient to accomplish this, and recently reported experiments have shown an improvement in thermal stability of z-cut $LiNbO_3$ waveguide switches using this method [17]. Decreasing the bulk resistivity of $LiNbO_3$ represents a more comprehensive (and technically challenging) solution, since charges arising both from the pyroelectric effect and the photorefraction on a $LiNbO_3$ chip could be screened or compensated [17].

## References

[1] Lefèvre H C 2014 *The Fiber-Optic Gyroscope* 2nd ed. (Artech House: Boston)
[2] Lefèvre H C 2016 *Proc. SPIE* **9852** 985203

IOP Conf. Series: Materials Science and Engineering **699** (2019) 012021 doi:10.1088/1757-899X/699/1/012021

[3] Khial P P, White A D and Hajimiri A 2018 *Nat. Photonics* **12** 671-5
[4] Pogorelaya D A, Smolovik M A, Aksarin S M, Strigalev V E, Shulepov V A and Muhtubaev A B 2017 *J. Phys. Conf. Ser.* **917** 072002
[5] Mao X, Tian Q, Zhang E, Sun L, Teng Y and Liu J 2005 *Proc. SPIE* **5826** 635
[6] Chunduru V, VaraLakshmi R et al 2010 *Proc. SPIE* **7604** 760415
[7] Bi R, She X, Huang T and Shu X 2019 *Opt. Eng.* **58** 067104
[8] Skeath P, Bulmer C H, Hiser S C, Burns W K 1986 *Appl. Phys. Lett.* **49** 1221
[9] Shafer K W et al 2000 US Patent 6044184
[10] Gampp L L et al 2000 US Patent 6128424
[11] Kostritskii S M, Korkishko Yu N and Fedorov V A 2013 *Proc. SPIE* **9065** 90650E
[12] Evdokimov S V, Shostak R I and Yatsenko A V 2007 *Phys. Solid State* **49** 1957-62
[13] Hossain A and Rashid M H 1991 *IEEE Transact. Indust. Appl.* **27** 824
[14] Yatsenko A V, Evdokimov S V, Palatnikov M N and Sidorov N V 2019 *Phys. Solid State* **61** 1211-7
[15] Becker R A 1985 *Opt. Lett.* **10** 417-9
[16] Higuma K, Hashimoto Y, Yatsuki M and Nagata H 2000 *Electronics Lett.* **36** 2013-4
[17] Bulmer C H, Burns W K and Hiser S C 1986 *Appl. Phys. Lett.* **48** 1036-8

IOP Conf. Series: Materials Science and Engineering **699** (2019) 012022 doi:10.1088/1757-899X/699/1/012022

# Kinetics of photo-stimulated adsorption of enzyme molecules onto *n*- and *p*-type silicon

**A V Kozlowski and S V Stetsyura**

Department of Nano- and Biomedical Technologies, Saratov State University, Saratov 410012, Russia

kozlowsky@bk.ru

**Abstract**. We report on a change in the surface density of glucose oxidase (GOx) molecules on surface of Si/$SiO_2$/polyethyleneimine structures depending on type of Si conductivity, adsorption time and white-light illumination of a substrate during GOx adsorption. It is clearly shown by AFM images that the dependence of GOx adsorption on adsorption time and illumination is different for *n*-Si and *p*-Si substrates. The results can be explained by the electrostatic interaction between GOx molecules and the surface charge of Si, which depends on illumination.

## 1. Introduction

Adsorption of enzyme molecules onto surface of a semiconductor transducer is critical for selectivity, sensitivity, and other important parameters of biosensors. Enzyme molecules (for example, glucose oxidase (GOx)) can be used to fabricate a receptor layer on the surface of an ion-sensitive semiconductor transducer. Previous experimental data of GOx adsorption and molecular simulation results of other enzymes [1, 2] have shown that electrostatic interactions play an important role in controlling the enzyme adsorption.

Despite the realization of the biological significance, enzyme adsorption mechanisms and kinetics are still not fully understood, and several important questions remain unanswered. For example, the kinetics of adsorption of enzyme molecules to a solid surface, depending on changes in the surface charge density (SCD), is poorly understood. Remote methods of influencing the SCD of a substrate during the immobilization of enzyme molecules are interesting. Semiconductors as a solid substrate allows us to remotely change SCD value affecting the electrostatic interaction between enzyme molecules and the substrate. Obviously, the remote change of the semiconductor SCD can be achieved by either a field effect using external electric field or a super band-gap illumination.

Previously, we have shown the effect of the ionic strength of a solution and the photoelectron processes in Si during GOx adsorption on the electrostatic interaction between enzyme molecules and the semiconductor surface [3, 4]. However, studies on the enzyme adsorption kinetics onto a semiconductor, depending on its conductivity type and changes in the surface charge density caused by illumination, have not been carried out before.

Weighing or indirect methods characterizing adsorption onto a solid substrate by changing solution parameters are often used to measure the adsorption kinetics. However, these methods allow us to obtain information on the amount of adsorbed substance only, but not on the distribution of adsorbed molecules on the substrate surface, their conformation, and the processes of molecules coagulation. AFM measurements can clearly show the kinetics of adsorption of enzyme molecules and allow us to draw conclusions about coagulation processes and the regular/random distribution of GOx molecules using photo-stimulated adsorption (PSA).

Thus, in our work, AFM was used to study the PSA kinetics of GOx molecules onto the surface of both *p*- and *n*-type silicon.

## 2. Experimental section

### *2.1. Materials and methods*

Experiments were performed with single-crystalline Si wafers of *n*-type ($\rho \cong 4\ \Omega$ cm) and *p*-type ($\rho \cong 8\ \Omega$ cm). Initially, the wafers were boiled in a peroxide–ammonia solution and rinsed in deionized water. This treatment leads to "reconstruction" of a native oxide layer to $SiO_2$. It can be noted that above pH = 2-3 the surface is negatively charged [5]. Afterwards, wafers were cut into substrates of 8×8 $mm^2$.

GOx molecules from *Aspergillus niger* were used as enzyme molecules. At pH > 4.2 the GOx molecule has an effective negative charge [6]. The size of the GOx molecule is 6.0×5.2×7.7 $nm^3$ [6]. A cationic polyelectrolyte polyethylenimine (PEI) with a molecular weight of 25 kDa was used to increase the adsorption of negatively charged GOx onto the silicon substrates. The PEI molecules were adsorbed onto the silicon substrates from the 1 mg/ml aqueous solution during 10 min followed by rinsing in deionized water ($\rho \cong 18.2$ M$\Omega$ cm) during 10 min and drying in nitrogen flow.

The photo-stimulated layer-by-layer adsorption technique suggested in [7] was used to adsorb GOx from the 0.5 mg/ml aqueous solution onto silicon covered with PEI. Two silicon substrates were placed in GOx solution. One of the substrates was illuminated during GOx adsorption by white-light of a halogen lamp (Philips 13186 EPX/EPV) with illuminance level of about 20000 lx. Adsorption time was varied from 10 to 180 min. The optical radiation of the halogen lamp was not absorbed by the GOx solution and did not heat the solution and substrate. However, the radiation of the halogen lamp absorbed by Si induced the photogeneration of nonequilibrium charge carriers. Thus, the surface charge density of Si substrate was changed. The PSA results of the GOx onto a silicon substrate were compared with the results of GOx adsorption in the dark.

### *2.2. Atomic Force Microscopy measurements*

The topography of the surface was measured using atomic force microscopy (AFM) by NTEGRA Spectra (NT-MDT Spectrum Instruments, Russia) in a tapping mode with scan frequencies of 0.3-1 Hz using HA_NC/$W_2C$ cantilevers of ETALON series with resonance frequency of 140 ± 10% kHz. The Gwyddion software was used for statistical analysis of AFM data. The measurements have not been corrected for the convolution with the tip shape. Therefore, the lateral size of enzyme molecules observed by AFM does not correspond to their real size due to the artifacts appearance ("profile broadening" effect due to the tip-sample convolution). However, using results from [8], we can recognize individual molecules of GOx on the scan.

## 3. Results and discussion

Figure 1 shows AFM-images of adsorbed GOx molecules onto the surface of *p*-Si structure either in the dark or under illumination for each value of the adsorption time (10, 30, 60, and 180 min), similar images were obtained also for *n*-Si (Fig. 2). It is seen that adsorption of GOx molecules onto the surface of Si/$SiO_2$/PEI structure increases with adsorption time and this does not depend on the silicon conductivity type or illumination level. The GOx adsorption efficiency can be estimated using a mean surface coverage calculated as the relative part of area covered with GOx molecules. Table 1 displays the mean surface coverage (±standard dev.) dependence on adsorption time in the dark and under illumination.

In the case of GOx adsorption onto *p*-Si/$SiO_2$/PEI, significant adsorption rate during the first hour and its acceleration by illumination were observed. In the case of GOx adsorption onto *n*-Si/$SiO_2$/PEI, a very low adsorption rate during the first hour of adsorption and illumination of the substrate during adsorption further slowed down the adsorption process. However, after 3 hours, the surface coverage values with GOx molecules for both *p*-type and *n*-type were comparable.

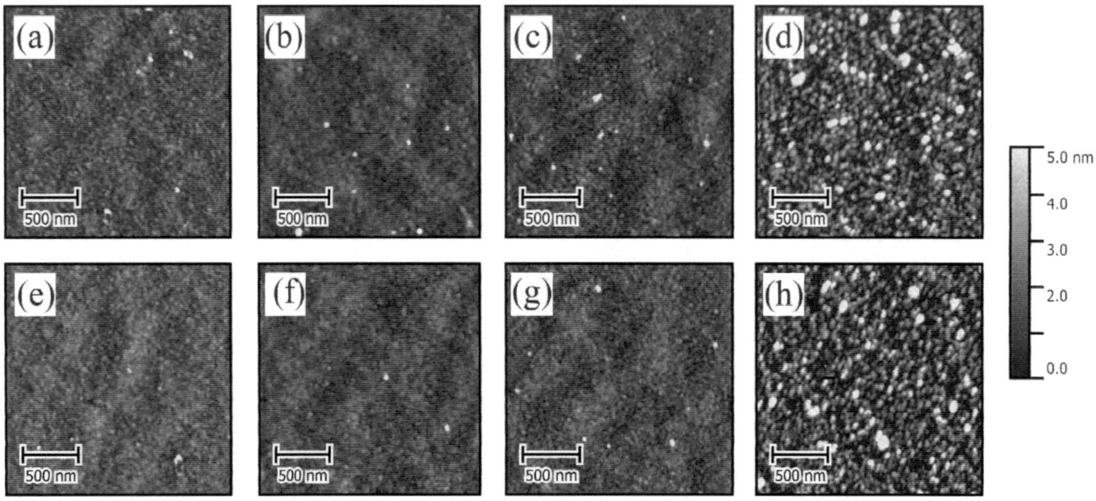


**Figure 1**. AFM-image of $p$-Si/$SiO_2$/PEI/GOx structures, which GOx adsorbed during 10 min (a, e), 30 min (b, f), 60 min (c, g) and 180 min (d, h). Top line corresponds to GOx adsorption in the dark, while down line corresponds to PSA.

Thus, in order to achieve a high value of surface coverage with GOx molecules in case $p$-Si/$SiO_2$/PEI substrate, the adsorption time is crucial. In case of $n$-Si/$SiO_2$/PEI, photo-stimulation of the substrate during GOx deposition allows one to achieve the same surface coverage, but in a shorter time. It is an important practical result.

Different kinetic dependences for $n$-Si and $p$-Si can be explained by different electrostatic interactions between negatively charged in solution enzyme molecules and the substrate surface charge. This leads to different values of rate constants for adsorption and desorption processes. These constants can also be controlled by illumination, because illumination induces changes in the surface charge density of the substrate.

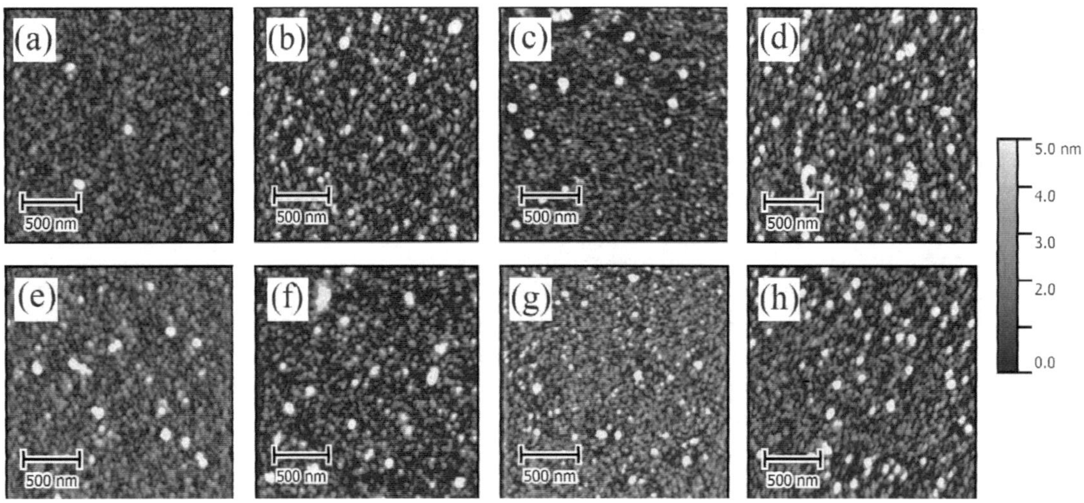


**Figure 2.** AFM-image of $n$-Si/$SiO_2$/PEI/GOx structures, which GOx adsorbed during 10 min (a, e), 30 min (b, f), 60 min (c, g) and 180 min (d, h). Top line correspond to GOx adsorption in the dark, while down line correspond to PSA.

**Table 1.** The mean percentage of the surface coverage (±standard deviation) after GOx adsorption in the dark and under illumination depending on the GOx adsorption time.

| GOx adsorption conditions | *p*-$Si/SiO_2$/PEI/GOx | | | | *n*-$Si/SiO_2$/PEI/GOx | | | |
|---|---|---|---|---|---|---|---|---|
| | 10 min | 30 min | 60 min | 180 min | 10 min | 30 min | 60 min | 180 min |
| in the dark | 0.30±0.1 | 0.29±0.1 | 0.48±0.1 | 4.85±0.5 | 0.40±0.1 | 2.21±0.5 | 3.26±0.5 | 5.4±0.5 |
| under illumination | 0.10±0.1 | 0.09±0.1 | 0.25±0.1 | 4.9±0.5 | 0.58±0.1 | 3.27±0.5 | 4.47±0.5 | 5.6±0.5 |

The same maximum values of the surface coverage with GOx molecules for *n*-Si and *p*-Si can be explained by the charge of GOx molecules, which depends only on values of pH and ionic strength of the solution. In our experiments, pH and ionic strength of the solution did not specifically change and the charge of individual GOx molecule was the same in all experiments. At a certain surface coverage with GOx molecules (ca. 5%), an electric field blocks further immobilization of enzyme molecules (the desorption rate constant prevails).

## 4. Conclusion

In the present study, we have investigated the kinetics of photo-stimulated adsorption of GOx molecules onto silicon substrates of different conductivity types. It was found that illumination increased the adsorption rate of GOx during the first hour in case of *n*-$Si/SiO_2$/PEI structure by about 50%. In the case of GOx adsorption onto *p*-$Si/SiO_2$/PEI structure, significant covering of the surface with enzyme molecules is possible only after 3 hours. In the first hour, illumination reduces the enzyme adsorption onto a *p*-type semiconductor. The obtained results were explained by illumination-induced amplification (for *n*-Si-based substrate) or attenuation (for *p*-Si-based substrate) of the electrostatic interaction between the enzyme molecules and semiconductor substrate.

## Acknowledgements

The work was supported by the German-Russian Interdisciplinary Science Center (G-RISC) funded by the German Federal Foreign Office via the German Academic Exchange Service (DAAD), Project No. P-2018a-10.

## References

[1] Ratner B D, Castner D G, Horbett T A, Lenk T J, Lewis K B and Rapoza R J 1990 *J. Vac. Sci. Technol. A* **8** 2306-17
[2] Xie Y, Li Z and Zhou J 2018 *Phys. Chem. Chem. Phys.* **20** 14587-96
[3] Kozlowski A V, Kiseleva E D, Maslennikova A A and Stetsyura S V 2018 *IOP Conf. Ser.: Mater. Sci. Eng.* **443** 012016
[4] Stetsyura S and Kozlowski A 2017 *Tech. Phys. Lett.* **43** 285-8
[5] Cloarec J-P, Chevalier C, Genest J, Beauvais J, Chamas H, Chevolot Y, Baron T and Souif A 2016 *Nanotechnology* **27** 295602
[6] Hecht H J, Kalisz H M, Hendle J, Schmid R D and Schomburg D 1993 *J. Mol. Biol.* **229** 153-72
[7] Malyar I, Gorin D, Santer S and Stetsyura S 2017 *Appl. Phys. Lett.* **110** 133104
[8] Makky A, Viel P, Chen S W, Berthelot T, Pellequer J and Polesel-Maris J 2013 *J. Mol. Recognit.* **26** 521-31

# Charge injection into the Ni-phyllosilicate nanoscrolls with reduced Ni nanoparticles using Kelvin force probe microscopy

**T S Kunkel[1], A A Krasilin[1,2], E K Khrapova[1], E A Straumal[3], A Nomine[2,4], J Ghanbaja[4] and A V Ankudinov[1]**

[1]Ioffe Institute, 194021 Saint-Petersburg, Russia
[2]ITMO University, 197101 Saint-Petersburg, Russia
[3]IPAC RAS, 142432 Chernogolovka, Russia
[4]Université de Lorraine, Institut Jean Lamour, UMR CNRS 7198, Nancy, France

tatyana_kunkel@mail.ru

**Abstract:** The evolution of the electrical charge injected into Ni-doped phyllosilicate nanoscrolls composites deposited on a conductive substrate was studied by Kelvin probe force microscopy. The nanoscrolls were synthesized by hydrothermal method and then annealed in $H_2$/Ar flow at 400-900°C in order to reduce Ni up to metal. A typical agglomerate of the $Ni_3Si_2O_5(OH)_4$ nanoscrolls accumulated the charge on one nanoscroll followed by a subsequent partial destruction of the agglomerate. For the $Ni_2MgSi_2O_5(OH)_4$ nanoscrolls, the whole agglomerate could be charged, and the injected charge was retained for hours. The shortest charge relaxation time was revealed for the nanoscrolls annealed at the highest temperature, when Ni was completely reduced.

## 1. Introduction

Currently, 1:1 phyllosilicates and clays with tubular morphology gained increased attention due to a number of peculiar properties [1,2]. Size mismatch together with difference in chemical composition between the two sheets [3] endue the phyllosilicate layer with ability of spontaneous curling. Resulting nanotubes and nanoscrolls possess a high specific surface area (from one to several hundreds of $m^2/g$ [4-6]), aspect ratio and completely different outer and inner surfaces. Tubular hydrosilicates like halloysite, chrysotile, pecoraite, and related compounds can be found in mineral form [1,7,8] or synthesized involving hydrothermal process [3,5,9,10]. One of the frequently mentioned nanoscrolls application related with their natural abundance and specific morphology is adsorption and encapsulation. Tubular particles can serve for wastewater remediation [4,11-13] or for controlled release of drugs and corrosion inhibitors [1].

Doping of hydrosilicate structure by *d*-elements like Fe, Co, and Ni opens up additional perspectives in the fields of magnetic materials [14-16] and catalysts [6,17-20]. A remarkable feature is that those dopants can be reduced by hydrogen up to metal nanoparticles, which are strongly coupled with the residual oxide matrix. Heterogeneous nanostructured metal catalysts possess great interest for application in a wide range of chemical process [21,22]. To create many active sites, it is necessary not only to form catalytically active nanoparticles, but also to find a way to separate and isolate them from each other. Traditional methods assume depositing metal nanoparticles upon initially prepared support using laser-assisted techniques [23-25] or various chemical methods [26-28]. Here, we study Ni-doped phyllosilicate nanoscrolls [11,29] with partially or completely reduced metallic nickel. A little has been known about the properties of such composite nanoscrolls and their ability to accumulate and retain the injected charge. One of the most suitable methods for studying the kinetics of charge accumulation and relaxation in nanoparticles is Kelvin probe force microscopy (KPFM) [30-32].

## 2. Experimental

The nanoscrolls were prepared by hydrothermal treatment (350 °C, 20 MPa, 12 hours, 0.25 M NaOH water solution) of Mg and Ni hydroxides coprecipitated in the presence of amorphous $SiO_2$ (hydrothermal synthesis is described in details elsewhere [29]). Phyllosilicates of two compositions were synthesized: nickel, $Ni_3Si_2O_5(OH)_4$, and nickel-magnesium, $Ni_2MgSi_2O_5(OH)_4$, ones.

In order to form metal nanoparticles inside the nanoscroll matrix, the samples were annealed in argon-hydrogen flow at various temperatures for 1 hour. For the nickel phyllosilicate samples, the annealing temperatures were 540, 670, and 900°C, whereas for the nickel-magnesium phyllosilicate ones – 400, 600, and 900°C.

Powder X-ray diffraction (PXRD) patterns of synthesized Ni-doped phyllosilicate nanoscrolls and the products of the reduction reaction were acquired using Rigaku Smartlab SE powder diffractometer ($CuK_\alpha$ radiation $\lambda$=0.15406 nm). A PDF-2 database was used for peak indexing.

Scanning transmission electron microscopy (TEM/STEM) was performed on JEOL ARM 200 F Cold FEG electron microscope operating at 200 kV and equipped with a spherical aberration (Cs) probe corrector (point resolution 0.19 nm in TEM mode and 0.078 nm in STEM mode).

Atomic force microscopy (AFM), KPFM, and scanning spreading resistance microscopy (SSRM) [33] measurements were performed on NTEGRA Aura system (NT-MDT SI, Russia). Conductive HA_C/W2C+ cantilevers (NT-MDT SI, Russia) with a probe tip radius <35 nm, a resonant frequency of about 37 KHz, and a stiffness coefficient of about 0.65 N/m were used. For the measurements, the nanoscrolls were deposited from isopropanol suspension on Si substrate (n-type conductivity, doping level $10^{16}$ $cm^{-3}$) covered by the 100-nm-thick nickel layer. All experiments were conducted in air at the room temperature and a relative humidity of about 30%.

Charge injection into the nanoscrolls and the investigation of the injected charge retention were carried out as follows. First, the topography and surface potential data of a sample area with an agglomerate of the nanoscrolls were collected in the tapping AFM and noncontact KPFM modes. Then, the probe touched the agglomerate in the AFM contact mode with the force of about 15-20 nN, and a charge was injected into the nanoscrolls by applying a rectangular positive pulse ($\tau$ = 1.5 min, amplitude + 10 V) between the grounded probe and the sample. After that, multiple scans of the surface potential of the sample region were performed, and the changes occurred were recorded. Then, the surface potential of the agglomerate relative to the surrounded substrate was analysed as a function of time.

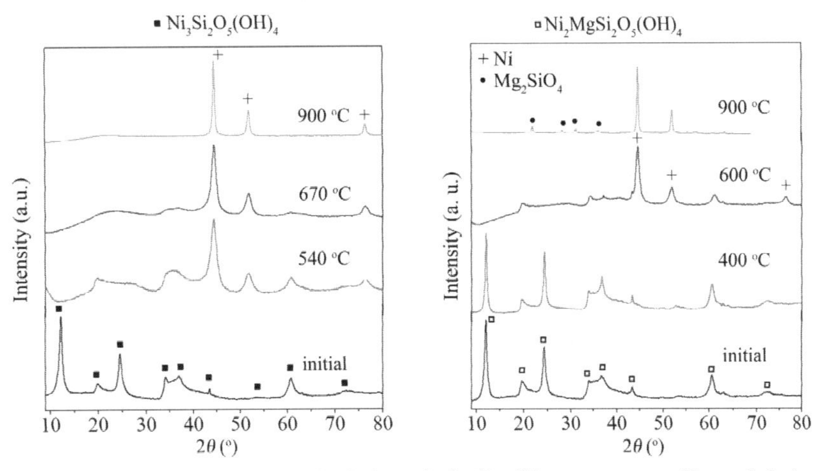


**Figure 1.** PXRD patterns of Ni-doped phyllosilicate nanoscrolls and their annealing products.

IOP Conf. Series: Materials Science and Engineering **699** (2019) 012023 doi:10.1088/1757-899X/699/1/012023

## 3. Results and discussion

Figure 1 shows changes in the crystal structure of Ni-doped phyllosilicate nanoscrolls after the annealing in the argon-hydrogen atmosphere. Temperature increase leads to a gradual degradation of the tubular phyllosilicate structure with simultaneous formation of the crystalline nickel phase. At 900°C, all Ni cations are supposed to be reduced to metal. Residual matrix phase composition depends on the nanoscrolls initial chemical composition. If there are magnesium cations in the system, then they form $Mg_2SiO_4$ crystal phase, otherwise $SiO_2$ stays in amorphous form at 900°C.

Figure 2 reflects morphological changes that occur with phyllosilicate nanoscrolls annealed at various temperatures. For the $Ni_2MgSi_2O_5(OH)_4$ nanoscrolls, the annealing at the lowest temperature, 400°C, was inefficient: it is difficult to say whether the metal has been reduced in this case (Fig. 2d). At the highest annealing temperature, 900°C, separate crystalline metallic particles embedded in the oxide matrix are visible for both samples (Fig. 2c,f). Typical sizes of the nanoparticles reduced out from $Ni_3Si_2O_5(OH)_4$ are 10-30 nm, and around 25-75 nm for the $Ni_2MgSi_2O_5(OH)_4$ case. The annealing at intermediate temperatures resulted in the Ni reduction in the nanoscrolls of both compositions. However, the resulted nanoparticles are smaller and not so clearly shaped.

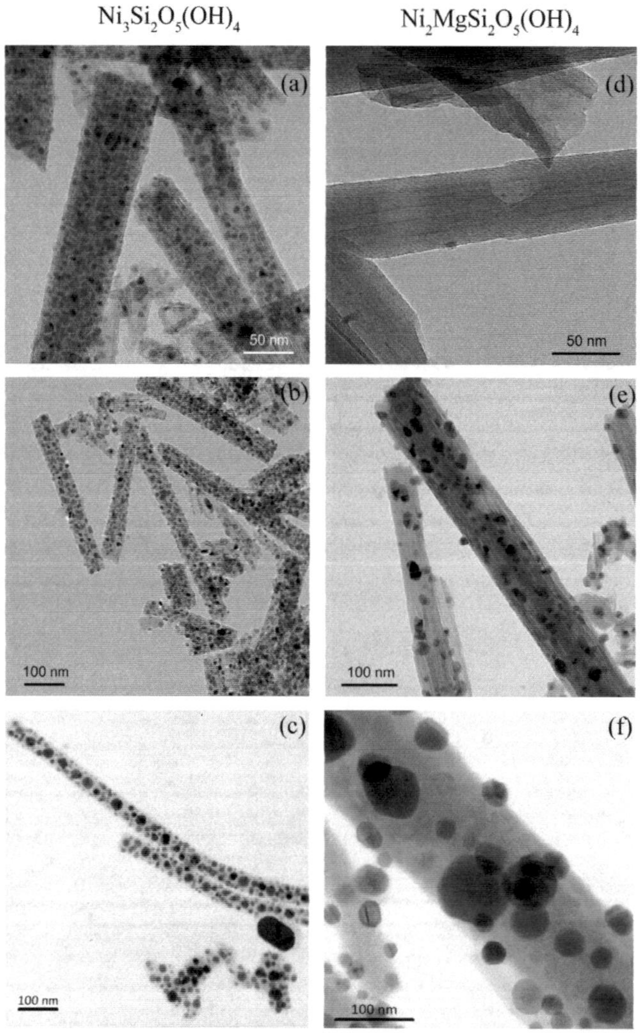


**Figure 2.** TEM/STEM micrographs of nanoscrolls after annealing: (a-c) $Ni_3Si_2O_5(OH)_4$, (d-f) $Ni_2MgSi_2O_5(OH)_4$. Annealing temperatures: (a-c) 540, 670, 900°C, respectively; (d-f) 400, 600, 900°C, respectively.

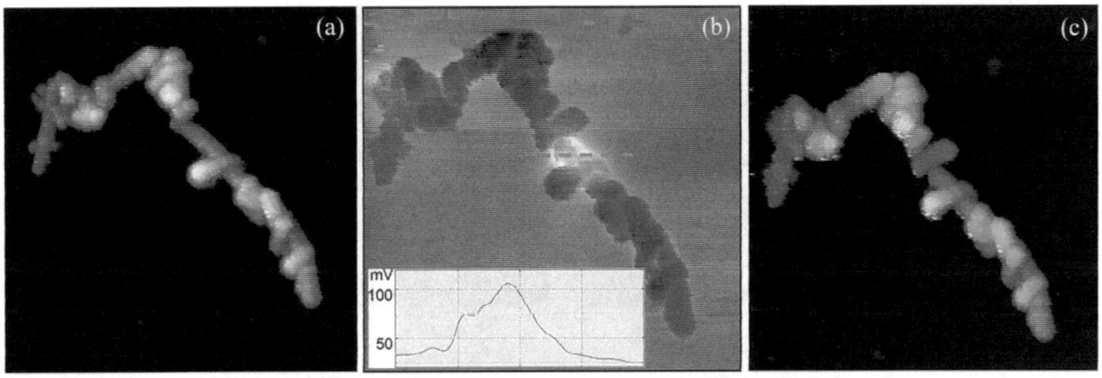


**Figure 3.** $Ni_3Si_2O_5(OH)_4$ nanoscrolls annealed at 900°C. AFM topography images of the same agglomerate (a) before and (c) after the charge injection. (b) KPFM surface potential image of the agglomerate. (insert) surface potential profile along the dashed line in the image. Frame size is 2.5x2.5 $\mu m^2$, profile length 1 μm. Scanning parameters: (a,c) free/set point oscillation amplitude 8 nm/5 nm; (b) probe-sample distance 10 nm, the probe is grounded.

First, the local conductivity measurements of the samples were made. Using the SSRM mode, the local current-voltage characteristics of nanoscrolls were recorded: the probe was pressed against the nanoscroll, a bias voltage was applied, and the current was recorded. The substrate surface areas free of the nanoscrolls were conductive, the conductivity of the nanoscrolls was not detected for all the samples studied (for probe-sample biases in the range of +/- 10 V and the current noise level of about 10 pA).

After the verification described above, the KPFM studies of the nanoscrolls were carried out. The difference in the properties of the nanoscrolls of various compositions was already observed at the charge injection stage. Independently on the annealing temperature, the $Ni_3Si_2O_5(OH)_4$ samples could be easily modified by this process: at the place of the voltage pulse application, changes in the topography were observed. As an example, Figure 3 shows results detected before, (a), and after the charge injection, (b) and (c), for the same agglomerate consisting of the $Ni_3Si_2O_5(OH)_4$ nanoscrolls annealed at 900 °C. The charge injection increased the surface potential locally by 120 mV (an insert in Figure 3b) and the one nanoscroll turned around across its initial orientation in the agglomerate. This nanoscroll was loosely coupled with the rest ones. It received the main injected charge and turned around due to the electrostatic repulsion from the neighbouring nanoscrolls, which was detected in the topography data. The observed weak nanoscroll-nanoscroll binding inside the agglomerates, as well as the low nanoscroll-substrate adhesion, facilitated the easy capture of some nanoscrolls onto the probe, even if the AFM worked in the tapping mode. In the charge injection experiments on the $Ni_3Si_2O_5(OH)_4$ samples, such arbitrary probe modifications made the time dependences of the KPFM data unreliable.

The charge injection tests of the $Ni_2MgSi_2O_5(OH)_4$ samples produced different and more informative results. First, the charge injection was not local: the voltage pulse charged either the entire conglomerate or most of its parts. Figure 4 represents the results of the charge injection into the $Ni_2MgSi_2O_5(OH)_4$ nanoscrolls annealed at various temperatures: topography (insert), surface potential before (a-d) and after the voltage pulse application (e-h).

To explore the charge leakage from the nanoscrolls, we analysed $\Delta\varphi(t)$, the potential of the nanoscrolls relative to the uncharged substrate. To determine a $\Delta\varphi(t)$ value, we used the profile data taken along the lines shown in Figure 4 and subtracted the background signal on the uncharged substrate from the signal averaged over the nanoscroll. Then, a nanoscroll normalized potential was determined as $\Delta\varphi(t)/\Delta\varphi(0)$; the $\Delta\varphi(0)$ values corresponded to the KPFM data obtained immediately (a few minutes) after the charge injection. Figure 5 shows the evolution of the nanoscroll normalized potential for four different samples.

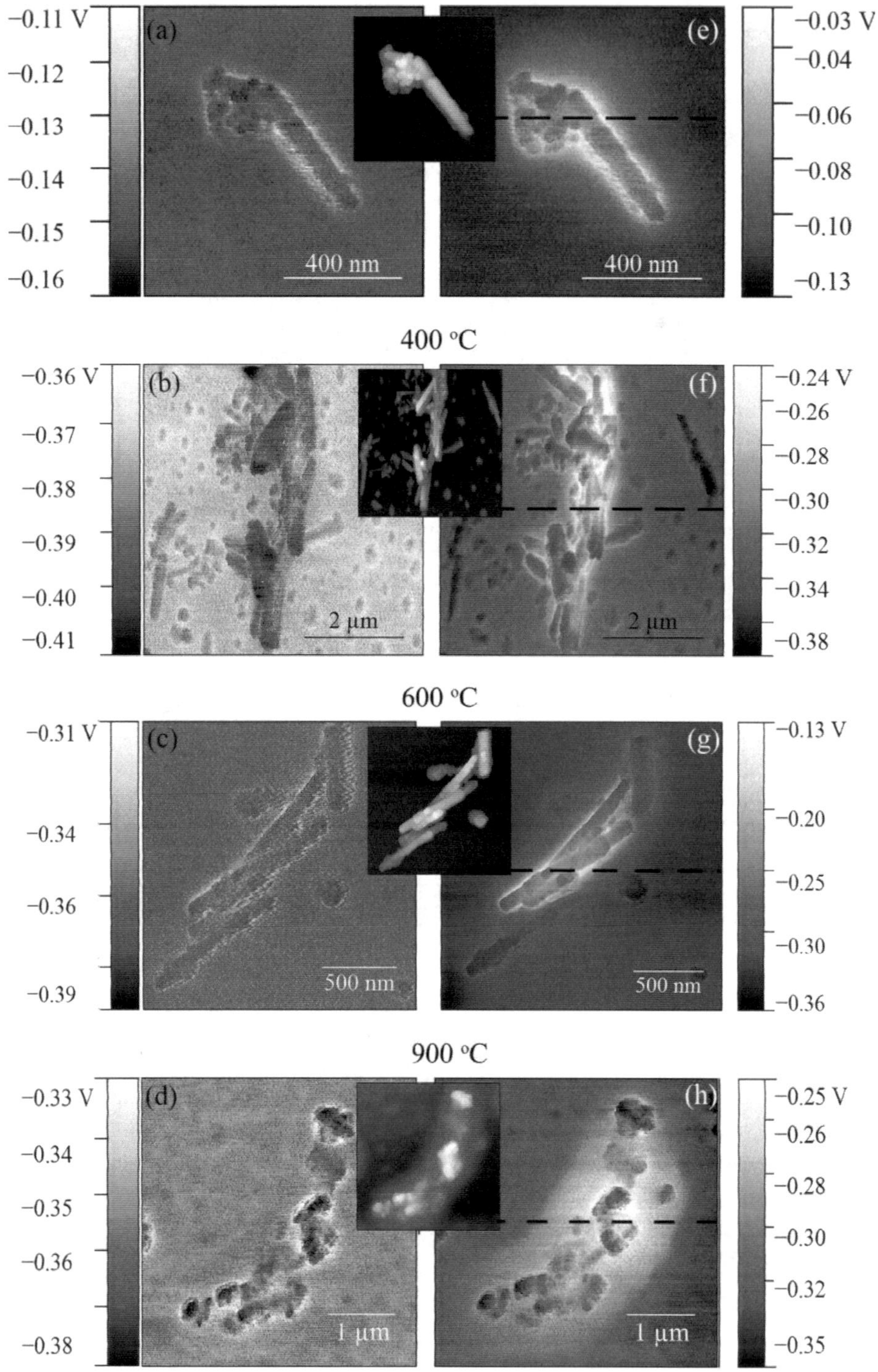


**Figure 4.** KPFM surface potential images of $Ni_2MgSi_2O_5(OH)_4$ nanoscrolls annealed at different temperatures: before (a-c) and after (d-f) the charge injection. Corresponding topography images are on the inserts. The dashed lines show the sections, along which the profiles were taken to prepare the data presented in Figure 5. Scanning parameters are close to Figure 3.

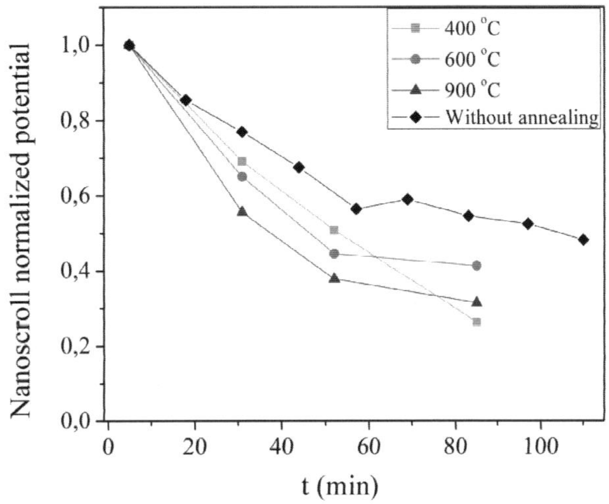


**Figure 5.** Injected charge relaxation for the $Ni_2MgSi_2O_5(OH)_4$ nanoscrolls annealed at different temperatures.

Using the data of Figure 5, we estimated the characteristic relaxation times (the time at the half value of the signal was used): for the sample without annealing – 100 min, for the sample annealed at 400°C – 45 min, 600°C – 45 min, and 900°C – 35 min. Hence, one can conclude that the metal reduction leads to a decrease in the characteristic relaxation time.

## 4. Conclusion

We used scanning probe microscopy to perform the electrical charge injection experiments on the nanostructured samples of complex composition. The accumulation and retention of the injected charge for Ni-doped phyllosilicate nanoscrolls with reduced metal nanoparticles were investigated.

The charge injection processes for the $Ni_3Si_2O_5(OH)_4$ and the $Ni_2MgSi_2O_5(OH)_4$ samples were significantly different: in the first case, only separate nanoscrolls were charged; in the second case, the charge was distributed over the nanoscroll agglomerate. This result clearly shows the significant difference between the levels of adhesive interaction of nanoscroll-nanoscroll and nanoscroll-substrate for the two studied compositions.

Unlike the $Ni_3Si_2O_5(OH)_4$ sample, the charge injected into the agglomerates of $Ni_2MgSi_2O_5(OH)_4$ nanoscrolls spread over the whole agglomerate (or most of it), and the agglomerate morphology did not change. The stability of the $Ni_2MgSi_2O_5(OH)_4$ samples morphology upon charge injection was used to study the charge relaxation time. It was found that this time decreases with increase in annealing temperature growth. It could be directly related with the efficiency of Ni reduction in the nanoscrolls at different temperatures.

We believe that these results could be in demand when characterizing metal inclusions distributed inside the dielectric nanostructures.

**Acknowledgements**
The research was supported by Russian Science Foundation (grant 19-13-00151).

**References**

[1] Lvov Y, Wang W, Zhang L and Fakhrullin R 2016 *Adv. Mater.* **28** 1227-50
[2] Smith R J, Holder K M, Ruiz S, Hahn W, Song Y, Lvov Y M and Grunlan J C 2017 *Adv. Funct. Mater.* **28** 1703289
[3] Krasilin A A, Nevedomsky V N and Gusarov V V 2017 *J. Phys. Chem. C* **121** 12495-502

[4] Belkassa K, Bessaha F, Marouf-Khelifa K, Batonneau-Gener I, Comparot J-D and Khelifa A 2013 *Colloids & Surfaces A: Physicochem. Eng. Asp.* **421** 26-33
[5] White R D, Bavykin D V and Walsh F C 2012 *J. Phys. Chem. C* **116** 8824-33
[6] Ashok J, Bian Z, Wang Z and Kawi S 2018 *Catal. Sci. Technol.* **8** 1730-42
[7] Dódony I and Buseck P R 2004 *Int. Geol. Rev.* **46** 507-27
[8] Villanova-de-Benavent C, Nieto F, Viti C, Proenza J A, Galí S and Roqué-Rosell J 2016 *Am. Mineral.* **101** 1460-73
[9] Korytkova E N, Semyashkina M P, Maslennikova T P, Pivovarova L N, Al'myashev V I and Ugolkov V L 2013 *Glas. Phys. Chem.* **39** 294-300
[10] Lafay R, Fernandez-Martinez A, Montes-Hernandez G, Auzende A L and Poulain A 2016 *Am. Mineral.* **101** 2666-76
[11] Krasilin A A, Danilovich D P, Yudina E B, Bruyere S, Ghanbaja J and Ivanov V K 2019 *Appl. Clay. Sci.* **173** 1-11
[12] Cataldo S, Lazzara G, Massaro M, Muratore N, Pettignano A and Riela S 2018 *Appl. Clay. Sci.* **156** 87-95
[13] Maslennikova T P and Korytkova E N 2010 *Glas. Phys. Chem.* **36** 345-50
[14] Krasilin A A, Semenova A S, Kellerman D G, Nevedomsky V N and Gusarov V V 2016 *EPL* **113** 47006
[15] Borghi E, Occhiuzzi M, Foresti E, Lesci I G and Roveri N 2010 *Phys. Chem. Chem. Phys.* **12** 227-38
[16] Yang Y, Liang Q, Li J, Zhuang Y, He Y, Bai B and Wang X 2011 *Nano. Res.* **4** 882-90
[17] Sarvaramini A and Larachi Mössbauer F 2011 *J. Phys. Chem. C* **115** 6841-8
[18] Zhang C, Zhu W, Li S, Wu G, Ma X, Wang X and Gong J 2013 *Chem. Commun.* **49** 9383-5
[19] Bian Z, Suryawinata I Y and Kawi S 2016 *Appl. Catal. B Environ.* **195** 1-8
[20] Bian Z and Kawi S 2017 *J. $CO_2$ Util.* **18** 345-52
[21] Bulut A, Yurderi M, Ertas İ E, Celebi M, Kaya M and Zahmakiran M 2016 *Appl. Catal. B: Environ.* **180** 121-9
[22] Bukhtiyarov V I and Slin'ko M G 2001 *Russ. Chem. Rev.* **70** 147-59
[23] Golubina E V, Lokteva E S, Maslakov K I, Rostovshchikova T N, Shilina M I, Gurevich S A, Kozhevin V M and Yavsin D A 2017 *Nanotechnologies in Russia* **12** 19
[24] Zhang J, Chaker M and Ma D 2017 *J. Colloid Interface Sci.* **489** 138-49
[25] Marzun G, Levish A, Mackert V, Kallio T, Barcikowski S and Wagener Ph 2017 *J. Colloid Interface Sci.* **489** 57-67
[26] Venta K, Wanunu M and Drndic M 2013 *Nano Lett.* **13** 423-29
[27] Yashtulov N A, Lebedeva M V and Flid V R 2015 *Russ. Chem. Bull.* **64** 1837-41
[28] He L, Weniger F, Neumann H and Beller M 2016 *Angew. Chem. Int. Ed.* **55** 12582-94
[29] Krasilin A A, Khrapova E K, Nominé A, Ghanbaja J, Belmonte T and Gusarov V V 2019 *Chem. Phys. Chem.* **20** 719-26
[30] Escasain E, Lopes-Elvira E, Baro A M, Colchero J and Palacios-Lidon E 2011 *Nanotechnol.* **22** 375704
[31] Dunaevskiy M S, Alekseev P A, Girard P, Lahderanta E, Lashkul A and Titkov A N 2011 *J. Appl. Phys.* **110** 84304
[32] Dement'ev P A, Dunaevskii M S, Aleshin A N, Titkov A N and Makarenko I V 2014 *Phys. Solid State* **56** 1054-7
[33] De Wolf P, Snauwaert J, Clarysse T, Vandervorst W and Hellemans L 1995 *Appl. Phys. Lett.* **66** 1530-2

IOP Conf. Series: Materials Science and Engineering **699** (2019) 012024 doi:10.1088/1757-899X/699/1/012024

# Evaluation of mechanical and electrical parameters of individual polyaniline nanoparticles

**I A Lobov[1], N A Davletkildeev[1,2], D V Sokolov[1] and E Yu Mosur[1,2]**

[1]Omsk Scientific Center SB RAS, Omsk 644040, Russia
[2]Dostoevsky Omsk State University, Omsk 644077, Russia

LI__87@mail.ru

**Abstract**. The work function, the static dielectric constant, and the elastic modulus of individual polyaniline nanoparticles in the base form deposited on HOPG were determined. Electrostatic force microscopy was used to determine the electrical parameters. The value of the work function is 4.66 eV. The dielectric constant is 2.2. Contact atomic force microscopy was used to determine Young's modulus. The modulus of elasticity is 4.42 GPa. The obtained values differ from the typical values obtained on bulk samples of the base form polyaniline. In the case of electrical parameters, this is due to contact phenomena between the conductive substrate and the dielectric nanoparticle. The Young's modulus of a polyaniline nanoparticle depends on the packing density of the macromolecule, while the Young's modulus of a compressed tablet depends mainly on the intermolecular forces.

## 1. Introduction

The desire to reduce the size of electronic devices toward submicron and even nanoscale lengths generates new requirements to the materials they are made of. The accurate measurement of electronic materials properties with high spatial resolution has become increasingly important. During the transition from the bulk structure to a separate nanoparticle, the properties of the material change significantly. Since standard characterization techniques are not capable of addressing the electronic properties of thin films or nanoparticles, new techniques are needed to achieve both higher accuracy and higher spatial resolution.

Polyaniline (PANI) is one of the most important conductive polymers, which is widely used in various electronic applications. The most important characteristics of conductive polymers are mechanical and electrical parameters, the magnitudes of which largely depend on the production conditions. The properties of PANI have been investigated basically with respect to its bulk state. Measurements, as a rule, are carried out on the polymer massive layers. Therefore, the results are influenced by the contact interactions between PANI molecules.

Scanning probe microscopy (SPM) is widely used to study the morphology and physical properties of 2D, 1D, and 0D materials. The method is based on the physical (mechanical, electrical, etc.) interaction of the probe with the sample. A variety of techniques that allow extracting different characteristics of the sample have been developed. Thus, mechanical interaction allows obtaining 3D surface morphology, elastic and adhesive properties. Electrical interaction allows identifying the sample portions with different resistance, revealing magnetic and ferromagnetic features, and calculating electronic properties. Most of the information obtained by SPM methods is the result of modelling and requires an understanding of the physics and the interaction nature in order to apply various models correctly. In respect of PANI nanoscopic methods of research, SPM is not so widespread. However, they can clarify many controversial issues, for example, related to the conduction mechanism.

The mechanical properties of PANI are mainly measured on compressed tablets [1]. The elastic properties of the polymer depend on the preparation conditions. Therefore, in order to obtain the mechanical properties of PANI itself, it is necessary to study them on a native sample. The measurement of PANI base form Young's modulus is the most appropriate, since PANI salts with various acids can affect the packing density of polymer molecules, hydrophobic properties (including the water content in the polymer), etc. SPM allows measuring Young's modulus both on films and on nanoscale particles. For this purpose, the measured indentation curve can be used in different models. The choice of a certain model depends, for example, on the shape of the probe or on the ductility of the material.

The electron work function (WF) is a fundamental thermodynamic property of the material. SPM allows measuring this parameter by various methods, such as scanning tunnelling microscopy, Kelvin probe force microscopy, or electrostatic force microscopy (EFM) [2,3]. PANI WF can vary from 4 to 5 eV. WF depends on protonation level, type of dopant acid, film preparation, and others.

Broadband Dielectric Spectroscopy is usually used to determine the dielectric characteristics of polymers. This technology allows measuring both static and dynamic permittivity. It can be used both for bulk materials and for thin films [4]. However, the study of 1D and 0D objects is not possible. SPM allows measuring the static dielectric constant of not only thin films, but also nanoscale objects [5].

This paper presents the results of measurements of the electron work function, dielectric constant, and elastic modulus of individual PANI nanoparticles (PANI-NP) in the emeraldine base form containing a small number of molecules. The measurements were carried out in a single experiment using two methods of scanning probe microscopy.

## 2. Experimental

### *2.1. Preparation of PANI-NP*

PANI was obtained via chemical oxidative polymerization of aniline [6]. Ammonium persulphate (APS) was used as an oxidizing agent. Two solutions were prepared. The first of them contained 3 ml of aniline, 25 ml of 3M HCl, and 40 ml of isopropyl alcohol. The second one included 7.5 g of APS and 15 ml of 3M HCl. The molar ratio of aniline monomer to APS is 1:1. Both solutions were cooled to 0°C in an ice bath. Then a solution containing an oxidizing agent was added dropwise to the first solution with constant stirring. The reaction flask was placed in an ice bath to maintain a constant synthesis temperature. Synthesis time is 40 min. Then PANI was transferred in an emeraldine base form by stirring a 1 M $NH_3$ for 24 h. The PANI-NP was separated from the bulk polymer by sonication in ethanol. A suspension of nanoparticles was deposited on a freshly split HOPG.

### *2.2. Characterization*

Measurements of individual PANI-NP using EFM and contact atomic force microscopy (AFM) were obtained using AFM MFP-3D (Asylum Research). Measurements were performed at room temperature in air. We used a HA_FM/Pt probe (TipsNano) with a resonant frequency of 108 kHz and a radius of curvature of 35 nm. Second pass height was 100 nm. The voltage was varied in the range from -9 to +9V. The electron work function was determined from the magnitude of the positive phase shift in the EFM images using the technique described in [7]. We are using the negative phase shift measured from EFM images and the model described by [5] to calculate the dielectric constant for PANI-NP. In the model used, the probe was taken as a sphere, and the sample was taken as a flat disk. The magnitudes of the Young's modulus were determined by fitting the model curve constructed using the Hertz model for a conical probe to the force-indentation curves using the Asylum Research MFP-3D Hertz analysis tool.

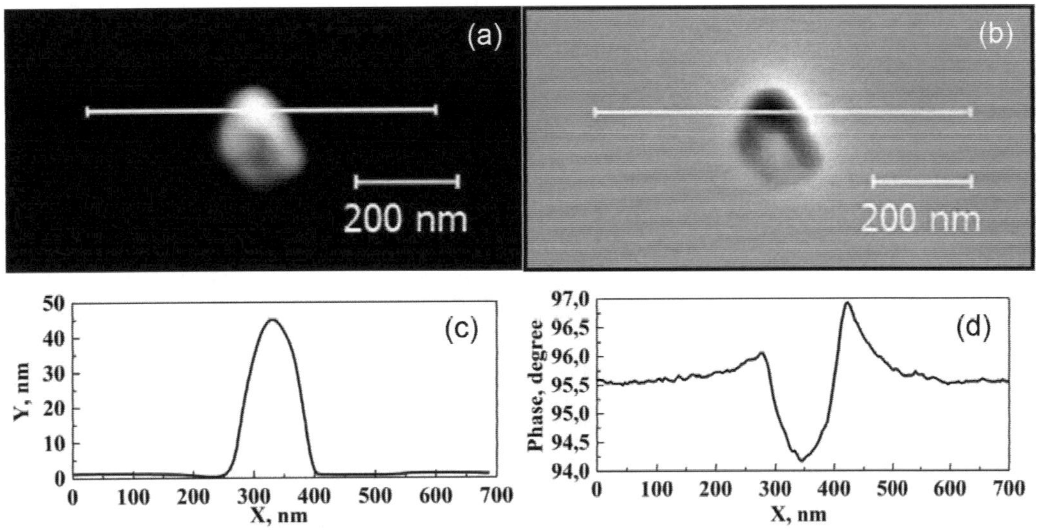


**Figure 1.** (a) AFM image, (b) EFM image of PANI-NP, and (c, d) their cross-sections along the lines.

## 3. Results and discussion

Figures 1a,b show an example of AFM and EFM images of a PANI-NP consisting of several polymer macromolecules, as well as their cross-sectional profiles showing the characteristic particle size of PANI (Fig. 1c) and the contrast feature of its EFM image (Fig. 1d). PANI-NP sizes range from 100 to 300 nm, and heights from 40 to 65 nm. The light contour (positive phase shift) around the nanoparticle and the negative phase shift inside it are clearly visible in the EPM image.

The electron work function was determined from the magnitude of the positive phase shift [7]. HOPG with a known work function of 4.6 eV was used in the work. Before measuring, the probe was calibrated on pure freshly split HOPG for goal determinate of tip WF (4.82 eV). Figure 2a shows the parabolic dependence of the tangent of the EFM positive phase shift on the magnitude of the tip voltage. The position of the dependence minimum on the voltage axis corresponds to the contact potential difference between the tip and the PANI-NP. Value of this difference $\Delta U_{PANI\text{-}NP/tip}$=0.16 V. The average electron work function of PANI-NP calculated on the basis of such dependences was $W_{PANI\text{-}NP}$=4.66 eV. Work function PANI measured on bulk structures lies in the range from 4 to 5 eV [2,8]. Such a large variation is due to the fact that many parameters influence the value of the work function. The obtained WF value is close to the values of the protonated forms of PANI. This is possible due to the presence of contact phenomena between the conductive substrate and the sample. Depending on the ratio of the WF PANI and HOPG, the formation of an ohmic contact or an energy barrier for charge carriers is possible.

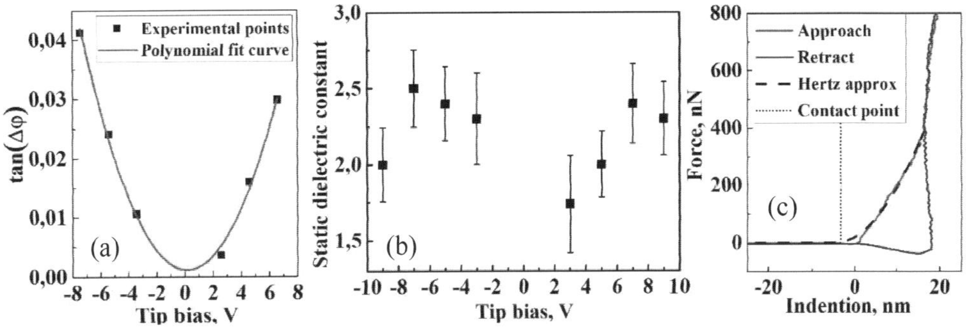


**Figure 2.** (a) Typical dependences of the tangent of the EFM positive phase shift and (b) the dielectric constant of the PANI-NP on the applied tip voltage; (c) experimental indentation curve of a PANI-NP and its approximation by the Hertz model.

We are using the negative phase shift measured from EFM images to calculate the dielectric constant of PANI-NP. The phase shift of the EPM signal is associated with the capacitance according to the formula given in [9]. In our model, the probe was taken as a sphere, and the sample was taken as a flat disk. Figure 2b shows the dependence of the calculated value of the static dielectric constant of the PANI-NP on the applied voltage. The average dielectric constant of PANI-NP is $\varepsilon_{PANI}$=2.2±0.6. The value of static dielectric constant for volume PANI is about 4 [10]. The small value of $\varepsilon_{PANI}$ in our experiment confirms the presence of contact phenomena.

Figure 2c shows a typical PANI-NP indentation curve with an AFM probe, built on the basis of force curves, its approximation by a model curve constructed on the basis of solving the Hertz model for a conical probe. When calculating the elastic modulus, we used the Poisson's ratio for PANI $\nu$=0.38 from [11]. The average value of the Young's modulus of the PANI-NP was E = 4.42 GPa. Typical values of the elastic modulus of the base form PANI layers is 1 GPa [1,12]. When measuring Young's modulus on pressed tablets or on polymer films, the main contribution is made by intermolecular interactions. When a probe is identified in a nanoparticle consisting of one or two layers of polymer molecules, the packing density of the macromolecule makes the main contribution to the Young's modulus.

## 4. Conclusion

In the work in a single experimental cycle, the electrical and mechanical properties of PANI-NP are investigated. It is shown that due to the small size of the object of study and the need to use a conductive substrate, the electrical properties of the nanoparticles differ from the similar parameters of the bulk PANI samples. To reduce or eliminate the influence of contact phenomena, it is necessary to use a different type of substrate. Further work will be aimed at studying this issue. The obtained value of Young's modulus corresponds to the elastic properties of an individual polymer macromolecule.

**Acknowledgements**

The work was carried out according to the state task of the Omsk Scientific Centre SB RAS (project registration number AAAA-A17-117041210227-8).

## References

[1] Valentová H and Stejskal J 2010 *Synth. MET* **160** 832-4
[2] Park J, Bang D, Jang K, Haam S, Yang J and Na S 2012 *Nanotechnology* **23** 365705
[3] Lobov I A, Davletkildeev N A and Sokolov D V 2018 *IOP Conf. Ser.: Mater. Sci. Eng.* **443** 012021
[4] Schönhals A and Kremer F 2003 *Broadband Dielectric Spectroscopy* (Berlin: Springer) 1-33
[5] Staii C, Johnson Jr A T and Pinto N J 2004 *Nano Lett.* **4** 859-62
[6] Lobov I A and Shulus I K 2018 *Omsk scientific bulletin* **3** 98
[7] Davletkildeev N A, Stetsko D V, Bolotov V V, Stenkin Y A, Korusenko P M and Nesov S N 2015 *Mater. Lett.* **161** 534-7
[8] Pandey S S, Misra S C K, Malhotra B D and Chandra S 1992 *J. Appl. Polym. Sci.* **44** 911
[9] Bolotov V V, Davletkil'deev N A, Stets'ko D V and Lobov I A 2014 *Tech. Phys. Lett.* **40** 965
[10] Javadi H H S, Cromack K R, MacDiarmid A G and Epstein A J 1989 *Phys. Rev. B* **39** 3579
[11] Pereira J N, Vieira P, Ferreira A, Paleo A J, Rocha J G and Lanceros-Méndez S 2012 *J. Polym. Res.* **19** 9815
[12] Pereira J N, Gholami H, Shakeri A and Saadattalab V 2017 *Curr. Chem. Lett.* **6** 151-8

IOP Conf. Series: Materials Science and Engineering **699** (2019) 012025 doi:10.1088/1757-899X/699/1/012025

# Structural, electronic, and magnetic properties of ferroelectric/dielectric heterostructures

**Yu V Lysogorskii[1,2], I I Piyanzina[1,2], D P Pavlov[1], A V Leontyev[1], R F Mamin[1,2], D A Tayurskii[2] and V V Kabanov[1,3]**

[1]Zavoisky Physical-Technical Institute, FRC KazanSC of RAS, 420029, Kazan, Russia
[2]Kazan Federal University, 420008, Kazan, Russia
[3]Department for Complex Matter, Jozef Stefan Institute, 1000 Ljubljana, Slovenia

mamin@kfti.knc.ru

**Abstract**. We present the results of structural, electronic, and magnetic properties of $BaTiO_3/LaMnO_3$ heterostructure by means of first-principles calculations. We demonstrate the possibility of highly conducting layer formation at the heterostructure composed of perovskite ferroelectric $BaTiO_3$ and antiferromagnet manganite $LaMnO_3$. We also analyse magnetic properties and an impact of ferroelectric polarization onto the conducting properties. Experimental results of magnetic field effect on $Ba_{0.8}Sr_{0.2}TiO_3/LaMnO_3$ heterostructure are presented as well. Effect of illumination of the $Ba_{0.8}Sr_{0.2}TiO_3/LaMnO_3$ heterostructure is also demonstrated.

## 1. Introduction

A high-mobility electron gas was found at the interface between two nonmagnetic wide-band gap insulating oxides $LaAlO_3$ (LAO) and $SrTiO_3$ (STO) in 2004 [1]. After that, similar heterointerfaces involving two insulating nonmagnetic oxides were comprehensively studied [1-12]. In particular, it was found that the metallic phase, so-called quasi-two-dimensional electron system (q2DES), was formed in the STO layers next to the LAO/STO interface, when the number of LAO overlayers exceeded three [2,3]. Such a system underwent a transition to a superconducting state below 300 mK [4]. Remarkably, this superconducting state coexisted with a magnetic state being stable up to the room temperature. It was concluded, that the primary mechanism responsible for the q2DES formation was the electronic reconstruction followed by structural reconstruction. Later, q2DES was found in other non-magnetic dielectrics. Their common feature is that the creation of q2DES might be due to either the polar nature of one of the components or due to defects or dopants. The polarity can be due to a sequence of the differently charged layers as in the LAO, or due to the ferroelectric material used as an overlayer [3,4]. The main advantage of using ferroelectric material in the heterostructure is a possibility to change the direction of the polarization in the ferroelectric layer and thus to control properties of the interface.

The most common mechanism for describing the q2DES formation is the polarization discontinuity model [1,12]. The important feature related to the 2DES formation is the local polarity of layers inside the LAO slab. LAO can be considered as a sequence of $(LaO)^+$ and $(AlO_2)^-$ layers. The conductivity occurs as soon as the field towards the surface exceeds the field, which arises due to the buckling, or in other words, when the number of LAO overlayers exceeds the critical value. In the present work, we have chosen the $BaTiO_3/LaMnO_3$ (BTO/LMO) heterostructure, where BTO has $(BaO)^0$ and $(TiO_2)^0$

neutral layers. Thus, there is no polarity due to charge sequence, but there is a ferroelectric polarization due to Ti atoms displacements out of the octahedron centre in the BTO slab. Moreover, the direction of this polarization might be switched by an external electric field. Based on first-principles band structure calculations, we demonstrate the possibility of conducting layer formation at the interface composed of perovskite ferroelectric BTO and antiferromagnet manganite LMO. We briefly introduce the employed computational method and first results of modelling of structural, electronic, and optical properties of BTO/LMO heterostructure with different numbers of layers. We also present the experimental results of magnetic field effect and the effect of illumination onto the $Ba_{0.8}Sr_{0.2}TiO_3/LaMnO_3$ heterostructure.

## 2. Simulation and experimental details

For the density of states (DOS) calculations and structural optimization, we use density functional theory (DFT) [12]. Exchange and correlation effects were accounted by generalized gradient approximation (GGA) [13]. Kohn-Sham equations were solved using the plane-wave basis set (PAW) [14] realized within the VASP code [15], which is a part of the MedeA® software of Materials Design [14]. The cut-off energy was chosen to be 400 eV. The force tolerance was 0.5 eV/nm and the energy tolerance for the self-consistency loop was $10^{-5}$ eV. The Brillouin zones were sampled including 5×5×1 **k**-points. Since there are strong correlations between *d* and *f*-electrons in our system, the GGA+*U* correction were included to our computational scheme [16]. The *U* parameter was added to La 4*f*, Ti 3*d* and Mn 3*d* orbitals ($U$ = 8 eV, 2 eV, and 4 eV, respectively). The choice of *U* for Ti and La atoms was based on our previous research [17]. In the case of BTO/LMO, heterointerface components represent two insulating oxides with different structures and parameters of elementary cell. Due to a high mismatch between BTO and LMO (≈30 %), the BTO unit cell was rotated by a 45° before merging with LMO substrate, since the $\sqrt{2}a_{BTO}$=0.5657 nm, resulting in less than 1 % mismatch with LMO substrate's in-plane cell parameter. It means that the substrate is compressive with respect to the film. As a result, the polarization axis of the ferroelectric BTO will be directed perpendicular to the interface. For modelling, the heterostructure the LMO central slab was enlarged by a factor of 1.5 in *z*-axis direction and bounded by a varying number of BTO layers with interface BaO or $TiO_2$ layers on both sides. Such a unit cell guarantees the absence of artificial dipoles and additional polarity, which might arise due to a non-symmetric structure. In order to avoid interaction of surfaces and slabs with their periodic images, a 20 nm vacuum region was added. In plane, *a* and *b* cell parameters were fixed to lattice parameters of LMO substrate, whereas the rest atom positions were allowed to relax during the optimization procedure. Since the substrate is compressive with respect to the film, the polarization axis of the ferroelectric BTO will be directed perpendicular to the interface.

The temperature dependence of the electrical resistance has been studied for heterostructures formed by antiferromagnetic $LaMnO_3$ single crystals with epitaxial films of ferroelectric $Ba_{0.8}Sr_{0.2}TiO_3$ (BSTO) deposited on it by reactive sputtering of stoichiometric targets using RF plasma (RF-sputtering) method [18, 19] at 650°C. In this work, we present a review of the investigation of conducting state and show the effect of illumination on the interface resistance between ferroelectric oxide and insulating oxide in the BSTO/LMO heterostructure.

## 3. Results and discussion

The DOS spectrum of BTO/LMO heterostructure with three BTO overlayers deposited on top of the LMO substrate is shown in Figure 1. In this case, Ti states cross the Fermi-level with sufficiently large DOS intensity (about 10 $eV^{-1}$). The band gap vanishes for the spin-up component, and we expect that the system is a good metal. A distinctive feature of the DOS with two and three BTO overlayers is the fact that in the first case the resulted magnetic state is antiferromagnetic, whereas in the second case it is ferromagnetic. Possibly, that is related to the number of free charge carriers. In the case of three BTO overlayers, the DOS at the Fermi-lever and charge carriers' concentration are higher. Consequently, indirect ferromagnetic exchange through charge carriers becomes efficient for the ferromagnetic ordering at the interface. It can be well observed from Figure 1, where DOS with spin up and down differ significantly (mainly due to Mn atoms shown by green colour).

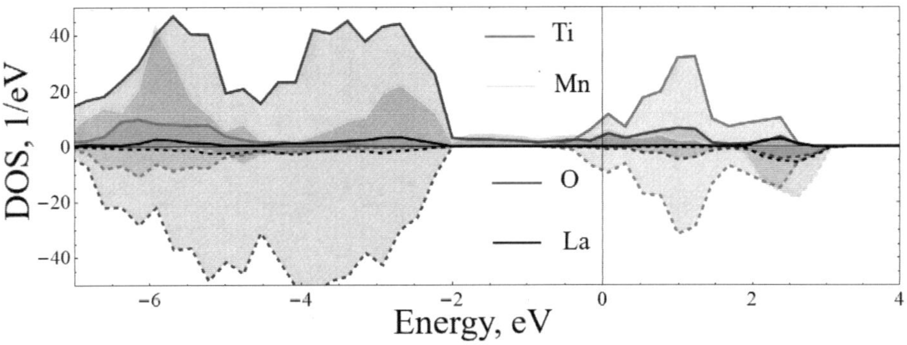


**Figure 1.** DOS spectrum of the BTO/LMO heterostructure with three BTO overlayers on top of the LMO substrate. Solid and dashed curves correspond to spin-up and down components.

It also follows that band gap vanishes for the heterostructure with relatively small number of BTO overlayers. Thus, we expect the interface area to absorb the light having the energy less than the band gap of LMO, in other words, photon energies considerably less than 1.8 eV. Besides, there is a metallic state, which might lead to the absorption in the terahertz spectral region. Thus, light absorption experiments in the terahertz spectral region might shed the light on the conductive properties. It is utterly important during such experiments to distinguish two contributions: the first one is related to the absorption by charge carriers in the conductive area, the second one is related to the optical phonon absorption in the region of 10-20 THz, if such phonons exist. Note, that the aforementioned line of reasoning is based on our assumption and so far not on experimental evidence. We applied magnetic field of different values and measured the temperature dependences of the heterostructure resistance by four contact method in the magnetic field and after switching it off.

After each cycle of the magnetic field application the resistance changed irreversibly. Temperature dependences of the resistance R(T) for the BSTO/LMO heterostructure after application of magnetic field are shown in Figure 2. Blue triangles show R(T) after few cycles of application of the magnetic field parallel to the *c*-axis of LMO and red diamonds show R(T) after few cycles of application of the magnetic field perpendicular to the *c*-axis of LMO. The inset shows a space distribution of the surface charge in the vicinity of the interphase. When the magnetic field is applied parallel to the *c*-axis of LMO, the resistance of the BSTO/LMO heterostructure sharply decreases after few cycles of the magnetic field application, but when the magnetic field is applied perpendicular to the *c*-axis, the resistance increases after few cycles of the magnetic field application.

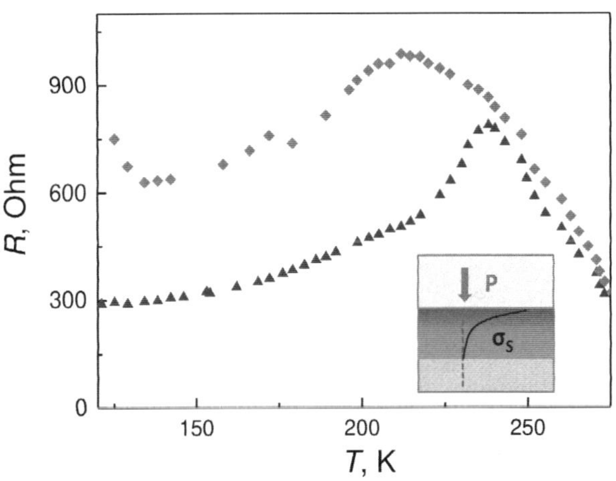


**Figure 2.** Temperature dependences of the resistance R(T) for the BSTO/LMO heterostructure after application of magnetic field: blue triangles show R(T) after few cycles of application of the magnetic field up to 1 T parallel to the *c*-axis of LMO and red diamonds show R(T) after few cycles of application of the magnetic field perpendicular to the *c*-axis of LMO. Space distribution of surface charge in the vicinity of the interphase is shown in the insert.

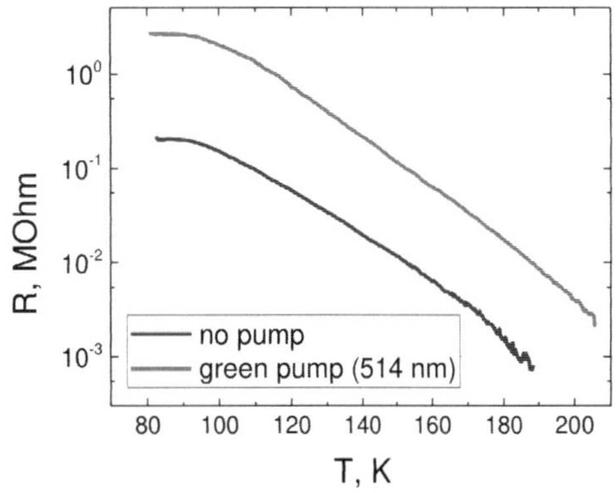


**Figure 3.** Temperature dependences of the resistance R(T) for BSTO/LMO heterostructure on the condition light switch off (blue line) and light switch on (red line).

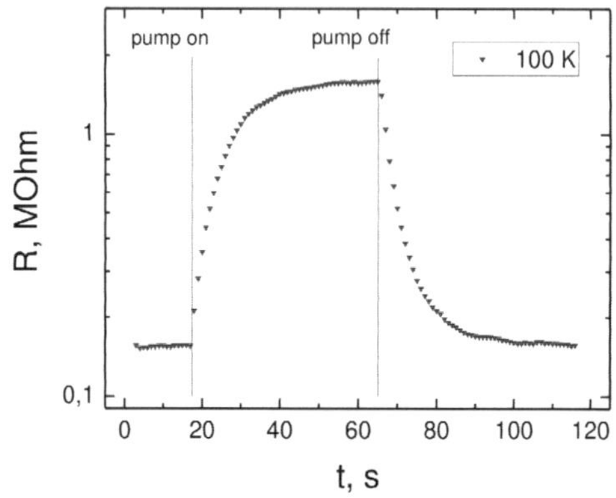


**Figure 4.** Time dependences of the resistance R(t) for BSTO/LMO heterostructure on the light switch on - light switch off regimes for different temperature – marked on the figure.

When the magnetic field is applied perpendicular to the *c*-axis, the resistance of the BSTO/LMO heterostructure shows semiconducting behaviour. The resistance measurements for these samples are shown in Figures 3 and 4. In particular, Figure 3 shows temperature dependences of the resistance for BSTO/LMO heterostructures under light radiation and without it. The resistance of the sample under radiation is more than 10 times higher than the resistance measured without radiation. The light absorption by the film affects the charge distribution in the BSTO/LMO heterostructure. Thus, we observe a negative photoconductivity. To understand the nature of resistivity increasing under the light radiation, we have measured time characteristics R(t). As it is shown in Figure 4, the resistance saturates in less than 10 seconds after switching the light on. The resistance recovers with the same relaxation time after switching the light off.

Since we found rather slow relaxation processes, we assume that charge dynamic is associated with relaxation processes in the ferroelectric film (relaxation processes in the metallic regions should be much faster). Under illumination, the free electron-hole pairs are generated in ferroelectric film. These pairs screen the polarization. This process leads to the depletion of the interface region. To clarify all these issues, further experiments with switching the polarization of the film in the strong electric field are necessary.

## 4. Conclusion

The results of structural, electronic, and optical properties calculations of $BaTiO_3/LaMnO_3$ heterostructure composed of different number of layers based on first-principles calculations and theoretical consideration were presented. A possibility of a conducting state at the interface formation was shown. The transition to a metallic state in $Ba_{0.8}Sr_{0.2}TiO_3/LaMnO_3$ heterostructure was observed when the polarization of the ferroelectric film was perpendicular to the surface of the heterostructure. The resistance of the heterostructure strongly decreased, when a magnetic field was applied perpendicular to the surface of heterostructure. The effect of illumination of the $Ba_{0.8}Sr_{0.2}TiO_3/LaMnO_3$ heterostructure was observed. Therefore, the resistance of the heterostructure may be controlled by the light. This example shows that highly conductive state can be created at the ferroelectric/dielectric interfaces using the relatively simple technique of preparation.

IOP Conf. Series: Materials Science and Engineering **699** (2019) 012025 doi:10.1088/1757-899X/699/1/012025

**Acknowledgements**
The reported study was funded by Russian Scientific Foundation according to the research project No. 18-12-00260. The authors from Kazan Federal University acknowledge support by the Russian Government Program of Competitive Growth of Kazan Federal University.

**References**

[1] Ohtomo A and Hwang H Y 2004 *Nature* **427** 423-6
[2] Niranjan M K, Wang Y, Jaswal S S and Tsymbal E Y 2009 *Phys. Rev. Lett.* **103** 016804
[3] Thiel S, Hammerl G, Schmehl A, Schneider C W and Mannhart J 2006 *Science* **313** 1942-5
[4] Reyren N, Thiel S, Caviglia A D, Kourkoutis L, Hammerl G, Richter C, Schneider C W, Kopp T, Ruetschi A-S, Jaccard D, Gabay M, Muller D A, Triscone J-M and Mannhart J 2007 *Science* **317** 1196-9
[5] Brinkman A, Huijben M, Van Zalk M, Huijben J, Zeitler U, Maan J C, van der Wiel W G, Rijnders G, Blank D H A and Hilgenkamp H 2007 *Nat. Mater.* **6** 493-6
[6] Kalabukhov A, Gunnarsson R, Borjesson J, Olsson E, Claeson T, Winkler D 2007 *Phys. Rev. B* **75** 121404
[7] Moetakef P, Cain T A, Ouellette D G, Zhang J Y, Klenov D O, Janotti A, Van de Walle Ch G, Rajan S, Allen S J and Stemmer S 2011 *Appl. Phys. Lett.* **99** 232116
[8] Jackson C A and Stemmer S 2013 *Phys. Rev. B* **88** 180403
[9] Biscaras J, Bergeal N, Kushwaha A, Wolf T, Rastogi A, Budhani R C and Lesueur J 2010 *Nat. Commun.* **1** 89
[10] Nakagawa N, Hwang H Y and Muller D A 2006 *Nat. Mater.* **5** 204-9
[11] Fredrickson K D and Demkov A A 2015 *Phys. Rev. B* **91** 115126
[12] Hohenberg P and Kohn W 1964 *Phys. Rev.* **136** B864-71
[13] Perdew J P, Burke K and Ernzerhof M 1996 *Phys. Rev. Lett.* **77** 3865-8
[14] MedeA®-2.20 2015 Materials Design, Inc., San Diego, CA, USA
[15] Kresse G and Furthmuller J 1996 *Comp. Mater. Sci.* **6** 15-50
[16] Dudarev S L, Botton G A, Savrasov S Y and Humphreys C J 1998 *Phys. Rev. B* **57** 1505-9
[17] Piyanzina I I, Kopp T, Lysogosrkiy Yu V, Tayurskii D and Eyert V 2017 *J. Phys.: Condens. Mat.* **29** 095501
[18] Pavlov D P, Piyanzina I I, Muhortov V I, Balbashov A I, Tauyrskii D A, Garifullin I A and Mamin R F 2017 *JETP Lett.* **106** 460-4
[19] Mukhortov V M, Golovko Y I, Tolmachev G N and Klevtzov A 2000 *Ferroelectrics* **247** 75-83

IOP Conf. Series: Materials Science and Engineering **699** (2019) 012026 doi:10.1088/1757-899X/699/1/012026

# Tailoring quasi-two-dimensional high conductivity and superconductivity areas at the interfaces of ferroelectric/dielectric heterostructures

**R F Mamin[1,2], D P Pavlov[1], I I Piyanzina[1,2], Yu V Lysogorskiy[1,2], V M Mukhortov[3], T Adachi[4], T Kawamata[5], Y Koike[5], A M Balbashev[6], D A Tayurskii[2] and V V Kabanov[1,7]**

[1]Zavoisky Physical-Technical Institute, FRC KazanSC of RAS, 420029 Kazan, Russia
[2]Kazan Federal University, 420008 Kazan, Russia
[3]Southern Scientific Center, Russian Academy of Sciences, 344006 Rostov-on-Don, Russia
[4]Department of Engineering and Applied Sciences, Sophia University, 102-8554 Tokyo, Japan
[5]Department of Applied Physics, Tohoku University, 980-8579 Sendai, Japan
[6]Moscow Power Engineering Institute, 111250 Moscow, Russia
[7]Department for Complex Matter, Jozef Stefan Institute, 1000 Ljubljana, Slovenia

mamin@kfti.knc.ru

**Abstract**. The results of the investigations of high conducting and superconducting areas at the interfaces between ferroelectric oxide and insulating oxide are presented. The numerical simulations of $BaTiO_3/LaMnO_3$ and $BaTiO_3/La_2CuO_4$ heterostructures have been performed. It is found that in the samples of the $Ba_{0.8}Sr_{0.2}TiO_3/LaMnO_3$ heterostructure the electrical resistance decreases significantly and exhibits metallic behaviour at low temperatures for the case when the *c* axis of ferroelectric film is directed along the normal to the surface of the single crystal. The superconducting behaviour with transition temperature $T_c$ about 30K has been found at the interface of the $Ba_{0.8}Sr_{0.2}TiO_3/La_2CuO_4$ heterostructure. The proposed concept promises the ferroelectrically controlled interface conductivity and superconductivity.

## 1. Introduction

Unique properties of functional materials are achieved due to the effects associated with the complex composition of the interface structure. New materials include heterointerfaces between two insulating oxides, in which unique transport properties are observed because of strong electronic correlations. A high-mobility electron gas was discovered at the interface between two oxide insulators $LaAlO_3$ (LAO) and $SrTiO_3$ (STO) in Ref. [1]. It has attracted significant attention [1-10] due to a wide range of physical phenomena observed in this system. This type of heterointerfaces was comprehensively studied. In particular, it was found that the metallic phase (quasi-two-dimensional electron gas, q2DEG) was formed in the STO layers at the LAO/STO interface, when the number of LAO layers exceeded three [2]. The density of the charge carriers in this heterostructure reached the value of 3 $10^{13}$ $cm^{-2}$. Subsequently, the coexistence of a two-dimensional superconductivity and ferromagnetism was discovered in this system [3,4]. Such a system underwent a transition to a superconducting state at temperatures below 300 mK [3,4]. The creation of high-$T_c$ quasi-two-dimensional superconductivity at the interface was impossible up to now without tailoring the atomically perfect interfaces [11,12].

The polarization discontinuity at the interface is usually discussed [1,2,9,10] in order to describe the q2DEG in the LAO/STO heterostructure. Along the [001] direction, LAO slab represents an alternation of the differently charged layers of $(LaO)^{+1}$ and $(AlO_2)^{-1}$. As it was shown experimentally, in the heterostructure with the $TiO_2$ interface layer the electric potential along the [001] direction appears due to the polarity disruption at the interface. Thus, the atomically flat quality of the interface between two components is utterly necessary since the effect is related to the strictly defined sequence of layers inside each slab. In this work, we study resistivity properties of two kinds of heterostructures: $Ba_{0.8}Sr_{0.2}TiO_3/LaMnO_3$ (BSTO/LMO) and $Ba_{0.8}Sr_{0.2}TiO_3/La_2CuO_4$ (BSTO/LCO). In the first case, we use $LaMnO_3$ (LMO) as a substrate, because antiferromagnetic LMO might be transferred to ferromagnetic state by increasing the concentration of free carriers. In the second case, we use $La_2CuO_4$ (LCO) as a substrate, because antiferromagnetic LCO might be transferred to a superconducting state by increasing the concentration of free carriers. The ferroelectric oxide $Ba_{0.8}Sr_{0.2}TiO_3$ (BSTO) was used as an overlayer in both heterostructures. For numerical simulations, we use $BaTiO_3$ instead of BSTO, because both materials have similar properties, but the simulations for $BaTiO_3$ are less complicated. As a result, we show that it is possible to use relatively simple techniques of sputtering of ferroelectric films in order to create q2DEG, which exhibits high conductivity and superconductivity at relatively high temperatures. It opens an opportunity to study the switching of such states by changing the polarization direction in the ferroelectric film.

## 2. Results and discussion

The numerical simulations [13-17] of the structural and electronic characteristics of the $BaTiO_3/LaMnO_3$ (BTO/LMO) and $BaTiO_3/La_2CuO_4$ (BTO/LCO) ferroelectric-antiferromagnet heterostructures have been performed. For density of states calculations and structural optimization, we use density functional theory [13]. Exchange and correlation effects were accounted by generalized gradient approximation [14]. Kohn-Sham equations were solved using the plane-wave basis set (PAW) [15] realized within the VASP code [16], which is a part of the MedeA® software of Materials Design [15]. We use the cut-off energy 400 eV. The force tolerance was 0.5 eV/nm and the energy tolerance for the self-consistency loop was $10^{-5}$ eV. The Brillouin zones were sampled including 5×5×1 k-points. Since there are strong correlations between *d* and *f*-electrons the GGA+*U* correction were included to computational scheme [17]. The *U* parameter was added to La 4*f*, Ti 3*d*, and Mn 4*d* orbitals ($U$ = 8 eV, 2 eV, and 4 eV, respectively). The choice of *U* for Ti and La atoms is based on Ref. [18]. Due to a high mismatch between BTO and LMO or LCO (≈30 %), the BTO unit cell was rotated by a 45° before merging with LMO and LCO substrate. Because the substrate is compressive with respect to the film, the polarization of the ferroelectric BTO is directed perpendicular to the interface. Thus, the polarization has a discontinuity near the interface. Therefore, uncompensated charge is created at the interface on the ferroelectric side. This charge is screened by conduction charged carriers in thin layer on the surface of the substrate, forming the q2DEG state in this heterostructure.

The density of states of BTO/LMO heterostructure for the case when two layers of BTO were deposited on top of LMO substrate is shown in Figure 1. It is obvious that even two layers of BTO result in 0.3 eV band gap with a finite occupation of states near the Fermi-level. That behaviour is a characteristic of a weak metal. When number of BTO layers becomes larger than two, the band gap vanishes, and we expect the system to be a good metal at the interface. The density of states of BTO/LCO heterostructure for the case when two layers of BTO were deposited on top of LCO substrate is shown in Figure 2. The density of states of BTO/LCO heterostructure at the Fermi-level becomes also sufficiently large, approximately 5 $eV^{-1}$. The band gap vanishes, and we expect the system to behave as a good metal at the interface. Thus, BTO/LCO heterostructure can be superconductive.

The temperature dependence of the electrical resistance has been studied for heterostructures formed by antiferromagnetic LMO and LCO single crystals with epitaxial films of ferroelectric BSTO deposited on them by reactive sputtering of stoichiometric targets using RF plasma (RF-sputtering) method [19,20]

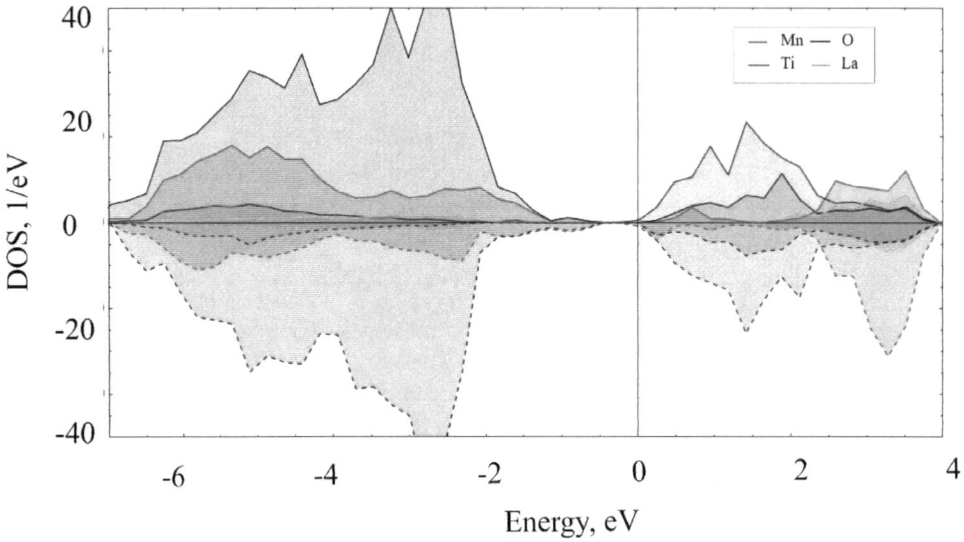


**Figure 1.** Density of states of $BaTiO_3/LaMnO_3$ heterostructure for the case of 2 $BaTiO_3$ layers on the top of $LaMnO_3$. Solid and dashed curves correspond to spin-up and down components.

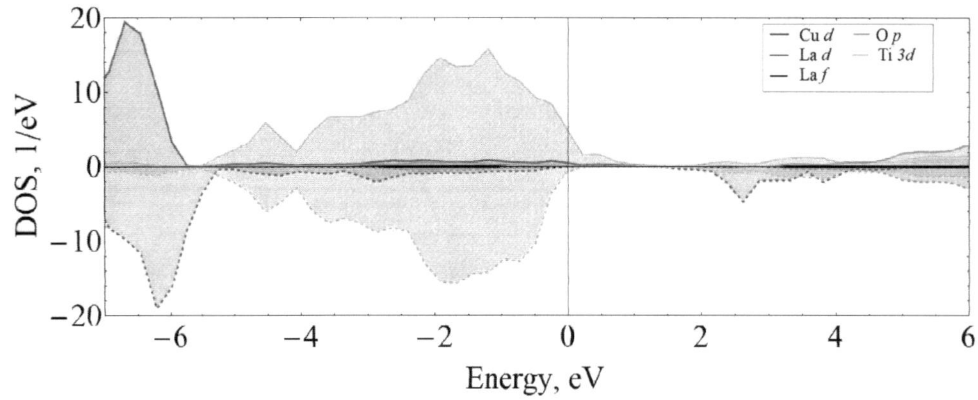


**Figure 2.** Density of states of $BaTiO_3/La_2CuO_4$ heterostructure for the case of 2 $BaTiO_3$ layers on the top of $La_2CuO_4$. Solid and dashed curves correspond to spin-up and down components.

at 650°C. Here, we present the results of investigations of q2DEG in the interfaces between ferroelectric oxide and insulating oxide in BSTO/LMO and BSTO/LCO heterostructures.

The results of electrical resistance measurement of the BSTO/LMO heterostructure are presented in Figure 3. The measured resistance was compared to that exhibited by LMO single crystal without the film. It is found that in the samples with *c*-axis of the ferroelectric film perpendicular to the interface the resistance decreases significantly with lowering the temperature below 165 K, exhibiting metallic behaviour (Fig. 3). But in the samples with the c-axis of the ferroelectric film parallel to the interface the q2DEG is not observed [19]. Thus, the transition to the state with q2DEG at the interface, which is connected to a sharp change in the perpendicular component of the polarization at the interface, is demonstrated. When the magnetic field was applied perpendicularly to the interface, the resistance of BSTO/LMO interface started to change irreversibly. First steps of such changes are shown in Figure 3 (blue triangles). When the magnetic field was applied parallel to the surface, the g2DEG state disappeared. Thus, it is shown that the magnetic field effect on the resistivity of the BSTO/LMO interface is very strong.

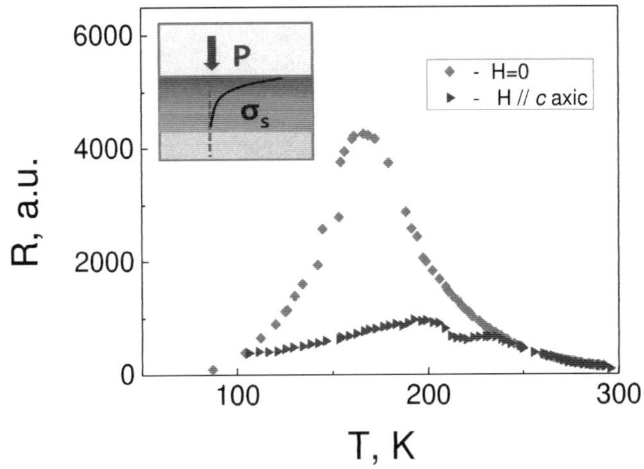


**Figure 3.** Temperature dependences of the resistance R(T) for $Ba_{0.8}Sr_{0.2}TiO_3/LaMnO_3$ heterostructure before and after application of the magnetic field: red diamonds show R(T) before application of magnetic field and blue triangles - after few cycles of application of magnetic field up to 1 T. Space distribution of surface charge near the interphase is shown in the insert.

The results of resistance measurements of the BSTO/LCO heterostructure are shown in Figure 4. The temperature dependence of the resistance in the wide temperature range shows that above 50 K it has semiconducting behaviour. At temperatures above 50K the main current flows through the substrate. The temperature dependence of the resistivity of LCO crystal is shown in Figure 4 by blue triangles. Below 50 K, the current flows mainly in the interface region. At low temperatures, the resistance decreases rapidly and superconducting behaviour is observed. It is confirmed by the observation of diamagnetic response in the susceptibility measurements in the BSTO/LCO heterostructure. The beginning of the transition to the superconducting state occurs around 40 K, similar to that observed in optimally doped $La_{2-x}Sr_xCuO_4$ single crystals [11]. Thus, the interface between the ferroelectric and insulating oxides shows superconducting behaviour with $T_c$ higher than in LAO/STO heterostructure.

These results allow relatively simple interpretation. The abrupt change of the polarization $P$ at the interface leads to the formation of the surface charge with the density $\sigma_s = P_n$, where $P_n$ is the component of polarization perpendicular to the interface (see insert in Fig. 3). This charge must be screened; therefore, the charged layer forms at the opposite side of the interface. This charged layer may be considered as a surface doping. The characteristic charge density in this layer can be estimated as $n_s = \sigma_s/e = P_n/e$, where $e$ is the elementary charge. If we assume characteristic value of polarization as is in the bulk of ferroelectric ($P_n \approx 30$ μC/cm), we obtain $n_s \approx 10^{14}$ cm$^{-2}$. We estimate that the characteristic thickness of this layer is about 1-2 nm.

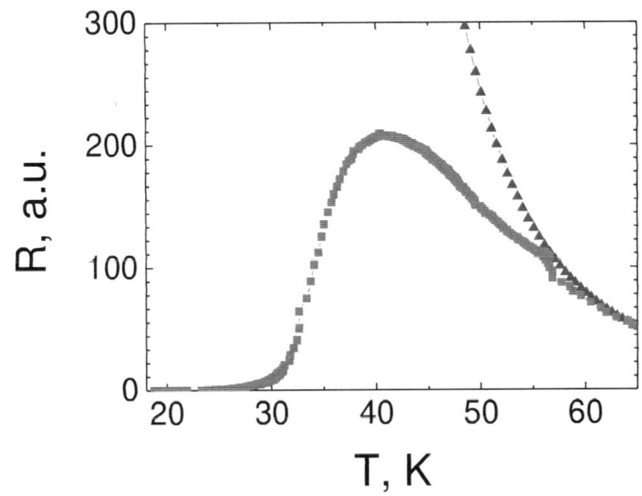


**Figure 4.** Temperature dependences of the resistance R(T) for $Ba_{0.8}Sr_{0.2}TiO_3/La_2CuO_4$ heterostructure for polarization directed perpendicular to ferroelectric film (red squares). Blue triangles show R(T) for $La_2CuO_4$ single crystal without film.

## 3. Conclusion

In conclusion, the transition to q2DEG state in BSTO/LMO heterostructure was found below 165 K in the samples with polarization of the ferroelectric film perpendicular to the interface. It was shown that the resistance of the interface changed irreversibly, when the magnetic field was applied perpendicular to the interface. BSTO/LCO heterostructure was found to be superconductive. We showed that quasi-two-dimensional highly conductive and superconductive layers might be realized at the ferroelectric/dielectric interfaces obtained by relatively simple preparation techniques.

## Acknowledgements

The reported study was funded by Russian Scientific Foundation, research project No. 18-12-00260. The authors from Kazan Federal University acknowledge partial support by the Russian Government Program of Competitive Growth of Kazan Federal University.

## References

[1] Ohtomo A and Hwang HY 2004 *Nature* **427** 423-6
[2] Thiel S, Hammerl G, Schmehl A, Schneider C W and Mannhart J 2006 *Science* **313** 1942-5
[3] Reyren N, Thiel S, Caviglia A D, Fitting Kourkoutis L, Hammerl G, Richter C, Schneider C W, Kopp T, Ruetschi A-S, Jaccard D, Gabay M, Muller D A, Triscone J-M and Mannhart J 2007 *Science* **317** 1196-9
[4] Brinkman A, Huijben M, Van Zalk M, Huijben J, Zeitler U, Maan J C, van der Wiel W G, Rijnders G, Blank D H A and Hilgenkamp H 2007 *Nat. Mater.* **6** 493-6
[5] Kalabukhov A, Gunnarsson R, Borjesson J, Olsson E, Claeson T and Winkler D 2007 *Phys. Rev. B* **75** 121404
[6] Niranjan M K, Wang Y, Jaswal S S and Tsymbal E Y 2009 *Phys. Rev. Lett.* **103** 016804
[7] Jackson C A and Stemmer S 2013 *Phys. Rev. B* **88** 180403
[8] Biscaras J, Bergeal N, Kushwaha A, Wolf T, Rastogi A, Budhani R C and Lesueur J 2010 *Nat. Commun.* **1** 89
[9] Nakagawa N, Hwang H Y and Muller D A 2006 *Nat. Mater.* **5** 204-9
[10] Fredrickson K D and Demkov A A 2015 *Phys. Rev. B* **91** 115126
[11] Gozar A, Logvenov G, Kourkoutis L F, Bollinger A T, Giannuzzi L A, Muller D A and Bozovic I 2008 *Nature* **455** 782-5
[12] Gozar A and Bozovic I 2016 *Physica C* **521-522** 38-49
[13] Hohenberg P and Kohn W 1964 *Phys. Rev.* **136** B864-71
[14] Perdew J P, Burke K and Ernzerhof M 1996 *Phys. Rev. Lett.* **77** 3865-8
[15] MedeA® -2.20 2015 Materials Design, Inc., San Diego, CA, USA
[16] Kresse G and Furthmuller J 1996 *Comp. Mater. Sci.* **6** 15-50
[17] Dudarev S L, Botton G A, Savrasov S Y and Humphreys C J 1998 *Phys. Rev. B* **57** 1505-9
[18] Piyanzina I I, Kopp T, Lysogosrkiy Yu V, Tayurskii D and Eyert V 2017 *J. Phys.: Condens. Mat.* **29** 095501
[19] Pavlov D P, Piyanzina I I, Muhortov V I, Balbashov A I, Tauyrskii D A, Garifullin I A, Mamin R F 2017 *JETP Lett.* **106** 460-4
[20] Mukhortov V M, Golovko Y I, Tolmachev G N and Klevtzov A 2000 *Ferroelectrics* **247** 75-83

IOP Conf. Series: Materials Science and Engineering **699** (2019) 012027 doi:10.1088/1757-899X/699/1/012027

# In-situ Raman mapping during indentation

**I I Maslenikov and A S Useinov**

Technological Institute for Superhard and Novel Carbon Materials, Troitsk, Moscow, 142190, Russia

i.i.maslenikov@gmail.ru

**Abstract.** Transparent diamond tip described elsewhere allows combining mechanical indentation measurement with optical methods, including not only surface observation, but also spectroscopy measurements, in particular, Raman spectroscopy. Current work considers the possibilities of creating Raman maps, which give an information about the pressure distribution and ratio of material phases. Corresponding maps are presented for the case of indentation of DLC film on a silicon sample.

## 1. Introduction

Transparent diamond tip was invented for the instrumented indentation tester "NanoScan-4D" – a device that conducts a measurement of hardness and elastic modulus by means of instrumented indentation [1]. In contrast with traditional indentation, this method relies on force-displacement data, rather than force and direct area values, which allows conducting measurements faster and with a smaller depth. However, on the other hand, the method neglects a "pile-up" and "sink-in" effects, which are taken into account when direct optical imprint area observation is used. The transparent tip allows observing the sample's surface in the real-time regime: before, during, and after the indentation, which allows measuring the imprint area without additional movement of the sample to the optical microscope. Thus, a direct imprint observation becomes faster and the area defined by means of instrumented indentation can be corrected with optical image for "pile-up" effects.

Naturally, the transparent tip can be used not only for the residual imprint area measurements, but also for the location of indentation, measurement of surface deformation and crack propagation during the scratch and, which is probably more straightforward, for the spectroscopy measurement. In order to be more consistent in description of corresponding measurement, one must take into account the tip shape. A detailed explanation is presented in Ref. [2], a concise description can be given as follows. A diamond cylinder is faceted in the way that its both ends have the Berkovich-pyramid shape. Pyramids' heights are parallel to each other and lie along the same line. Pyramids are symmetrical with respect to the indenter's centre so that each face of a given pyramid has a parallel face on the opposite end. Described configuration of the faces allows each incident beam of light that propagates parallel to the pyramid's height and falls into the indenter in the vicinity of the tip to go through the indenter and propagate parallel to the previous direction with a shift in the perpendicular direction only. The shift value as well as the visible area size are dependent on the indenter's height [2].

Therefore, each image directly observed through the tip consists of three sectors and the connected picture can be obtained via their simple linear translation. All maps in the following sections are presented for only one sector, the rest parts of the image differ only due to the indenters manufacturing shortcomings.

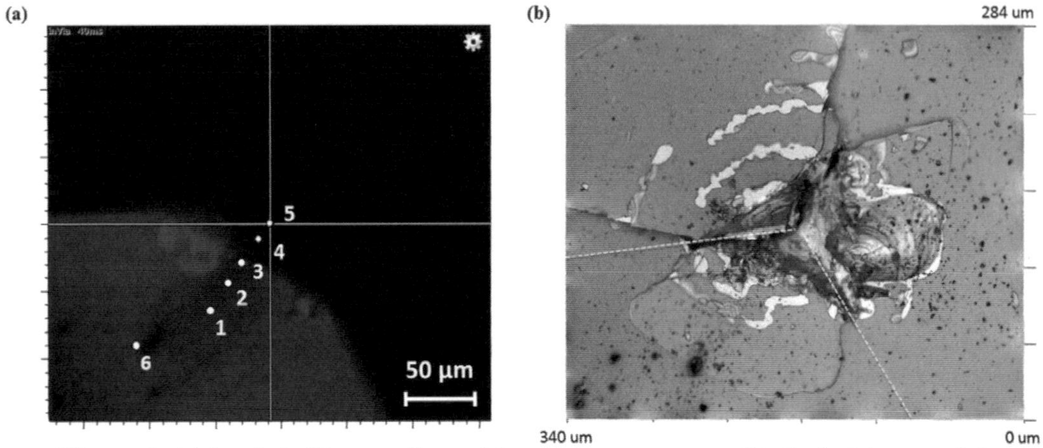

**Figure 1.** (a) Loaded silicon surface with measurement spots. (b) Optical image of the surface after the measurements.

The ability of the tip to transmit the light can be used to combine mechanical and spectroscopy measurements [3]. Similar simultaneous opto-mechanical measurements were known before, however, they didn't make use of the light passing through the tip. Work [4] considers an indentation of the transparent sample when spectra are collected from the bottom side of the sample, work [5] describes an indentation near the sample's edge with the collection of spectra in perpendicular direction. Current approach, which is described in [3], isn't limited to near-edge indentation and can be used for the opaque samples.

## 2. Experiment conditions

Raman mapping was performed using specially designed unit [3], that fits into the Renishaw InVia Raman spectroscope and allows applying the load by means of the stepper motor actuator and measuring the force acting on the indenter by means of the load cell. A silicon covered with a ~200 nm thick DLC coating was chosen as a sample for investigation. An indentation was performed up to the force of 40 N with further unloading up to the force of 30 N. When the load was achieved, the spectra were collected. A 514 nm laser, 3000 l/mm grating, and 20x objective were used. In the first experiment, five spectra were collected along the one of visible sectors diagonal, these points are shown in the Figure 1. Collection time was 60 s, each spectrum was collected five times and then averaged. In the second experiment, mapping over the visible sector's area was performed, the map contains 28x25 pts with 10 µm interpoint step. In each point 30 spectra were collected and averaged. Each spectrum had an acquisition time of 2 seconds. Optical image of the surface after the full unloading presented in the Figure 1b was obtained with the help of Sensofar S NEOX optical microscope in multicolour regime by stitching multiple captures in-focus regions.

## 3. Results

Figure 1a represents points of spectra collected in the first experiment, the picture was taken at the load of 30 N, after the unloading from 40 N load. In this sector, one can see black points in the left bottom corner, which correspond to the more evident debris seen in the Figure 1b, after the unloading. This picture also shows a number of cracks that occur mostly during the unloading cycle of the indentation. The section of Figure 1b, corresponding to the section visible in the Figure 1a is highlighted with yellow lines.

Figure 2b shows an example of spectrum collected during the mapping at the point 6 (Fig. 1a). This spectrum shows three peaks, which correspond to diamond (1332 $cm^{-1}$), DLC (1420 $cm^{-1}$), and Si-I (520 $cm^{-1}$). Spectra collected at the points 1-5 are shown in the Figure 2a. The detected frequencies are 50-900 $cm^{-1}$, that range was chosen to resolve peaks corresponding to the different phases of Si. The

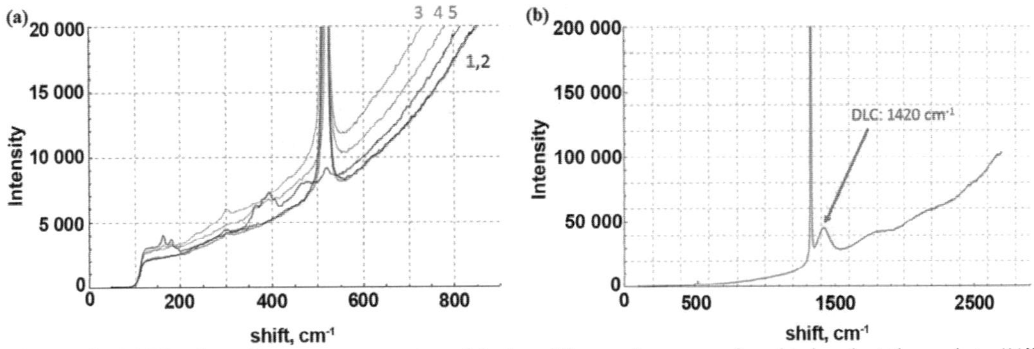


**Figure 2.** (a) The low-frequency spectra with the silicon phases peaks obtained at the points "1" to "5". (b) A full range spectrum obtained as a part of the mapping.

spectra are quite similar to the results presented in Ref. [3]. Each of them has an intensive peak that corresponds to Si-I phase (520 $cm^{-1}$). The range of intensities is limited to 20000 to underline the evidence of weaker peaks that correspond to the different phases. However, we can tell that, as it is partially seen from the Figure 2b, the most intensive peak 520 $cm^{-1}$ corresponds to the point "3", then intensities descend in the following order "2", "1", "4", and "5". Such behaviour can be most likely attributed to the following factors. Points "1" and "2" are located in the regions, where DLC coating is still present on the surface, points "4" and "5" are located in the region, where silicon bears a load and, since the unloading has already partially occurred, this silicon is transformed to different phases. Point "3" is located right at the edge of the indent – in the region where DLC coating is cracked and is likely to be at least partially chipped out, while uncovered silicon bears small load and thus significant part of the signal corresponds to the Si-I phase. Consideration presented above can be used to the explanation of the largest 520 $cm^{-1}$ peak intensity at the point "3". The peaks in the range below 520 $cm^{-1}$ can be attributed to Si phases as it is done in the work [3]. Peaks at 181 and 300 $cm^{-1}$ correspond to the a-Si, a peak at 161 $cm^{-1}$ corresponds to Si-III phase, while a "pile" around ~390 $cm^{-1}$ corresponds to the mixture of Si-III and Si-II phases [3].

Obtained spectra can be used for the properties mapping. Using the shift of the centre of the DLC line, one can obtain the pressure map. In order to obtain the corresponding picture, each spectrum was approximated by a fifth power polynomial in the range of frequencies from 1380 to 1600 $cm^{-1}$, then the argument of the fitted line maximum was used as a value that corresponds to the local pressure. Obtained map is shown on the Figure 3a. As can be seen from the picture the maximum pressure is observed in the contact region, in particular below the pyramid edges. Another feature seen on the figure is the low pressure in the sector centre, which can be attributed to the delamination of the DLC from the silicon.

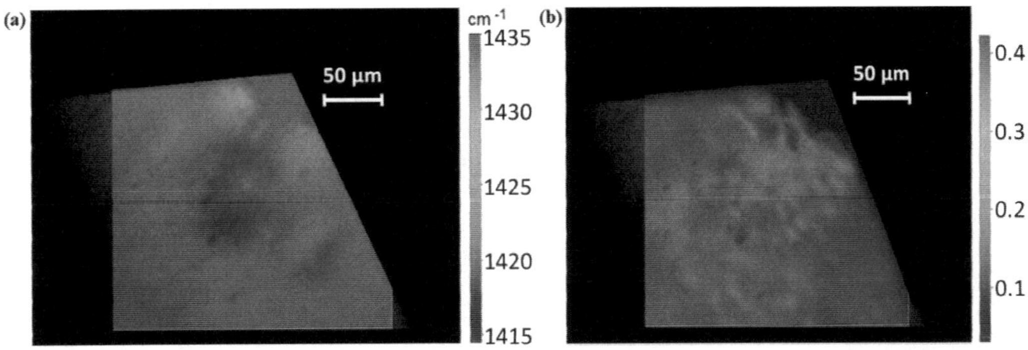


**Figure 3.** (a) Distribution of pressure (by the DLC-peak frequency). (b) Si-I to DLC peaks intensities ratio map.

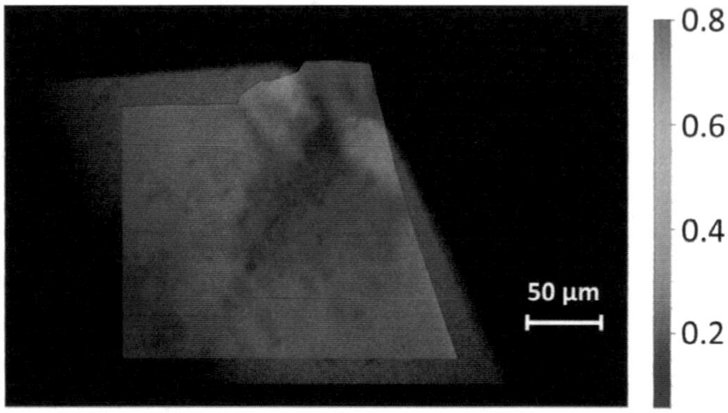


**Figure 4.** Intensities ratio I(340-410 $cm^{-1}$)/I(510-545 $cm^{-1}$) map.

Another example of the map, which represents the ratio of Si-I to DLC is shown in the Figure 3b. The map was obtained by dividing the maximum intensity within the range 510-540 $cm^{-1}$ to the maximum intensity within the range 1400-1460 $cm^{-1}$. The picture has a peak that is likely to be caused by the chipped Si crumbs lying over the DLC surface.

The map showing the ratio of maximum intensity within 340–410 $cm^{-1}$ to the maximum intensity within 510–545 $cm^{-1}$ frequency range is shown in the Figure 4. That map can be qualitatively and roughly attributed to the amount of Si-I transformed to the different phases. The map has a background level of 0.06 that corresponds to the case when there are no observable peaks within 340-410 $cm^{-1}$ range. Regions near the sector's edges are excluded from the picture as they have either a very small Si-I peak or no peaks within 340-410 $cm^{-1}$ range. Thus, these near-edge areas have phases ratio value about 0.6, which is highly caused by artefacts and, therefore, is inappropriate.

## 4. Conclusions

Presented results show the possibility to conduct an in-situ Raman mapping during the indentation measurements. It has been shown that combining mechanical testing with optical structure investigation is useful for local material characterisation. These methods implemented in a single instrument can be particularly useful for the heterogeneous materials and the materials that undergo the phase transformations.

## Acknowledgements

This study was supported by the Ministry of Science and Higher Education of the Russian Federation within the framework of the Agreement No. 14.577.21.0274 (EB 075-15-2019-024, the unique identifier of the project is RFMEFI57717X0274).

## References

[1] Oliver W C and Pharr G M 2004 Measurement of hardness and elastic modulus by instrumented indentation: Advances in understanding and refinements to methodology *J. Mater. Res.* **19** 3-20

[2] Maslenikov I I, Reshetov V N, Useinov A S and Doronin M A 2018 In situ surface imaging through a transparent diamond tip *Instrum. Exp. Tech.* **61** 719-24

[3] Maslenikov I I, Reshetov V N and Useinov A S 2019 Raman spectroscopy through the indenter working as an optical objective *Mater. Trans.* **60** 1433-5

[4] Gerbig Y B, Michaels C A, Forster A M, Hettenhouser J W, Byrd W E, Morris D J and Cook R F 2012 Indentation device for in situ Raman spectroscopic and optical studies *Rev. Sci. Instrum.* **83** 125106

[5] Manimunda P, Hintsala E, Asif S and Mishra M K 2017 Mechanical anisotropy and pressure induced structural changes in piroxicam crystals probed by in situ indentation and Raman spectroscopy *JOM* **69** 57-63

IOP Conf. Series: Materials Science and Engineering **699** (2019) 012028 doi:10.1088/1757-899X/699/1/012028

# Thromboresistant properties of nanostructured tantalum coatings on the stainless steel surfaces

**G B Melnikova[1], T A Kuznetsova[1], A S Petrovskaya[1], T N Tolstaya[1], V A Lapitskaya[1], S A Chizhik[1], A Zykova[2,3] and V Safonov[2,3]**

[1]A.V. Luikov Heat and Mass Transfer Institute of National Academy of Sciences of Belarus, 220072 Minsk, Belarus
[2]National Science Centre «Kharkov Institute of Physics and Technology», Kharkov, Ukraine
[3]Department of Physical Technologies, Kharkov National University, Kharkov, Ukraine

galachkax@gmail.com

**Abstract**. The nanostructured coatings of tantalum and its compounds on the stainless steel surfaces were investigated by atomic force microscopy. Their hydrophilic properties were studied by sessile drop and atomic force spectroscopy methods. The thromboresistant properties of samples were analysed *in vitro* by incubation of the samples in the rich platelets plasma of patients with acute coronary syndrome. The effect of surface properties on the thromboresistant properties was analysed.

## 1. Introduction

The development of metal stents is one of the main advances in the interventional cardiology. The requirements for stents are quite strong: high flexibility, plasticity, strength and rigidity, X-ray contrast, and biocompatibility. They should not be destroyed by surrounding tissue, induce tapping of electrolyte balance, toxicity, allergical, cancinogenic and immune reactions and thrombosis, disorders of cellular elements of blood, denaturation of plasma protein and enzymes of blood. For this purpose, the stent surfaces are covered with various types of ceramic coatings to prevent changing their properties. The biogrowth and thrombosis processes can occur on the surface of implants. For modern stents, both polymeric and metallic materials (steel or titanium and chromium alloys) are used. To improve materials performance, their surface is modified by different coatings [1]. One of promising coatings is tantalum, due to its good biocompatibility, mechanical strength, corrosion resistance, and radiopacity [2,3].

Modern methods of research, such as NMR-spectroscopy, scanning microscopy, and others allow receiving the data on the surface properties of materials [1-6]. Atomic force microscopy (AFM) gives the versatile characterization of the surface and at the same time can simultaneously examine and quantify different surface phenomena, such as roughness [7], adhesion [8,9], and morphological features after different treatment [10]. The study of mechanical properties of tantalum films and evaluation of their operational properties is an important task, because the surface of stents is often subjected to chemical and mechanical effects.

The work is aimed at studying the aggregation properties of thrombocytes before and after incubation with nanostructured coatings of tantalum and its compounds on the stainless steel substrates using AFM.

IOP Conf. Series: Materials Science and Engineering **699** (2019) 012028 doi:10.1088/1757-899X/699/1/012028

## 2. Experimental

The coatings were deposited using Direct Current planar magnetron sputtering. Before experiment, steel substrates of the 316 LSS were cleaned in an ultrasonic bath by the standard method and by ion beam of medium energies of ion source in crossed field of the Radical type in argon atmosphere (pressure is $5 \cdot 10^{-4}$ Torr, ion acceleration voltage – 3 kW, ion source current – 100 mA) during 5 min. Then stainless steel substrates were put into the camera, pumped to the vacuum less than $10^{-5}$ Torr, and Ta, $Ta_2O_5$, TaN, and TaON coatings were deposited on them. The distance between the magnetron and the sample was near 30 cm. The feature of this system is additional oxygen activation by discharge plasma source induction. Oxygen was puffed through plasma source for activation. For TaN deposition, only nitrogen was puffed without argon. The deposition parameters are presented in Table 1.

The surface microstructure was investigated by AFM method on NT-206 device (Belarus) with CSC 38 cantilevers (Mikromasch, Estonia). The stiffness of cantilevers was 0.03 N/m according to the manufacturer's passport, the radius of curvature was less than 10 nm. The adhesion characteristics were studied on Dimension FastScan (Bruker, USA) in PeakForce QNM regime with CSG10_SS cantilevers (Mikromasch, Estonia) with the curvature radius of 1.14 nm and stiffness of 0.365 N/m. Roughness values were determined by AFM method. The adhesion force was determined on the base of force-distance curves. The surface free energy at the interface between the probe and the sample was calculated by the formula for elastic adhesive contacts of the parabolic profile probe.

The hydrophilic properties were investigated by sessile drop method on the device DSA 100 E (KRUSS, Germany) by three test-liquids: water, glycerol, and diiodomethane. Then the free surface energy was calculated by the Jirifalko-Hood-Fawkes equation [11].

To determine the thromboresistant properties, the samples of steel with tantalum oxynitride coatings were cleaned with ethanol ($w_{mass} = 96\%$) and kept in platelet rich plasma during 15 min, obtained from the stabilized sodium citrate venous blood of patients with acute coronary syndrome. Then the samples were cleaned with phosphate buffer from non-adherent cell and dried in air. After that the optical microscopy (MICRO 200, Belarus) and atomic force microscopy (NT-206, Belarus) measurements were performed. The aggregation degree was estimated according to three morphological classes: (a) few activated (spherical and with pseudopodia); (b) highly activated (completely flattened); (c) aggregates. This classification was proposed in the article [12].

**Table 1.** The parameters of the magnetron sputtering for the coatings of tantalum and its compounds deposited on the steel substrates.

| Type of coating | $P_{Ar}$ (Torr) | $U_m$ (V) | $I_m$ (A) | Deposition time (min) | $Q$ (cm$^3$/min) |
|---|---|---|---|---|---|
| Ta | $8 \cdot 10^{-4}$ | 495 | 6.6 | 30 | - |
| $Ta_2O_5$ | $8 \cdot 10^{-4}$ | 500 | 6.4 | 20 | 25 ($O_2$) |
| TaN | $9 \cdot 10^{-4}$ ($N_2$) | 800 | 3.4 | 60 | 95 ($N_2$) |
| TaON | $11 \cdot 10^{-4}$ (general) | 620 | 4.0 | 30 | 55 ($O_2$) |

## 3. Results and discussion

### *3.1. Microstructure and properties of tantalum derivatives coatings on the steel surface*

Study of coatings structure of the tantalum and its compounds on the stainless steel 316 LSS by AFM showed that the polished steel surface had a cellular structure [13]. Cells were the surface areas rising above the main matrix. These parts might be phases with dopants, being harder than the rest matrix. Most of cells (90% of the total amount) have a diameter of 117 nm; the size of other formation is 61 nm. The AFM image of the initial steel and steel with the oxynitride tantalum coating is presented in Figure 1.

IOP Conf. Series: Materials Science and Engineering **699** (2019) 012028 doi:10.1088/1757-899X/699/1/012028

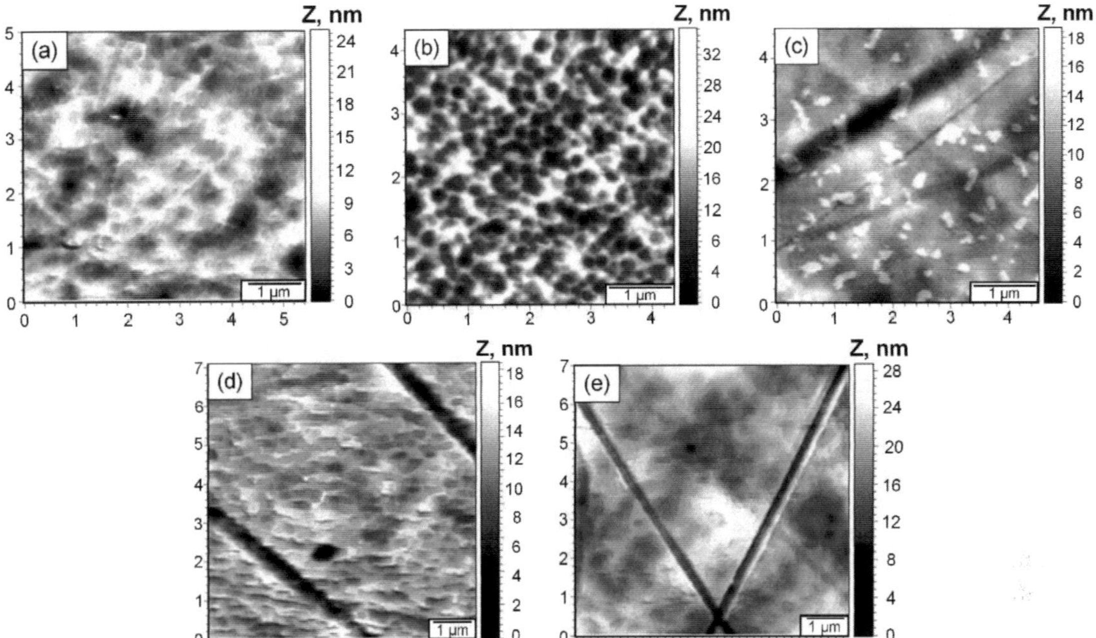


**Figure 1.** AFM image of (a) initial steel and steel with tantalum derivatives coatings: (b) tantalum; (c) tantalum oxide; (d) tantalum nitride; (e) tantalum oxynitride.

After the deposition of tantalum coating on the steel surface a cellular structure preserved (Fig. 1b), the values of Ra and Rq in 4x4 $\mu m^2$ scan increased, and in 7x7 $\mu m^2$ scan decreased, which indicates the increase in surface uniformity. This microstructure is rather typical for PVD coatings on the polished substrates [14]. The average cell size was 86.5 nm. The cellular surface of the tantalum coatings had more pronounced edges that bound the cells in comparison with the initial microstructure of steel. They formed even on the smallest cells and their height increased. Thereby, they played the role of crystallization centres, on which the tantalum atoms and clusters were initially fixed during magnetron sputtering. In the structure of tantalum oxide, the cells were not detected; the nature of the surface changed from a cellular structure to an islet. The height of the islands was 5.5 nm, and the size was from 100 to 500 nm. For the tantalum oxide coatings with the increase in the scanning area (Fig. 1c), the surface roughness Ra did not change, indicating the uniformity and equability of the applied coating.

For TaN (Fig. 1d) and TaON (Fig. 1e) coatings deposited onto the steel, the cellular structure also preserved. It should be noted that the presented protruding grains demonstrate predominant growth on the cells ledge. The cellular microstructure of TaON is significantly different from TaN. In this case, the cells were almost 2 times larger than those before deposition on the substrate. Rq values increased up to 11.4 in the case of TaN and up to 11.9 nm for TaON. The cells diameter for TaN was 96.3 nm and for TaON was 178.5 nm. The roughness values after tantalum coating deposition are presented in Table 2.

**Table 2.** The roughness values (Ra, Rq) of tantalum coatings on steel surface.

| Type of sample | Ra, nm | | Rq, nm | |
|---|---|---|---|---|
| | Scan area, $\mu m^2$ | | | |
| | 4 x 4 | 7 x 7 | 4 x 4 | 7 x 7 |
| Steel | 3.8 | 7.4 | 4.8 | 9.0 |
| Ta | 4.9 | 6.2 | 6.2 | 8.3 |
| $Ta_2O_5$ | 4.2 | 4.4 | 5.4 | 7.6 |
| TaN | 9.5 | 9.4 | 11.8 | 11.9 |
| TaON | 5.6 | 8.7 | 6.9 | 11.4 |

After the deposition of nanostructured tantalum coatings on the steel surface, as was shown in the article [15], all types of films, to a greater extent tantalum oxynitride, increased the hydrophobic properties of the surface, the contact angle increased from 75 to 94°. The tantalum and tantalum nitride coatings had the same hydrophilic surface properties, since they were characterized by the contact angle of 85°, and by the specific surface energies of 0.1 mJ/m$^2$, which was two times less than those for tantalum oxynitride coatings. The polar component of tantalum oxides surfaces was significantly reduced, and according to the data determined by two methods, they were characterized by pronounced hydrophobic properties.

*3.2. Thromboresistant properties of the nanostructured tantalum and its derivatives coatings*

It was shown, that on all of the researched samples platelets were adhered and aggregated to a greater or lesser extent, while their morphological characteristics significantly changed as compared with the initial state.

After incubating the samples of initial steel in platelet plasma, a large number of aggregates formed on the steel samples (Figs. 2a, 3a), while the diameter of an individual cells was 600-800 nm, which indicated about the exit of granules and the formation of clusters in the form of aggregates. From 2 to 5 platelets formed aggregates. We consider that process provided on the steel surface can be classified as the third class of platelet aggregation, described above.

According to the data of optical microscopy and AFM, a large number of pseudopodia are established on Ta coating. According to the AFM data, the granules with a diameter of 0.5 μm are presented; individual cells are spread over the surface of the samples. A large number of pseudopodia form a dense network (Figs. 2b, 3b). The thickness of the lamellas in the cross section is 400 nm. A reversible aggregation process is observed (class 2). These processes are due to hydrophilic properties of tantalum coating surfaces. The tantalum and TaN coatings (Figs. 2d, 3d) are characterized by similar properties. On the nanostructured surface of tantalum nitride, the degree of platelet aggregation is weak. In this case, platelets are stuck on the protrusions surface. According to the Table 2, the highest values of Rq = 11.6 nm were obtained for tantalum nitride. Platelets are unevenly adhered on the surface. As a rule, they are observed in hollows. The platelets behaviour can be classified as class 1.

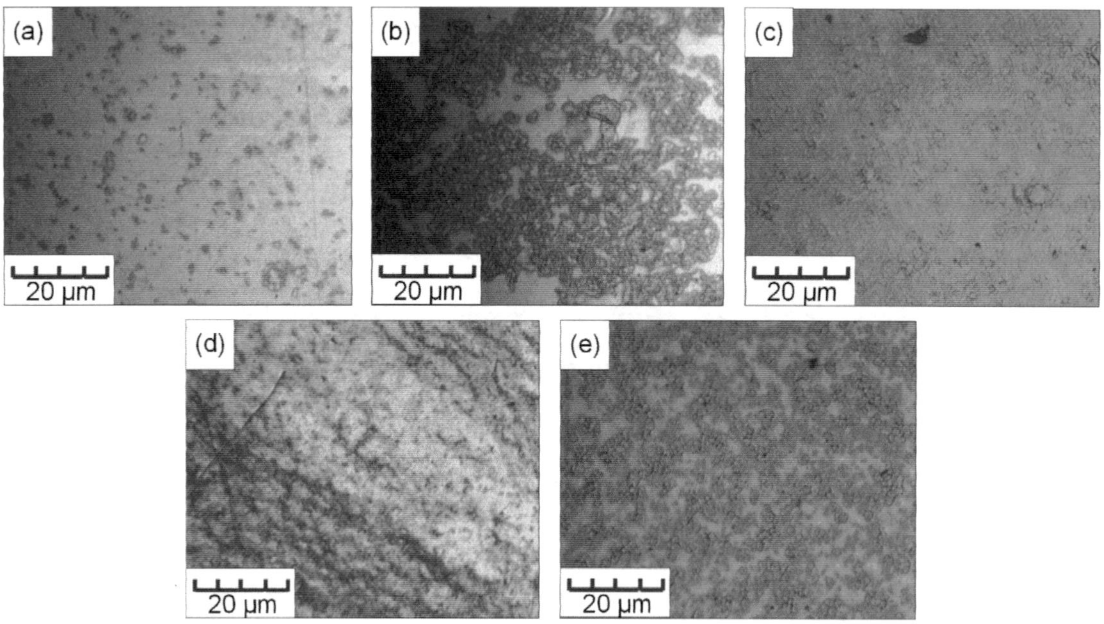


**Figure 2.** Optical images of steel surfaces with different coatings after the platelets aggregation: (a) initial steel; (b) tantalum; (c) tantalum oxide; (d) tantalum nitride; (e) tantalum oxynitride.

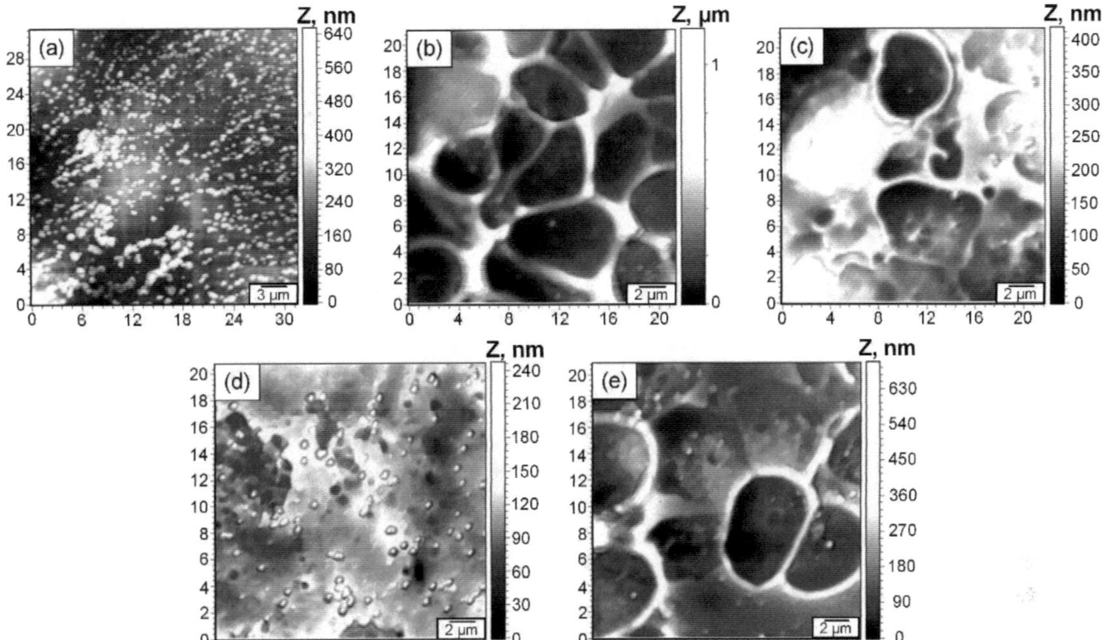

**Figure 3.** AFM-images of steel surfaces with different coatings after the platelets aggregation: (a) initial steel; (b) tantalum; (c) tantalum oxide; (d) tantalum nitride; (e) tantalum oxynitride.

In the same way, the individual platelets sank on the surface areas with a high roughness on the oxynitride coating surface (Figs. 2e, 3e). But for this type of surface the dispersion component of free surface energy significantly increases, which leads to the aggregation incremented. The number of aggregates is less compared to the tantalum coating. A dense network of pseudopodia is established. According to the classification, oxynitride coating surface can be classified as the second group.

On the surface of nanostructured tantalum oxide films, the adherent platelets are unevenly distributed, the platelets are weakly activated, spherical, and the number of pseudopodia is minimal (class 1). A large number of aggregates on these samples are not identified (Fig. 2c, 3c).

## 4. Conclusion

Thus, the type, nature of the material, and its hydrophilic properties are defined the thromboresistant properties of coatings. On the surfaces with hydrophilic properties (stainless steel), irreversible aggregation is provided. An increase in the hydrophobicity of the surface (tantalum nitride) should to increase thrombotic resistance, but additional adsorption of blood plasma proteins on the surface of the material is possible.

## Acknowledgements

This research was supported by the Belarusian Republican Foundation for Fundamental Research (BRFFR) (grant No. F18UKA-015), the National Academy of Sciences of Ukraine (grant No. 17-3-18), and by the World Federation of Scientists.

## References

[1] Papirov I I, Shkuropatenko V A, Shkurov V S and Pikalov A I 2010 *Medical Stent Materials* (Kharkov: NSC KIPT) p 40
[2] Sidorenko E S 2005 *Bull. RUDN Univ.* **1** 109-11
[3] Orlovsky P I, Gritsenko V V, Yukhnev A D, Evdokimov S V and Gavrilenkov V I 2007 *Artificial heart valves* (Moscow) p 448

[4] Volova T G, Volova T G, Shishatskaya E I and Mironov P V 2009 *Materials for medicine, cell and tissue engineering* (Krasnoyarsk: IPK SFU) p 116
[5] Sevastyanova V I 1999 *Biocompatibility* (Moscow) p 368
[6] Donkov N, Walkowicz J, Zavaleyev V, Zykova A, Safonov V, Dudin S and Yakovin S 2018 *Phys.: Conf. Ser.* **992** 012034
[7] Kuznetsova T A, Zubar T I, Lapitskaya V A, Sudzilouskaya K A, Chizhik S A, Didenko A L, Svetlichnyi V M, Vylegzhanina M E, Kudryavtsev V V and Sukhanova T E 2017 *IOP Conf. Ser.: Mater. Sci. Eng.* **256** 012022
[8] Kuznetsova T A, Lapitskaya V A, Chizhik S A, Uglov V V, Shymanski V I and Kvasov N T 2018 *IOP Conf. Ser.: Mater. Sci. Eng.* **443** 012018
[9] Lapitskaya V A, Kuznetsova T A, Chizhik S A, Sudzilouskaya K A, Kotov D A, Nikitiuk S A and Zaparozhchanka Y V 2018 *IOP Conf. Ser.: Mater. Sci. Eng.* **443** 012019
[10] Lapitskaya V A, Kuznetsova T A, Chizhik S A and Rogachev A A 2018 *IOP Conf. Ser.: Mater. Sci. Eng.* **443** 012020
[11] Kloubek J 1992 *Adv. Colloid Interface Sci.* **38** 99-142
[12] German E A, Younes R A, Khudoshin A K, Gabrielyan N I and Sevostyanov V I 2012 *Bull. Transplantol. Artif. Org.* **1** 66-71
[13] Melnikova G B, Petrovskaya A S, Kuznetsova T A, Chizhik S A, Zykova A, Safonov V and Yakovin S 2019 *Int. J. Nanosci.* **18** 1940078
[14] Warcholinski B, Kuznetsova T A, Gilewicz A, Zubar T I, Lapitskaya V A, Chizhik S A, Komarov A I, Komarova V I, Kuprin A S, Ovcharenko V D and Goltvyanytsya V S 2018 *J. Mater. Eng. Perf.* **27** 3940-50
[15] Petrovskaya A S, Lapitskaya V A, Melnikova G B, Kuznetsova T A, Chizhik S A, Zykova A V and Safonov V I 2019 *J. Phys.: Conf. Ser.* **1281** 012061

# The modification of PETF-membranes by Langmuir-Blodgett films of Nafion

**G B Melnikova[1], O G Ledvig[1], S A Chizhik[1], A L Kozlovskiy[3,4] and M V Zdorovec[3,4]**

[1]A.V. Luikov Heat and Mass Transfer Institute of National Academy of Sciences of Belarus, 220072 Minsk, Belarus
[2]Belarusian State Pedagogical University named after Maxim Tank, 220050 Minsk, Belarus
[3]L.N. Gumilyov Eurasian National University, 010000 Astana, Kazakhstan
[4]Institute of Nuclear Physics, Astana, Kazakhstan

galachkax@gmail.com

**Abstract**. In this research, the surfaces of track PETF-membranes were modified by Langmuir-Blodgett (LB) films of Nafion. The changes of structure of modified membranes were studied by atomic force microscopy (AFM). The LB-film of Nafion monolayer was found to be 30-100 nm thick. The incorporation of $SiO_2$ nanoparticles in LB-layer in small quantities was structuring the monolayer. The performance of water of PET-200 membranes with a Nafion film was increased to 401.3 ml/(m$^2$·h) from the initial values of 342.1 ml/(m$^2$·h). The air performance was increased for all modified PETF-200 membranes. The maximum performance was achieved by modifying the membranes with pure Nafion; the productivity was 154.3 ml/min compared with 96.4 ml/min for the initial membrane. It was shown that LB-films were stable during monitoring filtration characteristics.

## 1. Introduction

Membranes are used in various ways of water treatment [1], in micro-, ultra-, and nano-filtration and reverse osmosis, as well as in the processes of electro-deionization. Membranes hold up particles, bacteria, colloids, and solute (reverse osmosis and desalination processes). The development of an advanced technology for effective water treatment is an actual problem. The track membranes (TM) are the most interesting since they are characterized by a through pores system in a film with a thickness from 10 to 25 μm, of a uniform size, an identical shape, and a flat, smooth surface.

Due to these characteristics, TM have special properties and advantages over known polymer, metal, and ceramic membranes [2] used in microfiltration processes in medicine, biotechnology, microbiology, and virology [3]. PETF membranes are a commercial product with a well-developed production technology [4].

Their modification provides the principle possibilities to regulate TM properties, such as high throughput, selectivity, hydrophilicity, hydrophobicity, bactericidal activity, and allows enhancing the application area. One of the promising methods of surface modification is the sol-gel method. Organo-silicon compounds of various compositions [5] do not close the pores, but form a thin waterproof film on the membrane surface. The modified membranes are dried at room temperature and then in an oven for fixation of applied layer. If the modifier concentration in the solution increases, then the adsorption

IOP Conf. Series: Materials Science and Engineering **699** (2019) 012029 doi:10.1088/1757-899X/699/1/012029

capacity of the TM decreases; that is explained by the formation of a hydrophobic layer on the membrane surface and leads to a decrease in throughput of the modified membrane in the water. In this case, the air performance of the TM after modification does not change, which is explained by the preservation of the pore space of the membrane. The Langmuir-Blodgett (LB) method makes it possible to incorporate various molecules and molecular complexes, including biologically active ones, into the monolayer [6]. Using this method, it is possible to change the surface properties of the monolayer and, as a result, to form the high-quality film coatings due to precise control of the thickness of the resulting film, uniformity of coating, low roughness, and high adhesion of the film to the surface [7]. The properties of the films can also be easily varied by changing the structure of polar head of amphiphilic molecule, composition of monolayer, and also conditions of release – composition of subphase and surface pressure. The formation of thin nanometre-thick coatings allows providing the surface of macroobjects with the required hydrophilic properties. The LB method is used to form the ion-exchange membranes, e.g., based on Nafion [8], which can be used as ion-selective electrodes, sensors, in electrodialysis, and in water treatment systems [9]. Nafion-modified membranes are also used in the development of fuel cells [10]. It was shown [11] that incorporation of silicon oxide nanoparticles in the Nafion structure allowed improving current-voltage characteristics, increasing thermal stability of the material, and decreasing the coefficient of thermal expansion.

The aim of this work is to modify the surface of PETF-membranes by LB-films of Nafion and nanostructured films with silicon dioxide nanoparticles to increase their water and gas throughput.

## 2. Experimental

In the experiment, track PETF-membranes with pores diameters 100 and 200 nm were modified. PETF-membranes were prepared in Kazakhstan by method described in article [12]. The initial material for production of TM was a 12 μm thick PET film irradiated with 1.75 MeV / nucleon of Kr-ion beams with a fluency of $1 \cdot 10^6$ $cm^{-2}$ on a DC-60 cyclotron. All samples were additionally sensitized with a UV-lamp for 12 h on one side to accelerate etching of the irradiated material. UV-radiation led to photooxidation of the surface layer of the polymer. Transformation of highly defective regions (latent tracks) formed as a result of irradiation into pores was carried out by two-sided chemical etching in a 2.2 mol/l solution of NaOH at a temperature of 85 ± 1°C.

Nafion layers on the water sub phase were formed by horizontal LB method using device LB-Trough (Belarus). Nafion films were formed from initial Nafion suspension in low aliphatic alcohol (Nafion 1100 EW, Sigma-Aldrich) and diluted in two times with ethanol. The incorporation nanoparticles were provided by mix suspension of Nafion and suspension of silicon dioxide nanoparticles (d=10-20 nm, Sigma-Aldrich) in ethanol with concentration 1 mg/ml. Final suspension was kept in ultrasonic bath for 30 min to ensure that nanoparticles have uniform distribution. The ratio of the components in suspension is 1 g of Nafion to $3.9 \cdot 10^{-4}$ mol of $SiO_2$ nanoparticles. The pressure of thin films formation was defined on the base of isotherm of monolayer compression and presented in the Table 1. Thickness of Nafion films is 100 nm, thickness of films of diluted Nafion and nanostructured with silicon dioxide nanoparticles is 30 nm.

The structure of samples was investigated by atomic force microscopy (AFM) method using device NT-206 (Belarus) with NSC 11 A silicon cantilevers (Mikromasch, Estonia). The stiffness of cantilevers is 3 N/m, curvature radius is 10 nm. The contact angle of thin films was measured by sessile drop method using device DSA 100E (KRUSS, Germany). For this purpose, the drop of water with volume 2 μl was formed using automatically system of dosing.

**Table 1.** The pressure of films formation.

| The type of film | ($\pi$ +/-1), mN/m |
|---|---|
| Nafion | 43 |
| Nafion diluted (1 : 1 V) | 33 |
| Nafion + $SiO_2$ (1 g : $3,9 \cdot 10^{-4}$mol) | 24 |

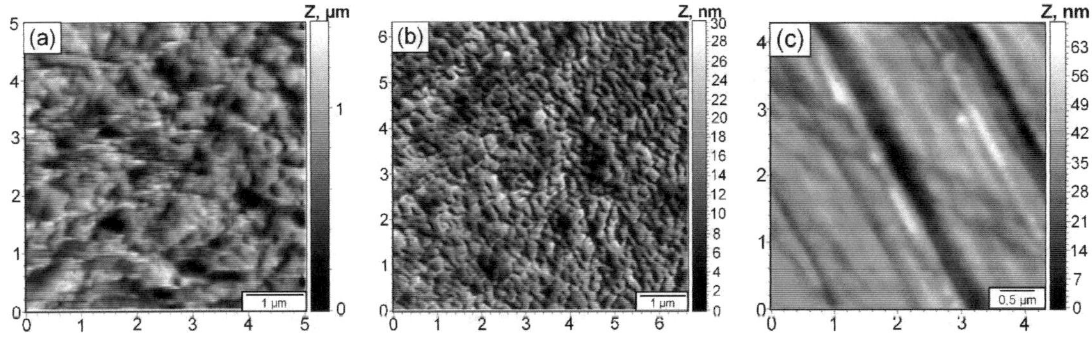


**Figure 1.** AFM-structure of LB-films: (a) Nafion, (b) diluted Nafion, and (c) Nafion with nanoparticles on silicon substrate.

The air throughput of initial and modified membranes was defined at a pressure drop of 0.02 MPa. The size of sample is 1 cm$^2$. The effective pore diameter was calculated by the gas performance, according to the formula:

$$r^3 = \frac{Q \cdot 3l}{\sqrt{\frac{2\pi}{R \cdot T \cdot M}} \cdot \Delta p \cdot 4n}, \quad (1)$$

where r – the radius of pores; Q – air performance; l – thickness of film; Δp – pressure; R – universal gas constant; M – molecular mass of air; n – radiation density; T – temperature.

The water performance was determined using deionized water (18 MΩ) and the pressure difference of 0.4 MPa. The size of samples is 2.54 cm$^2$. Performance was calculated using the Hagen-Poiseuille equation:

$$Q = \frac{n \cdot \pi \cdot r^4}{8\mu \cdot l} \cdot \Delta p, \quad (2)$$

where r – the radius of pores; l – thickness of film; Δp – pressure; n – radiation density; T – temperature, μ – dynamic fluid viscosity.

To identify the strength properties of the samples, we used the method of checking the tensile strength by increasing the flow of passing air through the membrane.

## 3. Results and discussion

As a result of studies of the structure of LB-films on the silicon substrates by AFM, it was shown that Nafion formed a dense and reticulate layer. The nanostructured films with nanoparticles had a directed structure determined by silicon dioxide nanoparticles (Fig. 1).

These surfaces were characterized by water contact angle of 70°. Free surface energy was 44 mJ/m$^2$ with prevailing polar component.

According to AFM-data, for track PETF-membranes a system of through pores with a diameter of 100 nm (Fig. 2a) and 200 nm (Fig. 3a) prevailed. In addition, interwoven fibres of the polymer were also visible on the surface. The surface of PETF membranes demonstrated the hydrophobic properties; the contact angle did not depend on pores diameter and was about 90°.

The structure of the modified membranes was similar to the initial surface. A dense layer of polyelectrolyte enveloped pores of the original membrane, thereby reducing their diameter by 10-20 nm. Wherein, the mean-square roughness values corresponded to those for the initial membranes. In comparison with the initial samples, the presence of thickened formations around the pores of the modified membrane was established. The Nafion monolayer obtained from a dilute suspension was less dense than the initial Nafion. As a result of modifying TM with Nafion films, the contact angle increased to 94°, thereby increasing the hydrophobic properties of the sample surface.

IOP Conf. Series: Materials Science and Engineering **699** (2019) 012029 doi:10.1088/1757-899X/699/1/012029

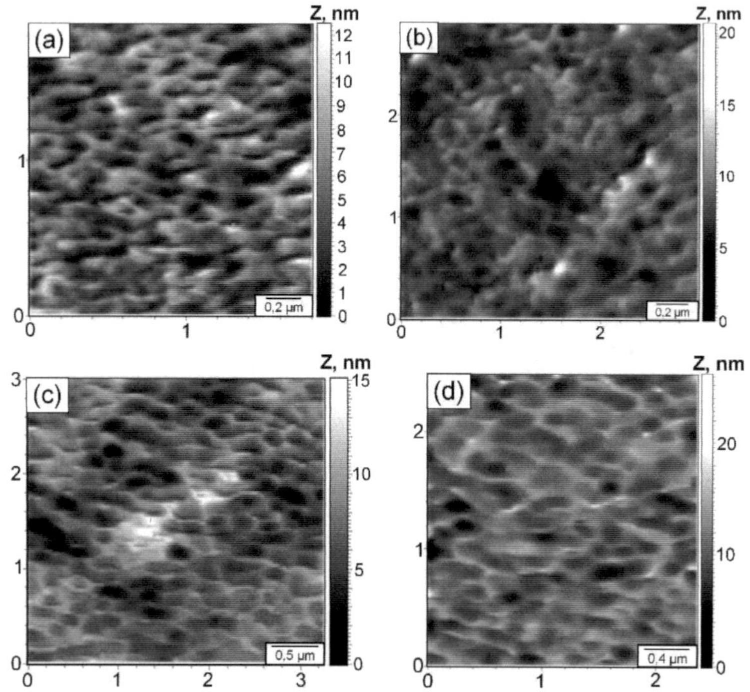


**Figure 2.** AFM images of the samples: (a) PETF-100; (b) PETF-100/Nafion; (c) PETF-100/Nafion$_{dil.}$; (d) PETF-100/(Nafion + npSiO$_2$).

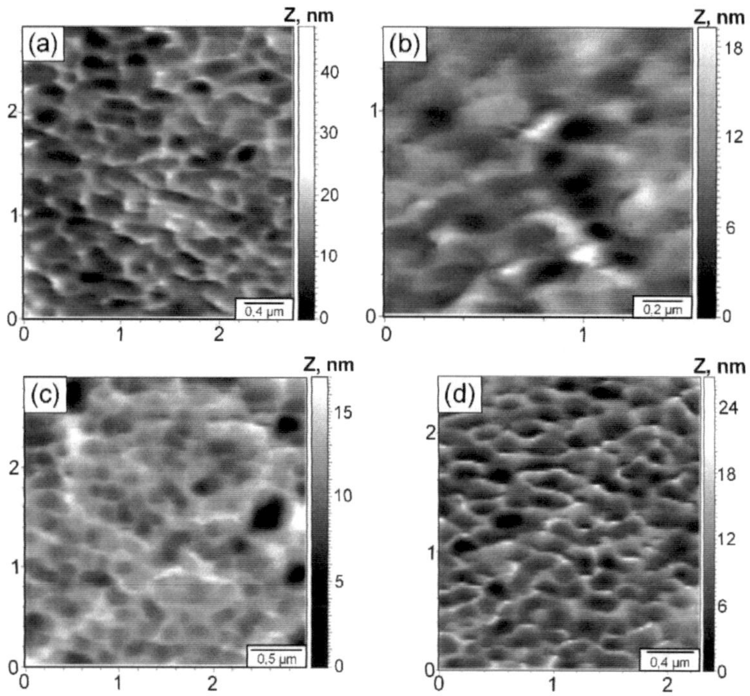


**Figure 3.** AFM images of the samples: (a) PETF-200; (b) PETF-200/Nafion; (c) PETF-200/Nafion$_{dil.}$; (d) PETF-200/(Nafion + npSiO$_2$).

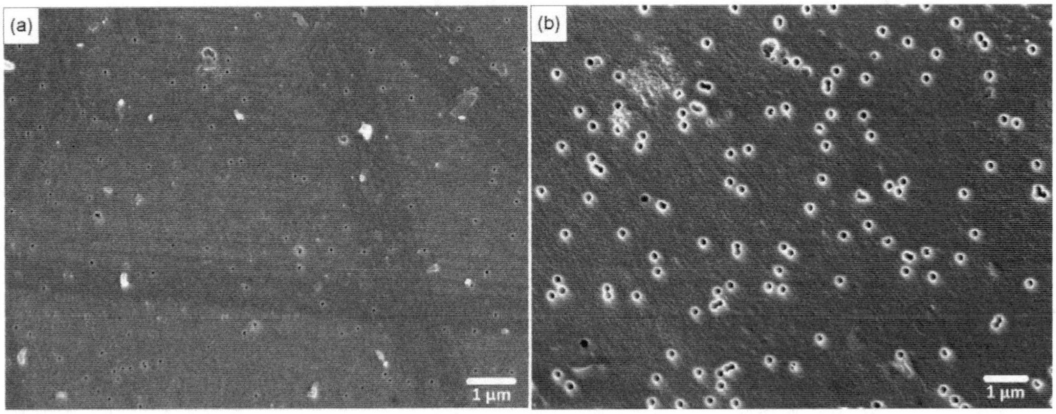


**Figure 4.** SEM-images of structure of surfaces PETF-membranes with diameter of pores of (a) 100 nm and (b) 200 nm, modified by layer of Nafion with $SiO_2$ nanoparticles.

In the case of the formation of composition structures (Nafion +$SiO_2$ np) on the surface of hydrophobic TM with dimensions of 100 nm, two regions of films were formed – island non-uniform films of Nafion and a homogeneous film with characteristic conglomerates of nanoparticles (Fig. 4a). This result can be explained by the high pressure of the film deposition near collapse of monolayer. When a film is formed on a TM with a pore diameter of 200 nm, silicon dioxide nanoparticles are structured around the pores of the membrane, forming characteristic rings (Fig. 4b).

After modification, the filtration properties of PETF100 membranes did not change (Table 2). However, the tensile strength of modified samples slightly increased. The modification of PET-200 membranes with a Nafion film showed the best results. The performance of water increased to 401.3 ml/·($m^2$·h) from the initial values of 342.1 ml/($m^2$·h). The air performance was increased for all modified PETF-200 membranes. The maximum performance was achieved by modifying with pure Nafion, the productivity was 154.3 ml/min compared with the initial membrane of 96.4 ml/min. These changes are due to an increase in the root mean square surface roughness by 1.2 times. The modification of PETF 200 membranes by a composite Nafion film with silicon dioxide nanoparticles led to the formation of characteristic ring-shaped structures along the membrane pore contour, which increased the adsorption of water on the surface. The LB-film not only modified the working surface of the membranes, but also could penetrate deep into the pores from 2 nm and more. As a result, hydrophilicity of the surface inside the pores increased; thereby, the water performance of the membranes also increased, despite the data obtained on the increase in surface hydrophobicity as a result of modification. The gas penetration changed as a result increasing air adsorption to modified surface of membrane.

**Table 2.** Flux parameters of initial PETF membranes and modified by Nafion films.

| **Samples** | **$Q_{air}$, ml/min** | **$Q_{water}$, ml/·($m^2$·h)** | **Tensile strength, MPa** |
|---|---|---|---|
| PETF 100 | 121.1 | 364.7 | 0.316 |
| PETF 100 / Nafion | 101.2 | 354.6 | 0.320 |
| PETF 100 / $Nafion_{dil}$ | 122.2 | 376.7 | 0.356 |
| PETF 100 / (Nafion+$SiO_2$) | 103.4 | 359.6 | 0.388 |
| PETF 200 | 96.4 | 342.0 | 0.224 |
| PETF 200 / Nafion | **154.3** | **401.3** | **0.460** |
| PETF 200 / $Nafion_{dil}$ | **132.1** | **397.5** | 0.268 |
| PETF 200 / (Nafion+$SiO_2$) | 123.1 | 276.7 | 0.340 |

Moreover, LB-films were stable during the experiment of monitoring filtration characteristic. The stability of the filtrate flow over time was also controlled. The values did not change, indicating the stability of the applied layers.

It should be noted that tensile strength increased significantly after modification of thin Nafion layer from 0.22 to 0.46 MPa.

## 4. Conclusion

The modification of LB-film of Nafion monolayer with thickness from 30 to 100 nm allowed improving both penetration properties of PETF-membranes and their strength characteristics. The incorporation of nanoparticles in LB-layer in small quantities was structuring the monolayer. It was shown that modification of only one monolayer significantly changed the properties of membranes.

## Acknowledgements

The investigation was performed within the Program of State Research "Energy systems, process and technologies", project 2.82, 2019-2020 yrs.

## References

[1] Kumar R U, Muthukrishnan M and Guha B K 2008 *Desalination* **230** 70-8
[2] He Z B and Guo S L 2015 *Phys. Procedia* **80** 131-4
[3] Wang R, Shi L, Tang C Y, Chou S, Qiu C and Fane A G 2010 *J. Membr. Sci.* **355** 158-67
[4] Zhdanov G S, Kitaeva N K, Bannova E A and Minyailo L V 2004 *Critical technologies* **2** 3-8
[5] Awasthi K, Soumyadip C, Hartmut K, Frank S and Manfred S 2014 *Int. J. Hydrogen Energy* **39** 9356-65
[6] Jung G-B, Weng F-B, Su A, Wang J-S, Yu T L, Lin H-L, Yang T-F and Chan S-H 2008 *Int. J. Hydrogen energy* **33** 2413-24
[7] Melnikova G B, Zhavnerko G K, Chizhik S A and Bildyukevich A V 2016 *Pet. Chem.* **55** 406-12
[8] Bertoncello P, Manoj K R, Notargiacomo A, Ugo P and Nicolini C 2002 *Phys. Chem. Chem. Phys.* **4** 4036-43
[9] Jakubik W P, Urbańczyk M, Maciak E and Urbanczyk M 2006 *J. De Physicue* **137** 99-102
[10] Kreuer K D 2001 *J. Membrane Sci.* **185** 29-39
[11] Shi J 2008 *Composite membranes for proton exchange membrane fuel cells* PhD Dissertation (Dayton, United States) p 127
[12] Kozlovskiy A L, Borgekov K K, Zdorovets M V, Arkhangelsky E, Shumskaya A E and Kaniukov E Yu 2017 *Proc. NAS of Belarus. Phys.-Tech. Ser.* **1** 45-51 (in Russian)

IOP Conf. Series: Materials Science and Engineering **699** (2019) 012030 doi:10.1088/1757-899X/699/1/012030

# Light polarization and intensity behaviour in aperture cantilevers with carbon tip created by focused ion beam

**T V Mikhailova[1], Yu E Vysokikh[2,3], S Yu Krasnoborodko[2,3], A S Kolomiytsev[4] and A A Fedotov[4]**

[1]V.I. Vernadsky Crimean Federal University, 295007 Simferopol, Russia
[2]Scientific and Technological Centre of Unique Instrumentation of the RAS, 117342 Moscow, Russia
[3]MTEON Ltd, Moscow, 124617 Zelenograd, Russia
[4]Southern Federal University, 344006 Rostov-on-Don, Russia

visokihy@gmail.com

**Abstract.** Modern atomic force microscopy (AFM) techniques and their combination with other methods give us a possibility to investigate the same samples from the new side at nano-scale. One of very promising combinations is polarization near field microscopy integrated into AFM tool. It allows one to make topography imaging by standard AFM tool and optical magnetic structure imaging by magneto-optical method simultaneously. Moreover, this combination gives synergy effect on both methods by improving their weak sides: elimination of magnetic influence of magnetic force microscopy on the sample, on the one hand, and overcoming the diffraction limit of the magneto-optical microscopy, on another hand. The cantilevers play a very important role in this combination of the methods since they interact with the surface and allow focusing the laser beam to the sample in near field mode. The geometry and other parameters of cantilevers are determined as a result of the combined methods implementation. Authors investigate the new type of carbon-tips aperture cantilevers for AFM and high resolution magneto-optical combination.

## 1. Introduction

Nowadays, atomic force microscopy (AFM) allows investigating not only morphology and physical properties [1,2] of the samples, but also optical data could be collected by the combination of AFM and near-field scanning optical microscopy (NSOM) methods. This combination allows one to study optical properties of the sample and morphology of same area simultaneously, which is important for new magnetic materials investigation [3-5]. One of the modern techniques obtained from such combination is polarization near-field optical microscopy, which allows performing polarization depended measurements with high spatial resolution [6].

There is a number of probes created for NSOM. Historically, optical fibre probes were used for this purpose, but they have the limitations of spatial resolution, light throughput, and reproducibility [7]. is One of the modern types of the probes for NSOM methods is silicon cantilevers. Authors propose to use focused ion beam technology to get carbon hollow pyramid tip on a silicon cantilever for creation high reproducible aperture cantilevers with high light throughput and high spatial resolution [8]. In this paper, authors study light polarization and intensity in aperture cantilevers with carbon tip created by focused ion beam.

IOP Conf. Series: Materials Science and Engineering **699** (2019) 012030 doi:10.1088/1757-899X/699/1/012030

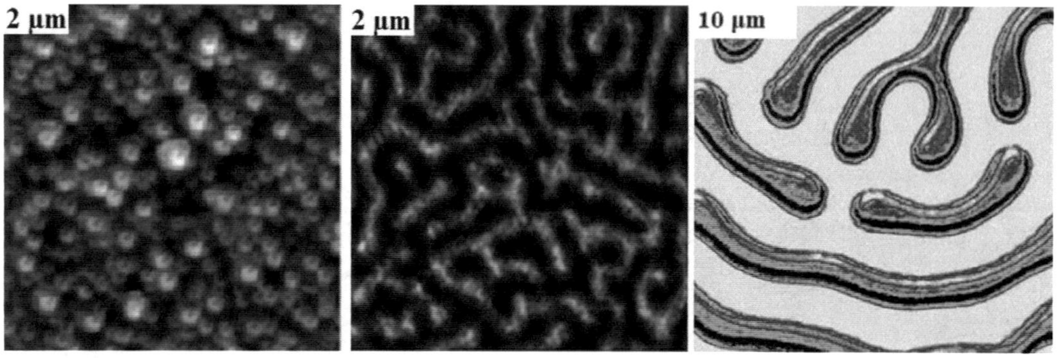


**Figure 1.** Example of polarization NSOM – magneto-optical polarization near field microscopy. Topography (left) and magneto-optical near field imaging of magnetic domain structures (centre and right).

## 2. Experimental

Images of Bi-substituted yttrium iron garnet (Bi:YIG) thin films were obtained using magneto-optical polarization near field microscopy (Fig. 1).

Carbon-tips cantilevers were made by focused ion beam to investigate light polarization and intensity behaviour. Authors propose a new technology of aperture cantilever creation. Tipless silicon cantilevers NSG-11 (TipsNano) were used for creation a new type of probes. Ion beam etching and ion beam assistance deposition were implemented using Nova NanoLab 600 tool. Gallium liquid metal was used as ion beam source. Round shape hole with diameter of 5 µm was created by ion beam etching. After the hole creation, carbon pyramid formation started. The trajectory of the ion beam was determined to deposit carbon in an orderly manner from the base of the cone (5 µm in diameter) to its tip. As a result, a hollow tip with a height of about 5-6 µm was formed. Scanning electron microscopy images of the carbon-tips cantilevers were obtained (Fig. 2).

Produced carbon-tip aperture cantilevers with aperture sizes from 83 to 264 nm were investigated by means of NTEGRA (NT-MDT) including AFM and NSOM methods. Optical images of apertures of the carbon-tip cantilevers (bottom illumination scheme) were collected (Fig. 3).

## 3. Results and discussion

Optical images were obtained by laser scanning (by mirror) of apertures of the carbon-tips cantilevers. Possibility to focus light spot to the aperture cantilevers is demonstrated (Fig. 4). Disturbances occurring in different sizes of aperture are investigated. It is shown that it may be caused by nonuniformity carbon rings due to some errors in technological process.

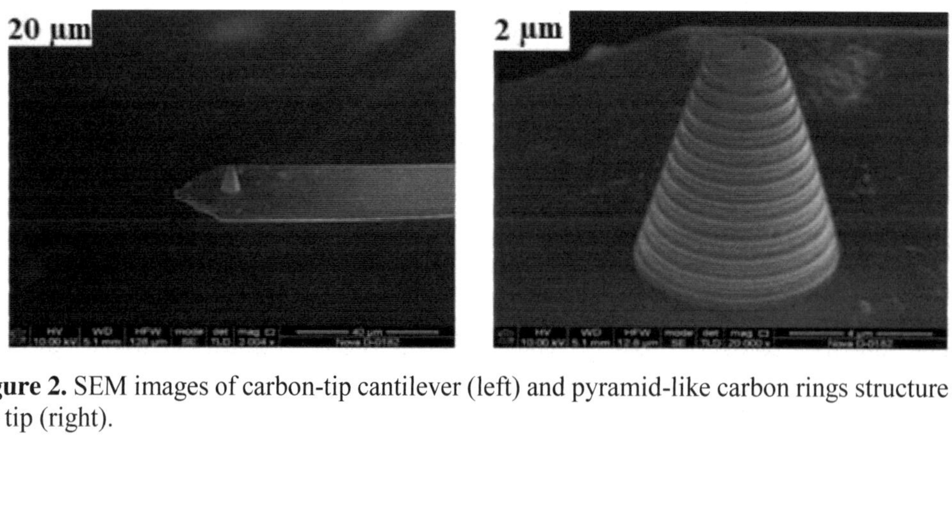


**Figure 2.** SEM images of carbon-tip cantilever (left) and pyramid-like carbon rings structure of the tip (right).

IOP Conf. Series: Materials Science and Engineering **699** (2019) 012030 doi:10.1088/1757-899X/699/1/012030

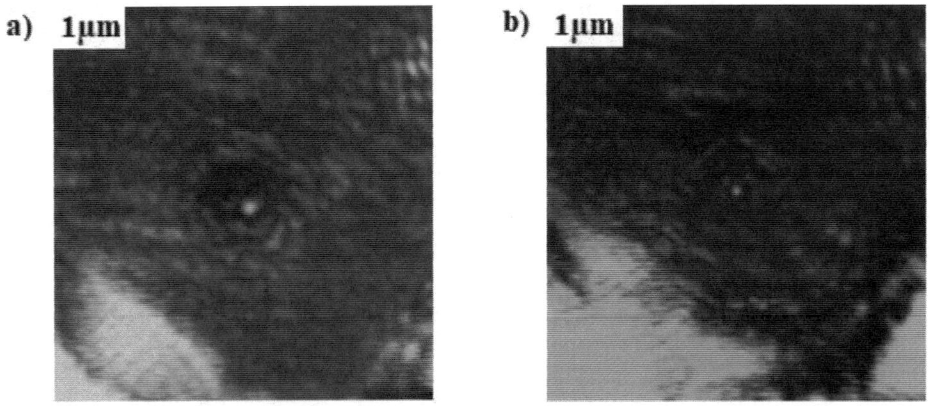


**Figure 3.** Optical images of apertures of the carbon-tip cantilevers (bottom illumination scheme): (a) 264 nm; (b) 90 nm.

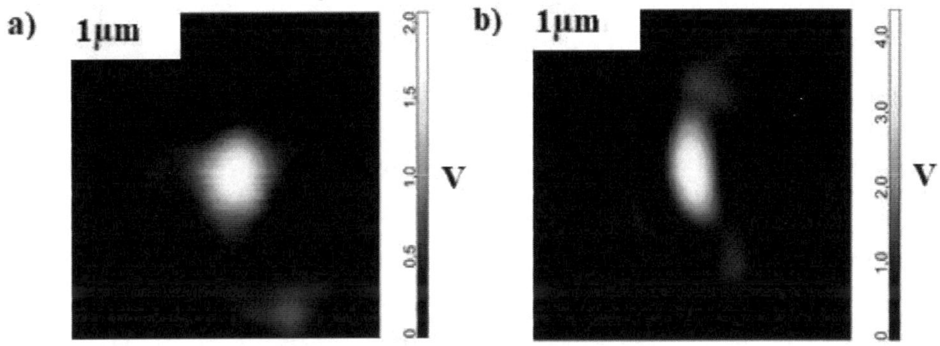


**Figure 4.** Optical images obtained by laser scanning (by mirror) of apertures of the carbon-tips cantilevers: (a) 264 nm; (b) 90 nm.

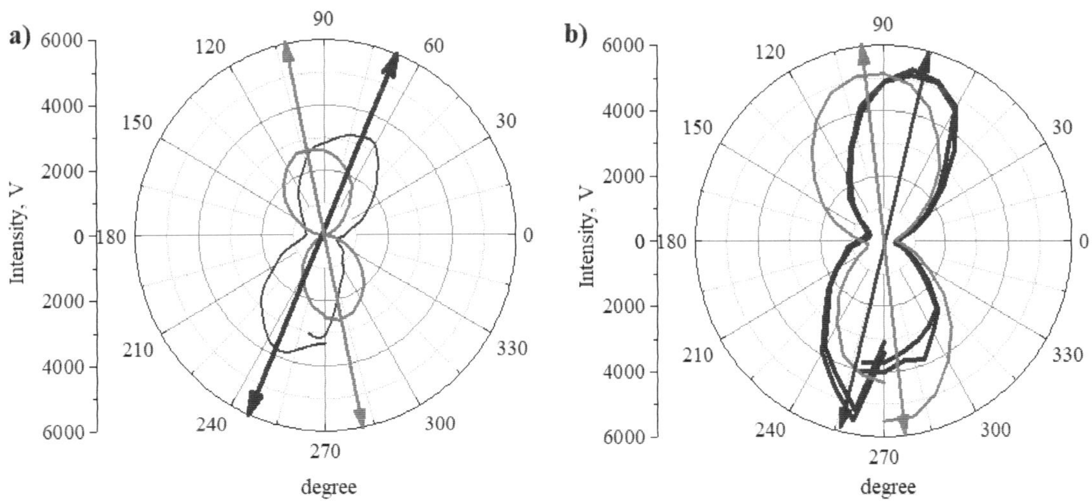


**Figure 5.** Polarization rotation in carbon-type aperture cantilevers. Red lines and arrows show polarization position of initial laser beam. Blue lines and arrows represent the polarization of light passed through the cantilever aperture. Aperture size: (a) 83 nm and (b) 90 nm.

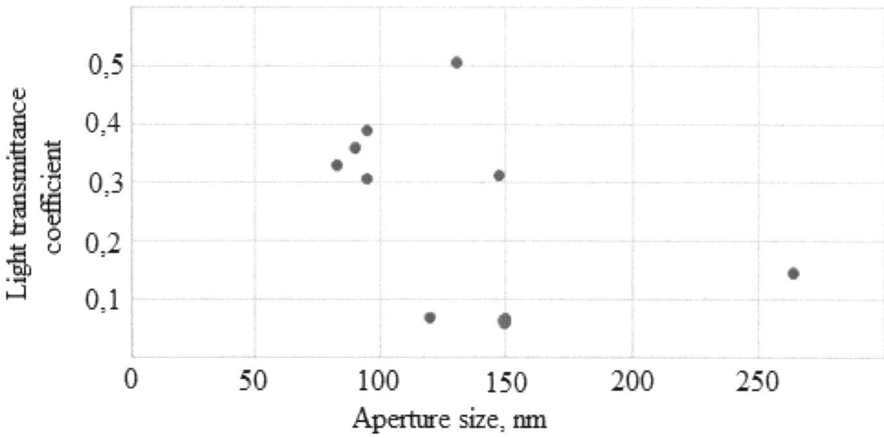


**Figure 6.** Coefficient of light transmittance/aperture size(nm) dependence.

Polarization rotation in carbon-tips aperture cantilevers was measured. The biggest value of polarization rotation was determined for cantilevers with the smallest aperture size: 35° and 21° for 90 and 83 nm aperture size, correspondently (Fig. 5).

Coefficient of light transmittance vs. aperture size(nm) dependence was calculated. Carbon-tips cantilevers (blue circles) showed good light transmittance compared to commonly used aperture type cantilevers (marked by red circles).

## 4. Conclusion

New carbon-tips aperture cantilevers for polarization near field microscopy were investigated. It was shown that these cantilevers allowed obtaining appropriate laser spot for near field microscopy, but disturbances appeared in some cantilevers should be eliminated by more precision technological process and changing pyramid geometry. Light transmittance coefficients of new type cantilevers were estimated to show good results compared with commonly used aperture cantilevers. Polarization rotation measured for different apertures showed necessity of analyser prealignment depending on cantilever aperture geometry and size.

## References

[1] Mironov V L 2004 *Fundamentals of Scanning Probe Microscopy* (Nizniy Novgorod: Inst. Fiz. Mikrostruktur)
[2] Vysokikh Yu E, Shaposhnikov A N, Prokopov A R, Berzhansky V N, Karavainikov A V, Gerasimenko N N and Smirnov D I 2017 *Mater. Res. Bull.* **95** 115-22
[3] Zvezdin A K and Kotov V A 1997 *Modern Magnetooptics and Magnetooptical Materials* (IOP Publishing Ltd)
[4] Bulatov M F and Churikov D 2019 *J. Surf. Invest.: X-Ray, Synchrotron Neutron Tech.* **13** 206-9
[5] Berzhansky V N, Karavainikov A V, Milyukova E T, Mikhailova T V, Prokopov A R and Shaposhnikov A N 2010 *Func. Mater.* **17** 120-6
[6] Vysokikh Yu E, Shelaev A V, Prokopov A R, Shevyakov V I and Krasnoborodko S Yu 2016 *J. Phys.: Conf. Ser.* **741** 012190
[7] Veerman J A, Otter A M, Kuipers L and Van Hulst N F 1998 *Appl. Phys. Lett.* **72** 3115
[8] Kolomiytsev A S, Shandyba N A, Panchenko I V and Lisitsyn S A 2018 *IOP Conf. Ser.: Mater. Sci. Eng.* **443** 12015

# Structural-mechanical AFM study of inhomogeneous stiff nanocoating of soft polymer substrate

**I A Morozov[1] and A S Kamenetskikh[2]**

[1]Institute of Continuous Media Mechanics UB RAS, 614013 Perm, Russia
[2]Institute of Electrophysics UB RAS, 620016 Ekaterinburg, Russia

ilya.morozov@gmail.com

**Abstract**. Modification of polymer with low-energy gas plasma is used to change the wettability of the surface and, as a consequence, to improve the interactions with biological objects. As a result of such treatment, a nanolayer, which is stiffer than the initial surface, is formed on a broad class of polymers. Elastic polyurethane is one of the widely used polymers. This is two-phase material consisting of soft and hard blocks. Its surface was treated by low-energy argon plasma and studied by an atomic force microscopy (AFM): the structure becomes rough and mechanically inhomogeneous depending on the part of the surface treated (hard or soft blocks of the material). The average value of elastic modulus increases with the treatment time, but the average thickness of the coating does not change, having a wide dispersion. All these features are related with the heterogeneous surface peculiarities of the initial polymer.

## 1. Introduction

Plasma treatment significantly changes the physicochemical, mechanical, and structural properties of the surface [1]. This is a promising way to modify the polymer surfaces. Polyurethanes, due to a wide range of thermomechanical properties from stiff plastics to soft elastomers, are used in many areas. Coatings formed by plasma modification of soft elastic polyurethanes have many potential applications, in particular in the development of biomedical products with improved properties. However, such substrates are easily deformed under the action of external loads, which often damages the coating [2] and the positive effect of the treatment falls [3]. Therefore, the low energy plasma modification of the polymer surface is of interest. Studies on the treatment of polyurethanes with oxygen [4], nitrogen [5], argon [6], or ammonia [7] low energy plasma have shown changes in surface hydrophobicity, which led to improved cell and protein adhesion.

Argon, being a noble gas, does not react with polymer atoms, but destroys C-C and C-H bonds of macromolecules; thus, free radicals are emerged and the surface activity increases [6]. At the same time, new C-C bonds are formed, and carbonization of the surface occurs [8]. The plasma-induced changes of the structural and mechanical properties of the surface should be taken into account, especially when creating coatings on soft polymers that are easily deformed. Note, that the polyurethane (PU) is a two-phase system consisting of soft and hard blocks of neat-like [9] or globular structure [10]. The influence of plasma on such an inhomogeneous material should lead to the formation of an inhomogeneous coating.

In this paper, the AFM methods are applied in study of structural-mechanical properties of the PU surface modified by the low energy argon plasma. The obtained results can be used to create deformable materials with improved biomedical properties.

IOP Conf. Series: Materials Science and Engineering **699** (2019) 012031 doi:10.1088/1757-899X/699/1/012031

## 2. Materials and methods

PU was made of pre-polymer (urethane based on simple polyester and toluene diisocyanate) and hardener (MOKA). The components were heated to 70°C, vacuumed for 5 min, mixed, then vacuumed again, and cured at 100°C for 20 h. The thickness of the obtained plates was 2 mm. An initial modulus of elasticity was 30 MPa and elongation at rupture – 800% (measured by uniaxial mechanical tests).

The samples (20x20 mm$^2$) were placed in a plasma unit in an electrically insulated holder cooled with water. The chamber was vacuumed to a pressure of $5*10^{-5}$ Torr. The pressure of argon was set to $2*10^{-3}$ Torr and a glow discharge with a current of 1 A was ignited in the source of low energy electron beam; the accelerating voltage of the electron source was 100 V. Injection of electrons into the plasma chamber provided generation of Ar plasma, which particles interacted with the surface of the samples. Treatment time $\tau$ was set to 6 or 8 minutes: preliminary study has shown that at these durations, the energy of the treated surface has a maximum and exceeds the value for raw material, i.e. the hydrophilicity increases.

Structural-mechanical study of surface properties was carried out by an AFM Ntegra Prima (by NT-MDT BV) in semi-contact regime, force modulation, as well as by nanoindentation. In all the cases, the scanning was performed by a sample. The geometry of the probe tip was evaluated by blind reconstruction method based on TipCheck samples: probe tip radii ranged from 6 to 15 nm, and the stiffness of $k$ cantilevers was estimated by the Sader method. For quantitative comparison of the force measurements, the sensitivity of the probes was calibrated using a silicon substrate.

In semi-contact regime, the probe oscillates near the sample surface and slightly taps it, causing changes in amplitude (determines the height) and phase (shows qualitatively the heterogeneity of the surface composition) of oscillations. In this mode, stiff probes ($k \sim 30$ N/m) with high resonance frequency were used.

In force modulation, the probe scans the sample in the contact mode: it is pressed against the surface with certain constant force. Additionally, the scanner with the sample performs harmonic modulation (the frequency was set to 3 kHz and the amplitude – to 0.05 V), so the probe begins vibrating. The amplitude of these vibrations is proportional to stiffness of the surface. Soft probes ($k \sim 0.5$ N/m) were used in this mode.

Nanoindentation was carried out to determine a thickness of the coatings, as well as to estimate mechanical properties of the surfaces. In all cases, several areas (3x3, 5x5 μm$^2$) of the sample were studied for statistical purposes. The indentation onset was considered to be the first point after the probe jumps to the surface; depth of indentation: $u = z - d$, where $z$ is the movement of the probe base; $d$ is the cantilever deflection. Typical curve of indentation of PU and corresponding dependence of applied load $F = kd$ vs. the depth of indentation are shown in Figure 1a. In this case, soft probes ($k = 0.5$ nN/nm) were used. The modulus of elasticity was estimated using the approximation of the $F(u)$ curves by the Hertz model.

The thickness of the coatings was determined by stiff probes ($k \sim 35$ N/m). Surface relief before and after the indentation was studied in semi-contact mode. It was found that the indentation imprints remained on the surface (Fig. 1c) after the indentation with a significant load; they ranged from 80 to 400 nN, depending on the sample. A typical indentation curve of the coating on elastic PU is shown in Figure 1b: the change in slope (indicated by an arrow) corresponds to onset of piercing the coating. The surface of untreated PU is elastically recovered after the indentation to the same depth. Thus, the thickness $t$ of the coating could be estimated by the depth of the indentation imprint. Relief irregularities in the vicinity of the imprint make it difficult to clearly determine $t$ by the height $h$ of the profile (Fig. 1e). Therefore, the cross-sections of the imprint depending on the depth (Fig. 1d) were analysed. The thickness of the coating was assumed to be the height of the imprint, where the ratio of the cross-section areas at the given and the previous $h$ was minimal (Fig. 1f).

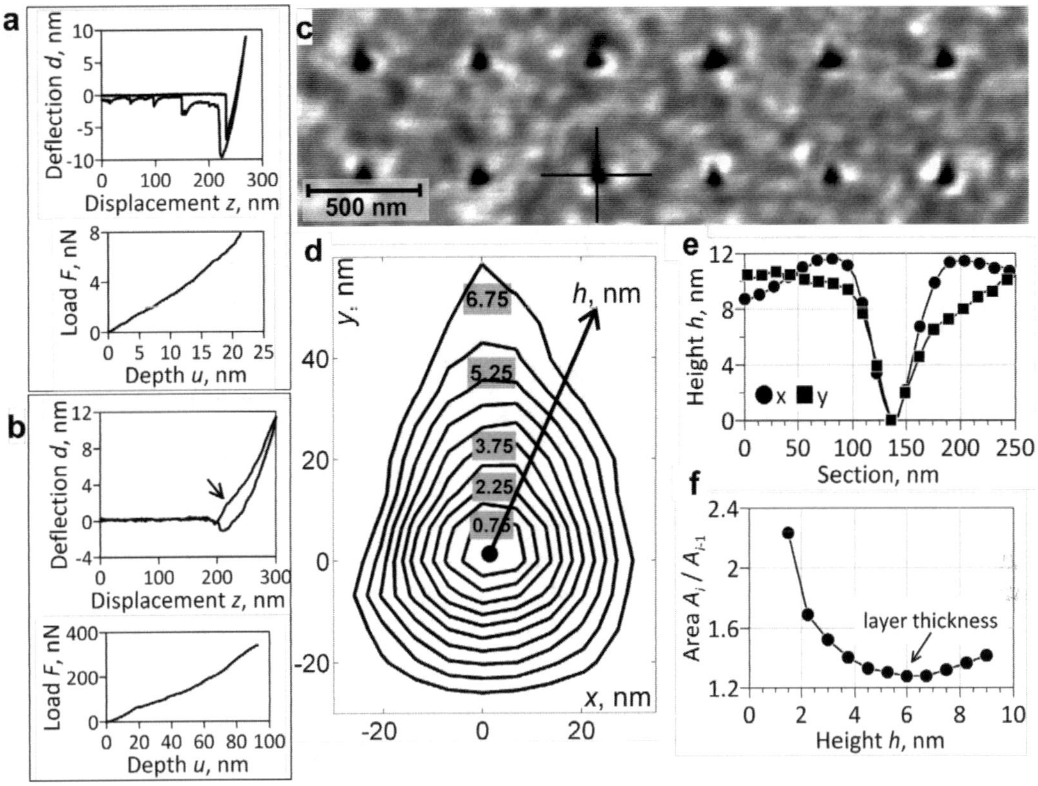


**Figure 1**. Indentation of (a) raw and (b) plasma-treated elastic PU: in coordinates of deflection vs. displacement of the cantilever base and the applied load vs. indentation depth. (c) The indentation imprints; the cross-sections of the marked imprint are given (d) along height and (e) along axes x and y. (f) The ratio of the contour areas vs. height.

## 3. Results and discussion

The original untreated PU is a complex system consisting of soft and hard phases. Its surface is not perfectly smooth (Fig. 2a). The hard phase looks like a mesh of threads clearly visible on the phase contrast of the AFM-image (Fig. 2a). These threads are distributed heterogeneously and form dense agglomerates with sizes from 20 nm to 4 μm, making bumps on the surface. Roughness of the surface will be discussed further. On the distribution (statistical probability density) of elastic modulus (Fig. 2b), three areas can be distinguished: 1) indentation of soft phase (modulus ~20 MPa); 2) indentation of mixed regions; 3) indentation of hard phase, which is the main part of the material (modulus ~40 MPa).

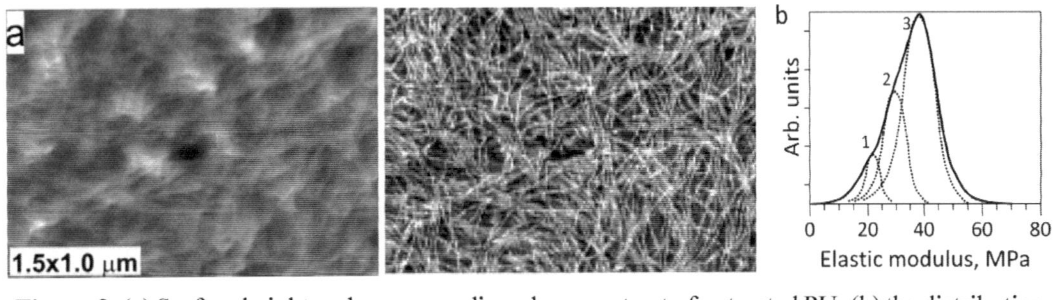


**Figure 2.** (a) Surface height and corresponding phase contrast of untreated PU; (b) the distribution of elastic modulus.

IOP Conf. Series: Materials Science and Engineering **699** (2019) 012031 doi:10.1088/1757-899X/699/1/012031

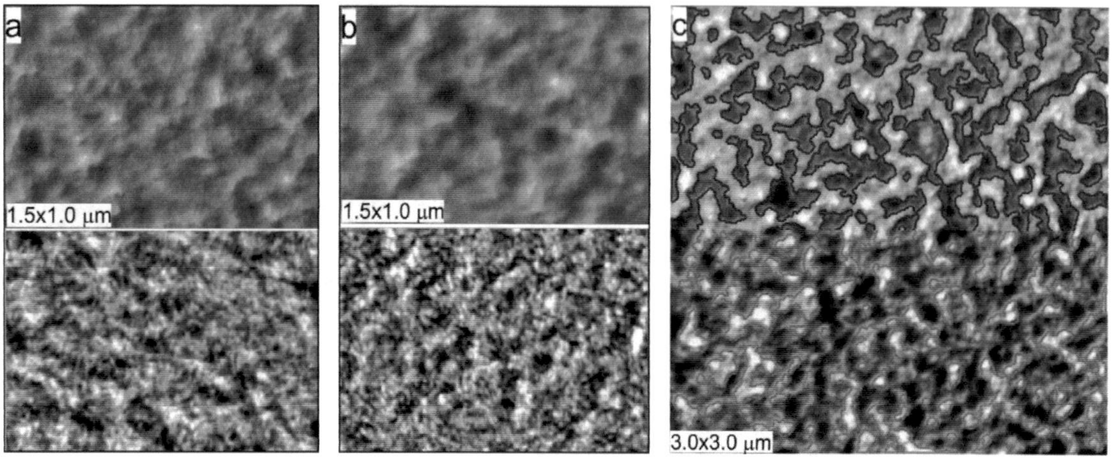


**Figure 3**. Surface height and corresponding phase contrast of PU treated with argon plasma for (a) 6 min and (b) 8 min; (c) illustration of the relief segmentation: contours of lowlands are shown in upper part (blue lines) and highlands in bottom part (red lines) of one representative image.

The interaction of argon plasma with the PU surface causes the destruction of C-C and C-H bonds on the polymer surface, the formation of free radicals, and increase in surface hydrophilicity [6]. At the same time, carbon bonds are being rearranged and the so-called surface carbonization [8] occurs.

As a result of argon treatment, the surface changes: the neat-like structures of the hard phase at $\tau = 6$ min are visible on the phase contrast of AFM-images (Fig. 3a) only partially and disappear completely at $\tau = 8$ min (Fig. 3b). RMS roughness (Fig. 4a) of the surface depends on the size of the observation: small areas could include flat regions or rough fragments of agglomerates of the hard polymer phase: the roughness increases on the scale up to 4 µm, but further reaches an asymptote.

The heterogeneous relief of the surface was segmented by the watershed method into highland areas ($z > 1.1z_{avg}$, where $z_{avg}$ is the mean height of the AFM image) and lowland ($z < 0.9z_{avg}$) ones. As an illustration, lowland contours are shown in the top of Figure 3c; highlands – in the bottom. The size of the segmented object was defined as $A^{0.5}$, where $A$ is the area of contour.

The roughness increases after 6 min of plasma treatment (Fig. 4a); further treatment flattens the relief below the level of the untreated material. Such behaviour can be explained by examining lowlands and highlands areas separately, the latter being agglomerates of the hard phase of PU.

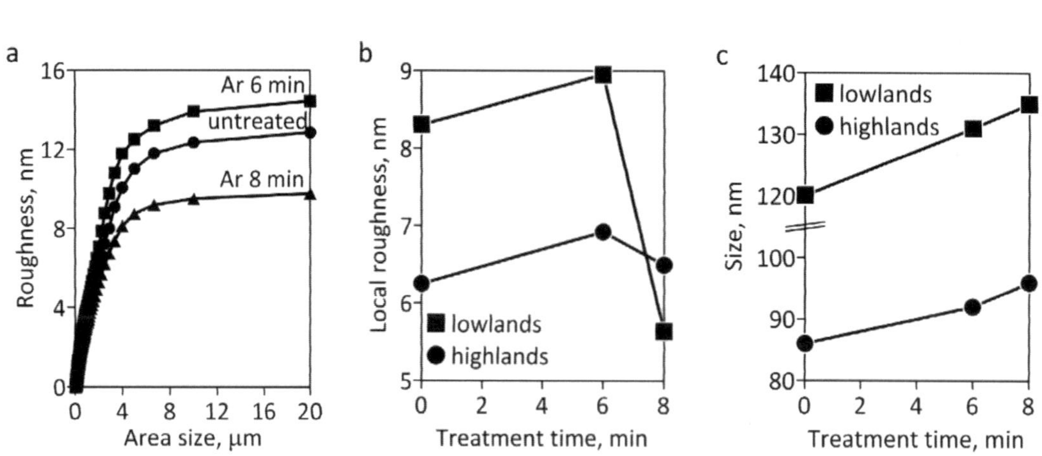


**Figure 4.** (a) Roughness depending on observation size; (b) local roughness and (c) size of lowlands and highlands.

IOP Conf. Series: Materials Science and Engineering **699** (2019) 012031 doi:10.1088/1757-899X/699/1/012031

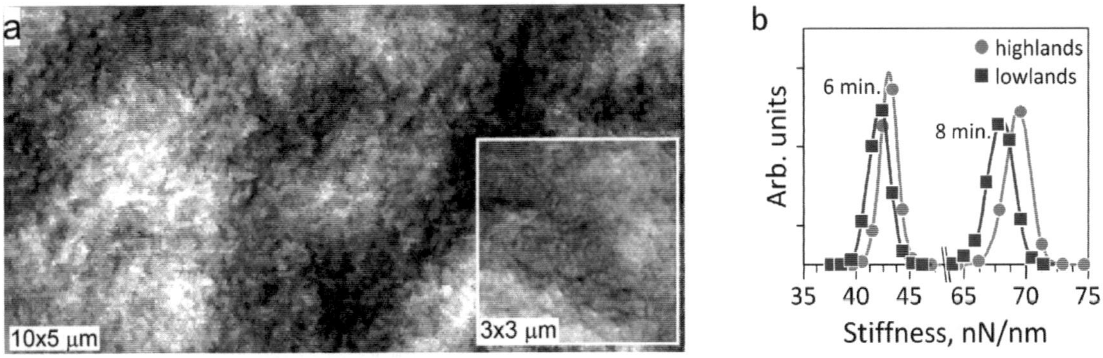


**Figure 5.** (a) AFM image of the surface stiffness obtained in force modulation regime and (b) the distribution curves.

Low-energy argon plasma treatment mainly affects the structure of lowlands: the areas, where the soft phase of the initial PU prevails. Local roughness (measured separately in the areas of highlands and lowlands) increases after the treatment for 6 min, and then falls (Fig. 4b). Changes in the roughness of the highlands are insignificant, while the roughness of the lowlands has fallen by almost a half. The average size of the lowlands increased by 15 nm after the treatment, and that of the highlands – by 7 nm (Fig. 4c); it means the etching of low (soft) areas and growth of high (hard) areas. At the same time, both types of surface segments became flatter; the total roughness (Fig. 4a) at $\tau = 8$ min fell. Thus, in the terms of microstructure, the given plasma treatment primarily affects the soft phase of PU.

The stiffness of the treated surfaces is heterogeneous. For example, Figure 5a shows a map of stiffness obtained in force modulation regime for the treated PU ($\tau = 8$ min): the stiffness of the highlands is 10% higher than that of the lowlands. The statistical distributions of the stiffness (obtained by the force modulation) are given in Figure 5b; longer activation leads to the significant rise of stiffness.

Estimated values of the average elastic modulus of the treated surfaces calculated from the initial part of the indentation curves (Fig. 1b), were 60 and 300 MPa for 6 and 8 min, respectively. The deformation of the soft elastic substrate was not accounted, so the modulus of the coating itself was higher.

The coating thickness was 5±2 nm regardless the treatment time, i.e. at $\tau \leq 6$ min the limiting thickness of the surface modification in this regime is reached; further treatment changes the structure and stiffness of this layer. Wide scatter of thickness from the mean value is explained by inhomogeneous properties of the original substrate.

## 4. Conclusion

Low energy argon plasma treatment of soft elastic polyurethane caused structural and mechanical changes of the surface. The initial substrate is a two-phase system: thread-shaped hard blocks (elastic modulus ~40 MPa) in a softer matrix (elastic modulus ~20 MPa). Dense clusters of hard phase form agglomerates, which look like bumps on the surface. Plasma treatment caused degradation of the threads of the hard phase, significantly changing the microstructure (size and roughness) of the surface lowlands, where the concentration of the hard phase is relatively low. The structure of the highlands (agglomerates of the hard polymer phase) has changed insignificantly.

The duration of plasma treatment (6 or 8 minutes) did not affect the thickness of the modified surface layer (non-uniform thickness from 3 to 7 nm), but significantly affected the average surface stiffness, which increased compared to untreated material by 2 and 10 times, respectively. The local stiffness is heterogeneous and depends on the surface area: treated agglomerates of the hard polymer phase are 10% harder than the surrounding surface.

It is known that such treatment increases the hydrophilicity of the surface and promotes sorption of certain blood proteins, which is a promising way to improve the biomedical properties of the material

IOP Conf. Series: Materials Science and Engineering **699** (2019) 012031 doi:10.1088/1757-899X/699/1/012031

[6]. However, a significant increase of coating stiffness in relation to the substrate can lead to undesirable consequences when deforming the treated material: under the influence of external loads the surface may be damaged. These processes require further study.

**Acknowledgements**
The work was supported by Russian Science Foundation, grant No 17-79-20042.

**References**
[1] Stüber M, Niederberger L, Danneil F, Leiste H, Ulrich S, Welle A, Marin M and Fischer H 2007 *Adv. Eng. Mater.* **9** 1114-22
[2] Tsubone D, Hasebe T, Kamijo A and Hotta A 2007 *Surf. Coat. Tech.* **201** 6423-30
[3] Morozov I A, Kamenetskikh A S, Beliaev A Yu, Scherban M G, Lemkina L M, Eroshenko D V and Korobov V P 2019 *Biophysics* **64** 527-35
[4] Ozdemir Y, Hasirci N and Serbetci K 2002 *J. Mater. Sci.-Mater. Med.* **13** 1147-52
[5] Sanchis M R, Calvo O, Fenollar O, Garcia D and Balart R 2008 *Polym. Test.* **27** 75-83
[6] Alves P, Pinto S, de Sousa H C and Gil M H 2011 *J. Appl. Polym. Sci.* **122** 2302-8
[7] Wilson D J, Rhodes N P and Williams R L 2003 *Biomaterials* **24** 5069-81
[8] Calcagno L, Compagnini G and Foti G 1992 *Nucl. Instr. Meth. B* **65** 413-22
[9] Klinedinst D B, Yilgör I, Yilgör E, Zhang M and Wilkes G L 2012 *Polymer* **53** 5358-66
[10] Zheng J, Ozisik R and Siegel R W 2006 *Polymer* **47** 7786-94

# Microscopy of composite materials based on carbon fibre

**T I Muravyeva[1], O O Shcherbakova[1], A G Shpenev[1] and D L Zagorskiy[1,2]**

[1]Ishlinsky Institute for Problems in Mechanics RAS, 119526 Moscow, Russia
[2]Gubkin Russian State University of Oil and Gas, 119991 Moscow, Russia

shcherbakovaoo@mail.ru

**Abstract.** The work is devoted to a composite material consisting of carbon fibres in a carbon pitch-based matrix. Due to its high wear and heat resistance, this material is widely used in aviation brake systems. Two samples with different impregnation of fibres were studied. It is shown that their tribological properties are different. A complex method of microscopy (optical, scanning electron, and scanning probe microscopies) was used to study the friction surface and inner structure of the samples. In the sample with the best properties, the fibres have a regular ribbed side surface, fibre bundles are located parallel to the friction surface, matrix around the fibres has a denser structure. It is also shown that in this sample wear rates of the fibres and the matrix are the same. Thus, a small change in the production process leads to a noticeable change in the structure, which determines the operational properties.

## 1. Introduction

The production and application of composite materials, and in particular, carbon-fibre materials, have significantly increased in recent years [1-3]. It is known that carbon fibre-based composites in a carbon matrix demonstrates high wear resistance and heat resistance, which makes them useful in the production of brake discs in aircraft industry [4,5]. Obviously, the properties of such materials depend on the structure of the composite materials [6]. It is known that the required characteristics are achieved only under certain conditions for the production of these materials, and even small changes in production technology can lead to a significant deterioration in properties [7].

This paper is devoted to investigation of the correlation of fibre manufacturing technology with the structure and tribological properties of the resulting composite material.

## 2. Materials and methods

### *2.1. Samples*

Composite materials based on graphitized fibre in a carbon matrix were studied. Samples were provided by JSC "AK Rubin". The original fibres had a diameter of about 7 microns, and were grouped into parallel bundles. The standard technology used includes impregnation of fibres with a special compound (so-called "apret"), high-temperature annealing (graphitization of fibres), followed by pouring by a carbon pitch-based matrix.

Using this technology, two types of samples were prepared (No. 1 and No. 2). The difference in their preparation consisted in different impregnation compound composition and lower fibre annealing temperature for sample No. 2.

Tests for wear resistance were carried out on the IM-58 friction machine. A study of mechanical properties was also carried out using bending tests.

### *2.2. Microscopy*
A complex method of microscopic studies was used to investigate the surface of the samples [8,9]. An optical microscope (OM) "Neophot-2" with an image observation system in digital format "Image Scope" was used in this work. Electron microscopic studies were carried out on a scanning electron microscope (SEM) "FEI" Quanta-650 "(an accelerating voltage of 25 kV) with a secondary electrons detector. The cuts (slides) of the samples were also studied by SEM. It should be noted that in this work the energy dispersive X-ray spectral analyser was not used, since all the samples had almost the same elemental composition. Moreover, the main chemical element of the composite was carbon, which could be determined by this method with a high error. For obtaining the high-resolution surface images, the scanning probe microscope (SPM) study was performed using a Smart SPM-TM. A taping mode was used (the resonant frequency – 250 kHz), the fpN10 cantilevers were used (the force constant about 10-20 N/m and the curvature of the tip about 20 nm). This microscope was equipped by the integrated optical camera for the preliminary selection of the area for investigation. The maximum scanning field was 100x100 microns.

## 3. Results and discussion

### *3.1. Tribological properties*
The tribological tests were done by the sample manufacturer (JSC "AK Rubin") and showed the significant difference in wear resistance: for sample No. 1 this parameter was significantly higher than for sample No. 2. It can be assumed that small changes in the production process can affect the properties of the composite components and significantly change their tribological properties. At the same time, other operational parameters of the studied samples (mechanical properties of the initial carbon fibres, bending and strength properties) were approximately the same for both samples. The main part of the work was devoted to identifying the differences between these samples by microscopy methods.

### *3.2. Optical microscopy*
At the first stage, samples were studied with a low magnification. The resulting images are shown in Figure 1. Analysis of these images indicates that the fibres are grouped into bundles. It is found that the fibre bundles have different diameters – for example, in sample No. 1 it is 50-100 microns, while in sample No. 2 it is 150-200 microns. Such a difference can be explained by the conditions of obtaining: a change in the structure of apret leads to the fact that the fibre bundles of sample No. 2 are worse divided into narrow groups of fibres.

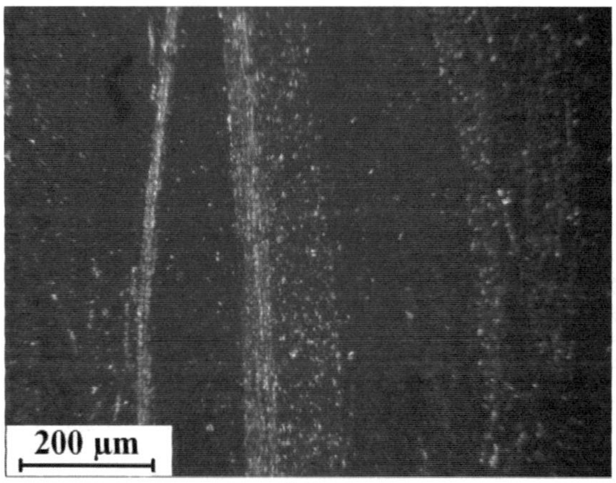

(a)

(b)

**Figure 1.** Optical microscopy of the sample surface: a - sample No.1; b - sample No.2.

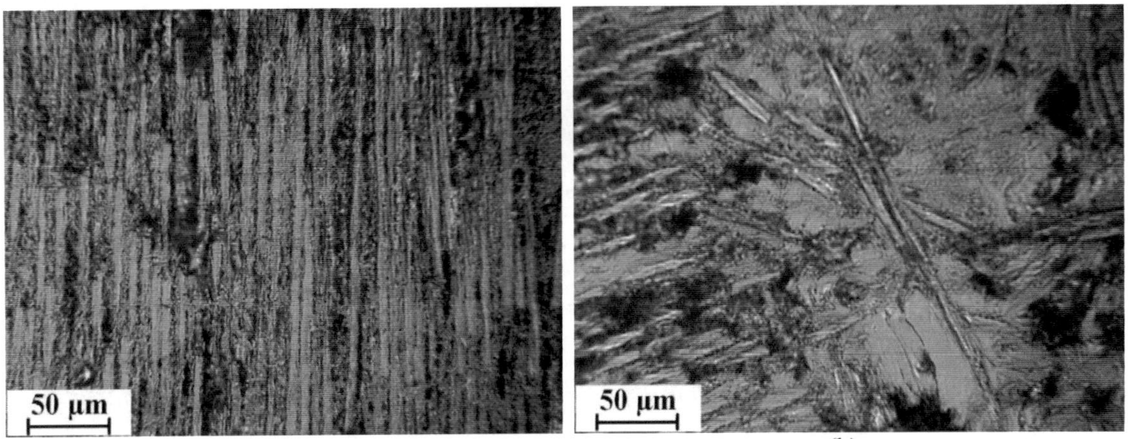


(a) (b)

**Figure 2.** Images (optical microscopy) of sample No. 1 surface after tribological tests.

Microscopic studies with high magnification were carried out for detailed study of the differences in the samples. The obtained results for different areas of the samples are shown in Figures 2 and 3.

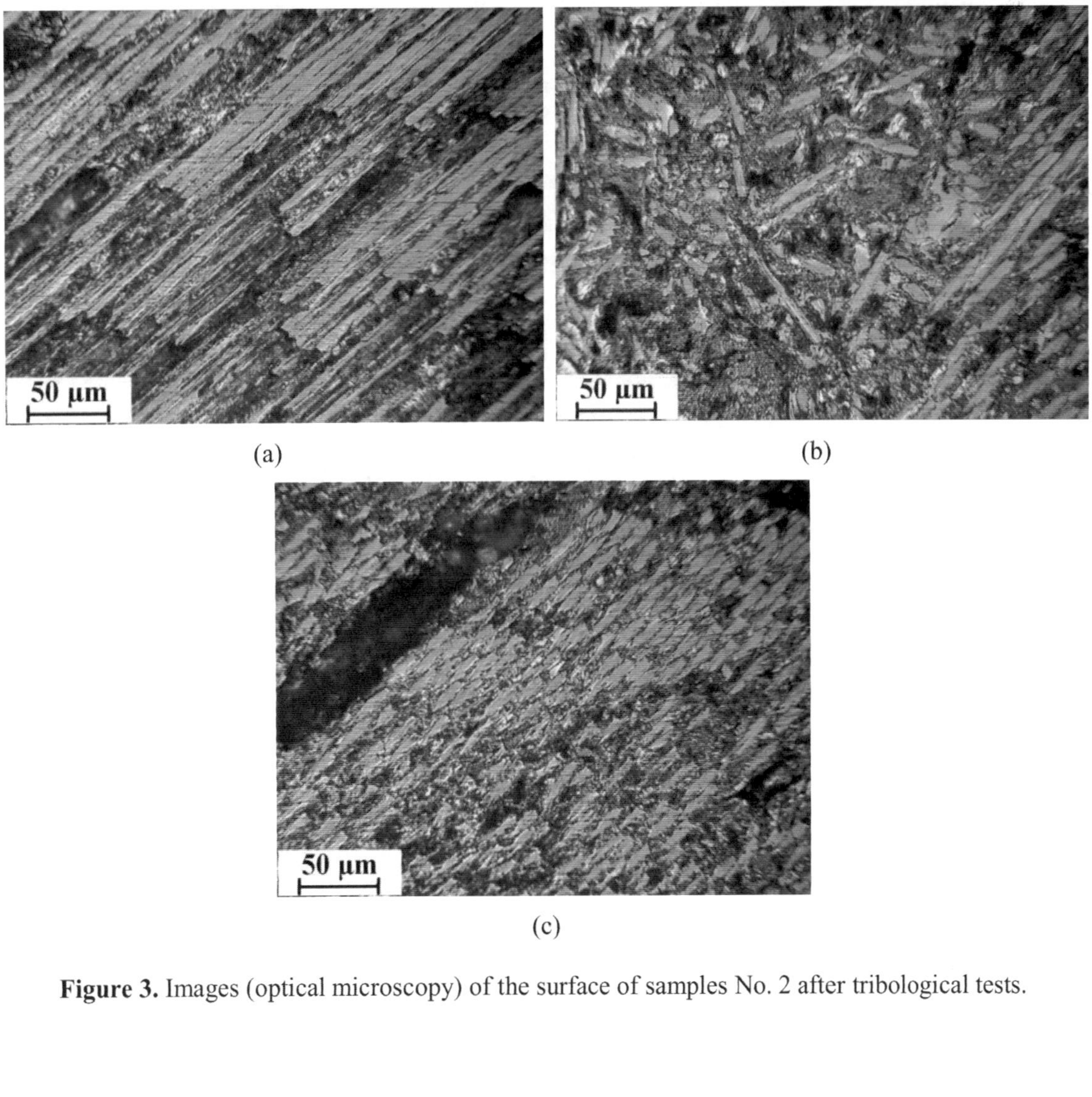


(c)

**Figure 3.** Images (optical microscopy) of the surface of samples No. 2 after tribological tests.

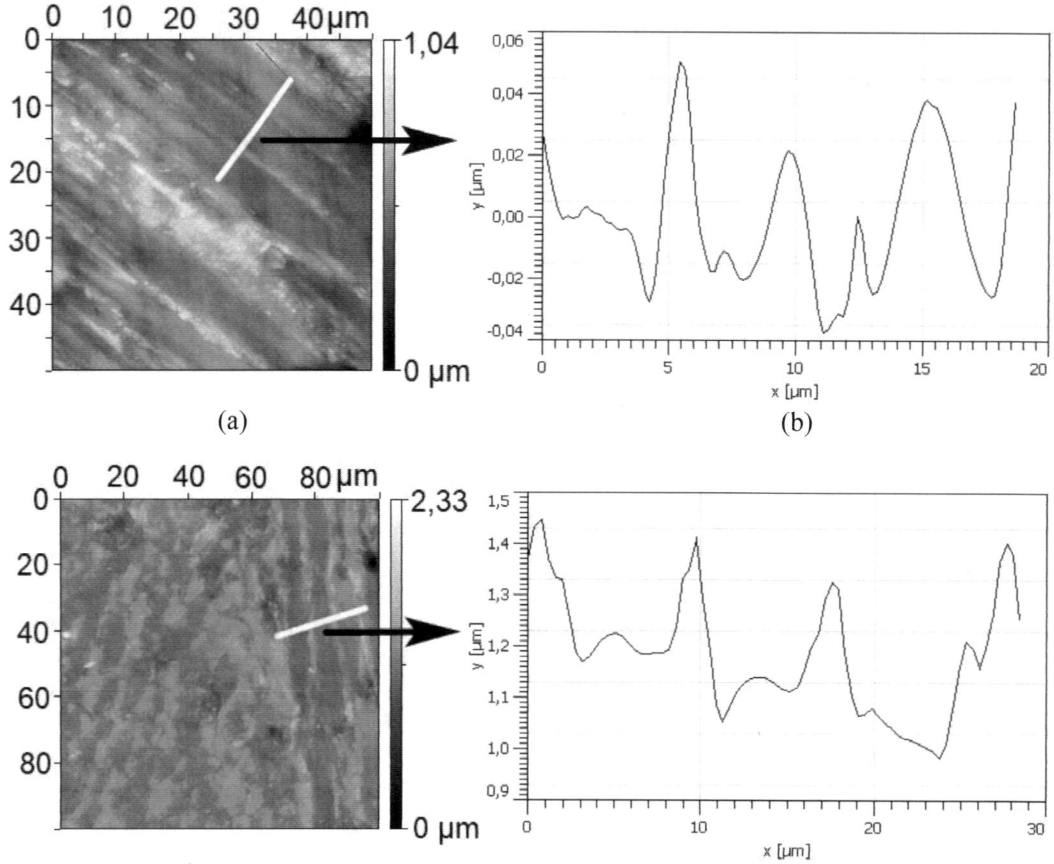


(a) (b)

(c) (d)

**Figure 4.** SPM images of the surface of samples No. 1 (a-topography, b-section profile) and No. 2 (in-topography, g-profile section). The profiles are indicated along the highlighted lines.

Analysis of these images showed that there are two types of fibre arrangement at the surface of sample No. 1. There are areas with fibres arranged parallel to the friction surface (regions of fibre bundles) and areas with a small number of randomly arranged fibres (regions of randomly reinforced matrix). At the same time, three types of fibre arrangement can be clearly distinguished at the surface of sample No. 2. Two types are the same as already described; the third type of fibre arrangement is their unparallel orientation to the friction surface. The appearance of areas with unparallel orientation of fibres is a consequence of the higher ability of the fibre bundles of sample No. 2 to bend (unlike bundles of sample No. 1, which split and break when flexed).

*3.3. Probe microscopy*

The SPM images of the surface and the surface profiles are shown in Figure 4. It is seen that the nature of the surface topography of the samples correlates with the data of optical microscopy. In addition, the method of SPM profilometry showed a significant difference in the surface topography of samples No. 1 and No. 2 and evaluated the pattern of surface wear in these cases.

Thus, the analysis and comparison of areas, in which the fibres lie parallel, showed that in samples No. 1 the surface relief is flat, the fibres and matrix surface are at the same level. So, it may be concluded that the fibres and the matrix wear at the same way. At the same time, sample No. 2 has a distinct relief: the fibres lie considerably below the level of the matrix surface (0.3-0.4 µm). The reason of this effect is lower wear resistance of the fibres.

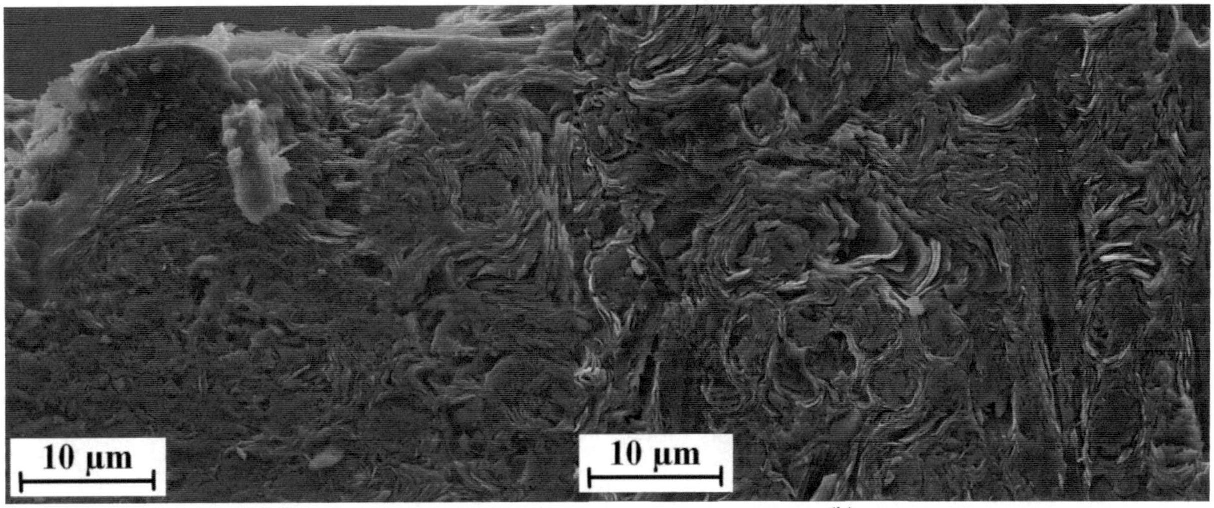


(a) (b)

**Figure 5.** SEM images of slices: a - sample No. 1, b - sample No. 2.

*3.4. Electronic microscopy*

Further studies were carried out using SEM. The images of the sections (slices) of samples No. 1 and No. 2 are shown in Figure 5. The fibres sections and the main matrix could be distinguished at both figures. It is easy to see that in sample No. 2 the fibres stand out in the matrix: they have a rather sharp border with the matrix, while the matrix around the fibres have a looser structure. For sample No. 1, the fibres and the matrix are practically merged, and the matrix itself is rather dense. The reason for this was identified in subsequent measurements.

Individual fibres for both samples were investigated using SEM with the highest magnification for samples No. 1 and No. 2. Figure 6 illustrates the difference of fibres. In sample No. 1, it consists of a large number of flat plates, which are layers of graphite. The side surface of these fibres consists of thin edges of graphite layers. Sample No. 2 fibre consists of a smaller number of thicker layers, having side surface, with cracks and deep cavities. These features have a great influence on the process of fibre/matrix interface formation and properties of matrix around fibres.

(a) (b)

**Figure 6.** SEM images of individual fibres: a - sample No. 1, b - sample No. 2

This difference in fibre structure may be explained by the features of the manufacturing process: different annealing temperatures and impregnation compounds. This difference in technology also leads to a change in the shape and number of crystallite particles covering the fibres (these crystallites are formed during the thermal decomposition of apret).

It can be assumed that the nature of the surface of the fibres, the state of the graphite layers strongly affects the properties of the fibres. These factors, as well as the presence of particles on the surface of the fibres, can greatly influence the interaction with the surrounding matrix. As a result of a change in the surface structure of the fibre, a change occurs in the layers of the matrix adjacent to the fibres. Obviously, the structure of the matrix layers surrounding the fibre depends on the structure of the fibre surface. This is clearly demonstrated by slice images of samples No. 1 and No. 2. It is clearly visible that the fibres in the cross section of the material in sample No. 2 have a sharp border with the matrix, and the layers of the matrix around the fibres have a loose structure. At the same time, in sample No. 1, the fibre-matrix boundary is practically indistinguishable, and the matrix itself has a rather dense structure. Obviously, all this also affects the tribological properties.

## 4. Conclusion

The features of sample preparation determine the shape of the fibre, the surface of the fibre, and its interaction with surrounding matrix and orientation of fibres in the matrix. Different orientations of the fibres in matrix determine sample`s friction properties. Friction properties are also connected with correlation of wear-resistances of fibres and matrix. Small changes in fibre production technology lead to significant changes in the structure and properties of the resulting material (fibre-matrix composite) [10-13].

**Acknowledgements**

This work was supported by the Russian Scientific Foundation, project No. 19-19-00548. Authors are grateful to JSC “AK Rubin” for the materials provided and their tribological testing.

## References

[1] Byrne C, and Wang Z 2001 *Carbon* **39** 1789-809
[2] Gomes J R, Silva O M, Silva C M, Pardini L C and Silva R S 2001 *Wear* **249** 240-5
[3] Sharma M, Rao I M and Bijwe J 2009 *Wear* **267** 839-45
[4] Shpenev A G, Kenigfest A M and Golubkov A K 2016 *Springer Proc. Phys.* **175** 551-9
[5] Schmidt D L, Davidson K E and Theibert L S 1999 *J. SAMPE* **35** 27-39
[6] Awasthi S and Wood J L 1988 *Ceram. Eng. Sci. Proc.* **9** 553-9
[7] Kasema H, Bonnamya S, Berthierb Y and Jacquemard P 2010 *Wear* **269** 104-11
[8] Stolyarova O O Muravyeva T I Zagorskiy D L and Belov N 2016 *Phys. Mesomech.* **19** 104-13
[9] Mezrin A M, Shcherbakova O O, Muravyeva T I, Zagorskiy D L and Shkalei I V 2019 *Front. Mech. Eng.* **5** 14
[10] Shpenev A G 2018 *Springer Proc. Phys.* **207** 459-68
[11] Ozcan S and Filip P 2005 *Wear* **259** 642-50
[12] Shin H K, Lee H B and Kim K S 2001 *Carbon* **39** 959-70
[13] Hutton T J, Johnson D and McEnaney B 2001 *Wear* **249** 647-55

IOP Conf. Series: Materials Science and Engineering **699** (2019) 012033 doi:10.1088/1757-899X/699/1/012033

# Uncertainty of tunnelling microscopy measurements of the field emission from multilayer nanostructures

**E V Panfilova[1], A B Syritskii[2] and A R Ibragimov[1]**

[1]Department of Electronic Technologies in Mechanical Engineering, Bauman Moscow State Technical University, 105005, Moscow, Russia
[2]Department of Metrology and Interchangeability, Bauman Moscow State Technical University, 105005, Moscow, Russia

panfilova.e.v@bmstu.ru

**Abstract**. The paper presents the results of scanning tunnelling microscopy investigation of electron field emission from multilayer nanostructures based on silica opal film Cr-$SiO_2$-Au-C. Opal film was deposited in colloidal solvent, while other films – by vacuum deposition methods. We discuss the problem of uncertainty of measurements of the field enhancement factor and carbon nano-edge sizes using the current versus applied voltage characteristics of the emission. Field enhancement factor is defined as the ratio of the electric field at the nano-edge to the field applied to the gap. We found that at well-controlled experimental conditions the combined uncertainty of field enhancement factor measurements is ~15% based on a 95% confidence interval and the combined uncertainty of nano-edge sizes measurements is ~18% based on a 95% confidence interval.

## 1. Introduction

The electron field emission (FE) structures (cold cathodes) can be applied in microwave, vacuum electronic and microelectronic devices, and various other devices and technologies [1,2]. The paper presents the results and metrological assurance [3] of scanning tunnelling microscopy (STM) investigation of the FE from multilayer nanostructures Cr-$SiO_2$-Au-C. Carbon a-C film was an emission active nanostructure. Gold film was a catalyst for the growth of carbon nanostructures. Silica $SiO_2$ was an opal film and had well-ordered structure. Opal film served as a matrix for formation of the topology of upper layers [4]. The Cr film provided the conductivity for the study on STM and acceptable adhesion of opal to the sitall substrate.

Opal film with diameter of silica microspheres of 220-250 nm and thickness of about 700 nm was deposited in colloidal solvent by vertical deposition method [5] on the Cr sublayer. The 100-nm thick Cr film and 80-nm-thick Au film were deposited by magnetron sputtering [6,7]. Carbon a-C structures with the thickness of about 50 nm were deposited by plasma enhanced chemical vapour deposition (PECVD). The Au and a-C films thicknesses were controlled on a witness sample – a sitall substrate. All films thicknesses were obtained by measuring the step height using atomic force microscopy (AFM). A schematic representation of the sample structure is shown in Figure 1.

In this study, we determine the uncertainties of measurements of the FE enhancement factor or gain coefficient $\beta$ and carbon nano-edges radius $r$ using the current $I$ versus applied voltage $V$ characteristics of the emission [8]. Factor $\beta$ is defined as the ratio of the electric field at the nano-edge to the one applied

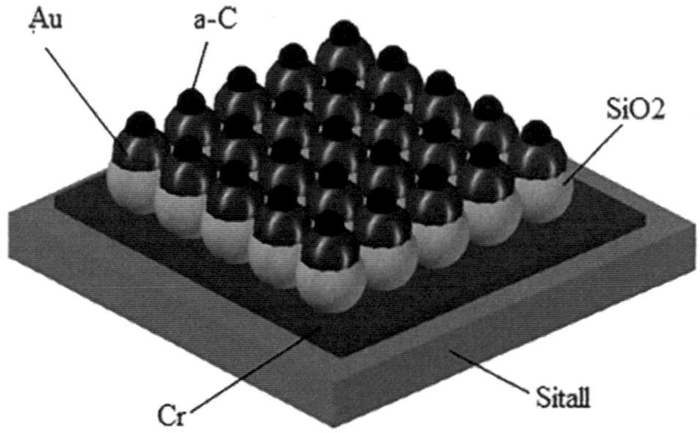


**Figure 1.** Scheme of the Cr-$SiO_2$-Au-C structure on the sitall substrate.

to the gap. A large $\beta$ factor allows emitters to operate at a small applied voltage. To achieve a low turn-on field and a high current density, field emitter materials should preferably have a low work function (or electron affinity) and be nanostructured with needle-like edges to enhance significantly the field at the edges [9]. While the factor $\beta$ is inversely proportional to the radius $r$, it is important to make the edges as sharp as possible to reduce the applied external field required for turning on the electron emission [10,11].

## 2. Method description

The samples were studied using scanning probe microscope (SPM) Solver P-47 (NT-MDT, Russia) by STM in topographic mode by direct current method and current spectroscopy mode. Measurements were operated at a controlled temperature and humidity of 293 K and 50 %, respectively. The SPM was set on a vibration isolation system. Minimum step STM-scanning was 0.006 nm. During scanning, the voltage varied from 0.1 to 1.0 V. It was found that the relief image of multilayer nanostructures on the metal sublayer is dependent on the tunnel voltage generated in the gap as shown in the Figure 2.

This effect was most likely not related to mass transfer and was explained by the presence of emission processes. In this way, three regimes can be defined in the current $I$ versus applied voltage $V$ in a Cr-$SiO_2$-Au-C structure: (1) tunnelling regime, when the applied voltage is lower than the work function, (2) the intermediate regime, and (3) the field emission or Fowler–Nordheim regime for higher voltages.

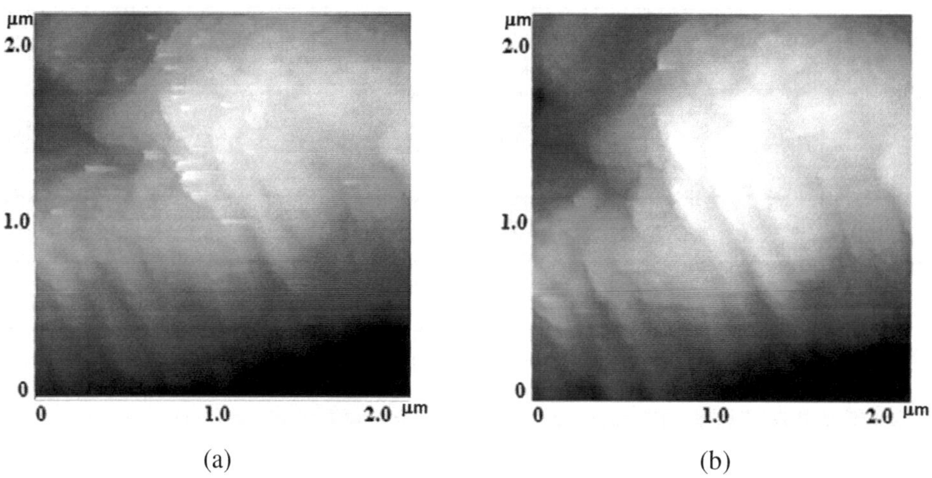


**Figure 2.** The STM-images of multilayer nanostructures Cr-$SiO_2$-Au-C for various regimes of the $I$-$V$ curve: (a) tunnelling regime, 0.1 V, (b) FE regime, 1.0 V.

IOP Conf. Series: Materials Science and Engineering **699** (2019) 012033 doi:10.1088/1757-899X/699/1/012033

We measured the current-voltage characteristics of the structures and analysed them by means of a Fowler-Nordheim (FN) plot. FN theory [12] describes the electric field induced electron tunnelling from a flat perfectly conducting planar surface through a potential-energy barrier. FN plot represents $\ln(I/E_0^2)$ on $E_0^{-1}$ dependence, where $I$ is a current and $E_0$ is a macroscopic field. The field enhancement factor $\beta$ can be deduced as $E/E_0$, where $E$ is the surface field of the emitter. The curve slope $\tan\alpha$ of $\ln(I/E_0^2)$ versus $E_0^{-1}$ gives the field enhancement factor [13], which allows calculating the characteristic size of carbon nano-edge radius.

The field enhancement factor $\beta$ is expressed as a function of the electron work function of material $\varphi$ and $\tan\alpha$:

$$\beta = \frac{B\varphi^{3/2}\lg e}{\tan\alpha}, \tag{1}$$

where $B$ is coefficient.

In certain cases, the field enhancement factor $\beta$ can be expressed as:

$$\beta = \frac{1}{\sqrt{2Rr}} \tag{2}$$

where $R$ is the distance between nano-edge and the probe. Consequently, nano-edge radius $r$ is expressed as a function of the electron work function $\varphi$, the slope $\tan\alpha$, and the distance $R$:

$$r = \frac{R\tan^2\alpha}{2B^2\varphi^3\lg^2 e} \tag{3}$$

In this case, equations (1) and (3) are the measurement models [14]. According to the Guide to the Expression of Uncertainty in Measurement (GUM) [15], a measurement results in a value of a well-defined physical quantity – the measurand – and a corresponding uncertainty, which characterizes the dispersion of the values attributed to the measurand [15]. It is widely recognized that any measurement result has an uncertainty range, within which the true value may be found with a certain probability [16]. The measurement uncertainty is determined by identifying, assessing, and combining all the sources of uncertainty associated with the quantity value.

We identified two types of uncertainty evaluations for inputs [15]. For equation (1) of type B, evaluation includes uncertainty in the electron work function $\varphi$. For equation (3) of type B, evaluation includes uncertainties in the electron work function $\varphi$ and the distance between nano-edge and the probe $R$. For both equations (1) and (3) of type A, evaluation is derived from repeated observations of the slope $\tan\alpha$. Based on these uncertainty evaluations, an uncertainty calculation scheme [15] is developed in this work.

## 3. Results

Fifteen observations of the current $I$ versus applied voltage $V$ characteristics were carried out for various dots on three samples. Typical $I$-$V$ curve is shown in Figure 3. The FE regime was observed for higher applied voltage.

The relative combined standard uncertainty $U_{rc}$ was calculated by collecting all aforementioned uncertainties. Finally, the relative expanded uncertainty $U_{exp}$ for the coverage factor $k$ at the confidence of level of 95% was obtained as follows:

$$U_{exp} = kU_{rc} = 2U_{rc} \tag{4}$$

To investigate the influence of the inputs on the uncertainty of the measurement, uncertainty budgets for the factor $\beta$ and the nano-edges radius $r$ were calculated. The uncertainties of the contributing variables for the factor $\beta$ are showed in Table 1. When the factor $\beta$ is 19.30 and the radius $r$ is 1.85 nm, the total results are reported in Table 2.

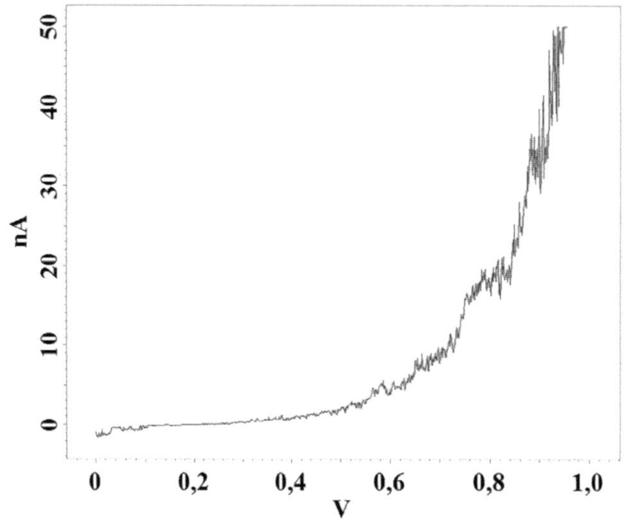


**Figure 3.** *I-V* curve of the Cr-$SiO_2$-Au-C structure on the sitall substrate.

**Table 1.** Uncertainty analysis.

| Uncertainty | Uncertainty evaluations type | Probability distribution | Degrees of freedom | Relative standard uncertainty (%) |
|---|---|---|---|---|
| $\varphi$ | B | normal | $\infty$ | 2.2 |
| $\tan\alpha$ | A | normal | 14 | 8.4 |

**Table 2.** Uncertainties of the factor $\beta$ and the nano-edge radius $r$ measurements.

| Measurand | The relative combined standard uncertainty (%) | The effective degrees of freedom | The coverage factor | The relative expanded uncertainty (%) |
|---|---|---|---|---|
| $\beta$ | 7.4 | $\infty$ | 2 | 14.8 |
| $r$ | 9.1 | $\infty$ | 2 | 18.2 |

Examination of $\tan\alpha$ measurements shows remarkable influence of this contamination on the detected factor $\beta$ and radius $r$.

## 4. Discussion

The reproducibility of the samples properties has the largest influence on the uncertainty of the *I-V* curve and $\tan\alpha$ . Multilayer film deposition is a very complex process. Study of films growth kinetics revealed that many factors were critical for reproducible deposition of the structure.

A fine tuning of structure properties was difficult, because growth parameters were very interdependent. It was shown that the opal film packing depended significantly on substrate pre-treatment, roughness of Cr film, colloid properties, and lifting velocity. The Au precursor resulting roughness depended on packing density of microspheres and was a dominant factor affecting a-C film growth. So it was hard to repeat structure growth process for several samples and various dots on the samples surface and reproducibility of the samples properties sometimes was not sufficient. More precise technologies and equipment could be further employed to reduce the uncertainty.

IOP Conf. Series: Materials Science and Engineering **699** (2019) 012033 doi:10.1088/1757-899X/699/1/012033

## 5. Conclusion

In this paper, the procedure and analysis of the uncertainty of tunnelling microscopy measurements of the electron field emission from multilayer nanostructures Cr-$SiO_2$-Au-C were given. The relative expanded uncertainties ($k$=2) of field enhancement factor measurements and of nano-edge sizes measurements were ~ 15 % and ~18 %, respectively. The result of our estimation of uncertainty is associated with determination of the factors influencing on the detected factor $\beta$ and radius $r$ in the measurement models (1) and (3). The standard uncertainty of the $\tan\alpha$ of a Fowler-Nordheim plot has the largest influence on the measurement results uncertainty. The $\tan\alpha$ uncertainty is strongly influenced by the reproducibility of the sample properties. Thereby, the phenomenon of uncertainty of STM measurements of factor $\beta$ and radius $r$ is mainly related to the reproducibility of the film deposition process and unevenness of multilayer structure. The main application of this methodology is laboratory testing of the emission nanostructures.

## References

[1] Saito Y and Uemura S 2000 *Carbon* **38** 169-82
[2] Silva S R P and Carey J D 2003 *Diamond and Related Materials* **12** 151-8
[3] Komshin A S and Orlova S R 2016 *Meas. Tech.* **59** 589-94
[4] Bunkin N F, Gorelik V S and Filatov V V 2010 *Phys. Wave Phenom.* **18** 90-5
[5] Kuleshova V L, Panfilova E V and Prohorov E P 2018 *Int. Russ. Automation Conf. (RusAutoCon)* (IEEE) 1-5
[6] Sidorova S V, Pronin M A and Isaeva A A 2018 *Int. Russ. Automation Conf. (RusAutoCon)* (IEEE) 1-4
[7] Panfilov Y V, Kolesnik L L, Ryabov V T and Sidorova S V 2017 *J. Phys.: Conf. Ser.* **872** 012010
[8] Syritskii A B and Panfilova E V 2018 *IOP Conf. Ser.: Mater. Sci. Eng.* **443** 012035
[9] Duan J L, Lei D Y, Chen F, Lau S P, Milne W I, Toimil-Molares M E, Trautmann C and Liu J 2015 *ACS Appl. Mater. Interfaces* **8.1** 472-9
[10] Xu N S and Huq S E 2005 *Mater. Sci. Eng.* R **48** 47-189
[11] Fang X, Bando Y, Gautam U K, Ye C and Golberg D 2008 *J. Mater. Chem.* **18.5** 509-22
[12] Jensen K L 2003 *J. Vac. Sci. Technol. B* **21** 1528-44
[13] Joag D S, More M A and Sheini F J 2011 *Nanowires-Implementations and Applications* IntechOpen Ch.24
[14] Possolo A 1900 Simple guide for evaluating and expressing the uncertainty of NIST Measurement Results *NIST Technical Note (*U.S. Department of Commerce: National Institute of Standards and Technology) p 103
[15] ISO/IEC Guide 98-3:2008 (JCGM/WG1/100) *Uncertainty of measurement - Part 3: Guide to the expression of uncertainty in measurement* (GUM:1995)
[16] Kreuzeder A, Santner J, Zhang H, Prohaska T and Wenzel W W 2015 *Environ. Sci. Technol.* **49** 1594-602

# Optimization of the photonic crystal colloidal films deposition by means of atomic force microscopy

**E V Panfilova[1], A B Syritskii[2] and A R Ibragimov[1]**

[1]Department of Electronic Technologies in Mechanical Engineering, Bauman Moscow State Technical University, 105005 Moscow, Russia
[2]Department of Metrology and Interchangeability, Bauman Moscow State Technical University, 105005 Moscow, Russia

panfilova.e.v@bmstu.ru

**Abstract**. This article describes the results of optimization of the colloidal photonic crystal opal films deposition by means of atomic force microscopy investigation. The main factors affecting the quality of the colloidal opal films are studied. Bragg diffraction occurs in the periodic distribution of colloidal microspherical particles, and ordered structure is critical for formation of the photonic band gap. It is found that polystyrene opal structures obtained in optimal conditions have a good periodicity, uniformity, and high packing density.

## 1. Introduction

Today it is already well known that the physical properties of nanoparticles and clusters, determined by their extremely high specific surface area, can differ significantly from the microscopic characteristics [1]. The ability to study nanostructures with high spatial resolution quickly and efficiently is an important question in the development of nanotechnologies. The atomic force microscopy (AFM) has been demonstrated to be capable to investigate almost any kind of samples without any sample preparation with nanometre resolution. This method is widely used in nanostructures research [2]. One of the perspective nanostructured materials is an artificial opal [3,4]. Because of high periodic regular structure, opal films can block the light spreading in wavelength range forbidden for this structure. Colloidal films with the structure of opal matrix can be used in photonics, optoelectronics, sensors, micro- and nanoelectronics, and other various devices and technologies [5,6]. While various strategies have been developed to direct the self-assembly of colloidal particles, fabrication of crack-free and transferrable colloidal film with controllable crystal structures still remains a major challenge [7]. This article describes the polystyrene opal film deposition process and its optimization procedure that was implemented using operating AFM control.

## 2. Experimental

The original method of vertical lifting was used as the opal film deposition method [6]. The substrate is placed vertically in a colloidal suspension of microspheres and is gradually exposed by evaporation or other slow solvent removal at a rate in the range from 0.1 to 10.0 mm/min [8-10]. While lifting, the capillary forces attract the microspheres nearest to the substrate to their surface. It makes possible to obtain sufficiently large-area colloidal opal films by vertical deposition method and to parametrize the deposition conditions for optimal film growth. Figure 1 shows the general scheme of the vertical deposition method.

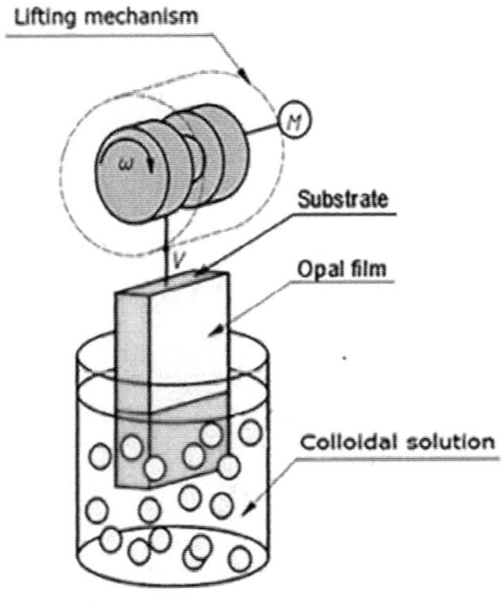


**Figure 1.** The schematic diagram of the vertical deposition method.

The structure of the obtained films was monitored by measuring the relative area of ordered structure by means Solver Next scanning probe microscope (NT-MDT, Russia) in the atomic force semi-contact mode [11]. The photonic properties of the obtained opal matrices were monitored by measuring the reflection coefficient of the photonic band gap (PBG) using Epsilon spectrophotometer (IZOVAC, Belarus).

## 3. Results and discussion

Colloidal polystyrene monodisperse latex was used as a material for the formation of opal matrix. A preliminary experiment was performed using 10 % colloidal solution with particles diameter 220 nm at lifting velocities of 1 and 0.8 mm/min. Films were deposited on the sitall CT-50-1 substrates with a working surface roughness of less than 0.032 μm.

Analysis of macroscopic images and topological analysis of the first obtained samples showed that the films did not have the required uniformity parameters. In this case, the photonic properties of the obtained structures were weakly expressed. Figure 2 presents the results of studies of obtained polystyrene films.

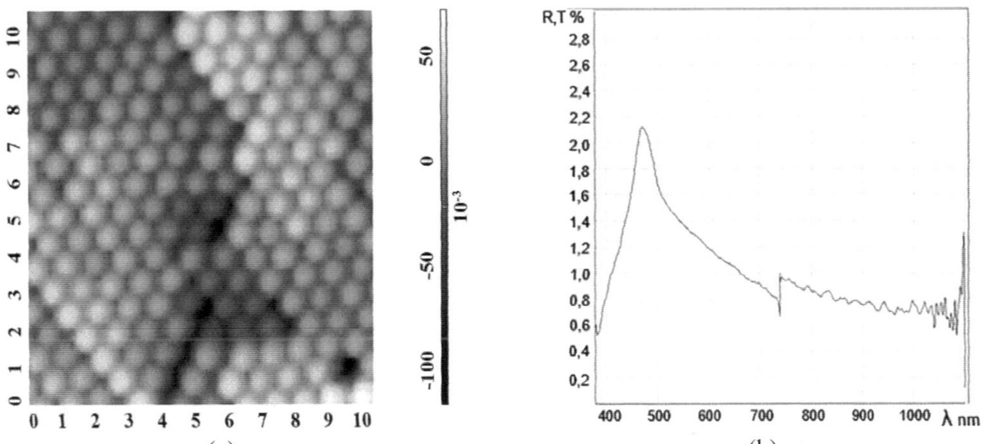


(a) (b)

**Figure 2.** (a) AFM image and (b) the reflection spectrum of the polystyrene film obtained during the preliminary experiment at the 1mm/min lifting speed.

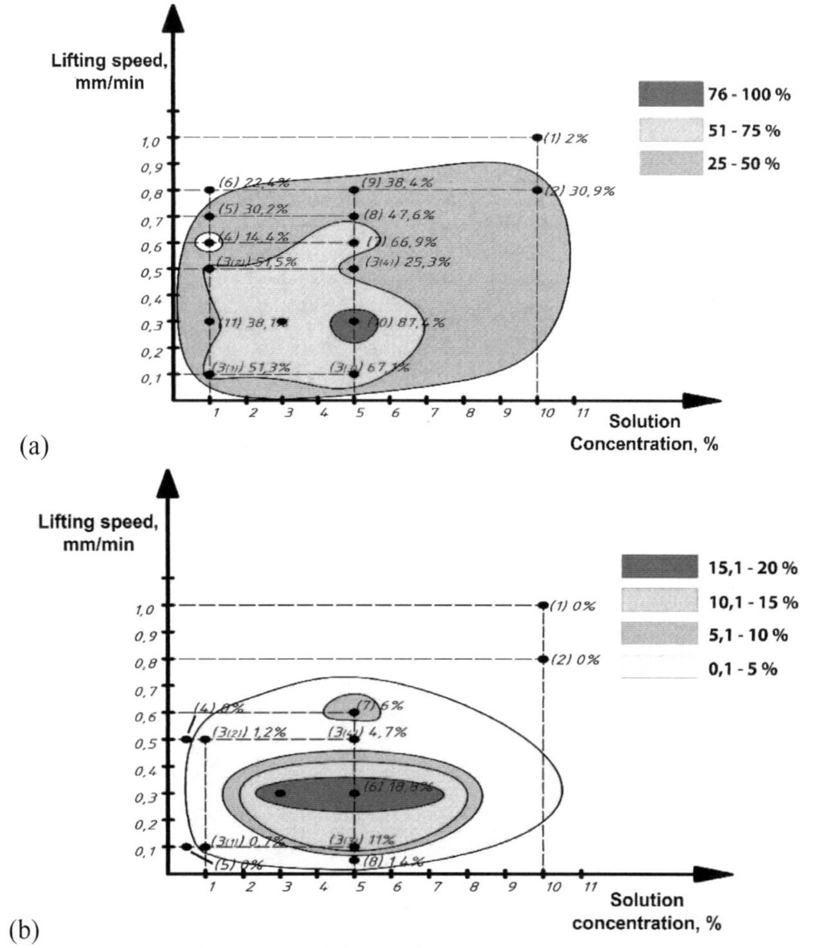


**Figure 3.** Optimization of the vertical deposition method according to: (a) the criterion of uniformly ordered coating, (b) the criterion of reflectance of the photonic band gap.

The process of obtaining colloidal films was optimized by the results of the full factorial experiment. Fifteen observations were carried out for every coordinate plane point. The lifting velocity was varied from 0.1 to 0.5 mm/min, the solution concentration – from 1 to 5%, and particles diameter – from 220 to 330 nm. The relative area of a structurally ordered film and the reflection intensity of the PBG were controlled. Regression equations were considered as joint estimates of linear effects and interaction effects [12]. The results of the full factorial experiment revealed areas of factor space, in which the films were well deposited and opal matrix had a good periodicity, uniformity, and high packing density. To optimize the process, the gradient method was used. Figure 3 shows optimization schemes for 220 nm particles film.

The optimization results revealed that for particles diameters varying from 220 to 330 nm lifting velocity of 0.3 mm/min and solution concentration of 5% make it possible to obtain the most uniform opal colloidal films with pronounced photonic crystal properties. The deposition process allows the opal films to have periodical structure and uniformity. Every particle connects with its neighbours without any defects. As a consequence of its structure periodicity, all obtained films show opalescence: reflected colours that come from the Bragg diffractions in the periodic distribution of microspheres. The greatest reflectance was 24%. The reflection spectrum and the surface structure for 220 nm particles film are shown in Figure 4.

IOP Conf. Series: Materials Science and Engineering **699** (2019) 012034 doi:10.1088/1757-899X/699/1/012034

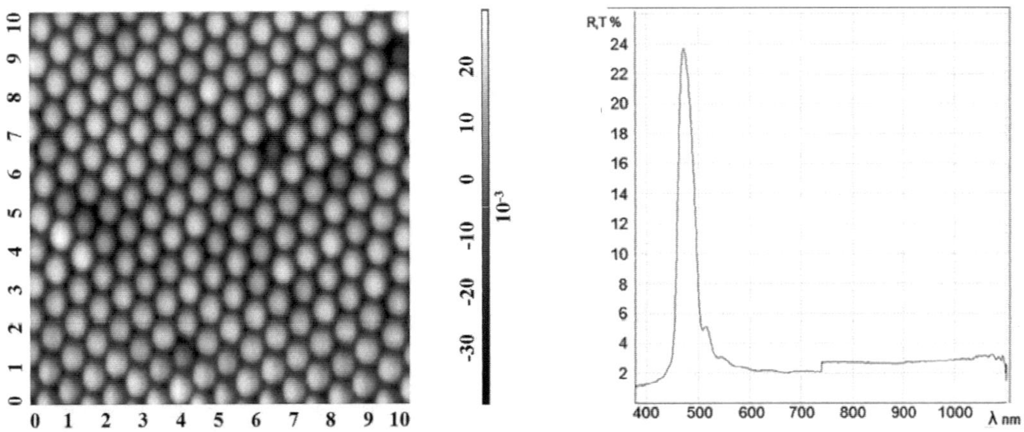


(a) (b)

**Figure 4.** Results obtained after optimization of the vertical deposition method: (a) AFM image of polystyrene film, (b) the reflection spectrum of the polystyrene film.

Concentration of more than 10% leads to agglomeration of the particles and appearance of large clusters of particles on the substrate. At velocities more than 1.5 mm/min, no deposition or empty plots of substrate were seen.

## 4. Conclusion

The vertical deposition process was optimized to allow depositing qualitative opal film. Operational quality control of the films was carried out using AFM. We have fabricated polystyrene opal matrix of particles with diameters ranging from 220 to 330 nm and less than 2% dispersion of Bragg stop band position. The results of the reflectance measurements and AFM investigation of deposited film show that polystyrene particles have been obtained using 5% colloidal solution and the vertical lifting velocity of 0.3 mm/min. The authors see the development of the work in the research of the phenomenon of self-assembly.

## References

[1] Garishin O and Lebedev S 2013 *PNRPU Mech. Bull.* **1** 68-80
[2] Pavliček N and Gross L 2017 *Nat. Rev. Chem.* **1** 0005
[3] Bunkin N F, Gorelik V S and Filatov V V 2010 *Phys. Wave Phenom.* **18** 90-5
[4] Moiseyenko V N, Brynza N P, Sal B A, Holze R and Gorelik V S 2018 *Inorg. Mater.* **54** 1250-5
[5] Panfilov Y V, Kolesnik L L, Ryabov V T and Sidorova S V 2017 *J. Phys.: Conf. Ser.* **872** 012010
[6] Sidorova S V, Pronin M A and Isaeva A A 2018 *Int. Russ. Automation Conf. (RusAutoCon)* (IEEE) 1-4
[7] Fan W, Chen M, Yang S and Wu L 2015 *Sci. Rep.* **5** 12100
[8] Kuleshova V L, Panfilova E V and Prohorov E P 2018 *Int. Russ. Automation Conf.(RusAutoCon)* (IEEE) 1-5
[9] Diao J J, Hutchison J B, Luo G and Reeves M E 2005 *J. Chem. Phys.* **122** 184710
[10] Meijer J 2012 *Langmuir* **28** 7631-8
[11] Syritskii A B and Panfilova E V 2018 *IOP Conf. Ser.: Mater. Sci. Eng.* **443** 012035
[12] Tamhane A C 2009 *Statistical analysis of designed experiments: theory and applications* **609** (John Wiley & Sons)

IOP Conf. Series: Materials Science and Engineering **699** (2019) 012035 doi:10.1088/1757-899X/699/1/012035

# High temperature superconductivity at the interface $Ba_{0.8}Sr_{0.2}TiO_3/La_2CuO_4$

**D P Pavlov[1], I I Piyanzina[1,2], N N Garig'yanov[1], T M Salikhov[1], A O Chibirev[1,2], I F Gilmutdinov[2], R V Yusupov[2], V M Mukhortov[3], T Adachi[4], T Kawamata[5], Y Koike[5], V V Kabanov[1,6] and R F Mamin[1,2]**

[1]Zavoisky Physical-Technical Institute, FRC KazanSC of RAS, 420029 Kazan, Russia
[2]Kazan Federal University, 420008 Kazan, Russia
[3]Southern Scientific Centre, Russian Academy of Sciences, 344006 Rostov-on-Don, Russia
[4]Department of Engineering and Applied Sciences, Sophia University, 102-8554 Tokyo, Japan
[5]Department of Applied Physics, Tohoku University, 980-8579 Sendai, Japan
[6]Department for Complex Matter, Jozef Stefan Institute, 1000 Ljubljana, Slovenia

mamin@kfti.knc.ru

**Abstract**. We present the results of the investigations of high conducting area and superconductivity at the interfaces between ferroelectric oxide and insulating oxide in heterostructures, isostructural to $BaTiO_3/La_2CuO_4$. The performed numerical simulations of $BaTiO_3/La_2CuO_4$ heterostructure show the possibility of high conductivity state at the interface. The temperature dependence of the measured electrical resistance of the $Ba_{0.8}Sr_{0.2}TiO_3/La_2CuO_4$ heterostructure interface have been studied and the superconducting behaviour with transition temperature $T_c$ about 30K has been found. Therefore, the transition to the state with 2DEG at the interface is demonstrated. The results offer the possibility to design novel electronic devices.

## 1. Introduction

The creation of high-mobility quasi-two-dimensional electron gas (q2DEG) at the interface and the ability to control such states by magnetic and electric fields is impossible without the use of new materials and without the development of new design of interfaces. Unique properties of functional materials are achieved due to the effects associated with the complex composition of the interface structure. Such new materials include oxide heterointerfaces between two nonconducting oxides, in which unique transport properties are observed due to strong electronic correlations [1-11]. A high-mobility electron gas has been discovered at the interface between two oxide insulators $LaAlO_3$ (LAO) and $SrTiO_3$ (STO) by Ohtomo and Hwang [1]. In particular, it was found that the metallic phase (quasi-two-dimensional electron gas, 2DEG) is formed in the STO layers at the LAO/STO interface, when the number of LAO layers is larger than three [2]. The density of the charge carriers in such heterostructure reaches the value of $3\ 10^{13}$ $cm^{-2}$. Subsequently, the coexistence of a two-dimensional superconductivity and ferromagnetism was discovered in this system [3,9] and in other systems with the atomically perfect interfaces [12,13]. The LAO/STO system undergoes a transition to a superconducting state at temperatures below 300 mK [3,9]. That is why it is essential to develop technical approaches to create quasi-two-dimensional superconductivity at higher temperatures and to study the processes of

IOP Conf. Series: Materials Science and Engineering **699** (2019) 012035 doi:10.1088/1757-899X/699/1/012035

superconductivity switching. In our work, we used the parent compound of a high-temperature superconductor (PCHTSC) $La_2CuO_4$ (LCO) as a substrate and ferroelectric oxide $Ba_{0.8}Sr_{0.2}TiO_3$ (BSTO) as an overlayer in the heterostructure. We chose the BSTO/LCO heterostructure, because we have a great experience in working with BSTO films [20,21]. Epitaxial films of ferroelectric BSTO were deposited by reactive sputtering of stoichiometric targets by RF plasma (RF-sputtering) [20,21] at 650°C on antiferromagnetic LCO single crystals. The polarization of the ferroelectric film was perpendicular to the surface of interface. Such heterostructure might provide an opportunity to the electron system formation in the substrate layers close to the interface as in the $LaAlO_3/SrTiO_3$ case. The difference is that the polar discontinuity at the interface is created not by the charge sequence in the upper slab, but by the bulk polarization. Therefore, we expect that quasi-two-dimensional superconductivity could be observed at much higher temperatures in the BSTO/LCO heterostructure. The temperature dependences of the electrical resistance and magnetic properties were studied. The numerical simulations were performed for the $BaTiO_3/La_2CuO_4$ (BTO/LCO) heterostructure.

## 2. Modelling and calculations

The $BaTiO_3$ (BTO) compound is close in its physical properties to ferroelectric oxide BSTO. Thus, the numerical simulations [14-20] of the structural and electronic characteristics have been performed for the $BaTiO_3/La_2CuO_4$ heterostructure, because the calculations become much simpler for this case. For density of states calculations and structural optimization, we use density functional theory (DFT) [14]. Correlation effects were accounted by generalized gradient approximation (GGA) [15]. Kohn-Sham equations were solved using the plane-wave basis set (PAW) [16] realized within the VASP code [17], which is a part of the MedeA® software of Materials Design [16]. The cut-off energy was chosen to be 400 eV. The force tolerance was 0.5 eV/nm and the energy tolerance for the self-consistency loop was $10^{-5}$ eV. The Brillouin zones were sampled including 5×5×1 k-points. Since there are strong correlations between *d* and *f*-electrons in our system, the GGA+U correction were included to our computational scheme [18]. The *U* parameter was added to La 4*f* and Ti 3*d* orbitals ($U = 8$ eV and 2 eV, respectively). The choice of *U* for Ti and La atoms was based on our previous results [19].

The density of states of BTO/LCO heterostructure for the case when 2 layers of BTO were deposited on top of LCO substrate is shown in Figure 1. Due to a high mismatch between BTO and LCO (≈30 %) the BTO unit cell was rotated by a 45° before merging with LCO substrate. Because the substrate is compressive with respect to the film, the polarization axis of the ferroelectric BTO will be directed perpendicular to the interface. The density of states of BTO/LCO heterostructure at the Fermi-level becomes also sufficiently large, approximately 5 $eV^{-1}$. The band gap vanishes, and we expect the system to be a good metal at the interface. Thus, the possibility of q2DEG state in this heterostructure has been shown. BTO/LCO heterostructure can be superconductive at low temperatures.

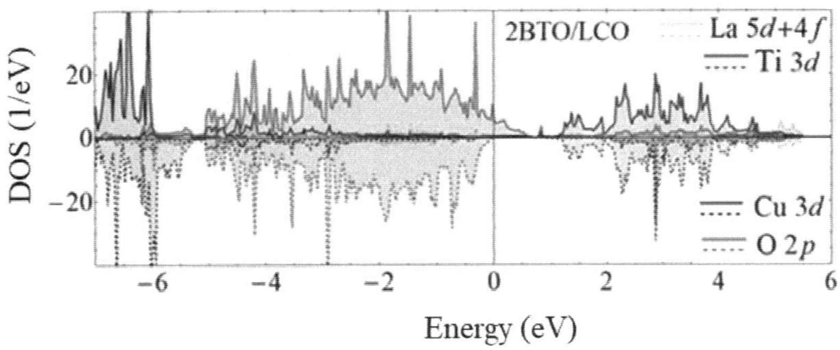


**Figure 1.** Density of states of $BaTiO_3/La_2CuO_4$ heterostructure for the case of 2 $BaTiO_3$ layers on the top of $La_2CuO_4$.

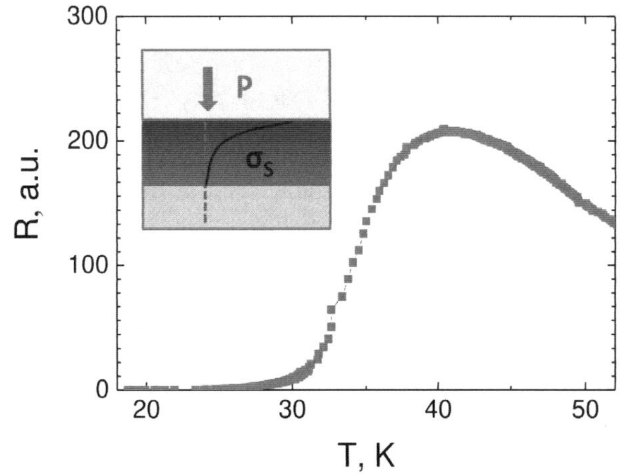


**Figure 2.** Temperature dependences of the resistance for $Ba_{0.8}Sr_{0.2}TiO_3/La_2CuO_4$ heterostructure for film polarization directed perpendicular to ferroelectric film (red squares). Space distribution of surface charge density $\sigma_s$ in the vicinity of the interphase is shown in the insert.

## 3. Experimental results and discussion

The electrical resistance of the BSTO/LCO heterostructure at the low temperatures has been measured. The results are shown in Figure 2. The temperature dependence of the resistance of the BSTO/LCO heterostructure above 50 K shows typical semiconducting behaviour. At temperatures above 50 K the main current flows through the substrate. And only a small part of the current is localized in the interface area. This cannot affect the current in the large volume of the substrate at high temperatures. At low temperatures, the resistance rapidly decreases and superconducting behaviour is observed. Most of the current flows in the interface region, because at these temperatures the resistivity of the substrate increases, but the interface is in the superconducting state. Therefore, the interface between the ferroelectric and insulating oxides demonstrates the superconducting behaviour with the critical temperature about 30 K (Fig. 2). The resistivity starts to drop around 40 K demonstrating the superconducting transition, similar to what is observed in $La_{2-x}Sr_xCuO_4$ (LSCO) single crystals at optimal doping [12].

In order to demonstrate the superconducting state in the heterostructure, we have measured the zero field cooled and field cooled magnetic susceptibility in different magnetic fields (Fig. 3). Observation of the diamagnetic susceptibility (Fig. 3) confirms the existence of superconducting phase at the interface. A very rough estimate of the superconducting layer thickness from diamagnetic susceptibility measurements provides value of 4-100 nm.

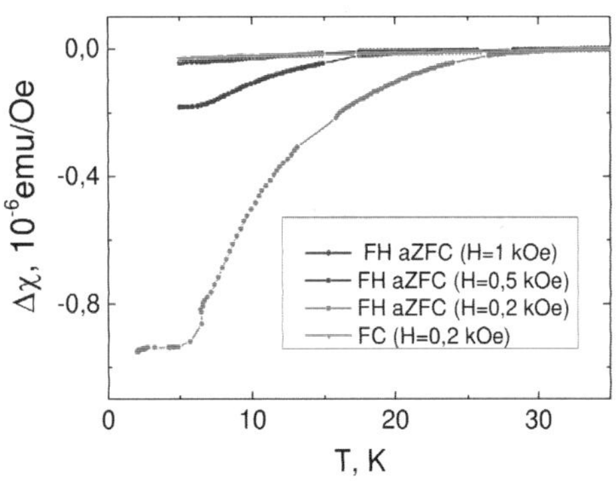


**Figure 3.** Temperature dependence of the magnetic susceptibility of $Ba_{0.8}Sr_{0.2}TiO_3/La_2CuO_4$ heterostructure.

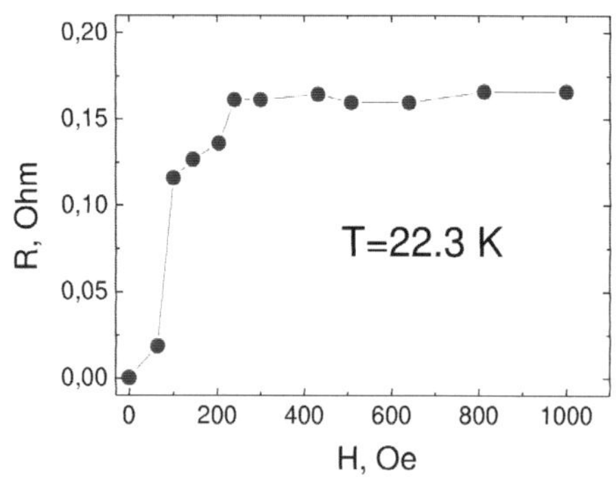


**Figure 4.** Magnetic field dependence of the resistance of the $Ba_{0.8}Sr_{0.2}TiO_3/La_2CuO_4$ heterostructure.

IOP Conf. Series: Materials Science and Engineering **699** (2019) 012035 doi:10.1088/1757-899X/699/1/012035

The magnetic field dependence of the resistance of the $Ba_{0.8}Sr_{0.2}TiO_3/La_2CuO_4$ heterostructure is shown in Figure 4. When a weak magnetic field is applied to the heterostructure in the direction perpendicular to the surface of the interface, the finite resistance of the interface appears and it increases with increasing the field (Fig. 4) as it was predicted [22]. The magnetic field was applied perpendicular to the surface and parallel to the c axis of the LCO substrate at T = 22.3 K. The magnetic field dependence of the heterostructure resistance shows that nonzero resistance appears at a very low field. The $H_{c1}$ for a thin layer of superconductor is very small and the magnetic field penetrates in the superconducting layer. In that case, the system demonstrates flux-flow resistance.

The polarization discontinuity model is most often discussed [1,2,10,11] to describe the q2DEG formation phenomena in the LAO/STO heterostructure. The polar discontinuity at the interface leads to the divergence of the electrostatic potential. The polar discontinuity for LAO/STO case exists due to an alternation of the differently charged layers of $(LaO)^{+1}$ and $(AlO_2)^{-1}$. Therefore, the atomically flat quality of the interface between LAO and STO components is utterly necessary in this case. For the case of BSTO/LCO heterostructure, a volume ferroelectric polarization exists, and it is independent on the sequence of layers. As a result, it is possible to use more simple techniques in order to create BSTO/LCO heterostructure and to obtain quasi-two-dimensional superconductivity at much higher temperatures.

Let us discuss the physical processes at the interface. We take into account that a sharp discontinuity of the polarization $P$ at the interface leads to the formation of the surface charge. The surface charge density $\sigma_s=P_n$, where $P_n$ is the component of polarization perpendicular to the interface (see insert in Fig. 2). This surface charge should be screened. Therefore, the regions with the excess charge are formed at the opposite side of the interface. This charged region from $La_2CuO_4$ side may be considered as a surface doping. The characteristic surface carrier density $n_s$ in this region can be estimated as $n_s = \sigma_s/e = P_n/e$, where $e$ is the elementary charge. Assuming that the characteristic value of polarization in the bulk of ferroelectric is $P_n \approx 30$ μC/cm$^2$ we obtain $n_s \approx 1{,}87\ 10^{14}$ cm$^{-2}$.

This charged region has a nonuniform charge distribution. The characteristic charge density decreases exponentially when we go away from the interface. The characteristic decaying length $d$ of the charge density is determined by the screening, and the screening length is defined as:

$$r_D = \sqrt{\frac{\varepsilon_0 E_F}{6\pi e^2 n_v}}, \qquad (1)$$

where $\varepsilon_0$ is the static dielectric constant, $E_F$ is the Fermi energy, and $n_v$ is the volume carrier density. Taking into account that $E_F$ is proportional to $n_v^{2/3}$ ($E_F=\hbar^2(3\pi^2 n_v)^{2/3}/2m$), the screening radius only weakly depends on carrier density $n_v$ ($r_D \sim n_v^{-1/3}$). In order to estimate the characteristic decaying length of the charge density, we assume that $d=r_D$. and, therefore, $n_v = n_s/r_D = n_s/d$. Substituting this density in Eq. (1) and in the expression for $E_F$ we obtain equation for $d$. Solving this equation, we obtain the following expression for $d$:

$$d = \left( \frac{\hbar^2 \varepsilon_0 (9\pi)^{1/3}}{12 e^2 m n_s^{1/3}} \right)^{3/5}, \qquad (2)$$

here $m$ is the effective mass of electron in parent compound, and $\hbar$ is the Planck constant. Taking into account that the effective mass is of the order of free electron mass and $\varepsilon_0$=30, we obtain $d = 4.5\cdot10^{-8}$ cm. It should be noted that this length is slightly shorter than the distance between electrons in 2D plane $r = n_s^{-0.5} \approx 10^7$ cm. Therefore, we believe that the characteristic thickness of the 2D electron gas would be more than 1-2 nm, which is slightly greater than the estimated value. The value of surface carrier density $n_s \approx 1{,}87\ 10^{14}$ cm$^{-2}$ exceeds the value $n_{ss} \approx 0{,}53\ 10^{14}$ cm$^{-2}$, which corresponds to the carrier density of the $CuO_2$ layer in superconducting state for optimal doping (x=0.15) in $La_{2-x}Sr_xCuO_4$. Thus, superconducting state can be realized on the one of the $CuO_2$ layers in the interval of 1-2 nm in the vicinity of the interface.

## 4. Conclusion

In conclusion, the numerical simulations have shown the possibility of q2DEG state in the $BaTiO_3/La_2CuO_4$ heterostructure. We observed the transition to quasi-two-dimensional electron gas state with superconducting behaviour below 40 K in the similar $Ba_{0.8}Sr_{0.2}TiO_3/La_2CuO_4$ heterostructure. The polarization of the ferroelectric film was perpendicular to the interface surface. It was shown that the resistivity properties of the superconducting state change irreversibly, when a magnetic field was applied perpendicular to the surface of interface. The observation of diamagnetic susceptibility confirmed the formation of the superconducting state of BSTO/LCO heterostructure. Therefore, our results demonstrate that quasi-two-dimensional highly conductive and superconductive regimes can be observed at the ferroelectric/dielectric interfaces.

**Acknowledgements**

The reported study was funded by Russian Scientific Foundation, research project No. 18-12-00260. The authors from Kazan Federal University acknowledge partial support by the Russian Government Program of Competitive Growth of Kazan Federal University.

**References**

[1] Ohtomo A and Hwang HY 2004 *Nature* **427** 423-6
[2] Thiel S, Hammerl G, Schmehl A, Schneider C W and Mannhart J 2006 *Science* **313** 1942-5
[3] Reyren N, Thiel S, Caviglia A D, Kourkoutis L, Hammerl G, Richter C, Schneider C W, Kopp T, Ruetschi A-S, Jaccard D, Gabay M, Muller D A, Triscone J-M and Mannhart J 2007 *Science* **317** 1196-9
[4] Brinkman A, Huijben M, Van Zalk M, Huijben J, Zeitler U, Maan J C, van der Wiel W G, Rijnders G, Blank D H A and Hilgenkamp H 2007 *Nat. Mater.* **6** 493-6
[5] Kalabukhov A, Gunnarsson R, Borjesson J, Olsson E, Claeson T and Winkler D 2007 *Phys. Rev. B* **75** 121404
[6] Niranjan M K, Wang Y, Jaswal S S and Tsymbal E Y 2009 *Phys. Rev. Lett.* **103** 016804
[7] Moetakef P, Cain T A, Ouellette D G, Zhang J Y, Klenov D O, Janotti A, Van de Walle Ch G, Rajan S, Allen S J and Stemmer S 2011 *Appl. Phys. Lett.* **99** 232116
[8] Jackson C A and Stemmer S 2013 *Phys. Rev. B* **88** 180403
[9] Biscaras J, Bergeal N, Kushwaha A, Wolf T, Rastogi A, Budhani R C and Lesueur J 2010 *Nat. Commun.* **1** 89
[10] Nakagawa N, Hwang H Y and Muller D A 2006 *Nat. Mater.* **5** 204-9
[11] Fredrickson K D and Demkov A A 2015 *Phys. Rev. B* **91** 115126
[12] Gozar A, Logvenov G, Kourkoutis L F, Bollinger A T, Giannuzzi L A, Muller D A and Bozovic I 2008 *Nature* **455** 782-5
[13] Gozar A and Bozovic I 2016 *Physica C* **521–522** 38
[14] Hohenberg P and Kohn W 1964 *Phys. Rev.* **136** B864-71
[15] Perdew J P, Burke K and Ernzerhof M 1996 *Phys. Rev. Lett.* **77** 3865-8
[16] MedeA® -2.20 2015 Materials Design, Inc., San Diego, CA, USA
[17] Kresse G and Furthmuller J 1996 *Comp. Mater. Sci.* **6** 15-50
[18] Dudarev S L, Botton G A, Savrasov S Y and Humphreys C J 1998 *Phys. Rev. B* **57** 1505-9
[19] Piyanzina I I, Kopp T, Lysogosrkiy Yu V, Tayurskii D and Eyert V 2017 *J. Phys.: Condens. Mat.* **29** 095501
[20] Pavlov D P, Piyanzina I I, Muhortov V I, Balbashov A I, Tauyrskii D A, Garifullin I A and Mamin R F 2017 *JETP Lett.* **106** 460-4
[21] Mukhortov V M, Golovko Y I, Tolmachev G N and Klevtzov A 2000 *Ferroelectrics* **247** 75-83
[22] Kabanov V V, Piyanzina I I, Tauyrskii D A and Mamin R F 2018 *Phys. Rev. B* **98** 094522

IOP Conf. Series: Materials Science and Engineering **699** (2019) 012036 doi:10.1088/1757-899X/699/1/012036

# Morphological and kinetic consequences of scratching of the growing crystal surface

**N N Piskunova**

Laboratory of Experimental Mineralogy, Institute of Geology of Komi Science Centre, Ural Branch of RAS, 167982 Syktyvkar, Russia

piskunova@geo.komisc.ru

**Abstract**. Using the atomic-force microscopy (AFM) it is shown that even a minor local contact of an AFM probe with a growing crystal surface at the nanoscale leads to important crystallogenetic (morphological and kinetic) consequences. These non-linear effects are the thickening of steps, the huge fluctuations, and a loss of morphological stability of the surface even at the distance of dozens of micrometres from the contact area. Near the equilibrium state, we observed the phenomenon of simultaneous growth and dissolution at the neighbouring points of one step.

## 1. Introduction

Most natural crystals have traces of selective dissolution, a mineralogist sees such traces not merely as an evidence of the destruction process, but as an imprint, which allows him to reconstruct one of the historical stages of the crystal and its medium. There were attempts of experimental studies of crystals' growth after their mechanical scratching, which led to a double increase of the growth rate. It was driven by the appearance of the more intensive growth centres [1]. Nowadays, the most informative method that enables to observe the growth and the dissolution of a crystal's surface at nanodimensional scale is atomic force microscopy (AFM). An AFM probe can also be used as a nanoindenter patterning the growing surface. This fact was already used in a number of works studying the processes on the surface after being affected by the probe. In particular, in Ref. [2] it was shown that thin steps in a protein crystal generated from a scratch gradually aggregated in a macroscopic step with their subsequent surface smoothing. Redistribution of solute material during relaxation of a KDP crystal scratch is driven by a reduction in surface free energy, as described in well-known laws [3]. However, the possibilities of using AFM not only for observing the surface evolution, but also for studying kinetic peculiarities of crystals' growth after the mechanical impact have not been fully investigated yet.

Therefore, in this work, using AFM, we have studied nanodimensional morphological and kinetic changes happening on the growing crystal's surface in the area of scratching. The results are compared to crystal's growth peculiarities without special external impacts.

## 2. Experimental

### *2.1. Model crystal*

As a model, we have used dioxidine's crystals (2,3-Bis(hydroxymethyl)quinoxaline 1,4-dioxide, $C_{10}H_{10}N_2O_4$). According to our X-ray structural analysis (Diffractometer Shimadzu XRD – 6000, radiation CuK$\alpha$, internal standard Si), dioxidine belongs to monoclinic system, its unit cell parameters are 8.79Å, 15.74 Å, and 7.99 Å, beta is 102.29, and cell volume is 1081.58 $Å^3$.

### *2.2. Exclusion of impurities impact*
A special attention was paid to the exclusion of impurities impact on the active processes. To achieve that, the dioxidine's solutions were thoroughly analysed using a mass-spectrometer with inductively coupled plasma Agilent 7700x. The concentrations were calculated with calibration lines set by standard solutions "High Purity Standards" with elements' concentration of 10 mg/l. All the dioxidine solutions contained impurity traces – hundredths in mg/l.

### *2.3. Instrument*
The investigations were carried out on an atomic-force microscope Ntegra Prima (NT-MDT, Russia) with full and intermittent contacts with a surface. We used standard silicon cantilevers (NanoProbe, Great Britain) with curvature edge radius of 5 nm. The investigations were carried out at room temperature (20 ± 2°C) in constant humidity. Snapshot "drift" from one shot to another is excluded as fixed points (dislocation channel exits) control was used.

### *2.4. Methods*
The crystals for our *in situ* studies were grown directly in the fluid AFM cell. When the feedback was switched off, an AFM probe-indenter was scratching the surface.

To calculate tangential rate, each image was overlaid by a special grid (based on gnomonic projection), affixed to certain points – the channels of dislocations. Then the position data of the intersection points of the grid meridian and the step contour were taken. Each two images required more than a thousand of coordinate readings. The accuracy of coordinate measure was discretization step, which was two hundredth nm for an area of 15×15 $\mu m^2$. The rate formula included a time correction, which was the function of the point *y*-coordinate on the step. Then we schemed the step rate distribution for every image pair. Each distribution was then approximated to the lognormal distribution, average tangential rates were determined as mathematical expectation values, and rate fluctuations were calculated as mean-square deviations.

## 3. Results and discussion
Observation of growth and dissolution of the surface after scratching has shown that even a minor local contact of an AFM probe with a growing crystal surface at the nanoscale leads to important crystallogenetic (morphological and kinetic) consequences. Apparently, strains appearing in the contact area are relaxed by defect formations, due to which local dissolution arises (Fig. 1). We carried out experiments with a mechanical deformation on topographically different parts of a growing and dissolving crystal. First, we scratched the place of layer-by-layer growth and then the top of the dislocation hillock.

### *3.1. Dissolution on the scratch*
On a flat surface in the undersaturated solution, we drew the letter *"O"* with the probe moving it counter-clockwise. Almost two hours of observations showed an intensive dissolution of the surface in the area of the scratch, which affected the surrounding surface very little (Fig. 1). In the mid of the experiment, there occurred enlargement of dissolving steps, and then they began to split into more thin ones. We have discovered anisotropy of dissolution rates, connected with one of the axes of the second order, along which the pits were eventually extended. It can be seen that the defects spread deep into the crystal.

### *3.2. Layer-by-layer growth area*
In this area, the tangential rate of equidistant steps almost of the same height was 0.9-1.8 nm/sec. (Fig. 2, snapshot 1). Using the AFM probe, two parallel scratches were drawn on the move and against of the steps movement. The dissolution began in spite of supersaturated condition, and soon, the width of the grooves increased – on the left up to 200 nm, on the right – more than 1 μm (Fig. 2, snapshot 2). Such

IOP Conf. Series: Materials Science and Engineering **699** (2019) 012036 doi:10.1088/1757-899X/699/1/012036

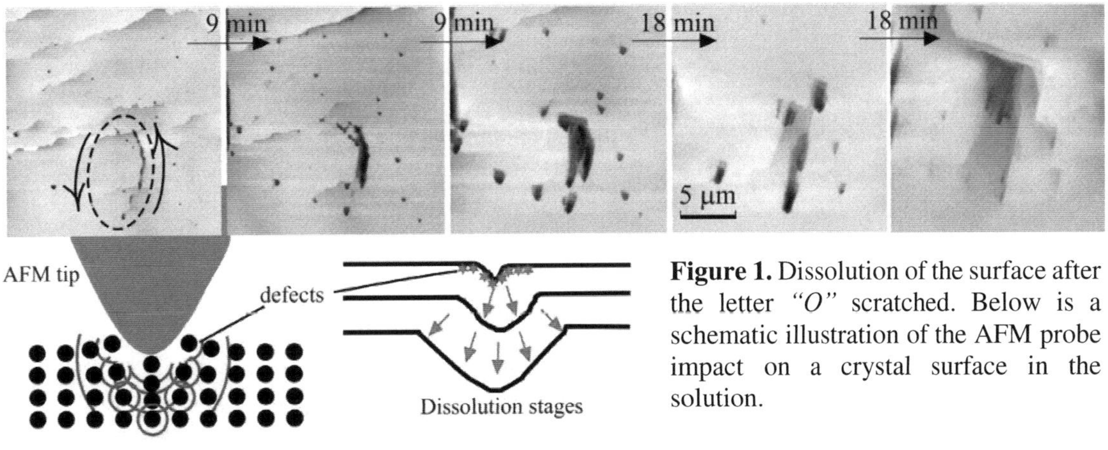


**Figure 1.** Dissolution of the surface after the letter *"O"* scratched. Below is a schematic illustration of the AFM probe impact on a crystal surface in the solution.

difference is caused by indenter's different impacts: on the move and against the direction of steps movement. In case of its going against the direction of the steps movement, more defects arise because of steps' resistance (more strain).

The steps growth rate around the scratches is 0.8 nm/sec. However, the overgrowing rate of the wide scratch is much higher: 1.2-3 nm/sec. At the same time, new growth centres do not arise. The analysis of the height distribution graphs shows that in 15 min after the surface being scratched, a lot of high steps, which were not there before, along with the original steps, arise (Fig. 2). Gradually, the original order of the steps' movement in the part of the surface gets disordered; the surface loses its morphological stability even in the areas without direct impact.

### *3.3. The screw dislocation hillock*

According to AFM data, the growing steps of the screw dislocation hillock were about 8 Å high, which agreed with the unit cell parameter. The tangential rate before the impact was 0.5-0.7 nm/sec. Two parallel furrows were drawn on two sides of the hillock top on the move direction and against of the steps movement (Fig. 3). This leads to the loss of morphological stability: by the 45$^{th}$ minute, the steps making large curves lose their original even profile and general law of collective movement. The hillock top becomes more and more flat and asymmetrical. We can observe the appearance of a certain crystallographic line, in which described processes occur more intensively. The effects of morphological instability cannot be a result of impurity damage, because there is very little amount of impurities in the solutions.

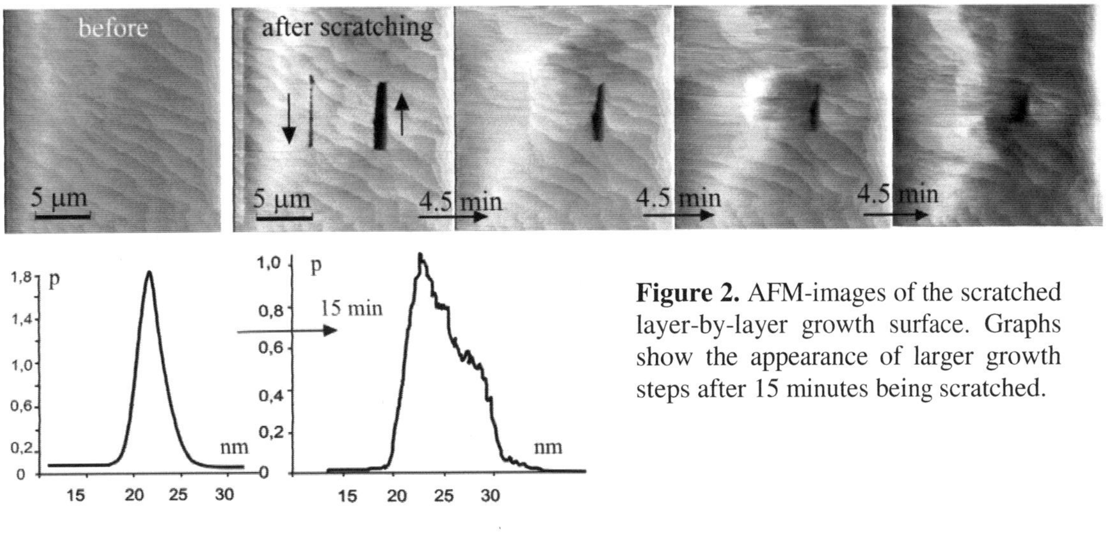


**Figure 2.** AFM-images of the scratched layer-by-layer growth surface. Graphs show the appearance of larger growth steps after 15 minutes being scratched.

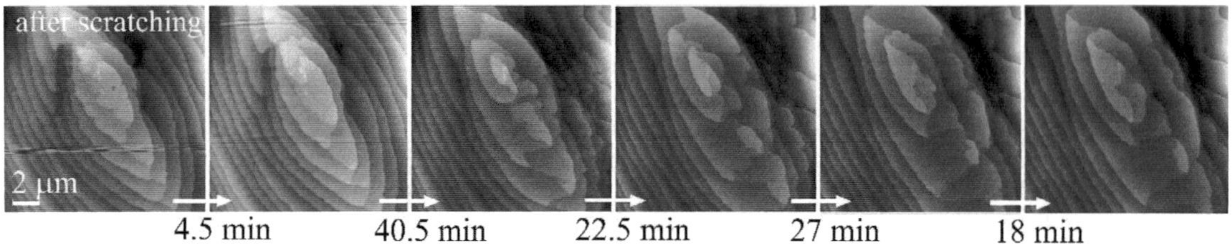


**Figure 3.** The loss of morphological stability of the screw dislocation hillock surface even at the distance of dozens of micrometres from the scratched area.

### *3.4. The phenomenon of simultaneous growth and dissolution*

At the top of the hillock, we observed a phenomenon of simultaneous growth and dissolution on the one step at the distance less than 1 μm. During the second hour of the experiment, some points of the steps grew with a rate 0.9 nm/sec, and the nearest ones dissolved with a rate up to 2 nm/sec (Fig. 4).

Experimental observation of the simultaneous growth and dissolution on the crystal surface is not unique. Sherwood and Ristic observed it on a millimetre scale using interferometry at the top of growth hillock in sodium chloride, which had been affected by radiation [4]. It is also worth mentioning a big number of points at the steps having zero rate at 1 h 43 min after scratching (Fig. 4). Probably, the phenomenon of simultaneous growth and dissolution at the neighbouring points of one step is also connected with approaching an equilibrium state.

### *3.5. The evolution of dislocation spiral without the mechanical impact*

Experiments in the same conditions, but without any special impact have shown that the morphological structure of such hillock growth on dioxidine is quite stable. Even after five-six hours of growth steps kept the original contours, their rate was monotonically decreasing by the end of the experiment (Fig. 5).

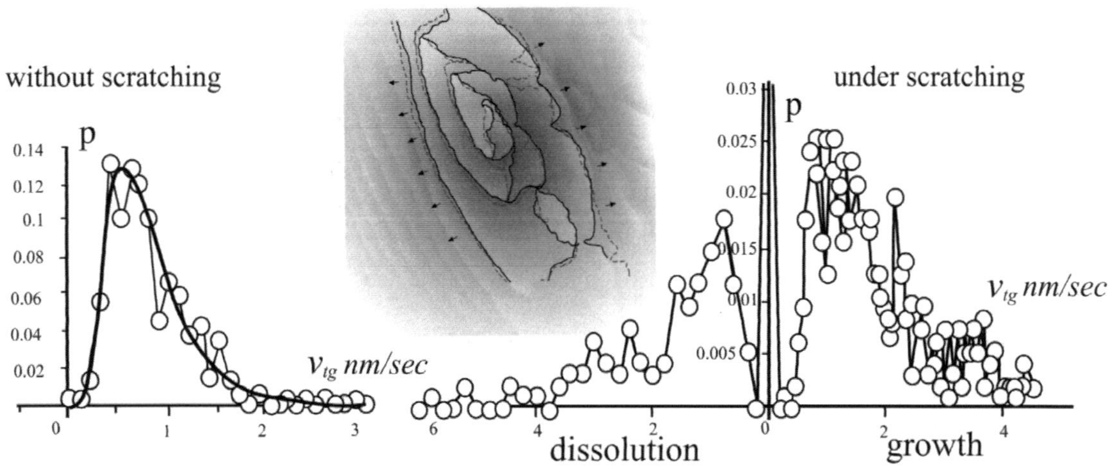


**Figure 4.** Empirical distribution (probability) of the tangential steps' rate before (left) and 1 h 43 min after scratching (right). Before: step – 0.1, mode – 0.7, dispersion – 0.22, average rate – 0.8 nm/sec and rate fluctuation 0.4 nm/sec. In 1 h and 43 min after scratching: step – 0.05, average rate – 1.8 nm/sec, rate fluctuation – 1.5 nm/sec. The average dissolution rate is 1.8 nm/sec, the fluctuation is 1.1 nm/sec. Picture: simultaneous growth and dissolution at the dislocation hillock. A solid line is the projection of the original step position; a dotted line is a profile of the same step after 4.5 minutes.

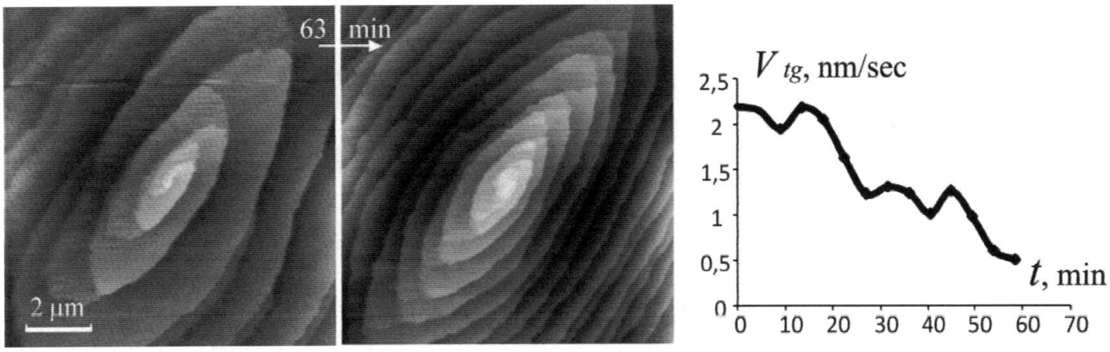


**Figure 5.** The evolution of dislocation spiral on dioxide without the mechanical impact and the tangential rate change within an hour.

## 4. Conclusion

Thus, we believe that kinetic effects of scratching at the nanoscale account for energy factors: mechanical impact initiates formation of multiple defects spreading towards stronger bonds on the surface. These defects decrease activation barriers so much that local dissolution can occur in the area of these defects even in the supersaturated solutions. Our direct observations by AFM showed the beginning of non-linear processes: steps thickening, appearance of huge fluctuation steps rates, a loss of morphological stability of the surface even at the distance of dozens of micrometres from the contact area, and also the simultaneous growth and dissolution in the neighbouring areas.

## References

[1] Bredikhin V I, Galushkina G L, Kulagin A A, Kuznetsov S P and Malshakova O A 2003 *J. Cryst. Growth* **259** 309-20
[2] Yanagiya S and Goto N 2012 *J. Surf. Eng. Mater. Adv. Technol.* **2** 210-4
[3] Elhadj S, Chernov A A and De Yoreo J 2008 *Nanotechnology* **19** 105304
[4] Sherwood J N and Ristic R I 2001 *Chem. Eng. Sci.* **56** 2267-80

# Shapes change of PbO nanoparticles produced by laser ablation in liquid

**V I Pryakhina, E V Gunina, B I Lisjikh, M A Osipova, E D Greshnyakov, E V Shishkina and V Ya Shur**

School of Natural Sciences and Mathematics, Ural Federal University, 620000 Ekaterinburg, Russia

viktoria.pryahina@urfu.ru

Production of non-spherical crystalline nanoparticles by laser ablation in liquid has been demonstrated recently. Here we studied the shape variety of PbO nanoparticles prepared by laser ablation in deionized water and IPA. The key role of water in the formation of PbO non-spherical nanostructures was demonstrated. Only spherical nanoparticles have been obtained in isopropyl alcohol (IPA). PbO non-spherical nanostructures have been formed only in presence of water, which supports the growth mechanism by oxidation. Using mixture of IPA and water in various proportions and revealing the time and temperature dependences of non-spherical nanostructure morphology can be used for controlled growth of PbO non-spherical nanostructures.

## 1. Introduction

Laser ablation in liquid and especially in water gives an opportunity to produce metal and metal oxide "clean" colloids [1]. Avoiding additional chemicals for stabilization is of great importance for applications in medicine and life science. The control of the shapes of metal oxide non-spherical nanostructures can increase their application potential in nanodevices. PbO non-spherical nanostructures are used already in direct conversion X-ray detectors [2], photovoltaic devices [3], and electrocatalysis [4]. It elucidates recent interest to controllable production of PbO non-spherical nanostructures with desired shapes by various methods: ultrasonic liquid phase exfoliation of nanosheets from bulk PbO [4], electrodeposition in presence of oxygen resulted in formation of nanoplates [6, 7] and nanowires [6].

Recently, we have shown that the laser ablation of Pb target in water results in formation of the non-spherical nanostructures at target surface and in colloid [8]. The obtained formation of 3D nanostructures was attributed to growth of seed oxide nanoparticles and 2D plates by seedless growth. Unavoidable hot water treatment during laser ablation was proposed as an additional accelerating mechanism in the formation of non-spherical nanoparticles [8].

Here, we studied the shape variety of PbO nanoparticles prepared by laser ablation in water and isopropyl alcohol. The key role of the water in the formation of PbO non-spherical nanostructures was demonstrated.

## 2. Experimental

### 2.1. *Laser ablation in liquid*

The Pb target (99.9999% purity) with 50 mm diameter and 2 mm thickness was used for laser ablation in liquid. The target exposed at ambient conditions was rinsed in ultrasonic bath by turns with acetone, isopropyl alcohol (IPA), and deionized water produced by Millipore Elix 10 (Merck, Germany) and finally purged by dry nitrogen. CCP oxygen plasma etching was used for homogeneous target surface preparations. Plasma etching was carried out for 30 min at 300 W power at the pressure 20 mTorr with 20 sccm oxygen flow (Plasmalab 80RIE, Oxford Instruments).

The Pb target was placed on the bottom of a dish and covered by 6.4 mm layer of deionized water (15 ml total volume) or 4.7 mm layer of 99.8% pure IPA (40 ml total volume). The glass Petri dish was used for IPA. The metal (anodized Al) dish used for water allowed realizing effective heat dissipation. The water temperature during laser ablation measured by IR thermal imaging camera H2640 (Nippon Avionics, Japan) was below 30°C.

Laser marking system Fmark-20 RL (LTC, Russia) equipped with pulse Ytterbium fibre laser (IPG Photonics, IRE-Polus, Russia) with 1062 nm wavelength, 100 ns pulse duration, and 21 kHz frequency was used for laser ablation. The focused laser beam (spot diameter 50 μm, fluence 51 J/cm$^2$) scanned over the target area about 18 cm$^2$ with linear velocity 270 mm/s (30 lines/mm). The typical duration of the laser ablation process was about 4 min.

### *2.2. Nanoparticles imaging*

The probe of the colloid (0.5 μl) was taken immediately after termination of the laser ablation process and dropped at Si plate. The duration of the drop drying process was about 3 min. The nanoparticles at the substrate were imaged by scanning electron microscope Merlin (Carl Zeiss, Germany).

## 3. Results and discussion

### *3.1. Laser ablation in water*

The PbO non-spherical nanostructures (plates and rods) was obtained after the laser ablation in water along with spheroids (Fig. 1). The diameter of spheroids was from 10 to 100 nm. The shape of the plates changed between circle and hexagon, the thickness – from 10 to 20 nm, and the diameter – from 200 to 2000 nm. The width of the rods was from 60 to 180 nm and length – from 1 to 5 μm.

It is necessary to point out that some PbO plates were obtained in water after Pb target soaking in water at room temperature.

As it was shown by us earlier [7], the nanoparticles of different shapes also differ by phase: plates are β-PbO (orthorombic) and rods are α-PbO (tetragonal). PbO non-spherical nanostructures (rods and octahedra) were found only at target surface, but not in the colloid. This fact is attributed to sedimentation of the large particles. In this study, the various shapes of nanoparticles were obtained in water after time interval below five minutes after laser ablation termination including water exposure during drop drying. This fact allows evaluating the time interval necessary for formation of the PbO non-spherical nanostructures.

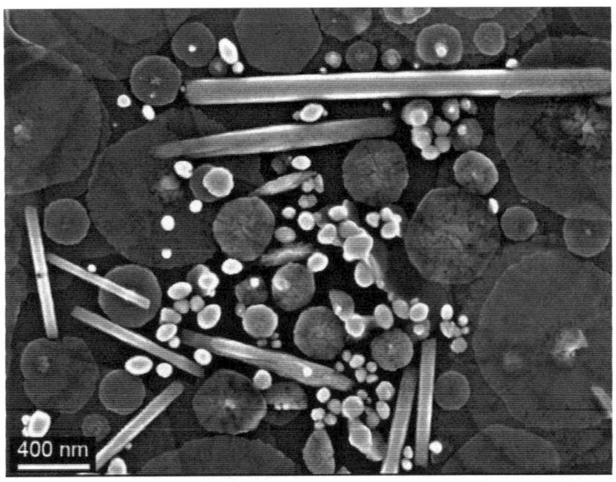


**Figure 1.** Plates, rods, and spheroids after laser ablation in water.

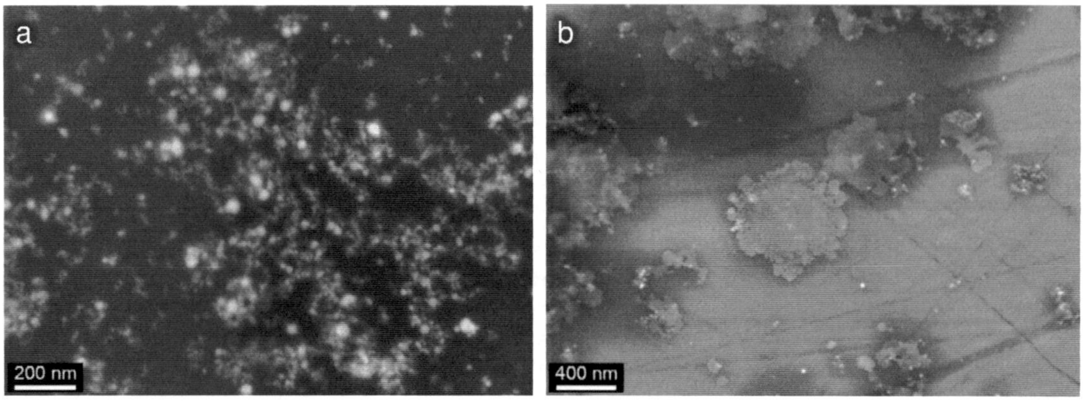


**Figure 2**. PbO nanoparticles: (a) produced by laser ablation in IPA, (b) after adding water.

### *3.2. Laser ablation in IPA*

According to the supposed model, water is important for oxidation of Pb and growth of oxide nanoparticles. The control experiment – laser ablation in IPA – resulted in formation of the spherical nanoparticles only with diameter ranging from 10 to 50 nm (Fig. 2a). The Pb target soaking in IPA did not result in formation of any nanostructures both in liquid and at the target surface.

Addition of equal amount of deionized water in the IPA with spherical nanoparticles after laser ablation resulted in formation of PbO plates (Fig. 2b). The plates had irregular shapes and effective diameter from 500 nm up to few microns. These facts confirm the proposed mechanism of PbO plates seedless growth in water.

The observed results allowed using mixture of IPA and water in various proportions for controlling PbO non-spherical nanostructures growth. The time and temperature dependences of sizes and shapes of non-spherical nanostructures need further investigations.

## 4. Conclusion

The shape variety of PbO nanoparticles prepared by laser ablation in deionized water and in IPA has been studied. Only spherical nanoparticles have been obtained in IPA. PbO non-spherical nanostructures have been formed only in presence of water, which supports the growth mechanism by oxidation.

Mixing IPA and water in various proportions could be used for controlling PbO non-spherical nanostructures growth. The time and temperature dependences of sizes and shapes of non-spherical nanostructures need further investigations. The next step in this research is production of nanoparticles by laser ablation in water-IPA mixture.

## Acknowledgements

The work has been supported in part by the Ministry of Science and Higher Education of the Russian Federation (projects 3.9534.2017/8.9 and 3.4993.2017/6.7) and by Government of the Russian Federation (act 211, agreement 02.A03.21.0006). The equipment of the Ural Centre for Shared Use "Modern Nanotechnology" Ural Federal University was used.

## References

[1] Zhang D, Gökce B and Barcikowski S 2017 Laser synthesis and processing of colloids: fundamentals and applications *Chem. Rev.* **117** 3990-4103

[2] Qamar A, LeBlanc K, Semeniuk O, Reznik A, Lin J, Pan Y and Moewes A 2017 X-ray spectroscopic study of amorphous and polycrystalline PbO films, α-PbO, and β-PbO for direct conversion imaging *Sci. Rep.* **7** 13159

[3] Cattley C A, Stavrinadis A, Beal R, Moghal J, Cook A G, Grant P S, Smith J M, Assender H and Watt A A R 2010 Colloidal synthesis of lead oxide nanocrystals for photovoltaics *Chem.*

*Commun.* **46** 2802

[4] Park S-E and Chang J-S 1992 Oxidative coupling of methane over a γ-alumina-supported lead oxide catalyst *Appl. Catal. A Gen.* **85** 117-28

[5] Xing C, Chen X, Huang W, Song Y, Li J, Chen S, Zhou Y, Dong B, Fan D, Zhu X and Zhang H 2018 Two-dimensional lead monoxide: Facile liquid phase exfoliation, excellent photoresponse performance, and theoretical investigation *ACS Photonics* **5** 5055-67

[6] Kwon Y, Lee H and Lee J 2011 Autonomous interfacial creation of nanostructured lead oxide *Nanoscale* **3** 4984

[7] Meng L, Ustarroz J, Newton M E and Macpherson J V 2017 Elucidating the cathodic electrodeposition mechanism of lead/lead oxide formation in nitrate solutions *J. Phys. Chem. C* **121** 6835-43

[8] Shur V Y, Gunina E V, Esin A A, Shishkina E V, Kuznetsov D K, Linker E A, Greshnyakov E D and Pryakhina V I 2019 Influence of hot water treatment during laser ablation in liquid on the shape of PbO nanoparticles *Appl. Surf. Sci.* **483** 835-9

IOP Conf. Series: Materials Science and Engineering **699** (2019) 012039 doi:10.1088/1757-899X/699/1/012039

# Determination of rate of epitaxial growth by means of combined usage of atomic force microscopy and light scattering data

**A E Rassadin**

Nizhny Novgorod Mathematical Society, Nizhny Novgorod, 603950, Russia

brat_ras@list.ru

**Abstract**. In the paper, a method of experimental estimation of epitaxial growth rate has been suggested. A corner stone of this method is obtaining image of initial shape of solid state surface by means of atomic force microscopy before the start of the process of epitaxial growth. These experimental data should be completed by measurements of special bistatic cross-section of visible light scattering on the sample surface both before the beginning of technological process and after its end. Mathematical model of epitaxial growth is based on simplification of the Kardar-Parisi-Zhang equation. The method of characteristics has been applied to solve this equation.

## 1. Introduction

Since the pioneer work [1] of Donald Eigler and Erhard Schweizer, words 'nanotechnology' and 'nanoengineering' were synonyms. But after appearance of paper [2], another sense of term 'nanoengineering' has taken place too; namely, one may understand nanoengineering as the problem of optimal control by distributed parameter system.

Let us describe epitaxial growth of solid state surface on the basis of the Kardar-Parisi-Zhang (KPZ) model [3]:

$$\frac{\partial H}{\partial t} = v + \frac{v}{2}\cdot(\nabla H)^2 + D\cdot\nabla^2 H + Q(\vec{x},t), \tag{1}$$

where $H(\vec{x},t)$ is the height of solid state surface, $\vec{x}=(x_1,x_2)$ is two-dimensional vector of transversal coordinates, $\nabla$ is two-dimensional gradient, $v$ is the rate of growth of the surface along the local normal to it, $D$ is the diffusion coefficient of sputtering matter and $Q(\vec{x},t)$ is additional source of external particles near the surface.

The KPZ-equation (1) proves to be highly adequate to the physical experiment during the simulation of manufacturing process of multilayer mirrors and gratings employed in X-ray optics [4]; therefore, verification procedure for model (1) of surface growth with support of this one by means of atomic force microscopy (AFM) is not required.

On the domain $\Omega \subset R^2$ equation (1) ought to be equipped by initial condition:

$$H(\vec{x},0) = H_0(\vec{x}), \tag{2}$$

corresponding to initial shape of the surface under investigation.

If the source $Q(\vec{x},t)\geq 0$ on the cylinder $\Omega\times[0,t_*]$, then one can consider this function as a control [2]. At last, let one impose on the solution of equation (1) with initial condition (2) the requirement that at the moment of time $t=t_*$ function $H(\vec{x},t)$ must provide the best mean square approximation of fixed shape $H_*(\vec{x})$:

$$\int_\Omega [H(\vec{x},t_*)-H_*(\vec{x})]^2 d^2x \to \min . \tag{3}$$

It is obvious that equation (1) and restrictions (2)-(3) form the problem of optimal control by distributed parameter system [2, 5]. But to construct the numerical solution of this problem, one has to find parameters $v$ and $D$ in input equation (1).

If $\operatorname{diam}\Omega >> \sqrt{D\cdot t_*}$, then there is a way to estimate these values; namely, in the absence of additional source $Q(\vec{x},t)$ of sputtering particles the approximate solution of the Cauchy problem (1)-(2) is equal to [3]:

$$H(\vec{x},t)\approx v\cdot t+\frac{2\cdot D}{v}\cdot\ln\left(\int_{-\infty}^{+\infty}\int_{-\infty}^{+\infty}\exp\left[-\frac{(\vec{x}-\vec{\xi})^2}{4\cdot D\cdot t}+\frac{v\cdot H_0(\vec{\xi})}{2\cdot D}\right]\cdot\frac{d^2\xi}{4\cdot\pi\cdot D\cdot t}\right). \tag{4}$$

Further height of the surface $H(\vec{x},t)$ on domain $\Omega$ must be measured by means of AFM both before the beginning of the technological process and after the finish of it. This information is enough to apply method of solution of inverse coefficient problem for determination of parameters $v$ and $D$ developed in paper [6]. The application of this method meets with difficulties in practice as it belongs to the advanced chapters of modern functional analysis [7]. Thus, more straightforward approaches for estimation of these parameters based on physical arguments are to be considered.

This paper deals with estimation for the rate $v$ of epitaxial growth of crystal surface in the situation when $v\cdot H >> D$, where $H$ is typical value of height of crystal surface roughness. In this case, one can neglect in input equation (1) by term with Laplasian. And after putting $Q(\vec{x},t)=0$ in equation (1), this one is reduced to the so-called equation simplified Kardar-Parisi-Zhang (sKPZ) equation:

$$\frac{\partial H}{\partial t}=v+\frac{v}{2}\cdot(\nabla H)^2 . \tag{5}$$

This equation differs sharply from the primordial KPZ one (1); moreover, equation (5) can be solved implicitly by means of the method of characteristics [8]:

$$H=v\cdot t+H_0(\vec{y})-\frac{v\cdot t}{2}\cdot(\nabla H_0(\vec{y}))^2, \qquad \vec{x}=\vec{y}-v\cdot t\cdot\nabla H_0(\vec{y}), \tag{6}$$

where $\vec{y}=(y_1,y_2)$ are Lagrange variables. To derive explicit dependence for height of the surface $H(\vec{x},t)$, one must find from the second equation of system (6) vector-function $\vec{y}=\vec{y}(\vec{x},t)$ and substitute it into the first equation of system (6). Generally speaking, it is impossible to obtain analytical formula connecting the initial shape of the surface (2) with the rate $v$. Nevertheless, absence of terms describing the surface diffusion of sputtering substance and the external source of sputtering particles in equation (5) gives one a possibility to determine the rate of growth of the surface.

In practice, initial condition (2) for equation (5) is often equal to superposition of smooth large scale function $h_0(\vec{x})$ and small perturbation $\mu\cdot u_0(\vec{x})$:

$$H_0(\vec{x})=h_0(\vec{x})+\mu\cdot u_0(\vec{x}), \tag{7}$$

where $\mu$ is small parameter: $0<\mu<<1$. As a rule, large scale function $h_0(\vec{x})$ is a result of preliminary technological processing of the surface before the start of epitaxial process and small perturbation is some technologically unremovable fluctuations. An example of such surface is shown in Figure 1.

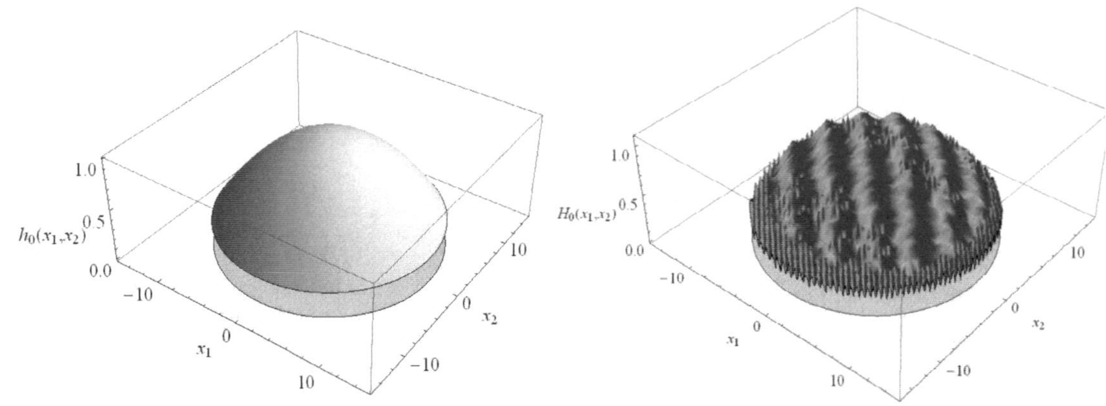


**Figure 1.** Initial condition as smooth large scale function (on the left) and as superposition of smooth large scale function and its small perturbation (on the right).

For further considerations, it is important that due to representation (7) of initial shape, specific bistatic cross-section of electromagnetic waves on this surface proves to be calculated.

The rest of the article is organized as follows: in Section 2 peculiarities of construction of asymptotic solution for the sKPZ-equation (5) for the special choice of function $h_0(\vec{x})$ are demonstrated. Section 3 deals with the estimation of specific bistatic cross-section of electromagnetic waves on this surface. Final Section is devoted to the discussion of results elaborated and perspectives of further investigations.

## 2. Derivation of asymptotic solution for the simplified KPZ-equation

Decomposition (7) for initial condition means that in order to construct solution of equation (5) one has to use perturbation theory:

$$H(\vec{x},t) = h(\vec{x},t) + \mu \cdot u(\vec{x},t) + \dots . \tag{8}$$

Substitution of expansion (8) into the sKPZ-equation (5) gives rise to a chain of the Cauchy problems; namely, for zero order approximation on $\mu$ in asymptotic series (8) the Cauchy problem is:

$$\frac{\partial h}{\partial t} = v + \frac{v}{2} \cdot (\nabla h)^2, \qquad h(\vec{x},0) = h_0(\vec{x}), \tag{9}$$

and for the first order approximation on $\mu$ in asymptotic series (8) the Cauchy problem is:

$$\frac{\partial u}{\partial t} - v \cdot \nabla h(\vec{x},t) \cdot \nabla u = 0, \qquad u(\vec{x},0) = u_0(\vec{x}). \tag{10}$$

In order to do the next step in solution of the system of equations (9)-(10) one has to render concrete function $h_0(\vec{x})$.

If one chooses regular shape $h_0(\vec{x})$ in initial condition (7) as follows:

$$h_0(\vec{x}) = h_* - \frac{\vec{x}^2}{2 \cdot L_*}, \tag{11}$$

where $h_*$ and $L_*$ are constant positive values, then in the framework of the method of characteristics for the first-order partial differential equation [9] it is easy to check that in this case exact solution of the Cauchy problem (8) is equal to:

$$h(\vec{x},t) = h_* + v \cdot t - \frac{\vec{x}^2}{2 \cdot (L_* + v \cdot t)}. \tag{12}$$

Substituting this function into equation (10) one can find that it is reduced to the next equation:

IOP Conf. Series: Materials Science and Engineering **699** (2019) 012039 doi:10.1088/1757-899X/699/1/012039

$$\frac{\partial u}{\partial t}+\frac{v\cdot\vec{x}}{L_*+v\cdot t}\cdot\nabla u=0\,. \tag{13}$$

This equation is linear partial differential equation with variable coefficients and one can derive exact solution of the Cauchy problem for equation (13) with initial condition $u_0(\vec{x})$ by the same method of characteristics [9]:

$$u(\vec{x},t)=u_0\left(\frac{L_*\cdot\vec{x}}{L_*+v\cdot t}\right). \tag{14}$$

Thus asymptotic solution of the sKPZ-equation (5) with regular part (11) of initial condition (7) is equal to:

$$H(\vec{x},t)=h_*+v\cdot t-\frac{\vec{x}^2}{2\cdot(L_*+v\cdot t)}+\mu\cdot u_0\left(\frac{L_*\cdot\vec{x}}{L_*+v\cdot t}\right)+O(\mu^2)\,. \tag{15}$$

Schematic graphs of section of function (15) under the assumption $h_*/L_* << 1$ at initial moment of time (below) and at arbitrary moment of time $t>0$ (higher) are presented in Figure 2. Temporal evolution of function (12) is presented on this figure by blue lines. From Figure 2 one can see that under growth of time roughness of surface shape becomes smaller in comparison with roughness of initial shape.

Such behaviour is determined by two circumstances. First of all, from formula (14) it is clear that spatial scale $l(t)$ of perturbation $u(\vec{x},t)$ grows as $l(t)=l_0\cdot(1+v\cdot t/L_*)$, where $l_0$ is spatial scale of initial perturbation $u_0(\vec{x})$. On the other hand, from formula (12) it is obvious that with increasing time function $h(x,t)$ becomes closer to plane $z=h_*+v\cdot t$. These planes are shown in Figure 2 by dashed lines. It means that if for some positive value $\lambda$ the following inequalities are true:

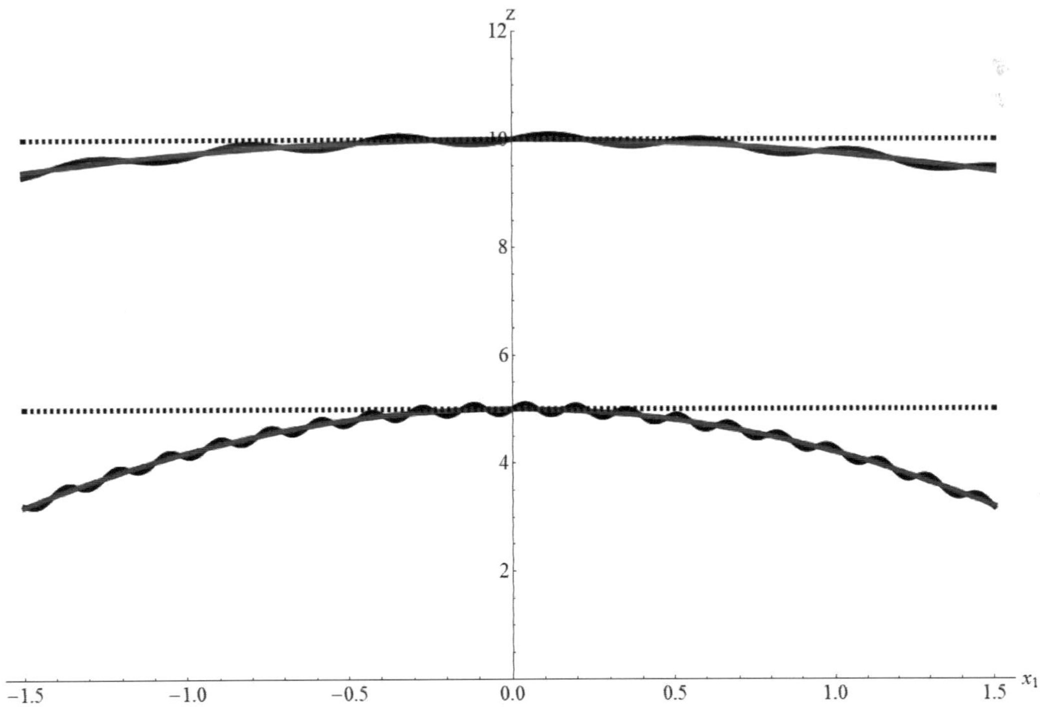


**Figure 2.** Temporal evolution of the section of the height by plane $x_2=0$.

$$\mu \cdot |u_0(\vec{x})| << \lambda, \qquad \mu \cdot |\nabla u_0(\vec{x})| << 1, \tag{16}$$

then with increasing time inequalities:

$$\mu \cdot |u(\vec{x},t)| << \lambda, \qquad \mu \cdot |\nabla u(\vec{x},t)| << 1, \tag{17}$$

are true too.

If one interprets value $\lambda$ as wavelength of monochromatic visible light: $\lambda = 380 \div 780$ nm, then under $\max_{\vec{x}\in\Omega} |u_0(\vec{x})| \sim 10$ nm inequalities (16) are valid. Therefore, inequalities (17) are valid too. Hence, one can describe scattering of monochromatic visible light on the surface (15) in the framework of a model of small-scale surface [10].

## 3. Specific bistatic cross-section of electromagnetic waves on the shape evolving in accordance with the simplified KPZ-equation

The simplest formulation for a model of small-scale surface can be done under the conjecture that perturbation of smooth surface $u_0(\vec{x})$ is stationary stochastic two-dimensional field with zero average: $< u_0(\vec{x}) >= 0$.

In this case, it is obvious from formula (14) that $< u(\vec{x},t) >= 0$. Therefore, from formula (15) one can easily find that $< H(\vec{x},t) >= h(\vec{x},t)$. It means that covariance function of total height $H(\vec{x},t)$ of solid state surface can be calculated directly:

$$B(\vec{\xi},t) =< (H(\vec{x},t) - < H(\vec{x},t) >) \cdot (H(\vec{x}+\vec{\xi},t) - < H(\vec{x}+\vec{\xi},t) >) >, \tag{18}$$

the two-dimensional Fourier-transform from function (18):

$$S(\vec{q},t) = \int B(\vec{\xi},t) \cdot \exp(-i \cdot \vec{q} \cdot \vec{\xi}) \cdot d^2\xi \tag{19}$$

being its spectral density.

Further, let one consider scattering on the surface (15) of monochromatic visible light with horizontal (H) and vertical (V) polarizations:

$$\vec{E}_H^{(i)} = (0,1,0) \cdot A(x_1, z), \qquad \vec{E}_V^{(i)} = (\cos\theta_i, 1, \sin\theta_i) \cdot A(x_1, z), \tag{20}$$

where $A(x_1, z) = E_0 \cdot \exp(i \cdot k \cdot \cos\theta_i \cdot x_1 - i \cdot k \cdot \sin\theta_i \cdot z)$ is complex amplitude of monochromatic wave, $k = 2 \cdot \pi / \lambda$ is wavevector corresponding to wavelength $\lambda$ and $\theta_i$ is input angle.

At last, because of validity of inequalities (17) specific bistatic cross-sections with polarizations $a,b = H,V$ of waves (20) on the surface under investigation are equal to [10]:

$$\sigma_{ab}^0(\vec{q},t) = \frac{4 \cdot k^4}{\pi} \cdot f_{ab}(\theta_i, \theta_s, \phi_s) \cdot S(\vec{q},t), \tag{21}$$

where

$$f_{ab}(\theta_i, \theta_s, \phi_s) = \begin{pmatrix} \cos^2\theta_i \cdot \cos^2\theta_s \cdot \cos^2\phi_s & \cos^2\theta_i \cdot \sin^2\phi_s \\ \cos^2\theta_s \cdot \sin^2\phi_s & (\sin\theta_i \cdot \sin\theta_s - \cos\phi_s)^2 \end{pmatrix} \tag{22}$$

is matrix of polarization coefficients, $\theta_s$ and $\phi_s$ are scattering angles in spherical system of coordinate and arguments in spectral density in (21) are equal to:

$$q_1 = k \cdot \sin\theta_s \cdot \cos\phi_s - k \cdot \sin\theta_i, \quad q_2 = k \cdot \sin\theta_s \cdot \sin\phi_s. \tag{23}$$

Using formula (14), it is easy to find that

$$B(\vec{\xi},t) = \mu^2 \cdot K_0\left(\frac{L_* \cdot \vec{\xi}}{L_* + v \cdot t}\right) + o(\mu^2), \tag{24}$$

where

$$K_0(\vec{\xi}) = < u_0(\vec{x}) \cdot u_0(\vec{x} + \vec{\xi}) > \tag{25}$$

is autocorrelated function of stochastic field $u_0(\vec{x})$.

Thus, spectral density for covariance function (24) is equal to:

$$S(\vec{q},t) = \mu^2 \cdot \left( \frac{L_* + v \cdot t}{L_*} \right)^2 \cdot S_0\left( \frac{L_* + v \cdot t}{L_*} \cdot \vec{q} \right) + o(\mu^2), \tag{26}$$

where $S_0(\vec{q}) = \int K_0(\vec{\xi}) \cdot \exp(-i \cdot \vec{q} \cdot \vec{\xi}) \cdot d^2\xi$ is spectral density corresponding to autocorrelated function (25).

In particular, at $t = 0$ formula (26) is reduced to:

$$S(\vec{q},0) = \mu^2 \cdot S_0(\vec{q}) + o(\mu^2). \tag{27}$$

Due to validity of inequalities (16) at $t = 0$ specific bistatic cross-sections also can be calculated in accordance with formulae (21)-(23).

Combining under $\vec{q} = 0$ formulae (26) and (27) with formula (21) one can derive that:

$$\frac{\sigma^0_{ab}(0,t)}{\sigma^0_{ab}(0,0)} = \left( 1 + \frac{v \cdot t}{L_*} \right)^2 + O(\mu). \tag{28}$$

Due to requirement $\vec{q} = 0$, formulae (23) give one position of receiver of monochromatic light with respect to transmitter:

$$\phi_s = 0, \quad \theta_s = \theta_i. \tag{29}$$

Substituting values (29) into formulae (22) one can obtain that $\sigma^0_{VH} = \sigma^0_{HV} = 0$ and $\sigma^0_{HH} = \sigma^0_{VV}$.

## 4. Conclusion

In the article, algorithm of determination of epitaxial growth rate $v$ in the framework of the sKPZ-equation has been presented. In order to find this parameter, one has to do the following operations:

- to prepare initial shape $h_0(\vec{x})$ of the surface under investigation with regular part (10) under condition $h_*/L_* << 1$;
- to measure specific bistatic cross-section $\sigma^0_{HH}$ (or $\sigma^0_{VV}$ ) of visible light before the start of technological process;
- to measure by means of AFM the real height $H_0(\vec{x})$ of the sample before the start of technological process;
- to perform for real initial shape $H_0(\vec{x})$ denoising procedure with the help of thresholding [11];
- to find the parameter $L_*$ of smoothed shape (11) in the vicinity of point $(0,0)$ by means of mean-square fitting;
- to fix the duration $t_*$ of technological process;
- to measure specific bistatic cross-section $\sigma^0_{HH}$ (or $\sigma^0_{VV}$ ) of visible light after the end of technological process;
- to calculate epitaxial growth rate $v$ in accordance with formula (28).

Thresholding is known to be a kind of denoising procedure in the framework of wavelet analysis [11]. To adjust this procedure to real AFM data, one has to do a number of computational experiments in order to find optimal type of wavelet and optimal level of wavelet decomposition.

In conclusion, it is necessary to underline that formulae (12) and (14) are rigorously valid only under suggestion that process of epitaxial growth develops only in small neighbourhood of point $(0,0)$. More careful mathematical treatment with boundary conditions for equation (5) is required in order to estimate its influence on exactness of formula (28).

**Acknowledgements**
The author prepared this article during work on the project № 18-08-01356-a of Russian Foundation for Basic Research.

**References**

[1] Eigler D M and Schweizer E K 1990 *Nature* **344** 524

[2] Rassadin A E, Sazanova T S, Stepanov A V and Fomin L A 2018 *IOP Conf. Ser.: Mater. Sci. Eng.* **443** 012027

[3] Kardar M, Parisi G and Zhang Y C 1986 *Phys. Rev. Lett.* **56** 889-92

[4] Goray L I and Lubov M N 2014 *J. Surf. Invest.* **8** 444-55

[5] Butkovsky A G 1965 *Theory of optimal control by distributed parameter systems* (Nauka) (in Russian)

[6] Boikov I V and Ryazantsev V A 2019 *Middle Volga Math. Soc. J.* **21** 149 (in Russian)

[7] Boikov I V 2012 *Differential equations* **48** 1308 (in Russian)

[8] Gurbatov S N, Rudenko O V and Saichev A I 2011 *Waves and Structures in Nonlinear Nondispersive Media. General Theory and Applications to Nonlinear Acoustics* (Springer and Higher Education Press)

[9] Polyanin A D, Zaitsev V F and Moussiaux A 2002 *Handbook of First Order Partial Differential Equations* (Taylor & Francis)

[10] Ishimaru A 1978 *Wave Propagation and Scattering in Random Media. Vol. 2* (Academic Press)

[11] Smolentsev N K 2008 *Foundations of Wavelet Theory. Wavelets in MATLAB* (DMC Press) (in Russian)

IOP Conf. Series: Materials Science and Engineering **699** (2019) 012040 doi:10.1088/1757-899X/699/1/012040

# Features of combining scanning probe microscope with optical and scanning electron microscopes

**I D Sapozhnikov, O M Gorbenko, M L Felshtyn, M V Zhukov and A O Golubok**

Laboratory of Scanning probe microscopy and Spectroscopy, Institute for Analytical Instrumentation RAS, 198095 St. Petersburg, Russia

aogolubok@mail.ru

**Abstract.** The designs of combining a scanning probe microscope (SPM) with an optical microscope (OM) or scanning electron microscope (SEM) have been presented. To reduce the size of the SPM units the self-sensing probes and a piezo-inertial displacement system are proposed to use. The peculiarities of working of such SPM units are discussed. The examples of visualization of different nature objects in combined systems SPM-OM and SPM-SEM have been presented.

## 1. Introduction
There are two main experimental approaches to visualizing the structure or mapping the properties of objects with nanometre spatial resolution. In one case, a focused beam of particles or radiation interacts with the sample under study. The beam source is far from the sample and the result of beam-sample interaction is detected. Otherwise, a sharp solid-state probe interacts with the sample under study. In this case, a sharp solid-state probe is located near the surface on a distance $d < \lambda$, where $\lambda$ - the probe-sample interaction decay length, and the probe-sample interaction is detected. Since the interaction of focused beams and sharp probes with the sample surface has a different nature, the combination of different experimental "beam" and "probe" approaches in one device [1-5] provides more complete information about the object under research.

## 2. Combined system of coarse approach and scanning
It is obvious that the compact design of the scanning probe microscope (SPM) assembly is optimal for combining it with an optical microscope (OM) or scanning electron microscope (SEM). To reduce the size of the SPM assembly, the commonly used mechanical design based on a stepper motor with complicated mechanical components has been excluded from the coarse approach system of the probe with the sample. Instead of a system on the base of a stepper motor, a simpler system on the base of a 1D piezo-inertial driver has been used.

In the Figure 1a, the principal scheme of 1D a piezo-inertial motor is presented. A stack multilayer piezo actuator (2) is attached to the fixed base (1) on one side. A titanium guide rod (4) is attached to the opposite side of the stack multilayer piezo actuator. The titanium carriage (3) is pressed against the surface of the titanium guide rod (4). Asymmetric electric pulses (Fig. 1b) with a sharp and smooth front are applied to the piezo actuator (2). Under the action of a sharp front, the piezo actuator (2) sharply stretches or contracts depending on the sign of the electric voltage. In this case, the inertial force acting on the carriage (3) is greater than the friction force acting between the surface of the guide rod (4) and the surface of the carriage (3). As a result, the carriage (3) slides along the surface of the guide rod (4). Under the action of a smooth front, the friction force is greater than the inertial one and

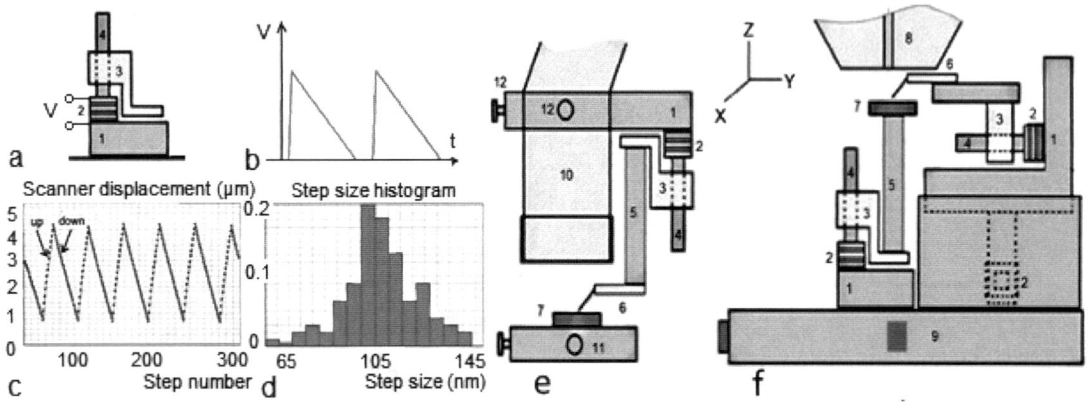

**Figure 1.** SPM-OM and SPM-SEM scheme: (a) principal scheme of 1D piezo-inertial motor, (b) control electric pulses, (c), (d) step measurement of piezo-inertial motor, (e) scheme of SPM-OM, (f) scheme of SPM-SEM. 1 – base, 2 – stack multilayer piezo actuator, 3 – carriage, 4 – guide rod, 5 – piezo tube X,Y scanner, 6 – probe sensor, 7 – sample, 8 – pole piece of the SEM objective lens, 9 – SEM goniometer, 10 – lens of optical microscope, 11 – 2D stage of optical microscope, 12 – screws for alignment of the probe.

slippage does not occur. The up or down moving direction of a carriage can be reversed by changing a sign of control voltage. Work reproducibility of a piezo motor with such design strongly depends on the quality of polishing of the titanium surfaces of friction pair. We applied mechanical polishing of titanium surfaces to the "orange peel" state.

The magnitude of the single step of the carriage movement depends on the force of the carriage pressing against the guide rod and the amplitude of the control pulse. It is clear that the application of a smoothly varying voltage to the piezo actuator provides a smooth movement of the carriage together with guide rod. In our case, the range of smooth movement (1D scanning) is ~7 μm, with a range of step by step movement of ~10 mm. On the carriage of the piezo motor, a probe sensor or a piezo tube scanner with a sample holder or a probe sensor were mounted. Also, it is clear that by combining 1D piezo motors it is possible to create 2D and 3D piezo drivers. Since the piezo motor can operate both as a linear stepper motor and as a smooth scanner, it is possible to use them both for coarse approach of the tip to the sample and for stabilization of the interaction between tip and sample in the constant force mode, when the SPM feedback loop is closed. Therefore, breaking and closing the feedback loop between the consecutive steps of the piezo motor, you can measure the step size by measuring the corresponding displacement of the scanner (Fig. 1c). When the guide rod is vertically positioned, the step up is smaller than the step down due to the action of gravity (Fig. 1c). As can be seen from Figure 1d, the average step of the piezo motor is ~ 100 nm and the minimum and maximum steps are ~ 50 nm and 150 nm, respectively.

## 3. Probe sensor

To reduce the size of the SPM unit, the quite bulky optical circuit usually used to measure the deflection of the cantilever has been excluded also from the SPM units design intended for SPM-OM and SPM-SEM combinations. Instead of usual Si cantilever, two types of self-sensing probe sensors have been used. One probe is a tuning-fork sensor fabricated from two piezo tubes with replaceable W tip [6] and another is a Si piezoresistive cantilever sensor [7].

The advantage of such kind of probe sensors compared with a standard Si cantilevers with optical registration of cantilever deviation is absence of any alignment after the probe's substitution. Tuning-fork sensor on the base of piezo tubes and replaceable W tip has a Q factor of about 30, while Q factor of Si piezoresistive cantilever is about 200. Therefore, tuning-fork sensor has a smaller sensitivity compared with Si piezoresistive cantilever.

Nevertheless, the sensitivity of tuning-fork sensor on the base of piezo tubes is enough for many experimental cases. For example, the tuning-fork sensor with W probe is preferable to a Si cantilever when operating in dynamic force nanolithography mode. From the practical point of view, a sensor on the base of piezo tubes with replaceable W tip is very convenient, because the users can fabricate the W tip themselves by electrochemical etching of tungsten wire with (100-150) μm diameter.

It should be note that at small tilt angle of cantilever (~10 degree) relative to the sample surface, which usually installed for a standard cantilever and at a standard value of a reference signal on input of feed-back loop of SPM, the amplitude of cantilever oscillations begins decreasing earlier than a needle starts tapping on the sample surface. We attribute this fact to the action on the cantilever of an air flow reflected from the sample surface. Indeed, the area of the piezo cantilever is larger than that of a conventional cantilever, since the Wheatstone bridge is located on its surface. It is experimentally found that if a reflected air flow entering the cantilever surface is decreased by increasing a tilt angle between sample surface and cantilever plane (up to 40 degrees), this effect disappears. This effect also disappears in a vacuum. Therefore, when working with a piezo resistive cantilever in tapping mode, it is necessary to select a reference signal for the feedback loop different from the signal corresponding to the case of a conventional cantilever. In this case, the operating point on the probe-sample approach curve should correspond to a larger value of the decreasing of the cantilever`s oscillations amplitude.

## 4. SPM-OM and SPM-SEM schemes

Figure 1e shows a scheme of SPM unit combined with the OM lens. Figure 1f shows a scheme of the SPM unit located on the goniometer of the SEM under the pole piece of the SEM objective electron lens. The gap between sample surface and the pole piece of the objective lens is 7 mm. In the case of a combination of SPM-OM (Fig. 1e), the probe moves in the (X, Y) plane under the OM light beam using screws. It is sufficient to use only 1D piezo-inertial mover to capture the interaction between the probe and the sample. To choose the place on the sample, a 2D OM stage is used. The piezo tube is used for scanning by the probe at X, Y directions. The stack multilayer piezo actuator is used both for fine scanning by the probe at Z direction and for stepwise coarse approach of the probe to the sample. In the case of a SPM-SEM combination (Fig. 1f), a 2D piezo-inertial mover is used to move the probe under the electron beam. To select a region on the sample surface, the SEM goniometer is used. In this case, the piezo tube is used to scan the sample at the X, Y directions, and the stack multilayer piezo actuator is used both for fine scanning the sample in the Z direction and for coarse approach of the sample to the probe using the 1D piezo-inertial mover.

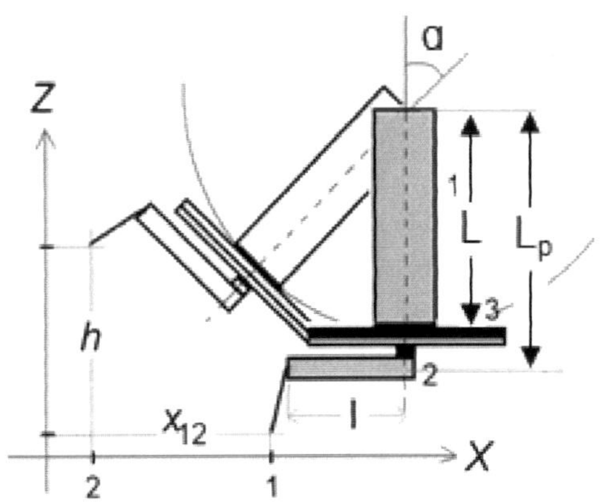


**Figure 2.** Probe sensor displacement along the Z direction during scanning: 1 – piezo scanner (X, Y) in the form of piezo tube with length L, 2 – probe sensor with length l, 3 – compensating piezo actuator in the form of bimorph membrane, α – small angle of piezo tube displacement during scanning, $x_{12}$ – the displacement of the end of the piezo tube along the X direction, h – the displacement of the end of the probe sensor along the Z direction.

It should be noted that for scheme shown in Figure 1e, a gap between the tip and sample surface is increased during scanning. Indeed, in frame of rough geometrical model, it can be assumed that the free end of the piezo tube (1) moves around a circle of a large radius and probe sensor (2) rotates at a small angle $\alpha \sim x_{12}/Lp$ during scanning, as it is shown in Figure 2, where $x_{12}$ is the probe displacement during scanning from point 1 to point 2 in X direction, Lp is the distance from the attachment point of the probe to the centre of the circle being approximately equal to the length L of the piezo tube (1). As a result, the gap between the tip and sample surface is increased by the amount $h \sim x_{12}\ l/L$, where l is the probe sensor length. For example, at L ~ 30 mm, l ~ 10 mm, and $x_{12}$ ~ 20 μm, the gap increasing h will be ~ 7 μm. This deviation is compensated by SPM feedback loop and looks like a slope on the SPM image. Of course, this slope can be deleted by mathematical treatment; however, the SPM working diapason along Z axis will be decreased. To compensate the working diapason decreasing, the extra piezo actuator in the form of bimorph membrane (3) is attached to the free end of the piezo tube. As a result, the working diapason on Z direction is ~ 20 μm. The SPM lens shown in Figure 1e, supplemented by a compensating piezo-bimorph (not shown in Fig. 1e), provides a spatial resolution of ~ 1 nm with a scanning range of 20x20 $\mu m^2$. The circuit shown in Figure 1f is devoid of the disadvantage described above, since on this scheme along the XY coordinates not the probe, but the sample moves. In this scheme, the scanning diapason is 20x20x7 $\mu m^3$ with the same spatial resolution.

## 5. Examples of SPM-OM and SPM-SEM images

Figures 3 and 4 present the results of SPM-OM and SPM-SEM application. Combining methods, firstly, makes it possible to move the SPM probe to the desired point under the control of OM or SEM and to carry out accurate measurements along the Z coordinate, which is rather difficult to do without

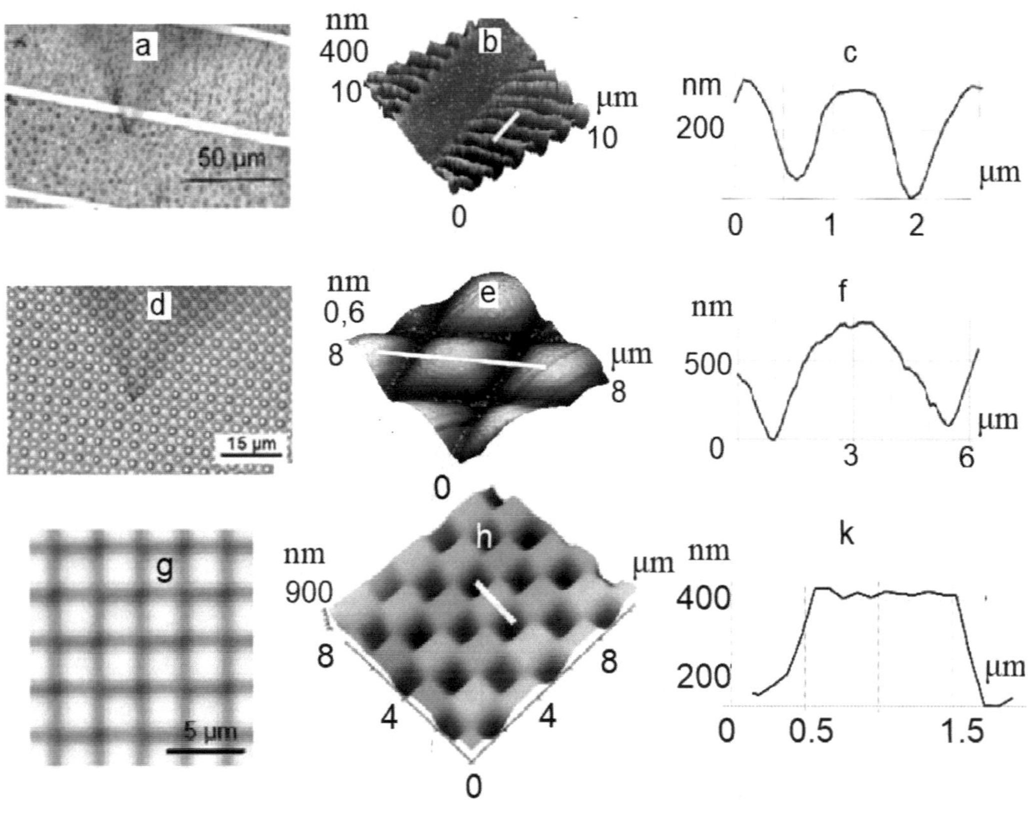


**Figure 3.** (a), (b), (c) SPM-OM images of polymer 2D-hologram; (d), (e), (f) semiconductor CCD-matrix; (g), (h), (k) diffraction grid on the Si surface. (a), (d), (g) OM images; (b), (e), (h) SPM images; (c), (f), (k) cross sections of SPM images.

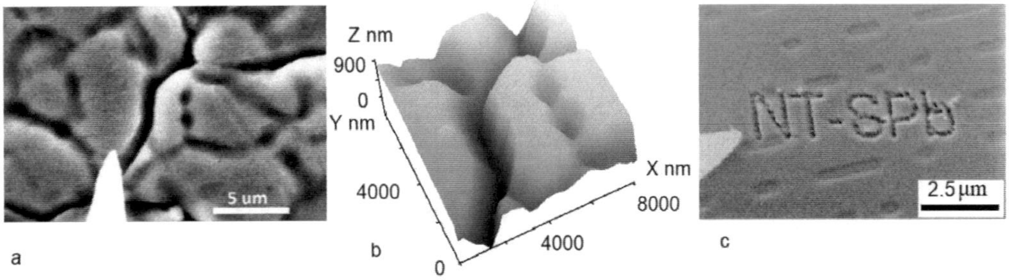


**Figure 4.** SPM-SEM images of steel surface and a result of a dynamic force nanolithography performed under an electron beam control; (a) SEM image of the steel surface, (b) SPM image received simultaneously with SEM image, (c) SEM image of microstructure created on the CD disk by dynamic force nanolithography.

using SPM. Secondly, the combination of different visualization methods in one experiment allows more accurate deciphering the micro- and nanostructure of the object. For example, on the optical image of the polymer hologram surface (Fig. 3a), periodically located large bands with a period of ~50 μm and a fine structure consisting of round objects with a typical size of ~1 μm are clearly visible. At the same time, on the SPM image both these large bands and a fine striped structure with a period of ~1 μm are observed (Fig. 3b,c), but the round objects with a typical size of ~1 μm are absent. Comparing optical and SPM images, it can be assumed that the fine striped structure on the surface of a polymer is not resolved by OM. It can be also assumed that the round objects on the optical image (Fig. 1a) are water micro drops condensed on the polymer surface and reflecting light. These micro drops of water are not visible on the SPM image, because the SPM probe penetrates through them and gives an image of the real surface of the polymer. It is interesting to compare the OM and SPM images of the diffraction grating (Fig. 3g,h).

In an optical image (Fig. 3g), the surface of the diffraction grid looks like a plane divided by narrow channels into square micro cells. From this image, the lattice period can be measured correctly. Although the correct data on the micro relief of the surface can be extracted only from the SPM image (Fig. 3h) and its cross section (Fig. 3k). This is due to the fact that in a vertical incidence, light is equally reflected from the surface of the protruding and recessed cells. SPM with nanometre accuracy allows measuring the height of the steps equal to 400 nm. Finally, SPM-OM-SEM combination fundamentally expands the possibilities of the experiment, for example, it becomes possible to measure local mechanical, electrical, and magnetic characteristics on the sample surface using a variety of SPM techniques.

## 6. Conclusion

Using piezo-inertial linear motors and self-sensing probes allows creating compact SPM modules for combining it with optical and electron microscopes. Due to the exclusion from the design stepper motors and an optical scheme for measuring the cantilever deflection, the SPM modules are easily placed on the lens of an optical microscope or goniometer of a scanning electron microscope. Since the scanning probe, scanning electron, and optical microscopes have a fundamentally different nature of contrast, their combination significantly expands the volume of information received in one experiment and increases the accuracy of nanodiagnostics of materials.

## Acknowledgements

The work was carried out with the support of the Ministry of Education and Science of the Russian Federation, State Project No. 075-00780-19-00 (Subject No. 0074-2019-0007).

## References

[1] Golubok A O and Timofeev V A 1992 *Ultramicroscopy* **42-44** 1558-63
[2] Stahl U, Yuan C W, de Lozanne A L et al. 1994 *Appl. Phys. Lett.* **65** 2878.
[3] Kreith J, Strunz T, Fantner E J, Fantner G E and Cordill M J 2017 *Rev. Sci. Instr.* **88** 053704
[4] AFM SEM/FIB Integrated Nanoscale Characterization, Nanonics Imaging Ltd., see http://www.nanonics.co.il/products/spm-for-sem-fib
[5] LensAFM - AFM for optical microscopes, Nanosurf company, see https://www.nanosurf.com/en/products/lensafm-afm-for-optical-microscopes
[6] Vasiliev A A, Kerpeleva S Yu, Kotov V V, Sapozhnikov I D and Golubok A O 2005 *Nauchnoe Priborostroenie* **15** 62-9
[7] Bausells J 2015 *Microelectron. Eng.* **145** 9-20

IOP Conf. Series: Materials Science and Engineering **699** (2019) 012041 doi:10.1088/1757-899X/699/1/012041

# Radial non-uniform piezoelectric response of perovskite islands in thin PZT films

**S V Senkevich[1,2], D A Kiselev[3,2], T S Ilina[3], V V Osipov[1], A V Elshin[4] and V P Pronin[2]**

[1]Ioffe Institute, 194021 St.-Petersburg, Russia
[2]Herzen State Pedagogical University of Russia, 191186 St.-Petersburg, Russia
[3]National University of Science and Technology (MISIS), 119049 Moscow, Russia
[4]Russian Technological University (MIREA), 119454 Moscow, Russia

SenkevichSV@mail.ioffe.ru

**Abstract**. A comparative study of the morphology and piezoelectric response of island and continuous perovskite thin PZT films deposited on a platinized silicon substrate $Si/SiO_2/TiO_2/Pt$ was carried out. It was shown that the self-polarization value of micron-size island films was about 1.4 times higher than that of continuous films. It is assumed that the difference is due to the relaxation of tensile mechanical stresses caused by silicon substrate at the island periphery.

## 1. Introduction

Thin $Pb(Zr,Ti)O_3$ (PZT) films are used in various silicon microelectronic devices – memory elements (FRAM), microelectromechanical systems (MEMS), and other devices [1-4]. The silicon substrates lead to certain difficulties in their manufacturing due to the difference in the temperature coefficients of linear expansion of silicon and PZT compositions (Zr/(Zr+Ti) = 0.52-0.54) corresponding to the region of the morphotropic phase boundary (MPB). This leads to the appearance of tensile mechanical stresses acting on the thin PZT layer from the substrate, reorientation of the significant part of the spontaneous polarization in directions close to the substrate plane, and decrease in the normally oriented macroscopic polarization, the value of the switched polarization, and the electromechanical coefficients, etc. [5-7].

One way to avoid the deterioration of the above characteristics is to reduce the local (transverse) dimensions of the perovskite regions. Previous studies revealed an increase in switched polarization for micron (submicron) sizes of perovskite regions, demonstrating decrease in the negative role of tensile mechanical stresses [8]. In this regard, the aim of this work was to study the microstructure and physical properties of perovskite islands arising during high-temperature annealing of amorphous thin PZT films.

## 2. Experimental

Thin films were prepared by ex-situ method of RF magnetron deposition by varying the working gas pressure [9]. The deposited amorphous layers of the composition corresponded to MPB with a thickness of 500-1000 nm were annealed in air at temperatures $T_{ann}$ = 550-600°C. The substrate was platinized silicon wafers with buffer ($SiO_2$) and adhesive ($TiO_2$) layers [$Si/SiO_2/TiO_2/Pt$]. Morphology and piezoelectric response were measured using Nterga Prima nanolaboratory (NT-MDT). The contact mode was used for studying the piezoelectric response with the application of AC voltage of 5 V

IOP Conf. Series: Materials Science and Engineering **699** (2019) 012041 doi:10.1088/1757-899X/699/1/012041

amplitude at a frequency about 50 kHz. The scanning area did not exceed 60×60 μm². The microstructure of thin films was also studied by the second harmonic generation (SHG) [10].

## 3. Results

The crystallization of the perovskite phase during high-temperature annealing occurred by nucleation and growth of perovskite islands in the low-temperature (intermediate) pyrochlore matrix. The volume of the perovskite phase depended on the temperature and duration of annealing, as well as pressure of the working gas during film deposition. A study of the piezoelectric response showed that both individual islands and films completely formed in the perovskite phase were characterized by the presence of macroscopic polarization (self-polarization effect). It was shown earlier that the reasons for its appearance were associated with formation of a space charge of oxygen defects near the lower interface of the film, which produced the bulk field leading to orientation of the ferroelectric dipoles [11-12]. A comparison of the properties of individual islands, merging islands, and continuous films allowed revealing several features.

1. The histograms of the distribution of the continuous films were well-formed peaks, the shift of which along the abscissa axis reflected the magnitude of self-polarization (Fig. 1a). In contrast, the histograms of the perovskite islands were characterized by a wide range of values with two maxima, i.e., were strongly inhomogeneous (Fig. 1b). In this case, the displacement of the right-hand side of the distribution significantly exceeded the self-polarization values characteristic of a continuous film.

2. Figure 2 shows an unusual signal of SHG distribution of two perovskite islands touching each other. By the magnitude of the SHG signal, one can quantitatively judge the distribution of polarization in the plane of a thin film. It can be seen from the figure that the signal intensity is radially inhomogeneous and reaches maximum values at the periphery ("crusts") of the islands. The width of the "crust" with a strong SHG signal region was ~ 3.5-4 μm. Another feature in the distribution was the absence of a strong signal at the boundary between two islands.

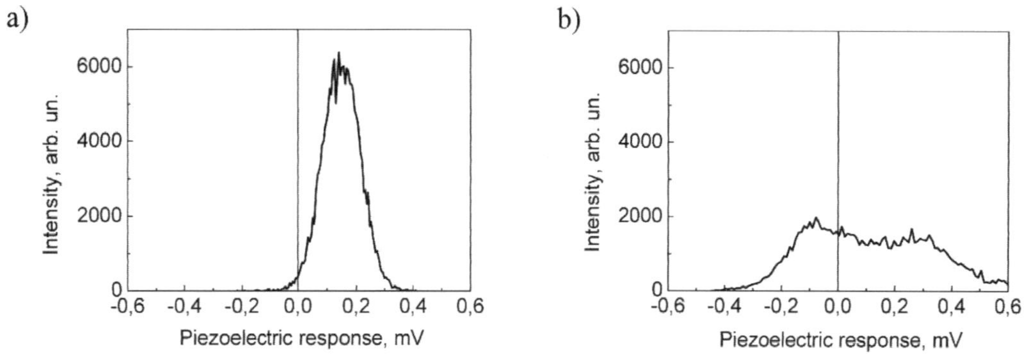


**Figure 1.** Histograms of the piezoelectric response distribution of (a) self-poled continuous and (b) island PZT films.

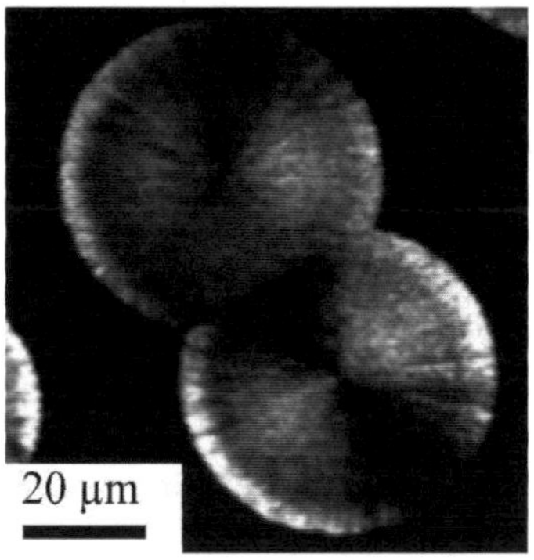


**Figure 2.** Nonlinear optical images (SHG) of two touching perovskite islands.

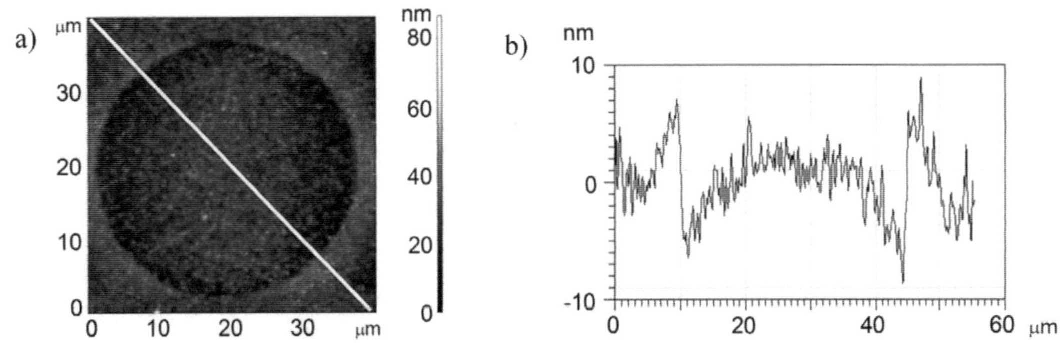


**Figure 3.** (a) The surface topography of the perovskite island; (b) the profile along the line.

3. The heterogeneity of surface morphology was the most pronounced on the periphery of the islands (Fig. 3). At the interface between the perovskite and pyrochlore phases, an increase in the film thickness was observed from the side of the pyrochlore phase, which sharply decreased upon transition to the perovskite phase. A sharp drop, as is well known, is associated with a difference in phase densities [13-14]. The reason for the increase in thickness on the side of the pyrochlore phase is not entirely clear. On the one hand, this may be due to the excess lead oxide present in the deposited film, which is displaced to the interface during crystallization of the dense perovskite phase. On the other hand, a feature of the solid-state phase transformation of the film on the substrate surface is the transition through an intermediate state, which is characterized by an increase in the porosity of the film, and, as a result, an increase in its thickness. In particular, excess lead oxide can just lead to an increase in porosity.

4. A study of the piezoelectric response of perovskite islands, reflecting the degree of unipolarity (self-polarization), revealed a radial inhomogeneity: the maximum signal was observed at the edges of the islands, and the signal significantly decreased towards the centre. It turned out that the degree of radial heterogeneity depended on the size of the islands: with a size (diameter) increase, the signal

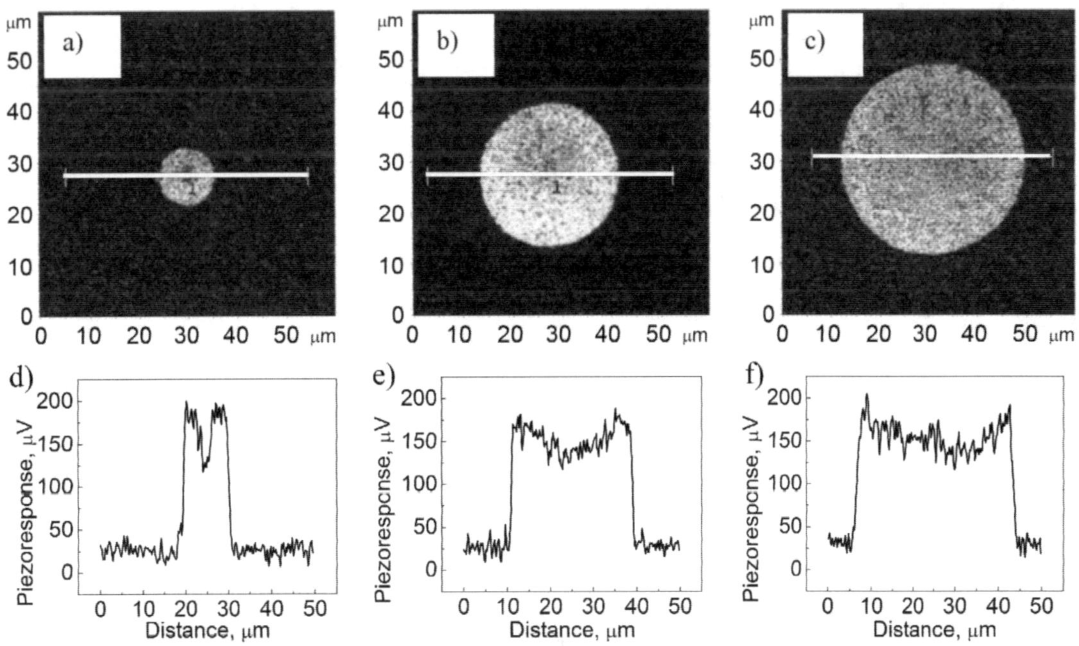


**Figure 4.** (a,b,c) Piezoelectric force microscopy images and (d,e,f) piezoelectric response profiles of perovskite islands of various diameters.

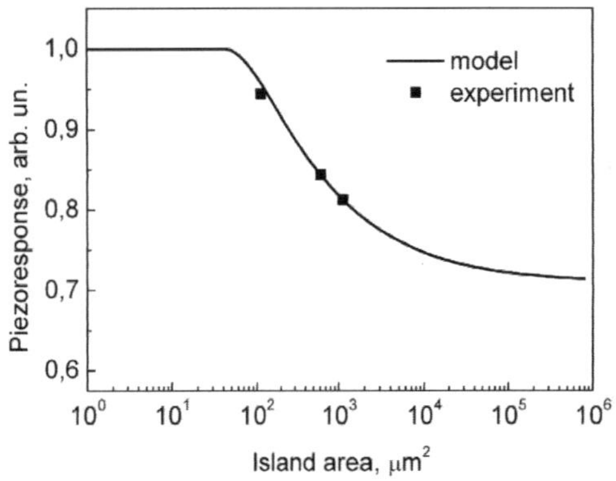


**Figure 5.** The change in the piezoelectric response of the island with increasing size: model and experiment.

changed more smoothly. It can be seen that the “crust” region was most clearly manifested in small islands (Fig. 4a), and with an increase in the diameter of the islands, the “crust” region was eroded (Fig. 4b,c). Nevertheless, the difference in the piezoelectric response in the centre of the island and at the periphery remained (Fig. 4).

## 4. Discussion

The analysis of the results obtained (Figs. 1-4) allowed us to make the assumption that the reason for the heterogeneity of the piezoelectric response of the islands, the signal of the second optical harmonic in them, and the difference in the distribution of the piezoelectric response in the continuous film and the island is related to the mechanical stresses acting on the PZT film from the silicon substrate [5-7]. The temperature coefficient of linear expansion of a thin film of the composition corresponding to MPB, as noted earlier, is greater than that of a silicon substrate, which leads to stretching of the film under the two-dimensional mechanical stresses. It is assumed that in the periphery of the perovskite island there is a relaxation of mechanical stresses on the pores, as a result of which the reorientation of the ferroelectric polarization is noticeably facilitated. We cannot exclude the greater likelihood of the formation of a monoclinic modification of the ferroelectric phase [15] in the absence of mechanical stresses [16]. This assumption is also supported by the results of studying the phase state of thin PZT films by the method of reflected electron diffraction, where the content (fraction) of the monoclinic phase increased in island perovskite films in comparison with continuous ones [17].

The results obtained indicate that the width of the “crust” with high piezoresponse values is equal 3.5-4.0 μm, where its value is approximately 1.4-1.6 times higher than in the continuous film. Based on these data, the dependence of the averaged piezoelectric response over the island area depending on its area is plotted in relative units (Fig. 5, solid curve). A constant value on the curve, equal to the unity, corresponds to the maximum signal of the piezoelectric response of the island, the radius of which does not exceed a critical size equal to the width of the "crust" ~ 3.7 μm. The decrease in the average piezoelectric response signal with an increase in the diameter of the island occurs to the level of the signal from the continuous film corresponding to the magnitude of the piezoelectric response signal in the central part of the island. For calculations, the signal magnitude was 1.4 times less than the piezoelectric response signal of the “crust”.

The points in Figure 5 correspond to the experimental data of the averaged values of the piezoelectric response signal over the area of islands normalized with respect to the maximum signal. It can be seen that a comparison of the model curve with the above parameters and experimental data is in good agreement with each other.

## 5. Conclusion

The results of the study suggest that, in order to obtain thin PZT films with maximum self-polarization, a periodic island structure should be constructed. A similar structure can be obtained, in particular, by femtosecond laser annealing [18-19]. Another chance for obtaining highly effective self-poled films is

IOP Conf. Series: Materials Science and Engineering **699** (2019) 012041 doi:10.1088/1757-899X/699/1/012041

the use of substrates with a higher temperature coefficient of linear expansion. In the latter case, the field of practical use of such films is limited by the complexity of integration with silicon microelectronics.

**Acknowledgements**
The study was partly funded by RFBR according to the research project № 18-32-00092 and was partly supported by the Ministry for Education and Science of Russian Federation (grant № 16.2811.2017/4.6).

**References**
[1] Scott J F 1998 *Ferroelectr. Rev.* **1** 1-129
[2] Muralt P 2008 *J. Am. Cer. Soc.* **91** 1385-96
[3] Eerenstein W, Mathur N D and Scott J F 2006 *Nature* **442** 759-65
[4] Izyumskaya N, Alivov Y-I, Cho S-J, Morkoç H, Lee H and Kang Y-S 2007 *Crit. Rev. Solid State* **32** 111-202
[5] Ogawa T, Senda A and Kasanami T 1991 *Jpn. J. Appl. Phys.* **30** 2145-8
[6] Bruchhaus R, Pitzer D, Schreiter M and Wersing W 1999 *J. Electroceram.* **3** 151-62
[7] Pronin I P, Kaptelov E Yu, Gol'tsev A V and Afanasjev V P 2003 *Phys. Solid State* **45** 1768-73
[8] Bühlmann S, Dwir B, Baborowski J and Muralt P 2002 *Appl. Phys. Lett.* **80** 3195-7
[9] Osipov V V, Kaptelov E Yu, Senkevich S V, Kiselev D A and Pronin I P 2018 *Ferroelectrics* **535** 78-82
[10] Firsova N Yu, Mishina E D, Sigov A S, Senkevich S V, Pronin I P, Kholkin A L, Bdikin I and Yuzyuk Yu I 2012 *Ferroelectrics* **433** 164-9
[11] Kholkin A L, Brooks K G, Taylor D V, Hiboux S and Setter N 1998 *Integr. Ferroelectr.* **22** 525-33
[12] Balke N, Bdikin I, Kalinin S V and Kholkin A L 2009 *J. Am. Cer. Soc.* **92** 1629-47
[13] Krupanidhi S B 1992 *Integr. Ferroelectr.* **1** 161-72
[14] Pronin I P, Kaptelov E Yu, Senkevich S A, Klimov V A, Zaitseva N V, Shaplygina T A, Pronin V P and Kukushkin S A 2010 *Phys. Solid State* **52** 132-6
[15] Noheda B, Cox D E, Shirane G, Guo R, Jones B and Cross L E 2001 *Phys. Rev. B* **63** 014103
[16] Yan L, Li J, Cao H and Viehland D 2006 *Appl. Phys. Lett.* **89** 262905
[17] Dolginzev D M, Kanareikin A G, Pronin V P, Kaptelov E Yu, Senkevich S V and Pronin I P 2015 *Nanomaterials and Nanostructures - XXI Century* **6** 21-8
[18] Elshin A S, Firsova N Yu, Marchenkova M A, Emel`yanov V I, Pronin I P, Senkevich S V, Mishina E D and Sigov A S 2015 *Tech. Phys. Lett.* **41** 418-21
[19] Elshin A S, Abdullaev D A and Mishina E D 2016 *Phys. Solid State* **58** 1154-9

IOP Conf. Series: Materials Science and Engineering **699** (2019) 012042 doi:10.1088/1757-899X/699/1/012042

# Novel technology for fabrication of probe tips for SPM using focused ion beam-induced deposition method

**N A Shandyba[1], A S Kolomiytsev[1], I V Panchenko[1] and S A Lisitsyn[2]**

[1]Southern Federal University, Institute of Nanotechnologies, Electronics, and Equipment Engineering, 347922 Taganrog, Russia
[2]State autonomous educational institution of the Astrakhan region of additional education "Regional school technopark", Kvantorium department, 414000 Astrakhan, Russia

shandyba@sfedu.ru

**Abstract**. The results of an experimental study of the fabrication of high aspect probe tips for atomic force microscopy (AFM) and critical dimension AFM (CD AFM) using the focused ion beam (FIB) milling and ion beam induced deposition of carbon are presented. It is shown that the probes formed by the FIB method have a higher resolution of surface scanning and allow studying nanoscale structures without distortion. In this work, a technique for the formation of AFM probes based on the use of a combination of ion-beam milling and ion-induced carbon deposition was developed. Based on the developed technique, high aspect probe with about 8 μm high, with a tip radius of about 40 nm and an aspect ratio of 30:1, were fabricated. A probe for CD AFM was also formed using FIB-induced deposition of carbon with about 320 nm of flare. The results obtained in the study can be used in the development of technological processes for the manufacturing and modification of special probes for AFM, including probes CD-AFM. The paper shows the possibility of forming the tip of the probes for specialized tasks of nanodiagnostics and nanometrology. The proposed technology allows reducing the artefacts of scanning probe microscopy and improving the adequacy of the results of AFM studies.

## 1. Introduction

Currently, atomic force microscopy (AFM) is one of the most promising methods of obtaining information on the structure and properties of the solid surface with nanometre spatial resolution [1]. Various AFM techniques allow obtaining information about the surface relief as well as its mechanical and electrical parameters under different conditions [2]. The quality of the obtained AFM images and the correctness of the measured properties of the sample directly depends on the shape of the probe and the radius of sharpening of its tip. Now, samples with a limited range of geometric parameters are presented on the market of probe cantilever sensors, which leads to the appearance of artefacts when scanning samples (Fig. 1). The figure shows, how the shape of the AFM probe can affect the image quality of specific structures. The application of new methods of local nanostructuring makes it possible to form probes of various geometric shapes for AFM and for specialized methods of scanning probe microscopy (SPM) [3]. One of the promising methods for forming probe tips with high accuracy and resolving power is the use of focused ion beams (FIB) [4].

The FIB method allows carrying out technological operations of local ion-beam milling and ion-induced deposition of materials from the gas phase under high vacuum conditions without need to use resists, masks, and chemical etchers. When chemically active gases are supplied to the FIB zone, their

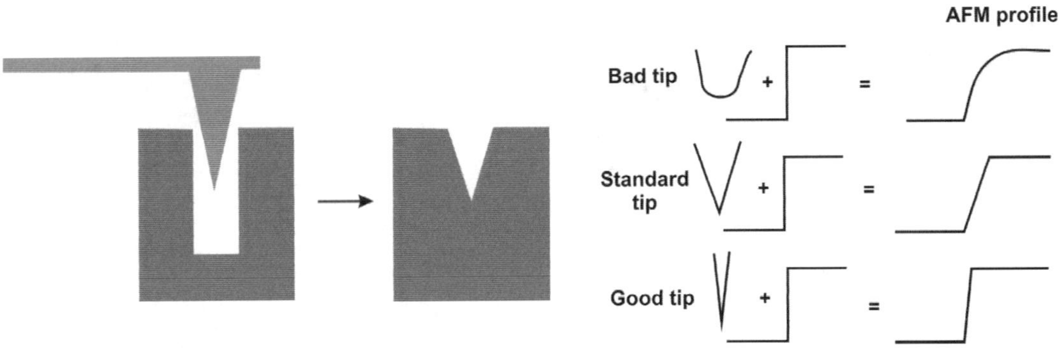


**Figure 1.** AFM artefacts obtained by scanning with a commercially available probe.

molecules dissociate. As a result, one of the gas components is deposited on the surface, and the volatile reaction products are removed by a vacuum system [5]. By setting the trajectory of the ion beam using control electronics, it is possible to form three-dimensional nanoscale structures of a complex profile. By adjusting the ion beam current, one can control the deposition rate and resolution. The wide range of materials deposited by this method allows using FIB in formation of nanoscale structures for electronics (quantum dots, field emitters, and point contacts) and nanodiagnostics (probe tips and relief measurements of the nanometre range for SPM) [6]. The main disadvantage of the FIB technology, hindering its widespread industrial implementation, is the relatively low productivity of the process, since only one ion beam is used. In recent years, the possibility of developing multi-beam process plants has been actively discussed, but in view of the high complexity of such equipment, it is advisable to search for alternative ways to combine the FIB method and traditional microelectronics processes.

The aim of this work is the formation of high-aspect AFM probes with a minimum radius of sharpening and probes for studying lateral surfaces using the FIB method and their experimental study. Using the flexibility and versatility of the FIB technology, it is possible to develop methods for forming the tips of probes of complex shape on any kind of cantilevers, both conductive and non-conductive, as well as to develop automation tools for the manufacturing process of the probe tips.

## 2. Experimental

Experimental studies were carried out using a FEI Company DualBeam system Nova NanoLab 600, combining a Ga+ FIB and a field emission scanning electron microscope (SEM) and the Ntegra Vita scanning probe microscope (NT-MDT, Russia). The standard AFM cantilevers NSG-10 (NT-MDT, Russia) with broken probes due to operation were used as a substrate [7].

At the initial stage of work on the formation of a high aspect probe for an AFM, the surface of a broken tip was cleaned and smoothed using a local ion beam etching method. This operation was performed with a current of FIB of 3 nA.

At the next stage, the tip of the probe was formed by the method of ion-stimulated deposition of carbon ($C_6H_{14}$) from the gas phase by a gallium FIB (Fig. 2b). In the Nova NanoLab 600 control program, a raster graphic template for ion-stimulated deposition was formed, being a stream file in ASCII format. Ion-stimulated deposition of the tip of the probe was performed at the following values of the FIB parameters: the time of the beam exposure at the point was 100 μs, the FIB current was 10 pA, and the accelerating beam voltage was 30 keV [8]. During the process, the vacuum in the working chamber was maintained at a level of $5x10^{-4}$ Pa. The process of forming the tip of the probe was monitored by SEM. Thus, a probe was formed with a height of about 7 μm, a tip radius of about 40 nm, and an aspect ratio of 30:1.

The FIB technology capabilities allow creating micro- and nanostructures of complex shape, as well as controlling their parameters during the formation process. Also, according to the method described above, a probe was formed to study the side walls of various structures. The technological route included cleaning the place from a broken tip, growing the base of the probe for CD AFM and forming flare using

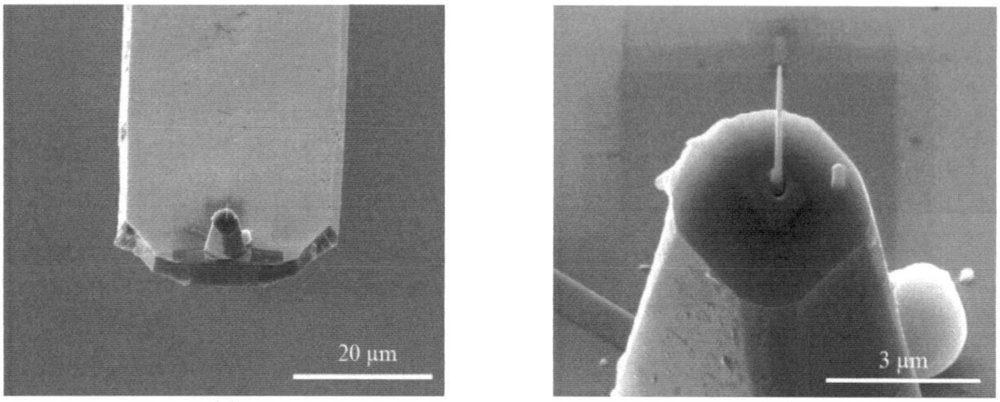


(a) (b)

**Figure 2.** SEM image of cantilever NSG-10 with a tip (with a radius of about 40 nm) formed by ion-induced carbon deposition on the site of a broken tip.

specially formed templates in the FIB control program (Fig. 3a). General parameters of ion-stimulated deposition of carbon from the gas phase by a FIB: accelerating voltage – 30 keV, beam current – 10 pA, beam dwell time – 150 μs, overlap value – 75%. Thus, a probe for CD AFM was formed with the diameter of flare about 320 nm (Fig. 3b).

## 3. Results and discussion

The study of the obtained probes was carried out using a Ntegra Vita scanning probe microscope (NT-MDT, Russia). The material of study was the standard calibration grating TGQ-1 (Fig. 4a), which was first studied in the tapping mode by a commercially available probe (Fig. 4b), and then by a probe formed by the FIB method (Fig. 4c). This object of study was chosen as a model structure of an object with vertical walls, such as, for example, a channel on the surface of an IC or the structure of an optical microresonator. The study of the parameters of such structures is extremely important for processing technology.

Studies have shown that the probe formed by the FIB methods has improved the accuracy of the study of the test object profile. So, the passport values of the TGQ-1 protrusion width are 1.5 ± 0.01 nm, values measured with a standard probe are 2.3 ± 0.2 nm, and with the modified probe – 1.5 ± 0.07 nm. The passport values of the lattice period are 3.0 ± 0.01 nm, values measured with a standard probe are

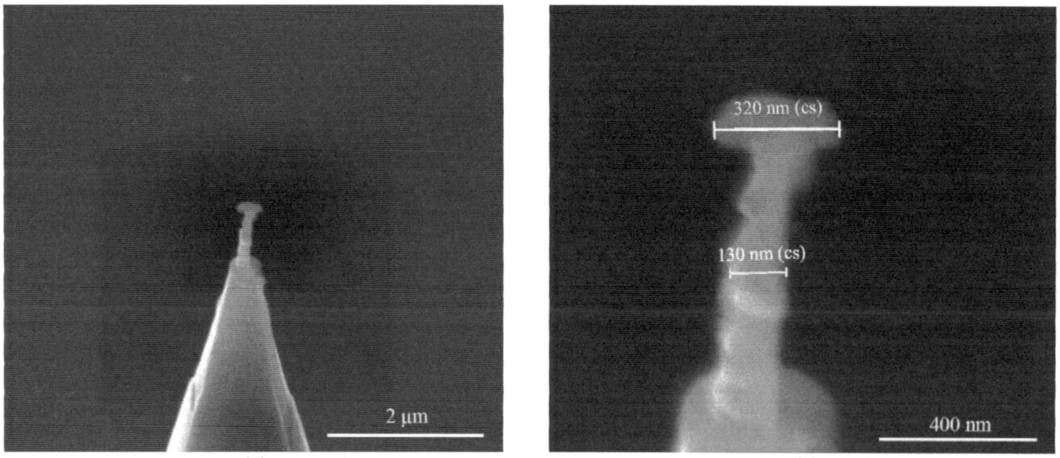


(a) (b)

**Figure 3.** CD AFM probe shaped using FIB-induced deposition of carbon.

IOP Conf. Series: Materials Science and Engineering **699** (2019) 012042 doi:10.1088/1757-899X/699/1/012042

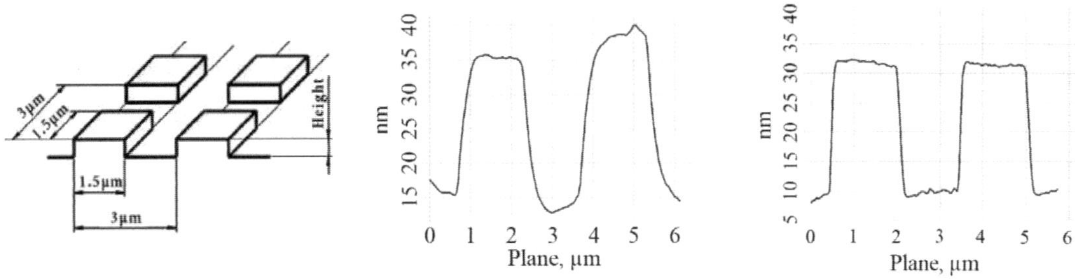


(a) (b) (c)

**Figure 4.** (a) Scheme of the TGQ-1 calibration grating and its profiles scanned using (b) the standard NSG-10 AFM probe and (c) the probe formed by ion-induced deposition.

2.9 ± 1.7 nm, and with modified probe – 3.0 ± 0.5 nm. Standard calibration gratings for AFM are not suitable for assessing the quality of manufacture of probes for CD AFM; therefore, a test sample in the form of a vertical step was fabricated by local ion-beam milling of silicon. The sample is a vertical step with a width of 20 µm and a height of 500 nm, with horizontal and vertical lines formed (Fig. 5a).

The study of probes for CD AFM formed by the method of local ion-stimulated deposition of FIB was carried out in the following way. The resulting image of the relief of the side surface using the fabricated probe is shown in Figure 5b. The diameter of the formed flare is about 320 nm. An analysis of the obtained images shows that those obtained by the probe formed by the FIB method are high-quality, have high resolution, and do not have any noticeable distortion.

In the work, an arithmetic mean roughness estimation technique was developed based on extracting the relief of the lateral face through the Grain Analysis tool of the Nova Image Analysis 3.0 software package (NT-MDT). Automated analysis was performed using the MathLab software package after importing the extracted roughness profile. The measured value of the arithmetic mean roughness of the side wall was 19.4±3.0 nm.

At the next stage, a probe manufacturing automation technique for sequential reproducing the procedure and parameters of ion-stimulated deposition was developed. A script file was generated containing a description and a sequence of all operations, as well as parameters of the ion-beam exposure, vacuum and gas-injection systems. After that, using the AutoFIB (FEI Company) software package, the coordinates of all the cantilevers to make the points on were set and the script was run. Thus, the tips were formed sequentially, according to the procedure, at the indicated points.

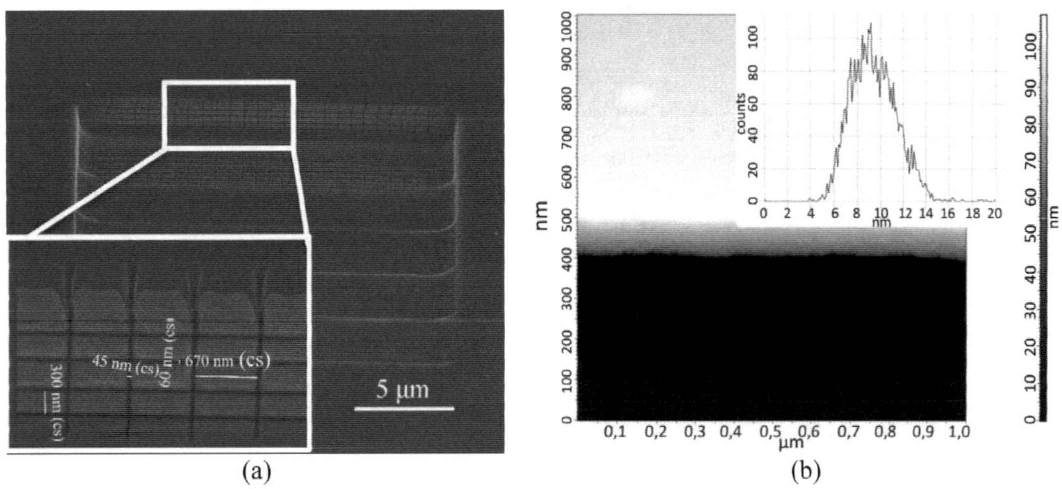


(a) (b)

**Figure 5.** The sample for CD AFM probes testing fabricated by FIB (a) and CD AFM image obtained by the fabricated probe (b).

The manufactured sensors have been tested to have a longer life as a standard AFM measurement tool. In addition, the ability to vary the shape of the probe tips allows the formation of optimized probes for the desired surface research methods. The ability to form probes of the required sizes and parameters mounted on broken cantilevers can significantly increase the economic efficiency of the proposed technology.

## 4. Conclusion

Thus, it has been shown that the use of the method of local ion-stimulated FIB deposition makes it possible to form conductive probes for nanodiagnostics of solid surfaces that are more accurate than commercially available, due to their high aspect ratio. Research conducted by such a probe contributes to obtaining surface data with less distortion and defects. In this work, a technique for the formation of AFM probes based on the use of a combination of ion-beam etching and ion-induced carbon deposition was developed. Based on the developed technique, high aspect probe about 8 μm high, with a tip radius of about 40 nm and an aspect ratio of 30:1 were fabricated. Moreover, a probe for CD AFM was also formed using FIB-induced deposition of carbon with about 320 nm of flare. A technique has been developed for automating the manufacture of probes, which makes it possible to increase the reproducibility of probe parameters and productivity of technology.

The results obtained in the study can be used in the development of technological processes for manufacturing and modification of special probes for AFM, including probes for CD-AFM, and in the development of procedures for the express monitoring of parameters of the technological process for manufacturing elements for micro- and nanoelectronics and micro- and nanosystems engineering.

## Acknowledgements

This work was supported by the Russian Science Foundation Grant No. 18-79-00175. The results were obtained using the equipment of the Scientific-Educational Centre and the Centre for Shared Use "Nanotechnologies" of the Southern Federal University (Taganrog).

## References

[1] Shandyba N and Panchenko I 2018 The formation of probes for specialized modes of scanning probe microscopy by the method of focused ion beams *Saint Petersburg OPEN Book of Abstracts* 606-7

[2] Ageev O A, Kolomiytsev A S and Konoplev B G 2011 Formation of nanosize structures on a silicon substrate by method of focused ion beams *Semiconductors* **45** 89-92

[3] Chang W S, Bauerdick S and Jeong M S 2008 *Ultramicroscopy* **108** 1070-5

[4] Lisitsyn S A, Kolomiytsev A S, Ilyin O I, Ilyina M V, Konoplev B G, Bykov A V and Ageev O A 2017 *Russ. Microelectron.* **46** 1-6

[5] Ageev O A, Kolomiitsev A S, Konoplev B G, Serbu N I and Smirnov V A 2012 *Russ. Microelectron.* **41** 41-50

[6] Knittel P, Hibst N, Mizaikoff B, Strehle S and Kranz C 2017 *Ultramicroscopy* **179** 24-32

[7] Kolomiytsev A S, Shandyba N A, Panchenko I V and Lisitsyn S A 2018 *IOP Conf. Ser.: Mater. Sci. Eng.* **443** 012015

[8] Bilbao-Guillerna A, Eachambadi R T, Cadot G B J, Axinte D A et. al. 2018 *J. Mater. Proc. Tech.* **252** 636-42

IOP Conf. Series: Materials Science and Engineering **699** (2019) 012043 doi:10.1088/1757-899X/699/1/012043

# Scanning probe microscopy investigation of iron garnet films for magnetoplasmonics

**A N Shaposhnikov[1], S D Lyashko[1], A S Nedviga[1], A V Karavainikov[1], E Yu Semuk[1], Yu E Vysokikh[2] and T V Mikhailova[1]**

[1]V.I. Vernadsky Crimean Federal University, Physics and Technology Institute, Simferopol 295007, Russia

[2]Scientific and Technological Centre of Unique Instrumentation of the RAS, Moscow 117342, Russia

tatvladismikh@cfuv.ru

**Abstract.** Topography and domain structures of Bi-substituted iron garnet films proposed for magnetoplasmonic applications are presented. Investigations were carried out by scanning probe microscopy methods, including polarization near-field optical microscopy. The most suitable films with less rough surface were chosen. The period of domain structure of sputtered film of composition $Bi_{2.8}Y_{0.2}Fe_5O_{12}$ were determined, $2w = 0.5$ μm.

## 1. Introduction

Synthesis and investigation of magnetoplasmonic (MP) nanostructures is relevant for construction of micro-dimensional optical isolators, modulators and switches, miniature integrated nanophotonic devices for fast multi-mode and multi-directional control, plasmonic circuit elements controlled by magnetic field [1,2]. For example, nanostructure "iron garnet (IG) – perforated Au film" experimentally demonstrated the effective modulation of transmission with contrast up to 98% using a weak external magnetic field [3]. Creation of metallic coating on IG films imposes specific surface requirements. Additionally, the films should have definite magnetic and magnetooptical properties [4]. In the work, authors present the investigation of topography and domain structures of IG films proposed for MP applications by scanning probe microscopy methods, including polarization near-field optical microscopy [5].

## 2. Experimental

Films of bismuth-substituted IG with micro- and nanoscale thicknesses were synthesized by liquid-phase epitaxy (LPE) and reactive ion beam sputtering (RIBS). For RIBS-films, the garnet phase was formed by crystallization annealing process in the air and high temperature $T_a$ (above 650 °C) after deposition on respective substrate from ceramic target [4,6-8]. Substrates of gadolinium gallium garnet (GGG), calcium niobium gallium garnet (CNGG), and fused quartz $SiO_2$ were used for RIBS-films. All LPE-films were synthesized on GGG substrates.

Grown LPE-films have the compositions of $(Gd,Bi,Lu)_3(Fe,Ga,Ge)_5O_{12}$ (No.1), $(Bi,Y,Eu)_3(Fe,Al,Ga)_5O_{12}$ (No.2), and $(Gd,Bi,Lu)_3(Fe,Ga,Ge)_5O_{12}$ (No.3). The thicknesses of films are 1.9, 5.1, and 11.7 μm, respectively.

Two sets of RIBS films were realized.

**Table 1.** Annealing regimes of double layer RIBS-films No. 6-9.

| No. | Layer of films | $T_a$, °C | $t_a$, min | Heating rate |
|---|---|---|---|---|
| 6 | M2 | 710 | 60 | low |
| | M3 | 680 | 60 | low |
| 7 | M2 | 700 | 20 | dipping |
| | M3 | 680 | 20 | dipping |
| 8 | M2 | 700 | 60 | low |
| | M1 | 650 | 25 | dipping |
| 9 | M2 | 700 | 60 | low |
| | M1 | 660 | 60 | low |

The first set contained the single layer films of composition $Bi_{2.8}Y_{0.2}Fe_5O_{12}$ (M1) on the substrates of GGG and CNGG. The films are denoted as GGG/M1 (No.4) and CNGG/M1 (No.5) hereinafter, respectively. The samples were synthesized at the same temperature $T_a = 650$ °C during annealing time $t_a = 20$ min. Garnet thickness is 100 nm.

The second set contained double layer films $SiO_2$/M2/M3 (No.6), GGG/M2/M3 (No.7), $SiO_2$/M2/M1 (No.8), and GGG/M2/M1 (No.9), where M2 and M3 are bismuth-substituted IG $Bi_{1.0}Y_{0.5}Gd_{1.5}Fe_{4.2}Al_{0.8}O_{12}$ and $Bi_{1.5}Gd_{1.5}Fe_{4.5}Al_{0.5}O_{12}$, respectively. The layers M2, M3, and M1 have the thicknesses of 63, 195, and 177 nm, correspondingly. The double layer films given successful results in magnetophotonic crystals and Tamm structures design [4,8-11]. Different annealing regimes were used to reduce the roughness of RIBS-layers. Annealing conditions are given in the Table 1.

Semicontact atomic force microscopy (AFM) was applied to describe the surface and measure its parameters – height range $R_{max}$, average roughness $R_a$, root mean square roughness *rms*, and polycrystalline size $Z$ (for RIBS-films only). To visualize the domain structure of films, semicontact two-pass or parallel magnetic force microscopy (MFM) and scanning near-field optical microscopy (SNOM) were used. The measurements were carried out by cantilevers of HA_HR ETALON, MFM01, and SNOM_NC with typical aperture diameter of (120 ± 25) nm. The scheme of used experimental equipment, scanning probe microscope (SPM) NTEGRA (NT-MDT) with configuration that allows combining the AFM and SNOM techniques, is illustrated in Figure 1.

## 3. Results and discussion

Table 2 lists the determined parameters of surface topography of investigated samples and total thicknesses of films $h$. Data were obtained by AFM measurements on frames with area of 2x2 $\mu m^2$ using standard AFM cantilever.

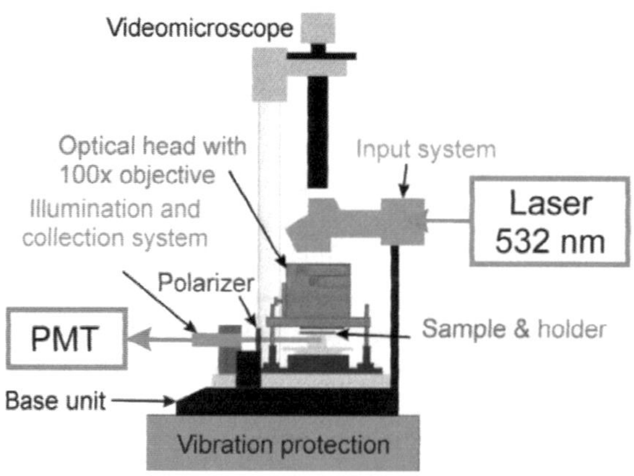


**Figure 1.** Scheme of used configuration of SPM NTEGRA (NT-MDT).

**Table 2.** Surface parameters of investigated LPE- (No.1-3) and RIBS-films (No.4-9).

| No. | $R_{max}$, nm | $R_a$, nm | *rms*, nm | *Z*, nm | *h*, nm |
|---|---|---|---|---|---|
| 1 | 12.6 | 0.6 | 0.8 | - | 1900 |
| 2 | 8.2 | 0.6 | 0.75 | - | 5100 |
| 3 | 13.4 | 1.0 | 1.47 | - | 11700 |
| 4 | 58 | 6.4 | 7.6 | 90 | 100 |
| 5 | 39.5 | 7.6 | 9.0 | 65 | 100 |
| 6 | 27 | 3.2 | 4.0 | 130 | 258 |
| 7 | 24 | 3.2 | 4.0 | 150 | 258 |
| 8 | 49 | 5.5 | 6.9 | 125 | 240 |
| 9 | 63 | 6.0 | 7.9 | 150 | 240 |

Measurements of various LPE-films showed that *rms* could reach values less than 0.9 nm. Figure 2 shows the typical domain structure of LPE-film No.1 with period of 16 μm. From obtained data we notice that width of domain wall of film No.1 is $w_{dw} = 895$ nm measured as distance from minima to maxima of optical intensity signal. Nevertheless, the structure of domain walls and domain edges is not obvious and may be more complicated as Figure 2b,c,d demonstrates.

Figure 3 shows the surface and domain structure of RIBS-film No. 4. MFM and SNOM data give the same domain period of the film that is $2w = 0.5$ μm. The film before measurements was heated. We calculated the parameters of surface from the images obtained by SNOM tip. The values of $R_{max}$, $R_a$, and *rms* decrease to 33.7, 3.9, and 4.8 nm in comparison to the same values in the Table 2. Average size of crystallites increases to $Z = 270$ nm. So, the surface parameters of polycrystalline films with the rough surface are not defined correctly by SNOM tip due to its pyramidal geometry, and development of SNOM tip with more appropriate forms is necessary [12, 13].

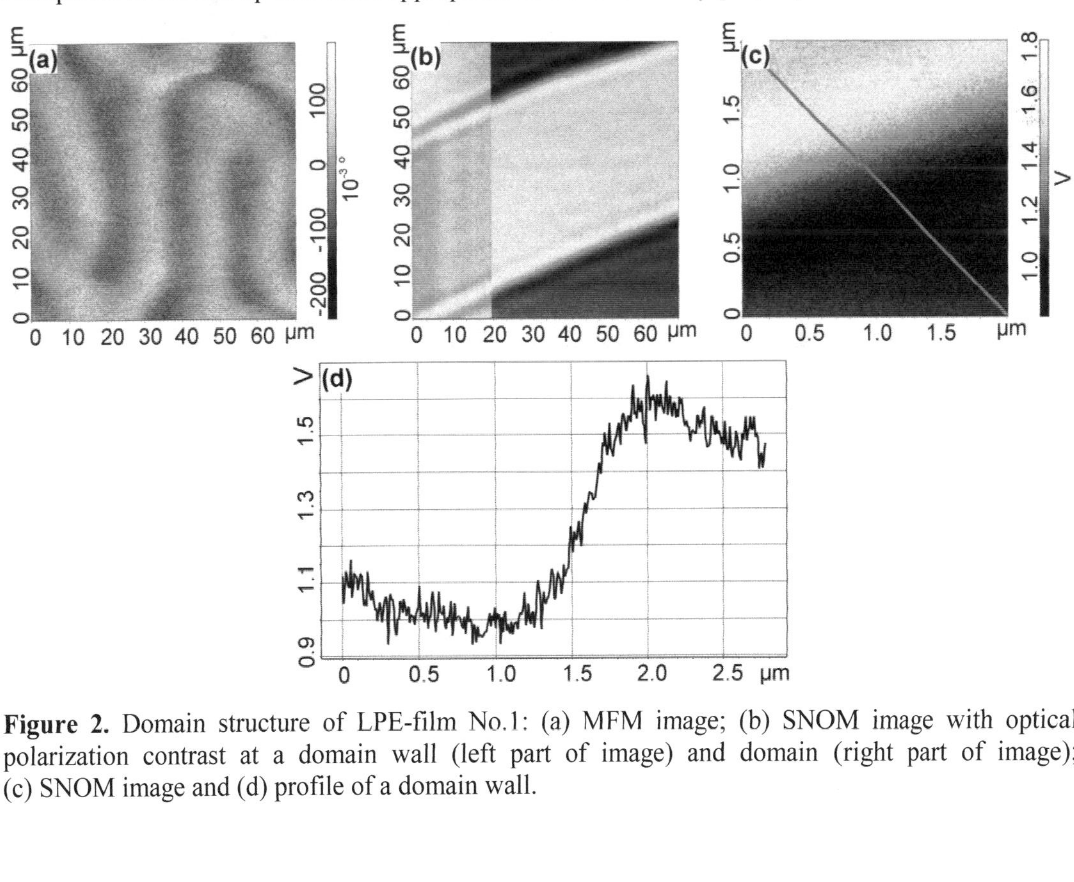


**Figure 2.** Domain structure of LPE-film No.1: (a) MFM image; (b) SNOM image with optical polarization contrast at a domain wall (left part of image) and domain (right part of image); (c) SNOM image and (d) profile of a domain wall.

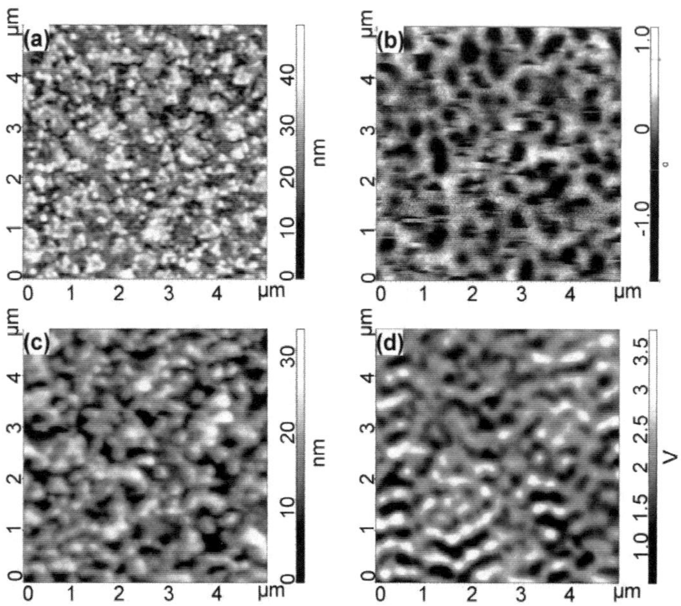


**Figure 3.** (a, c) Topography and (b, d) domain structure of RIBS-film No.4 of the surface area measured by (a, b) MFM and (c, d) combination of AFM and SNOM.

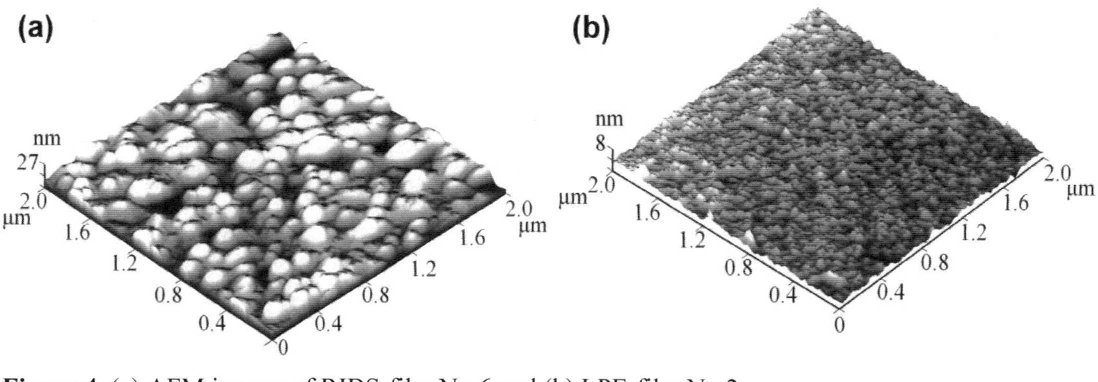


**Figure 4.** (a) AFM images of RIBS-film No.6 and (b) LPE-film No.2.

Figure 4 shows the surfaces of LPE-film No.2 and RIBS-film No.6 with the smallest roughness (Table 2). We propose these films for further experiments.

## 4. Conclusion

LPE and RIBS Bi-substituted IG films were investigated by AFM, MFM, and SNOM to choose the most suitable films with less rough surface for magnetoplasmonic applications. Typical value of *rms* of synthesized LPE-films is 0.8 nm. The *rms* of LPE-films increases with thickening the samples. The structure of domain wall of LPE-film with uniaxial anisotropy was presented.

Topography of RIBS-films strongly depended on annealing regimes. We reduced *rms* of RIBS polycrystalline films to 4.0 nm. The double layer films of composition $Bi_{1.0}Y_{0.5}Gd_{1.5}Fe_{4.2}Al_{0.8}O_{12}$/ $Bi_{1.5}Gd_{1.5}Fe_{4.5}Al_{0.5}O_{12}$ had the smallest values of *rms*.

The period of domain structure of RIBS-film of composition $Bi_{2.8}Y_{0.2}Fe_5O_{12}$ was determined as $2w = 0.5\ \mu m$.

**Acknowledgements**
Research performed by grant of Russian Science Foundation (project no. 19-72-20154).

**References**
[1] Inoue M, Baryshev A V, Goto T, Baek S M, Mito S, Takagi H and Lim P B 2013 Magnetophotonic crystals: Experimental realization and applications *Magnetophotonics* eds. Inoue M, Levy M, Baryshev A V (Springer-Verlag Berlin Heidelberg) chapter 7 p 163
[2] Armelles G, Cebollada A, García-Martín A and González M U 2013 *Adv. Optical Mater* **1** 10
[3] Wurtz G A, Hendren W, Pollard R, Atkinson R, Guyader L Le, Kirilyuk A, Rasing Th, Smolyaninov I I and Zayats A V 2008 *New J. Phys.* **10** 105012
[4] Berzhansky V N, Karavainikov A V, Mikhailova T V, Prokopov A R, Shaposhnikov A N, Shumilov A G, Lugovskoy N V, Semuk E Y, Kharchenko M F, Lukienko I M, Kharchenko Yu M and Belotelov V I 2017 *JMMM* **440** 175
[5] Vysokikh Yu E, Shelaev A V, Prokopov A R, Shevyakov V I and Krasnoborodko S Yu 2016 *J. Phys.: Conf. Ser.* **741** 012190
[6] Berzhansky V, Mikhailova T, Shaposhnikov A, Prokopov A, Karavainikov A, Kotov V, Balabanov D and Burkov V 2013 *Appl. Opt.* **52** 6599-606
[7] Berzhansky V N, Karavainikov A V, Milyukova E T, Mikhailova T V, Prokopov A R and Shaposhnikov A N 2010 *Funct. Mater.* **17** 120-6
[8] Berzhansky V N, Shaposhnikov A N, Prokopov A R, Karavainikov A V, Mikhailova T V, Lukienko I N, Kharchenko Yu N, Golub V O, Salyuk O Yu and Belotelov V I 2016 *J. Exp. Theor. Phys.* **123** 744-51
[9] Mikhailova T V, Berzhansky V N, Shaposhnikov A N, Karavainikov A V, Prokopov A R, Kharchenko Yu M, Lukienko I M, Miloslavskaya O V and Kharchenko M F 2018 *Opt. Mater.* **78** 521
[10] Mikhailova T, Shaposhnikov A, Prokopov A, Karavainikov A, Tomilin S, Lyashko S and Berzhansky V 2018 *EPJ Web of Conferences* **185** 02016
[11] Mikhailova T V, Lyashko S D, Tomilin S V, Karavainikov A V, Prokopov A R, Shaposhnikov A N and Berzhansky V N 2017 *J. Phys.: Conf. Ser.* **917** 062053
[12] Vysokikh Yu E, Krasnoborodko S Yu, Kolomiytsev A S, Fedotov A A, Shaposhnikov A N, Bulatov M F, Churikov D V and Shevyakov V I 2019 *BOOK of ABSTRACTS 6th International School and Conference on Optoelectronics, Photonics, Engineering and Nanostructures* (Saint Petersburg: St. Petersburg Academic University, 2019) 469-470
[13] Kolomiytsev A S, Shandyba N A, Panchenko I V and Lisitsyn S A 2018 *IOP Conf. Ser.: Mater. Sci. Eng.* 2018 **443** 012015

IOP Conf. Series: Materials Science and Engineering **699** (2019) 012044 doi:10.1088/1757-899X/699/1/012044

# Heterogeneous ferromagnetic state in small particles and its connection with ferroelectricity

**T S Shaposhnikova and R F Mamin**

Zavoisky Physical-Technical Institute, FRC Kazan Scientific Centre of RAS, 420029 Kazan, Russia

t_shap@kfti.knc.ru

**Abstract**. PTs in spherical particles of a cubic ferromagnetic was considered in the framework of the Ginzburg-Landau phenomenological theory. Concentrating on depolarizing field effects, we study the competition between states with homogeneous magnetization and vortex structures. For large sphere radii ($R>R_c$), a PT to a vortex state is possible, while for $R<R_c$ it can be a homogeneous state. We obtain the inhomogeneous distribution of the ferromagnetic order parameter in the form of 2D and 3D vortices, for which the absolute value of the local magnetization depends on the distance from the centre of the vortex. This approach can be used for nano-sized particles. In our work, we compute the polarization caused by ferromagnetic 2D and 3D vortices within the phenomenological approach.

## 1. Introduction

Magnetism at small length scales has in recent years provided a wealth of scientific interest and potential technological applications [1,2]. The advances in nanometre scale fabrication technology have opened up the possibility of studying artificially fabricated magnetic systems, in which all three dimensions are geometrically constrained on a nanometre length scale. “Single-domain particles” are very promising candidates for high-density hard disk data storage and integrated magnetoelectronic devices, such as nonvolatile magnetic memory. Discussion of ferromagnetic phase transitions (PTs) in small particles has become especially popular over the past two decades. The problem of phase separation has attracted considerable attention [3-7]. There are materials, where PTs are observed with various types of structural, magnetic, charge, and orbital ordering, for example, manganites, high-temperature cuprate superconductors, and multiferroics. Phase separation in these materials is often accompanied by an inhomogeneous charge distribution. The spatial size of single-phase regions depends on the ratio between the Coulomb energy and the energy gain due to the presence of a more ordered phase.

Formally, the phenomenological description of PTs in ferromagnets and ferroelectrics has the same form, although ferromagnetism and ferroelectricity have completely different sources in solids. We consider the problem of the PT of ferromagnetic regions of the nanoscale size and different forms. Therefore, the order parameter in our case is the magnetization vector, i.e. the magnetic moment averaged over the volume much larger than the unit cell, but much smaller than the volume of our nanoparticle, and not the magnetic moments of individual atoms. We assume that the absolute value of the local magnetization depends on the distance from the centre of the nanoparticle. Such a dependence of the magnetization amplitude was considered, for example, in Ref. [3] in the analysis of ferroelectric nanoparticle near the PT and in Ref. [4] in the analysis of ferromagnetic skyrmions near PT.

It was proposed that magnetic frustration and ferroelectricity had strong connections in multiferroics [5]. In multiferroics, the simultaneous existence of magnetic and electric dipoles does not always cause a strong interaction between them, since the microscopic mechanisms of ferroelectricity and magnetism are different. A probable microscopic mechanism to obtain ferroelectricity in magnetic spirals was discussed, for example, in Ref. [6]. It involves the antisymmetric Dzyaloshinsky-Moriya (DM) interaction. The DM interaction causes a non-collinear spin ordering. If the magnetic ordering is inhomogeneous, it can lead to nonzero polarization [5,7].

In our paper, we consider first the case of a 2D or infinite cylindrical particle of a ferromagnet with two polar axes in the plane and in an isotropic paramagnetic environment. Then we consider the ferromagnetic PT in a 3D spherical particle of a material using the same assumptions and approximation as 2D case. We compute the polarization caused by ferromagnetic 2D and 3D vortices within the phenomenological approach, based on Landau theory of the second order PTs.

## 2. Results and discussion

The effective magnetic field $\vec{H}^{eff}(\vec{r},t)$ is defined as the variation derivative of free energy $F$ by the magnetic moment

$$\vec{H}^{eff}\left(r,t\right)=-\frac{\delta F}{\delta \vec{M}}. \tag{1}$$

### 2.1. 2D case

Let us consider the 2D case. The linearized equations of state using the harmonic part of the free energy of the Landau-Ginzburg-Devonshire [8] has the form

$$F=\int \rho d\rho d\varphi\left(\frac{A}{2}\vec{m}^2+\frac{g}{2}\left\{\left(\frac{\partial m_\rho}{\partial \rho}\right)^2+\left(\frac{\partial m_\varphi}{\partial \rho}\right)^2+\frac{1}{\rho^2}\left[\left(\frac{\partial m_\rho}{\partial \varphi}-m_\varphi\right)^2+\left(\frac{\partial m_\varphi}{\partial \varphi}+m_\rho\right)^2\right]\right\}\right), \tag{2}$$

where $\rho$, $\varphi$ are polar coordinates, $A=A'(T-T_c)$, $T_c$ is PT temperature in bulk, $g=const$, and gradient term can be found in Ref. [9]. A similar expression for free energy was used to describe the ferroelectric PT in small particles in Ref. [3]. There are two equations for the ferromagnetic magnetization

$$H_\rho^{eff}=\left(A+\frac{g}{\rho^2}\right)m_\rho-g\left[\frac{1}{\rho}\frac{\partial}{\partial \rho}\left(\rho\frac{\partial m_\rho}{\partial \rho}\right)+\frac{1}{\rho^2}\frac{\partial^2 m_\rho}{\partial \varphi^2}\right]+\frac{2g}{\rho^2}\frac{\partial m_\varphi}{\partial \varphi},$$

$$H_\varphi^{eff}=\left(A+\frac{g}{\rho^2}\right)m_\varphi-g\left[\frac{1}{\rho}\frac{\partial}{\partial \rho}\left(\rho\frac{\partial m_\varphi}{\partial \rho}\right)+\frac{1}{\rho^2}\frac{\partial^2 m_\varphi}{\partial \varphi^2}\right]-\frac{2g}{\rho^2}\frac{\partial m_\rho}{\partial \varphi}.$$

For a vortex on a circle at $m_\rho=0$, $m_\varphi=m_\varphi(\rho)$, and $\vec{H}^{eff}=0$, the equation on $m_\varphi$ will have the form

$$\frac{1}{\rho}\frac{d}{d\rho}\left(\rho\frac{dm_\varphi}{d\rho}\right)-\left(\frac{A}{g}+\frac{1}{\rho^2}\right)m_\varphi=0 \tag{3}$$

with the boundary condition of the free surface $\left.\frac{dm_\varphi}{d\rho}\right|_{\rho=R}=0$ at $\rho=R$, where $R$ is the radius of the particle. The solution of the equation (3) has the form

$$m_\varphi(\rho)=CJ_1\left(\sqrt{-A/g}\rho\right), \tag{4}$$

where $J_1(x)$ is the first-order of the Bessel function of the first kind. The boundary condition is satisfied when $A=-g\left(\mu_n/R\right)^2$, where $\mu_n$ is the $n$-th root of equation $J_1'(x)=0$ ($\mu_1=1.8$ is the smallest value of the earliest loss stability).

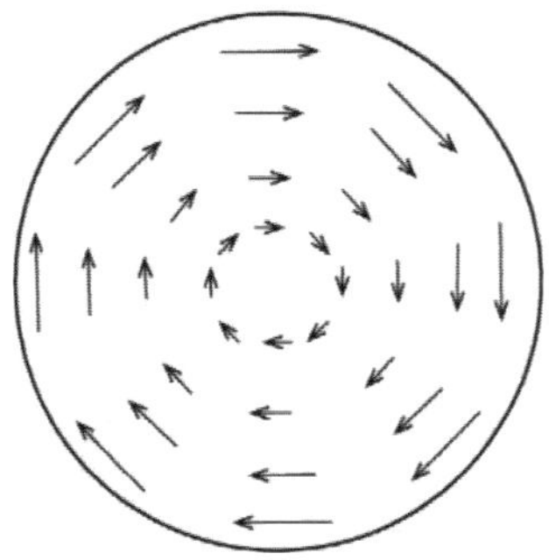

**Figure 1.** Magnetization distribution just after the paramagnetic-ferromagnetic PT in an infinite cylinder. 2D vortex. Vector plot.

The vector field and the form of $m_\varphi(\rho)$ arising at the stability loss are shown in Figure 1. $m_\varphi(\rho)$ is an increasing function of $\rho$ on the interval $0 \le \rho \le R$.

We have to compare the loss of stability with respect to vortex formation with other possibilities of stability loss. We mean specifically the loss of stability with respect to the homogeneously magnetized circle. The magnetic field inside the circle is $\vec{H} = -(\mu+1)^{-1}\vec{M}$, where $\mu$ is the relative magnetic permeability of the ferromagnetic. One can find that the loss of stability occurs at $A = -(\mu+1)^{-1}$. Since the temperature of this type of stability loss does not depend on *R*, the earliest stability loss for large *R* is with respect to vortex formation and the PT temperature is given by the earlier result. However, for small *R* it is possible that the PT into a homogeneous state occurs before and instead of vortex formation. The character of the PT changes at $R = R_c = \mu_1(g)^{1/2}(\mu+1)^{1/2}$.

### 2.2. *3D case*

Consider the 3D case. In spherical coordinates, the thermodynamic potential of the system near the second order PT has the form [9]

$$F = \int r^2\sin(\theta)drd\theta d\varphi\left(\frac{A}{2}\vec{m}^2 + \frac{g}{2}\left\{\left(\frac{\partial m_r}{\partial r}\right)^2 + \left(\frac{\partial m_\theta}{\partial r}\right)^2 + \left(\frac{\partial m_\varphi}{\partial r}\right)^2 + \frac{1}{r^2}\left[\left(\frac{\partial m_r}{\partial \theta} - m_\theta\right)^2 + \left(\frac{\partial m_\varphi}{\partial \theta}\right)^2 + \left(\frac{\partial m_\theta}{\partial \theta} + m_r\right)^2\right] + \right.\right.$$

$$\left.\left.\frac{1}{r^2\sin^2(\theta)}\left[\left(\frac{\partial m_r}{\partial \varphi} - m_\varphi\sin(\theta)\right)^2 + \left(\frac{\partial m_\theta}{\partial \varphi} - m_\varphi\cos(\theta)\right)^2 + \left(m_r\sin(\theta) + m_\theta\cos(\theta) + \frac{\partial m_\varphi}{\partial \varphi}\right)^2\right]\right]\right\}\right), \quad (5)$$

where *r*, $\theta$, and $\varphi$ are the spherical coordinates.

For a vortex on a sphere at $m_r = m_\theta = 0$ and $m_\varphi = m_\varphi(r,\theta)$ and $\vec{H}^{eff}(\vec{r},t) = 0$

$$F = \int r^2\sin(\theta)drd\theta d\varphi\left\{\frac{A}{2}m_\varphi^2 + \frac{g}{2}\left[\left(\frac{\partial m_\varphi}{\partial r}\right)^2 + \frac{1}{r^2}\left(\frac{\partial m_\varphi}{\partial \theta}\right)^2 + \frac{m_\varphi^2}{r^2\sin^2(\theta)}\right]\right\}. \quad (6)$$

Consider the variation $\delta F$ by $\delta m_\varphi$. This variation is zero, if $m_\varphi(r,\theta)$ is the solution of the equation

$$Am_\varphi + \frac{g}{r^2\sin^2(\theta)}m_\varphi - \frac{g}{r^2}\frac{\partial}{\partial r}\left(r^2\frac{\partial m_\varphi}{\partial r}\right) - \frac{g}{r^2\sin(\theta)}\frac{\partial}{\partial \theta}\left(\sin(\theta)\frac{\partial m_\varphi}{\partial \theta}\right) = 0. \quad (7)$$

Then the solution of (7) can be found as

$$m_{\varphi l}(r) = C\left(\frac{\pi}{2r}\right)^{1/2} J_{l+1/2}\left(\left(-\frac{A}{g}\right)^{1/2} r\right) = Cj_l\left(\left(-\frac{A}{g}\right)^{1/2} r\right), \quad (8)$$

where $j_l(x)$ is a spherical Bessel function of $l$–th order and $C$ is a constant. At $A>0$, the argument of the function $J_{l+1/2}(x)$ is imaginary, i.e. the solutions do not satisfy the boundary condition. When $A<0$ the argument of the function $J_{l+1/2}(x)$ is valid, the boundary condition is satisfied, when $A=-g(\mu_n^l/R)^2$, where $\mu_n^l$ is the $n$-th root of the equation $j_l'(x)=0$. Minimum value is $\mu_n^l=\mu_1^1\approx 2.1$. Then $J_{3/2}(x)=(2/(\pi x))^{1/2}(\sin(x)/x-\cos(x))$. The form of the magnetization distribution arising at the stability loss with respect to vortex formation is $m_r=0$, $m_\theta=0$ and $m_\varphi(r,\theta)=Cj_1(\mu_1^1 r/R)\sin(\theta)$ (Fig. 2). $m_\varphi(r,\theta)$ is an increasing function of $r$ on the interval $0\le r\le R$.

Small ferromagnetic particles are 'single-domain' uniformly magnetized formations. The criterion for their size is obtained by comparing the magnetic energy of a uniformly magnetized particle with the energy that would arise in the presence of a significant heterogeneity in the distribution of magnetization in the volume of the particle. The first energy is of order $M^2V$, the second one is of order $gM^2V/l^2$. Therefore, the size of single-domain particles is $R_c=C(g)^{1/2}$ [8]. Analogously to the 2D case, we find $\vec{M}=-(\mu+2)^{-1}\vec{H}$ for homogeneous polarization of the ball, and the stability is lost at $A=-(\mu+2)^{-1}$. Therefore, for the critical radius we get $R=R_c=\mu_1^1(g)^{1/2}(\mu+2)^{1/2}$.

Consider another situation with $m_r=0$, $m_\varphi=0$, and $m_\theta=m_\theta(r,\theta)$. Here, the thermodynamic potential of the system has the form

$$F=\int r^2\sin(\theta)drd\theta d\varphi\left\{\frac{A}{2}m_\theta^2+\frac{g}{2}\left[\left(\frac{\partial m_\theta}{\partial r}\right)^2+\frac{1}{r^2}\left[m_\theta^2+\left(\frac{\partial m_\theta}{\partial \theta}\right)^2\right]+\frac{m_\theta^2\cos^2(\theta)}{r^2\sin^2(\theta)}\right]\right\}. \quad (9)$$

The variation $\delta F$ by $\delta m_\theta$ is zero, if $m_\theta(r,\theta)$ is the solution of the equation

$$Am_\theta+\frac{g}{r^2\sin^2(\theta)}m_\theta-\frac{g}{r^2}\frac{\partial}{\partial r}\left(r^2\frac{\partial m_\theta}{\partial r}\right)-\frac{g}{r^2\sin(\theta)}\frac{\partial}{\partial\theta}\left(\sin(\theta)\frac{\partial m_\theta}{\partial\theta}\right)=0. \quad (10)$$

The solution of this equation has the same form as the solution of the equation (7), and $m_\theta(r,\theta)=Cj_1(\mu_1^1 r/R)\sin(\theta)$ (Fig. 3).

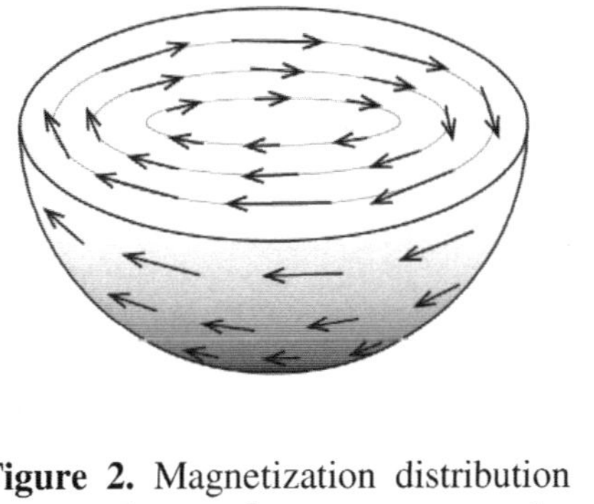

**Figure 2.** Magnetization distribution just after the paramagnetic-ferromagnetic PT in a sphere. 3D vortex. Vector plot.

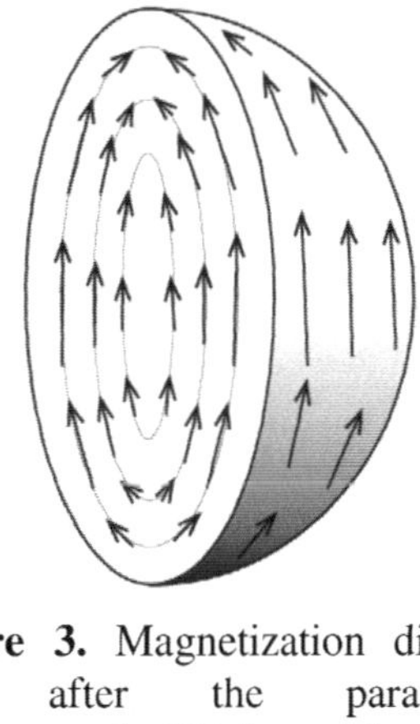

**Figure 3.** Magnetization distribution just after the paramagnetic-ferromagnetic PT in a sphere. Vector plot.

### *2.3. The polarization*

For cubic crystals, the allowed form of the magnetically induced electric polarization [7] is

$$\vec{P} \propto \left[ \left( \vec{m} \cdot \nabla \right) \vec{m} - \vec{m} \left( \nabla \cdot \vec{m} \right) \right]. \tag{11}$$

Then the expression for the polarization of the circle in the 2D case will have the form

$$\vec{P} \propto -\left( m_\varphi^2(\rho) / \rho \right) \vec{e}_\rho, \tag{12}$$

where $m_\varphi(\rho)$ has the form (4). The polarization has the form of 2D hedgehog with $P_\varphi = 0$ and $P_\rho = P_\rho(\rho)$.

For 3D vortex on a sphere at $m_r = m_\theta = 0$ and $m_\varphi = m_\varphi(r, \theta)$ the polarization has the form of 3D hedgehog with $P_\varphi = 0$ and

$$\vec{P}(r,\theta) \propto -\frac{j_1^2(\mu_1^1 \cdot r / \mathrm{R}) \sin(\theta)}{r} \left( \sin(\theta) \vec{e}_r + \cos(\theta) \vec{e}_\theta \right). \tag{13}$$

For 3D distribution of magnetization in the form $m_r = 0$, $m_\varphi = 0$ and $m_\theta = m_\theta(r, \theta)$, the polarization has the same form as for 3D vortex on a sphere (13).

## 3. Conclusion

In conclusion, we have studied the PT in 2D and 3D spherical particles of a cubic ferromagnetic in the framework of the phenomenological theory. A phenomenological model has taken into account the amplitude variations of magnetization. We have obtained expressions for the magnetization of these particles in the vortex phase and in the single-domain phase. We have shown that the existence of a vortex phase is possible for particles with size $R > R_c$ in a zero external magnetic field. The electrical polarization has been calculated for three different distributions of magnetization.

## References

[1] Hehn M, Ounadjela K, Bucher J-P, Rousseaux F, Decanini D, Bartenlian B and Chappert C 1996 *Science* **272** 1782-5
[2] Cowburn R P, Koltsov D K, Adeyeye A O, Welland M E and Tricker D M 1999 *Phys. Rev. Lett.* **83** 1042
[3] Levanyuk A P and Blinc R 2013 *Phys. Rev. Lett.* **111** 097601
[4] Rößler U K, Bogdanov A N and Pfleiderer C 2006 *Nature* **442** 797-801
[5] Cheong S-W and Mostovoy M 2007 *Nat. Mater.* **6** 13-20
[6] Sergienko I A and Dagotto E 2006 *Phys. Rev. B* **73** 094434
[7] Mostovoy M 2006 *Phys. Rev. Lett.* **96** 067601
[8] Landau L D and Lifshitz E M 1975 *Electrodynamics of Continuous Media* (New York: Pergamon Press)
[9] Hubert A and Schafer R 1998 *Magnetic Domains* (Berlin: Springer)

IOP Conf. Series: Materials Science and Engineering **699** (2019) 012045 doi:10.1088/1757-899X/699/1/012045

# The SPM study of oligonucleotides consisting of repeated nucleotide sequences

**T I Sharipov[1], R R Garafutdinov[2], I T Amangulova[1] and R Z Bakhtizin[1]**

[1]Bashkir State University, 450076 Ufa, Russia
[2]Institute of Biochemistry and Genetics RAS, 450054 Ufa, Russia

sha-t@yandex.ru

**Abstract.** Recently, the study of the dependence of the physical characteristics of DNA molecules on their structure has become of considerable interest. Various methods are used to conduct such research, among which scanning probe microscopy (SPM) is becoming increasingly important. In this work, the short single-stranded DNA molecules consisting of dC nucleotides are visualized and their conductivity is investigated by SPM methods. It has turned out that the distribution of DNA molecules on the surface of the substrates is uniform. The measured current-voltage characteristics have made it possible to estimate the differential electrical resistance of individual $d(C)_{12}$ molecules.

## 1. Introduction

The study of nucleic acid molecules – DNA and RNA – as physical objects is of considerable interest. The variety of possible structures of nucleic acids (single-stranded and double-stranded state, linear and ring molecules, various conformational forms, variability of nucleotide composition, etc.) determines the presence of a wide range of similar objects. Recently, new data on the dependence of the physical characteristics of DNA molecules on their structure have appeared. Thus, it has been shown that the conformation and hardness of DNA are influenced by the molecular environment [1,2], nucleotide composition and methylation [3-5]. The difference in the nature of DNA destruction under mechanical action (irradiation of aqueous solutions with ultrasound) has been shown [6,7].

One of the actively used methods for studying the physical parameters of nucleic acids is scanning probe microscopy (SPM), which, thanks to its high resolution, allows studying various objects at the molecular and submolecular levels, including DNA molecules. Synthetic single-stranded DNA molecules with a homonucleotide sequence, that is, consisting of nucleotides of the same type, are of particular interest. Such structures assume a special nature of the electron density distribution and charge transfer, so the interest in them is consistently high due to the prospects of using these molecules in nanoelectronics. Attempts to measure the electrical resistance of DNA give conflicting results [8-11]. The ambiguity of the results is influenced by the experimental conditions and the parameters of DNA molecules studied, namely length, nucleotide composition, different nucleotide sequence in the DNA chain, and the number of chains in the molecule [12]. The study of the surface topography and immobilization of the DNA molecules by SPM methods is also important in terms of the development of DNA microarrays [13-16].

The electrical properties of DNA molecules can be studied using one of the varieties of SPM – scanning tunnelling microscopy (STM), by obtaining the current-voltage curves. To do this, the molecule is placed between two electrical contacts, one of which is a conducting probe of the microscope, and the other is a fragment of the substrate surface of an electrically conductive material. The aim of this work was to visualize and study the conductivity of short single-stranded DNA consisting of nucleotides of only one type, the nitrogen base of which is cytosine.

## 2. Materials and methods

$d(C)_{12}$ oligonucleotide was synthesized on an automatic synthesizer ASM-800 ("Biosset", Russia) by amidophosphite method. Water of the highest quality category (>18 MΩ, "Millipore", France) was used to prepare the solutions, the concentration was determined by the optical density of the aqueous solution at 260 nm on the BioSpec-Mini spectrophotometer ("Shimadzu", Japan).

SPM study was carried out in air using Solver P47 and Ntegra-Prima devices ("NT-MDT Spectrum Instruments", Russia). For atomic force microscopy (AFM) studies, the probes were silicon cantilevers NSG10 ("NT-MDT Spectrum Instruments") with the curvature radius of 10 nm and force constant of 11.5 N/m. For STM studies, the probes were wolfram tips obtained by the method of electrochemical etching.

In STM experiments, silver thermally deposited on the mica surface in vacuum was used as a substrate. In AFM studies, the substrate was the surface of fresh mica cleavage.

The object of the study was the oligonucleotide molecules consisting of 12 identical units of deoxycytidine. A solution of $d(C)_{12}$ oligonucleotide at a concentration of 5 ng/µl was heated at a temperature of 75-80°C for 6-7 min for denaturation, and then a drop of this solution of 5 µl was deposited on a silver or mica substrate. The drop was dried up in 30-40 minutes.

## 3. Results and discussion

Previously, we have shown the possibility of studying of short single-stranded DNA using SPM [17-19]. In this work, the object was the $d(C)_{12}$ oligonucleotide molecules, which initially were imaged on the mica surface by AFM in the tapping mode. On one of such AFM-images (Fig. 1a), objects of spherical shape are clearly visualized, some of which are individual molecules of oligonucleotides, and others are aggregates representing clusters of single oligonucleotide molecules. These aggregates have different sizes both in normal and lateral directions, which is due to the difference in the number of molecules making them up.

Because the smallest objects observed on the AFM image, which in our opinion are single molecules, were of the greatest interest, a cross-section profile was obtained for them, by which the geometric parameters were determined (Fig. 1b). Thus, the average size of such objects in lateral direction was about 36±1.8 nm, height was about 0.9±0.05 nm, the calculation was carried out for 50 smallest objects observed on the AFM image.

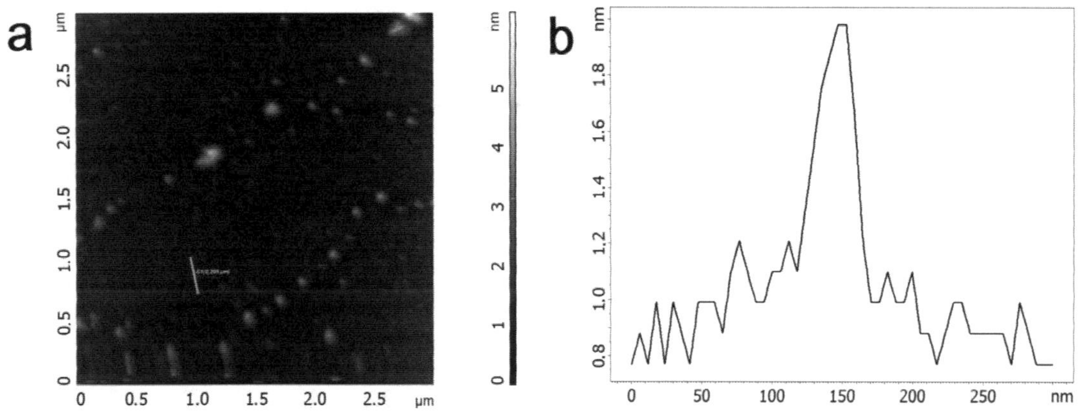


**Figure 1.** (a) AFM image of $d(C)_{12}$ oligonucleotides on mica surface; (b) cross-section profile along the line marked at (a).

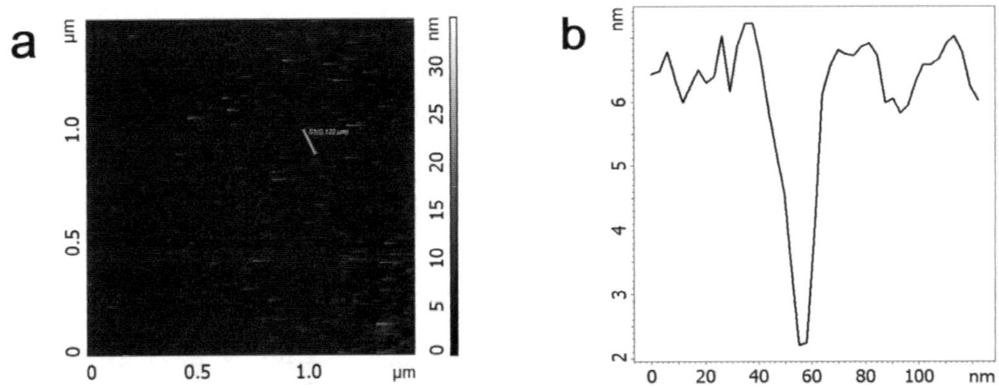


**Figure 2.** (a) STM imaging of $d(C)_{12}$ oligonucleotides on the silver surface; (b) $d(C)_{12}$ oligonucleotide cross-section profile along the line marked at (a).

Then $d(C)_{12}$ oligonucleotides were deposited on the atomically flat silver surface and previously studied by STM. Oligonucleotides were clearly visualized (Fig. 2a) and were located on the surface of the substrate relatively rare, which is very convenient for further spectroscopic study.

Usually, oligonucleotides on STM images are represented by dark objects with small lateral dimensions. This is explained by the fact that they have lower electrical conductivity compared to the electrical conductivity of silver. Cross-section profiles (Fig. 2b) estimated the lateral dimensions of these objects, which were on average equal to 17±0.9 nm. Measuring the depth of the observed objects gives a result of about 3.9±0.22 nm. This value is not true, because the vertical dimension values on STM images are indirect. It is a topographic display of the electrical conductivity at each individual point of the scan area. The above values were calculated for 30 smallest dark objects observed on the STM images.

After obtaining STM images and visualization of oligonucleotides on the substrate surface in the mode of scanning tunnelling spectroscopy, their current-voltage curves were measured in the areas, where individual oligonucleotide molecules were presumably located. Several current-voltage curves are measured at each point. Final current-voltage curve is obtained by averaging. By performing the described method for measuring the current-voltage curves of several oligonucleotides identified in the scan area shown in Figure 2a, we obtained the resulting averaged current-voltage curve (Fig. 3).

The kind of current-voltage curve was nonlinear. It has a symmetric appearance concerning zero values of both current and voltage in the voltage range from –1.5 V to +1.5 V. In the voltage range more than +1.5 V and less than –1.5 V it has an asymmetric form concerning zero values. For voltage applied between the substrate and the STM tip from –0.8 V to +1.2 V, the tunnelling current is close to

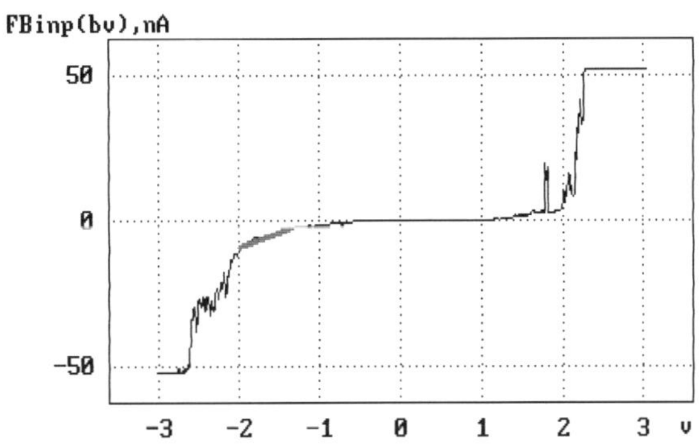


**Figure 3.** Averaged current-voltage curve of $d(C)_{12}$ oligonucleotides.

zero. Using current-voltage curve, the differential resistance of $d(C)_{12}$ oligonucleotides can be calculated. To do this, on the current-voltage curve we chose a site, where there were no zero current values and at the same time there were no significant current fluctuations. In Figure 3, it is highlighted with a thick red line. It was calculated that the differential electrical resistance of a single molecule was approximately equal to $R_{dif}=0.85\cdot10^8$ Ω.

## 4. Conclusion

Visualization of samples obtained after immobilization of $d(C)_{12}$ oligonucleotide on a solid substrate by scanning probe microscopy showed a uniform distribution of DNA molecules either in a single state or in the form of aggregates – clusters of several molecules, regardless of the substrate nature (mica or silver). In addition, the results of an experimental study of electrical conductivity are obtained. The current-voltage curves have been measured. Differential electrical resistance of the $d(C)_{12}$ oligonucleotide molecules has been estimated.

**Acknowledgements**
The work was performed with financial support of the grant of the Republic of Bashkortostan for the young scientists for 2019, contract № 29GR.

## References

[1] Ninaber A and Goodfellow J M 1999 *Radiat. Environ. Biophys.* **38** 23-9
[2] Vologodskii A and Frank-Kamenetskii M D 2013 *Nucleic Acids Res.* **41** 6785-92
[3] Elsner H I and Lindblad E B 1989 *DNA* **8** 697-701
[4] Perez A, Castellazzi C L, Battistini F et al. 2012 *Biophys. J.* **102** 2140-8
[5] Severin P M, Zou X, Gaub H E and Schulten K 2011 *Nucleic Acids Res.* **39** 8740-51
[6] Grokhovsky S L, Il'icheva I A, Nechipurenko D et al. 2011 *Biophys. J.* **100** 117-25
[7] Garafutdinov R R, Galimova A A, Sakhabutdinova A R and Chemeris A V 2016 *Molecular Biology* **50** 272-8
[8] Fink H W and Schonenberger C1999 *Nature* **398** 407
[9] Porath D, Bezryadin A, de Vries S and Dekker C 2000 *Nature London* **403** 635
[10] Sharipov T I and Bakhtizin R Z 2019 *University proceedings. Volga region. Physics and mathematics sciences. Astronomy* № **1 (49)** 133-41
[11] Sharipov T I and Bakhtizin R Z 2017 *IOP Conf. Ser.: Mater. Sci. Eng.* **256** 012009
[12] Iijima M, Kato T, Nakanishi S et al. 2005 *Chem. Lett.* **34** 1084
[13] Garafutdinov R R, Shepelevich I S, Chemeris A V and Talipov R F 2005 *Vestnik Bashkirskogo universiteta* **10** 49-54
[14] Tuktamysheva A Kh, Garafutdinov R R and Talipov R F 2012 *Vestnik Bashkirskogo universiteta* **17** 1247-52
[15] Ayupova A Kh, Garafutdinov R R, Talipov R F et al 2012 *Vestnik Bashkirskogo universiteta* **17** 1677-82
[16] Sharipov T I and Bakhtizin R Z 2017 *IOP Conf. Ser.: Mat. Sci. Eng.* **195** 012002
[17] Sharipov T I, Garafutdinov R R, Talipov R F and Bakhtizin R Z 2009 *Perspektivnye materialy* **7** 368-71
[18] Sharipov T I and Bakhtizin R Z 2013 *Vestnik Bashkirskogo universiteta* **18** 340-3
[19] Sharipov T I, Ghots S S and Bakhtizin R Z 2011 *Perspektivnye materialy* **12** 579-84

IOP Conf. Series: Materials Science and Engineering **699** (2019) 012046 doi:10.1088/1757-899X/699/1/012046

# Studying surface Fermi level pinning in GaP nanowires with gradient Kelvin probe microscopy

**V A Sharov[1,2], P A Alekseev[1], V V Fedorov[2], A D Bolshakov[2] and I S Mukhin[2]**

[1]Ioffe Institute, 194021 Saint-Petersburg, Russia
[2]Saint-Petersburg Academic University, 194021 Saint-Petersburg, Russia

vl_sharov@mail.ru

**Abstract**. In this work, we use scanning Kelvin probe gradient force microscopy to study surface Fermi level pinning in horizontal undoped nanowires with axial GaP/GaPAs heterojunction. The nanowires were separated from the growth substrate and dispersed on smooth surface of Ni/Si substrate. Then longitudinal and transverse surface potential profiles of several nanowires were captured. The study revealed a distinction between the work function of GaP and GaPAs regions. It was shown that the Fermi level in undoped GaP nanowires was pinned in the middle of the energy gap.

## 1. Introduction

Semiconductor nanowires (NWs) remain prospective building blocks for the next generation of nanoscale electronic devices. Mechanical strength and the ability to be grown on highly mismatched substrates have led to the development of NW-based optoelectronics [1] and photovoltaics [2,3]. In this case, understanding NWs transport properties is of interest. It is well known that surface effects in one-dimensional nanostructures are strong due to high surface to volume ratio. In particular, surface Fermi level pinning leads to a surface band bending and to a formation of the near surface depleted area. In this case, the conductivity of the III–V NWs with an oxidized surface is governed mainly by surface states. The depletion area width is governed by doping level, NW diameter, surface state density, and position of the Fermi level pinning [4]. To quantitatively analyse transport properties of NW, one should know a position of the surface Fermi level pinning (work function). Conventionally, NW surface is terminated by [110] facets. Fermi level in GaAs and in InAs [110] surface with native oxide layer is pinned in the midgap and near the conduction band, respectively [5,6], resulting in different transport properties [7]. Scanning Kelvin probe force gradient microscopy (KPFGM) proved to be an effective tool to directly investigate NW work function. Recently, in has been shown that excess arsenic is responsible for the Fermi level pinning in III-As NWs with native oxide [8]. However, surface Fermi level position in III-P NWs with native oxide layer is still poorly studied. GaP NWs are prospective for betavoltaic devices [9] and photodetectors [10], while wurtzite GaP is promising for green-range optoelectronics due to direct band gap [11]. For designing GaP NWs based devices with controllable electronic parameters, the position of the Fermi level pinning has to be obtained. In this work, we measure the work function on the [110] crystal plane of NWs with GaP/GaPAs axial heterojunction.

## 2. Experimental

Undoped Gallium phosphide NW arrays were synthesized by solid source molecular beam epitaxy using Veeco GEN-III machine on (111) silicon substrate (Fig. 1a). Self-catalytic vapour-liquid-solid mechanism with the use of gallium droplet acting as a catalyst was used. The average NWs length was approximately 5 μm. Arsenic source was turned on after 2 μm growth. The bottom part of NWs is 130 nm thick GaP with arsenic shell, while the top part is 150 nm thick GaPAs. NW crystal structure was studied with transmission electron microscopy (TEM) (Fig. 1b) and turned out to be cubic with a lot of stacking faults with 200 nm wurtzite insertion near GaP/GaPAs interface. The top 250 nm GaP region has a zincblende structure with small number of twin-defects. Phase features in the top of III-V NWs may be due to the contact angle behaviour during the catalytic drop disappearance in the end of the growth process [12]. Energy-dispersive X-ray spectroscopy (EDS) confirmed that NW array was separated from the growth substrate and dispersed on a Si/Ni substrate for subsequent atomic force microscopy (AFM) measurements. AFM topography of individual horizontal NW is shown in Figure 1f.

AFM measurements were carried out using Ntegra Aura setup (NT-MDT, Russia). We used silicon probes with $W_2C$ conductive coating, 3.5 N/m average force constant, and 30 nm tip curvature radius. Fermi level pinning position was determined by phase-modulated Kelvin probe force gradient microscopy (KPFGM) [13]. At each scanning point, this technique measures the surface potential, which is the difference between the work functions of a microscope probe and of a semiconductor surface. The latter value presents the energy gap between the vacuum level and the surface Fermi level pinning position. The scanning process is one-passed and includes two steps in each scanning point. First, the topography is measured in semi-contact regime. Second, the probe magnitude is being reduced by one order, the tip-sample distance is 20 nm, constant voltage $U_{sample}$ is applied on the sample, and the cantilever phase $\varphi$ is measured with respect to the $U_{sample}$. The phase is proportional to squared contact potential difference $U_{CPD}= U_{tip}-U_{NW}$. In each scanning point the parabolic fit $\varphi(U)$ is obtained. The voltage of the parabola apex is $U_{CPD}$. Inherently, KPGFM technique is very similar to electrostatic force microscopy (EFM), but output signals are different as EFM shows only phase of cantilever oscillations and does not show surface potential. More detailed KPGFM description can be found elsewhere [14]. The force gradient technique is much more sensitive than a regular KPFM [15] and allows studying low-dimensional nanostructures such as NWs. Before each measurement, the work function of the probe was calibrated using freshly cleaved highly oriented pyrolytic graphite (HOPG) with well-known work function ($\Phi_{HOPG}$) of 4.48 eV [16]. Then the NW work function $\Phi_{NW}$ can be expressed as $\Phi_{NW} = 4.48 + (U_{HOPG} - U_{NW})/e$, where $U_{HOPG}$ is the surface potential of the HOPG measured by the same probe as $V_{NW}$ and e is the electronic charge. The measurements were carried out in medium vacuum (around 1 Torr) in order to increase cantilever Q-factor up to ~2500 and decrease the role of potential shielding by the surface water layer.

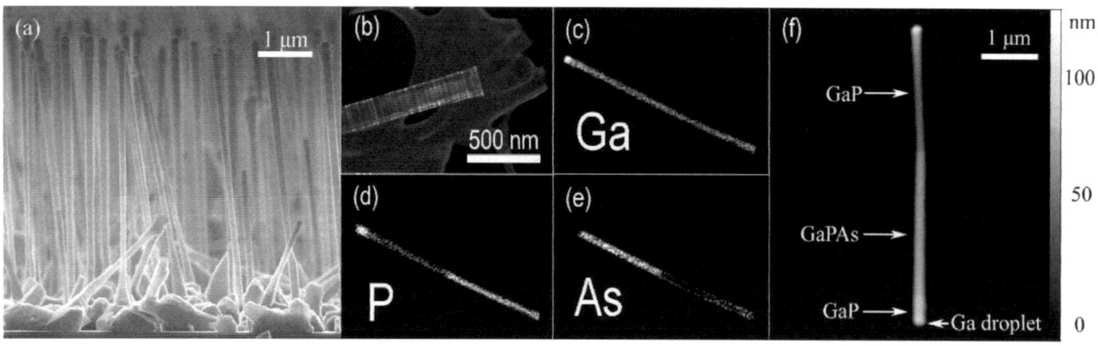


**Figure 1.** (a) Scanning electron microscopy image of GaP/GaPAs NWs array, (b)-(e) TEM and EDS study of individual NW, (f) AFM scan of individual NW.

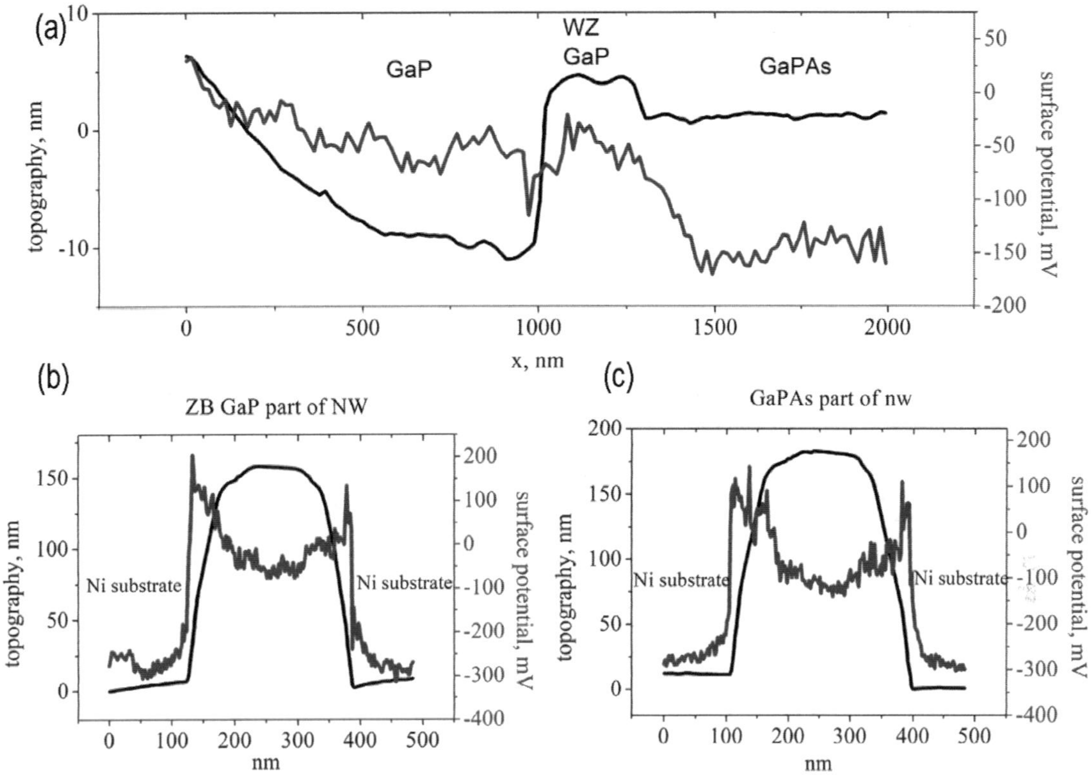


**Figure 2.** KPGFM scans of individual NW containing topography and surface potential: (a) longitudinal scan containing GaP/GaPAs interface; (b) transverse scan of GaP part; (c) transverse scan of GaPAs part.

## 3. Results and discussion

Figure 2 shows the main AFM results. The KPGFM profile along the NW is shown in Figure 3a. The GaP/GaPAs interface is in the middle of the profile, at 1000 nm. The right part is thicker; therefore, this NW region corresponds to GaPAs. Figure 2b, c shows KPGFM profiles across GaP and GaPAs regions, correspondingly. All profiles indicate that the surface potential is about 70 mV higher in GaP part. It means that GaPAs work function is 70-100 meV lower than that of GaP. This is confirmed by multiple measurements on several NWs. The transverse profiles overestimate the potential signal on NW sides, so only the signal from the central [110] facet (-50 mV on GaP and -120 mV on GaPAs) should be taken into account. The probe work function was checked out on freshly cleaved HOPG and turned out to be 4.9 eV. KPGFM lateral resolution is less than 50 nm for 30 nm probe radius and 20 nm tip-sample distance during the phase measurement [15]. Taking into account that the width of top [110] facet is 2-3 times larger, we assume that the measurement error is mainly due to the KPGFM noise that does not exceed 20 mV. Thus, GaP and GaPAs work functions are 4.95±0.01 and 5.02±0.01 eV. Taking into account that GaP electron affinity is 3.8 eV, one can conclude that the Fermi level of undoped NWs is pinned in the midgap. The obtained work function value is in a good agreement with Ref. [17]. The surface states responsible for the pinning have energy around 1.2 eV below the conductive band minimum. It is shown in Ref. [18] that such pinning position can be attributed to P-derived dangling bond states.

## 4. Conclusion

Surface electronic properties of undoped GaP/GaPAs nanowires were studied with Kelvin probe force gradient microscopy. The values of surface Fermi level pinning position at [110] facet turned out to be 4.95±0.01 and 5.02±0.01 eV in GaP and GaPAs, respectively. Fermi level of undoped GaP NWs was pinned in the midgap. The obtained data are important for calculation of the surface depletion layer thickness that is crucial for understanding the transport properties of nanowires.

**References**

[1] Zhao S, Nguyen H P, Kibria M G and Mi Z 2015 III-Nitride nanowire optoelectronics *Prog. Quant. Electron.* **44** 14-68
[2] Aberg I, Vescovi G, Asoli D, Naseem U, Gilboy J P, Sundvall C, Dahlgren A, Svensson K E, Anttu N, Björk M T et al. 2015 A GaAs nanowire array solar cell with 15.3% efficiency at 1 sun *IEEE J. Photovolt.* **6** 185-90
[3] Alekseev P, Sharov V, Geydt P, Dunaevskiy M, Soshnikov I, Reznik R, Lysak V, Lähderanta E and Cirlin G 2018 GaAs wurtzite nanowires for hybrid piezoelectric solar cells *Semiconductors* **52** 609-11
[4] Schmidt V, Senz S and Gösele U 2007 Influence of the $Si/SiO_2$ interface on the charge carrier density of Si nanowires *Appl. Phys. A* **86** 187-91
[5] Spicer W, Lindau I, Skeath P, Su C and Chye P 1980 Unified mechanism for Schottky-barrier formation and III-V oxide interface states *Phys. Rev. Lett.* **44** 420
[6] Smit K, Koenders L and Mönch W 1989 Adsorption of chlorine and oxygen on cleaved InAs (110) surfaces: Raman spectroscopy, photoemission spectroscopy, and Kelvin probe measurements *J. Vac. Sci. Tech. B: Microel. Proc. Phenom.* **7** 888-93
[7] Alekseev P A, Sharov V A, Dunaevskiy M S, Kirilenko D A, Ilkiv I V, Reznik R R, Cirlin G E and Berkovits V L 2019 Control of conductivity of $In_xGa_{1-x}As$ nanowires by applied tension and surface states *Nano Lett.* **19** 4463-9
[8] Alekseev P A, Dunaevskiy M S, Cirlin G E, Reznik R R, Smirnov A N, Kirilenko D A, Davydov V Y and Berkovits V L 2018 Unified mechanism of the surface Fermi level pinning in III-As nanowires *Nanotechnology* **29** 314003
[9] McNamee S, Wagner D, Fiordaliso E M, Novog D and LaPierre R R 2018 GaP nanowire betavoltaic device *Nanotechnology* **30** 075401
[10] Kuyanov P, McNamee S and LaPierre R 2018 GaAs quantum dots in a GaP nanowire photodetector *Nanotechnology* **29** 124003
[11] Assali S, Zardo I, Plissard S, Kriegner D, Verheijen M, Bauer G, Meijerink A, Belabbes A, Bechstedt F, Haverkort J et al. 2013 Direct band gap wurtzite gallium phosphide nanowires *Nano Lett.* **13** 1559-63
[12] Fedorov V, Bolshakov A, Kirilenko D, Mozharov A, Sitnikova A, Sapunov G, Dvoretckaia L, Shtrom I, Cirlin G and Mukhin I 2018 Droplet epitaxy mediated growth of GaN nanostructures on Si (111) via plasma-assisted molecular beam epitaxy *Cryst. Eng. Comm.* **20** 3370-80
[13] Ladutenko K, Ankudinov A and Evtikhiev V 2010 On the accuracy of quantitative measurements of the local surface potential *Tech. Phys. Lett.* **36** 228-31
[14] Portes L, Girard P, Arinero R and Ramonda M 2006 Force gradient detection under vacuum on the basis of a double pass method *Rev. Sci. Instr.* **77** 096101
[15] Dunaevskiy M, Alekseev P, Girard P, Lashkul A, Lahderanta E and Titkov A 2012 Analysis of the lateral resolution of electrostatic force gradient microscopy *J. Appl. Phys.* **112** 064112
[16] Hansen W N and Hansen G J 2001 Standard reference surfaces for work function measurements in air *Surf. Sci.* **481** 172-84
[17] Alonso M, Cimino R and Horn K 1990 Surface photovoltage effects in photoemission from metal-GaP (110) interfaces: Importance for band bending evaluation *Phys. Rev. Lett.* **64** 1947
[18] El-Dessouki M, Attia V, El-Deen M S and Gobrial F 1987 Surface photovoltage spectroscopy of real <111> GaP surfaces *Phys. Stat. Sol. (a)* **99** 171-9

IOP Conf. Series: Materials Science and Engineering **699** (2019) 012047 doi:10.1088/1757-899X/699/1/012047

# Conductive AFM study of the electronic properties of individual epitaxial GaN nanowires

**V Sharov[1,2], A Bolshakov[1], V Fedorov[1], K Yu Shugurov[1], A M Mozharov[1], G A Sapunov[1] and I S Mukhin[1]**

[1]Saint-Petersburg Academic University, 194021 Saint-Petersburg, Russia
[2]Ioffe Institute, 194021 Saint-Petersburg, Russia

vl_sharov@mail.ru

**Abstract**. In this work, we use conductive atomic force microscopy (CAFM) to study the impact of substrate surface preparation and buffer layer composition on the electrical transport properties of GaN nanowires (NWs). I-V curves of single NWs from seven differently prepared samples were obtained. The tip of atomic force microscope (AFM) was used as a top conductive electrode to create stable electric contact to NW free upper grain, while the bottom contact was established between the highly doped Si substrate and a grounded sample holder of the AFM device. Single NW I-V curves were compared to those of NW arrays. The difference between them was discussed.

## 1. Introduction

Development of one dimensional semiconductor structures remains one of the most perspective directions in modern nanoelectronics. In particular, GaN nanowires (NWs) are important for the development of the light-emitting diodes and optoelectronic devices operating in the blue and near-ultraviolet range as well as for nanogenerators and piezoelectric devices. The combination of beneficial mechanical, optical, and transport properties of NWs makes them promising building blocks for the next generation of optoelectronic [1], photovoltaic [2], sensing, and energy harvesting devices [3]. Thus, the understanding of their electrical properties is crucial.

One of the problems that III-N-based devices fabrication technology is facing today is the high cost of $Al_2O_3$ or GaN substrates promoting researchers to develop the nitride compounds growth techniques on more accessible wafers, such as silicon, and find new approaches for the defect free epitaxial growth of highly lattice mismatched systems [4]. Fortunately, quasi one-dimensional NW geometry provides a unique possibility of high crystallinity nanostructures growth on lattice mismatched substrates due to more effective strain relaxation in comparison to the thin film heterostructures [5]. In this work, we study influence of the Si (111) substrate surface preparation and buffer layer composition on the GaN NW/Si heterojunction electronic properties.

## 2. Experimental

The investigated GaN NWs were grown on boron doped Si (111) wafers by plasma assisted molecular beam epitaxy (PA-MBE) using Veeco GEN-III machine. The experimental setup description can be found elsewhere [6]. Equivalent pressures of nitrogen and gallium molecular beam fluxes were kept at $\sim 2\times 10^{-7}$ and $\sim 1\times 10^{-8}$ Torr, correspondingly. Substrate temperature for the NWs growth was set to

760°C in all the experiments. We studied seven samples that had been grown using different substrate surface preparation techniques after the oxide removal.

In the first growth experiment, GaN deposition was initiated immediately after the oxide removal. In the second one, a thin amorphous $SiN_x$ layer was formed on the substrate surface as a result of exposition under activated nitrogen flux. In the third sample, thin AlN buffer layer was deposited on the substrate. The fourth sample was covered with ~20 nm thick $GaO_x$ layer prepared by the plasma enhanced chemical vapour deposition (PECVD) prior to the placement into MBE growth chamber. In 5-7 samples, GaN seeding layer was obtained via nitridation of Ga predeposited on bare Si(111) surface with equivalent thicknesses of 0.3, 0.6, and 2 ML, respectively. All samples were Si-doped for the last two hours of the growth. More details about the growth process can be found in Ref. [7]. Figure 1 shows scanning electron microscopy (SEM) images of all the samples. Average length of NWs varied from 350 nm for the $1^{st}$ and the $6^{th}$ samples to 750 nm for the $3^{rd}$ one, with typical diameter from 50 to 150 nm.

Electrical properties of grown structures were characterized by obtaining IV curves. Two approaches were used. Firstly, we measured the curves from individual NWs. Then the curves from the whole array were obtained and the results were compared. To obtain the IV curve from an individual freestanding NW, two electric contacts were required. The first one was created using conductive atomic force microscope (AFM) tip being a nanosized electrode accurately positioned on the NW top. The second one was formed between the highly doped Si growth substrate and an AFM sample holder. The principal scheme of the experiment is shown in Figure 2a. This technique proved its effectiveness in a number of researches [8,9]. The experiments were performed at room temperature using NT-MDT Ntegra AURA setup with $W_2C$-coated probes having tip curvature radius of 30 nm. The measurements were carried out in medium vacuum (around 1 Torr) in order to decrease the role of the surface water layer.

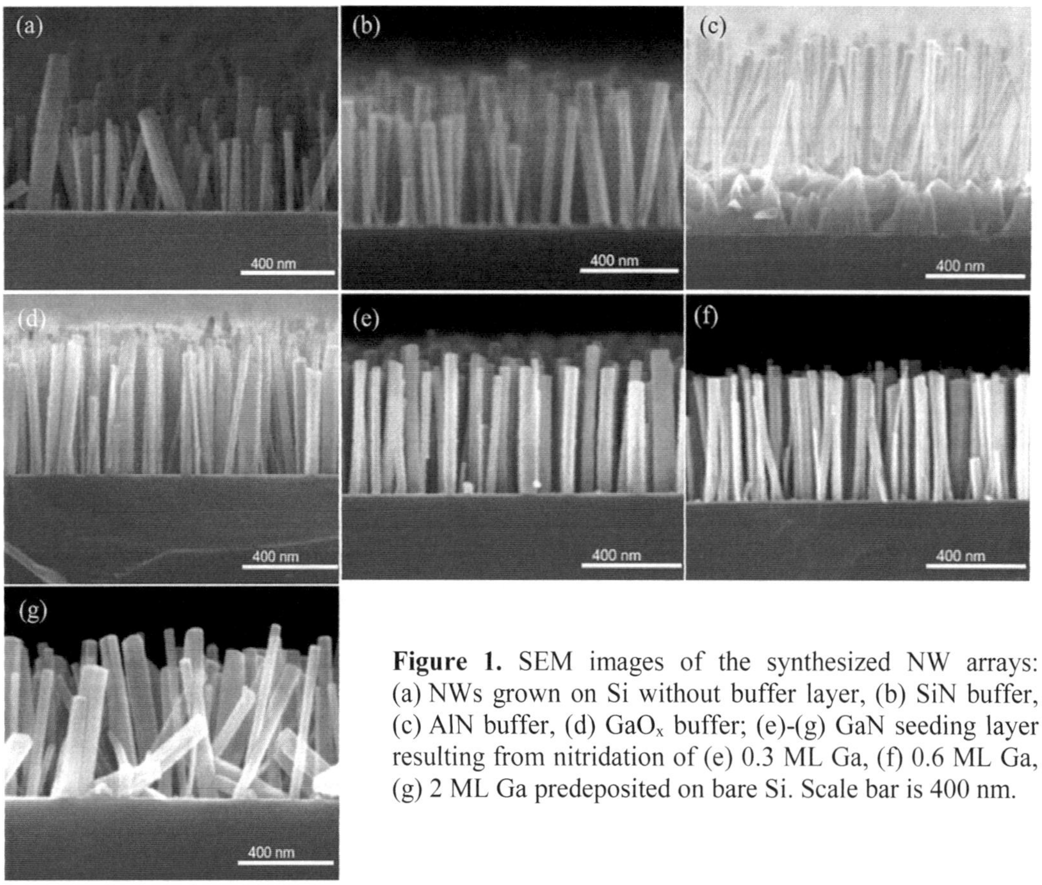


**Figure 1.** SEM images of the synthesized NW arrays: (a) NWs grown on Si without buffer layer, (b) SiN buffer, (c) AlN buffer, (d) $GaO_x$ buffer; (e)-(g) GaN seeding layer resulting from nitridation of (e) 0.3 ML Ga, (f) 0.6 ML Ga, (g) 2 ML Ga predeposited on bare Si. Scale bar is 400 nm.

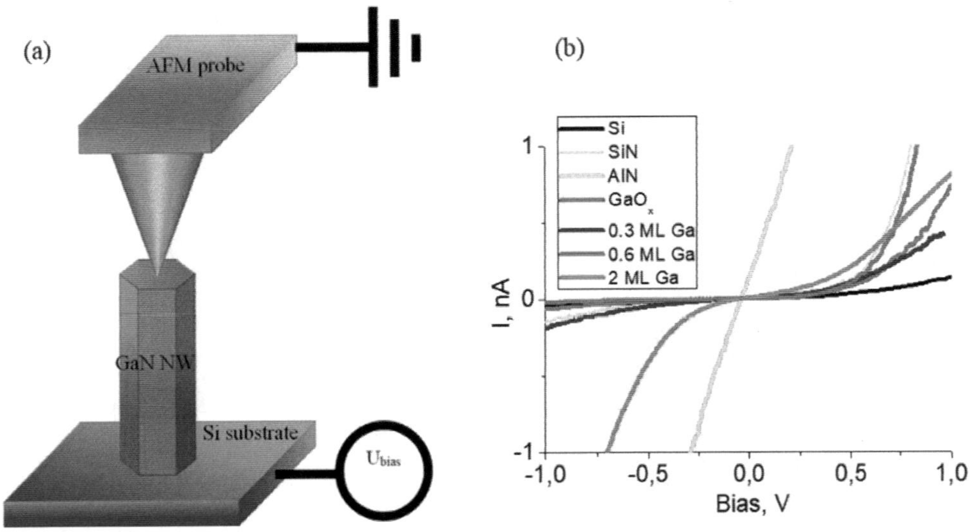


**Figure 2.** (a) Scheme of I-V curve measurement from an individual NW (not in scale), (b) averaged I-V curves from individual NWs.

To obtain the IV curves from the whole NW array, it was covered with a dielectric epoxy layer that was then etched with oxygen plasma to uncover the upper parts of the NWs. After that, the samples were covered with transparent conductive ITO layer and the bus bars were made of silver conductive paste. Use of transparent conductive ITO layer and bus bar contact geometry were dictated by the interest in further photocurrent measurements. The back side of each substrate was covered with Al. Thus, two electric contacts were created to obtain integral I-V characteristics from the grown arrays. The I-V curves were obtained using Keithley 2400 SourceMeter multimetre.

## 3. Results and discussion

In case of n-GaN/p-Si heterojunction, the forward current is attributed to hot electrons passing through the barrier. The hot-holes current is negligible due to high potential barrier [2]. Reverse current may be connected with defect-related leakage mechanisms [10] and band-to-band tunnelling [11]. Figure 2b shows I-V curve series from individual NWs obtained with C-AFM. Several NWs were studied on each sample with subsequent averaging. Sample 3 (AlN) demonstrates the highest current values in both forward and reverse bias. It is also the only sample that does not show exponential behaviour of the I-V curve. It is known that Al is a good p-type dopant for Si and easily incorporates in its lattice. Thus, sample 3 represents n-GaN/p-Si heterojunction with highly doped p-type region at the interface. Higher values of the reverse current are observed for this sample in comparison with the others. It is typical for a tunnel or backward diode operating according to the BBT mechanism. However, knee voltage in the forward region is not pronounced, which indicates contribution of the SRH mechanism and high defect density at the heterointerface. Reverse current is rather high for the 7$^{th}$ sample (2 ML Ga) that corresponds to high recombination rate at the heterointerface. Thus, we assume that unintentional doping of Si substrate occurs for the samples grown on AlN or GaN seeds prepared by the droplet epitaxy. However, growth on the nitridated 0.3 ML thick Ga layer has a negligible effect on the reverse region of the I–V curve similar to the case of the nitridated Si surface. The lowest obtained leakage current value and the most rectifying behaviour were found for the sample with GaO$_x$ buffer layer indicating that this interlayer effectively suppresses the GaN/Si interface recombination.

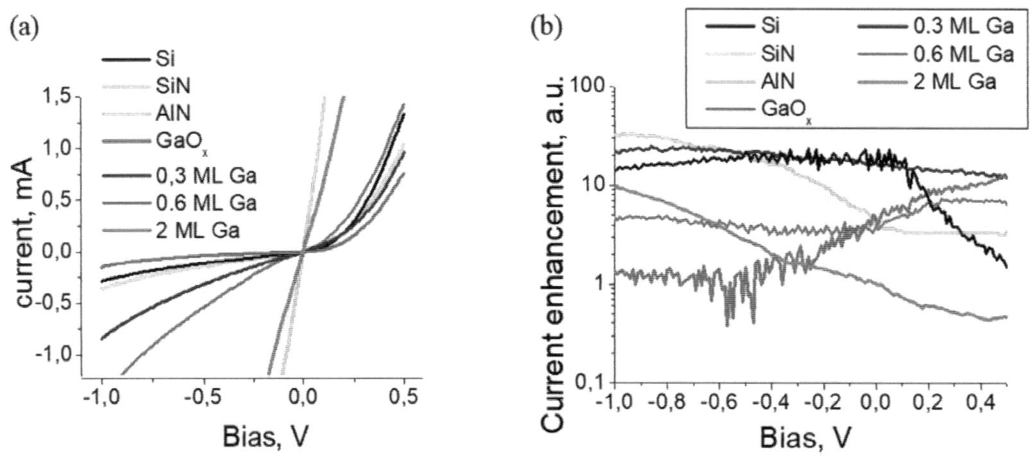


**Figure 3.** (a) Averaged I-V curves from individual NWs, (b) current enhancement.

The I-V curves for the samples with ITO top contact pad and Al-back contact measured at 300K are presented in Figure 3a. Enhancement of the current density (Fig. 3b) was observed for the I-V curves obtained with C-AFM at both forward and reverse biases. The ratio was calculated as $(I_{c\text{-}afm} - I_{1NW})/I_{1NW}$, where $I_{afm}$ – is the current density from the C-AFM I-V curve measurement, and $I_{1NW}$ – is the current density obtained from the I-V curves for the ITO-coated NWs. It can be seen that the C-AFM measurements with the use of conductive probe as a top contact demonstrates up to the 40 times current increase. We assume that such a phenomenon relates to only partial uncoverage of the NWs top facets after the epoxy etching resulted in lower effective contact area compared to the calculated value and, correspondingly, to a lower value of the calculated current density. Second, we assume that the current could be reduced due to the voltage drop in the ITO layer in the NWs array geometry. Third, the difference of the Schottky barrier for the different experiment geometries exists. That is clearly seen for the 6$^{th}$ sample (0.6 ML Ga) as the reverse current increased and for the 7$^{th}$ sample (2 ML Ga) as the curve behaviour changed from rectifying type (Fig. 2b) to Ohmic one (Fig. 3a). Taking into account abovementioned issues, we assume that for better understanding NW electronic properties, C-AFM measurements are preferable.

## 4. Conclusion

We investigated electrical properties of GaN NW/Si(111) heterostructures grown with various techniques of substrate surface preparation and buffer layer compositions. C-AFM measurements were carried out to compare the electrical transport properties of individual heterojunctions formed by a single NW on Si and that of NW arrays. Significant difference in the curves shape and the current density was observed. It was demonstrated that buffer layer preparation technique chosen for further GaN growth on Si(111) had an essential effect on the electronic transport properties of GaN/Si nanoheterostructures.

**Acknowledgements**

This work was supported by the Ministry of Science and Higher Education of the Russian Federation (grants 3.9796.2017/8.9 and 16.2593.2017/4.6) and the Russian Federation President Council (grants MK-1204.2019.2 and SP-2324.2018.1).

**References**

[1] Zhao S, Nguyen H P, Kibria M G and Mi Z 2015 III-Nitride nanowire optoelectronics *Prog. Quant. Electron.* **44** 14-68

[2] Mozharov A, Bolshakov A, Cirlin G and Mukhin I 2015 Numerical modeling of photovoltaic efficiency of n-type GaN nanowires on p-type Si heterojunction *Phys. Stat. Sol. Rapid Res. Lett.* **9** 507-10

[3] Huang C-T, Song J, Lee W-F, Ding Y, Gao Z, Hao Y, Chen L-J and Wang Z L 2010 GaN nanowire arrays for high-output nanogenerators *J. Am. Chem. Soc.* **132** 4766-71
[4] Bertness K A, Sanford N A and Davydov A V 2011 GaN nanowires grown by molecular beam epitaxy *IEEE J. Sel. Top. Quant.* **17** 847-58
[5] Bertness K, Sanford N and Schlager J 2011 Gallium nitride nanowires achieve crystalline perfection *SPIENewsroom*
[6] Fedorov V, Bolshakov A, Kirilenko D, Mozharov A, Sitnikova A, Sapunov G, Dvoretckaia L, Shtrom I, Cirlin G and Mukhin I 2018 Droplet epitaxy mediated growth of GaN nanostructures on Si (111) via plasma-assisted molecular beam epitaxy *Cryst. Eng. Comm.* **20** 3370-80
[7] Bolshakov A D, Fedorov V V, Shugurov K Y, Mozharov A M, Sapunov G A, Shtrom I, Mukhin M S, Uvarov A V, Cirlin G E and Mukhin I S 2019 Effects of the surface preparation and buffer layer on the morphology, electronic and optical properties of the GaN nanowires on Si *Nanotechnology* **30** 395602
[8] Alekseev P A, Sharov V A, Dunaevskiy M S, Kirilenko D A, Ilkiv I V, Reznik R R, Cirlin G E and Berkovits V L 2019 Control of conductivity of $In_xGa_{1-x}As$ nanowires by applied tension and surface states *Nano Lett.* **19** 4463-9
[9] Alekseev P, Sharov V, Geydt P, Dunaevskiy M, Soshnikov I, Reznik R, Lysak V, Lähderanta E and Cirlin G 2018 GaAs wurtzite nanowires for hybrid piezoelectric solar cells *Semiconductors* **52** 609-11
[10] Bessire C D, Björk M T, Schmid H, Schenk A, Reuter K B and Riel H 2011 Trap-assisted tunneling in Si-InAs nanowire heterojunction tunnel diodes *Nano Lett.* **11** 4195-9
[11] Nguyen H P T, Djavid M, Cui K and Mi Z 2012 Temperature-dependent nonradiative recombination processes in GaN-based nanowire white-light-emitting diodes on silicon *Nanotechnology* **23** 194012

# SEM and SPM investigations of the surface of antifriction aluminium alloys of the system Al-5%Si-4%Cu-6%Sn

**O O Shcherbakova[1], T I Muravyeva[1], D L Zagorskiy[1,2], I V Shkalei[1], V A Lapitskaya[3] and T A Kuznetsova[3]**

[1]Ishlinsky Institute for Problems in Mechanics RAS, 119526 Moscow, Russia
[2]Gubkin Russian State University of Oil and Gas, 119991 Moscow, Russia
[3]A.V. Luikov Heat and Mass Transfer Institute of NAS Belarus, 220072 Minsk, Belarus

shcherbakovaoo@mail.ru

**Abstract.** Two antifriction alloys based on the Al–5%Si–4%Cu–4%Sn system with the addition of bismuth, lead, and cadmium were investigated. The effect of these elements on the change in surface topography and the tribological properties of the samples was studied. Tribological tests simulating the work of the friction unit were carried out using "shoe-roller" scheme. After heat treatment in both alloys, the phase components acquired a globular shape. Mass transfer between the shoe and the roller led to the formation of a film of secondary structures of variable thickness on the roller surface. SEM and EBSD allowed observing a near-surface layer (50-100 µm thick) with a modified structure and grain shape. SPM made it possible to evaluate the difference in the mechanical properties of the surface layer at the nanoscale level. It was shown that the material was strengthened in the process of friction.

## 1. Introduction

The development of new antifriction materials is an important national economic task. In particular, it is of great importance for engineering to develop alloys used in friction mechanisms, for example, in plain bearings [1,2]. The service life of such nodes depends on the correct, scientifically based selection of the composition of the alloys. For the manufacture of such bearings expensive bronze are usually used [3, 4]; therefore, the urgent task is to create cheaper materials with high performance properties. Alloys based on aluminium, possessing antifriction characteristics comparable to bronze, meet these requirements. It is known that aluminium alloys are promising materials used in various fields of mechanical engineering [5,6].

Al–5%Si–4%Cu–6%Sn alloys were proposed as the optimal variant of antifriction materials for bearing assemblies [7, 8]. It is assumed that this alloy can be a basic composition. The addition of small amount of alloying fusible elements can effectively influence the structure and properties of the resulting alloy [8]. The idea of this work was the partial replacement of tin by bismuth, lead, and cadmium. It is known that the addition of cadmium to some aluminium alloys significantly improves their properties – for example, increases their hardness [9]. In this case, the use of cadmium seems promising (despite its toxicity). One of the aims of this work was to verify the effectiveness of cadmium alloying of the proposed system.

The change in the properties of the resulting alloys can be evaluated by various methods. Thus, tribological tests allow determining the main operational parameters (coefficient of friction, wear resistance, and abrasive resistance). The changes occurring on the contact surfaces and in the surface layers during friction were evaluated after these tests. These studies were carried out by a complex of different methods. So, scanning electron microscopy (SEM) studies of surfaces are carried out not only in secondary, but also in back-scattered electrons, which greatly increases the possibilities of research, allows more accurate studying the surface morphology and visualizing its phase composition. X-ray analysis is used for investigation of elemental composition. The SEM method is also used to estimate the near-surface layers on sample slices (cuts, sections).

The electron back-scattered diffraction (EBSD) method provides information on the surface crystal structure. As a rule, the electron diffraction patterns are processed and presented as an image of the orientation of the grains, which is very important for evaluating the properties of the surface.

Electron microscopic studies are often carried out in combination with the scanning probe microscopy (SPM) surface studies. The value of the SPM method consists both in obtaining the topographic images of the surface itself and in the possibility of using additional modes. Among the latter, the measurement of force curves can be noted.

The purpose of this work was to study the new antifriction alloys using a complex of the above-mentioned methods.

## 2. Experimental

### *2.1. Samples*

Two samples based on the Al–5%Si–4%Cu–4%Sn system were studied. Samples of the following composition were obtained: No. 1 – Al–5%Si–4%Cu–4%Sn–0.5%Bi–0.5%Pb–0.5%Cd and No. 2 – Al–5%Si–4%Cu–4%Sn–0.5%Bi–0.5%Pb.

The proposed composition of the alloy is determined by several factors. The alloy must have sufficient hardness, which provides by the basic composition of Al–5%Si–4%Cu. The addition of silicon improves casting properties. Also, the alloy must contain a sufficient number of low-melting elements, which is important for reducing friction and ensuring the resistance to tearing during dry friction or in lubrication conditions. It is known that under extreme conditions of friction on the contact surface, these elements are released, which protects the surface, performing the function of solid lubricant [10]. Previous studies have shown that the addition of Bi to the base alloy increases the fill rate of the form, and the addition of Cd increases the strength and hardness of the alloy [9]. Sample No. 1 differed from sample No. 2 by the presence of 0.5% cadmium.

The samples were subjected to heat treatment, which consisted in heating to 500°C followed by quenching in water and aging (at a temperature of 175°C for three hours). This mode of maintenance was previously selected for alloys of similar composition in Ref. [9].

Sample preparation for further studies (SEM, EBSD, and SPM) consisted in the preparation of thin sections (initial surface and sections after testing) and their polishing.

### *2.2. Investigation of hardness*

Surface hardness of the samples was measured. The Brinell measurements were carried out on thin sections (cuts) of the samples in the cast state and after heat treatment. The hardness was determined by the standard method on the Nemesis 9000 universal hardness tester with ball diameter 2.5 mm and load on the indenter about 600 N.

### *2.3. Tribological tests*

Tribological tests of alloys were carried out on the T-05 tribometre according to the shoe (from the material under study) – roller (counterbody made from steel St45), with step-by-step load changes. This contact pair was studied according to the previously tested method [8] at the pressures of 0.5 and 1 MPa. After tribological tests, the surface and cuts (slides) of the sample were investigated (see 2.4). Perpendicular slices were prepared by cutting the sample after tests and polishing the surface of the cut.

IOP Conf. Series: Materials Science and Engineering **699** (2019) 012048 doi:10.1088/1757-899X/699/1/012048

### *2.4. Microscopy*

In this work, a complex microscopic examination technique was used. Electron microscopic studies were performed on a Quanta-650 SEM (comp. FEI) with EDAX analytical equipment including EDS X-ray microanalyser and an EBSD electron back-scattered diffraction chamber (EBSD method). An accelerating voltage of up to 25 kV and two detection modes, secondary and back-scattered electrons, were used. Samples were examined before and after heat treatment, as well as before and after tribological tests. After tribological tests, the surface of the end sections was also studied.

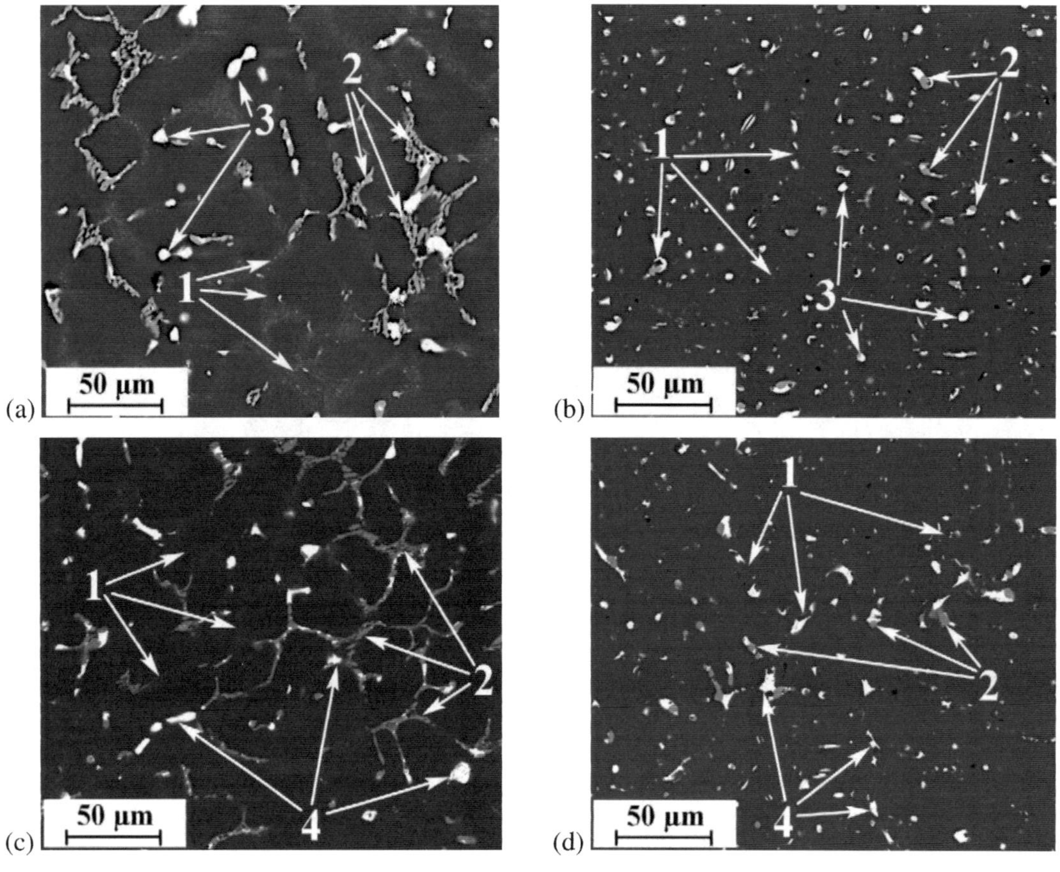


**Figure 1.** SEM image (secondary electrons) of the initial surface of the samples: (a) No. 1 in the cast state, (b) No. 1 in the heat-treated state; (c) No. 2 in the cast state, (d) No. 2 in the heat-treated state. The numbers indicate the various phases, determined by X-ray analysis: 1 – Al-Si; 2 – Al2Cu; 3 – Sn-Bi-Pb-Cd; 4 – Sn-Bi-Pb.

Nanoscale studies were performed using various methods of probe microscopy. SPM Dimension FastScan (Bruker). SPM studies of the alloys were carried out in the PeakForce Tapping QNM mode using standard silicon cantilevers of the NSC-11 type (MikroMasch) with a probe tip curvature radius of 29 nm and force constant of 99.72 N/m. SPM images of surfaces were obtained, and their local properties were investigated. At each point of the image, the probe approaches the surface of the sample with recording force curves. Obtaining and recording such curves is the basis of the PeakForce QNM mode, which automatically recalculates the values of mechanical properties (modulus of elasticity and adhesion), taking into account the characteristics of the probe used.

IOP Conf. Series: Materials Science and Engineering **699** (2019) 012048 doi:10.1088/1757-899X/699/1/012048

## 3. Results and discussion

### *3.1. The influence of heat treatment on the structure and elemental composition of the surface*

For both samples, the structure of the initial surface of the samples and its change as a result of maintenance were investigated by the SEM method. Figure 1 shows the obtained micrographs of the surface of alloys 1 and 2 – in the cast state and after heat treatment.

The analysis of previous optical images (not presented here), Figure 1, and other SEM images shows high surface uniformity and the absence of macroscopic defects. It can be seen that despite some difference in composition (alloy No. 1 contains Cd), both alloys are similar in structure and contain similar phases. Earlier, for alloys of similar composition, it was shown that heat treatment led to spheroidization of individual phases and to partial dissolution of copper in the aluminium matrix [11]. In the present work, the SEM method also shows that heat treatment leads to similar changes. So, after maintenance, the elongated phase constituents of the alloy – silicon and low-melting phases – acquired a more compact rounded shape. By the nature of the change in the phase distribution on the surface, it can be seen that copper dissolves in the aluminium matrix.

### *3.2. Measurement of hardness*

For both samples, the hardness values were determined. The measurement results are presented in Table 1 for cast alloys and after heat treatment. Cast alloys have hardness at the level of the base composition Al–5%Si–4%Cu, which is about 70 HB. After heat treatment, the alloys have a hardness of 125-130 HB, which is the same as in the alloy Al–5%Si–4%Cu–6%Sn. This indicator of hardness is not worse than that of the brand alloy AO3-7. The table shows that the values of hardness are almost the same for both samples.

### *3.3. The effect of tribological tests on the structure and elemental composition of the surface*

When testing for wear, it was found that both alloys had a score at a pressure of 1.0 MPa, while alloy No. 2 had a higher wear resistance. Figure 2 shows the SEM images of the surface of contact pairs (shoe and rollers for two alloys) after tribological tests at a pressure of 1 MPa. To identify the phases and assess their distribution on the surface of the roller after testing (Fig. 2f), the distribution of chemical elements (ElementOverlay) is presented. In Table 2, the elemental composition of the surfaces of the shoe at different pressures (0.5 and 1 MPa) is shown.

Studies have shown that the topography and chemical composition of the surface after testing at various pressures are different from the parameters of initial surface. Thus, at minimum pressures, iron was found on the surface of a shoe made of an alloy with a lower content of low-melting elements (Table 2). This is due to the fact that, in view of the small amount of lubricating component, the solid phase components of the alloy act as an abrasive, causing mass transfer of iron from the steel surface

**Table 1.** Hardness of the samples.

| Alloy | Hardness, HB | |
|---|---|---|
| | Cast | Heat-Treated |
| **No.1** (Al –5%Si–4%Cu–4%Sn–0.5%Bi–0.5%Pb–0.5%Cd) | 73 ± 0.9 | 129 ± 1.6 |
| **No.2** (Al–5%Si–4%Cu–4%Sn–0.5%Bi–0.5%Pb) | 71 ± 1.5 | 125 ± 1.2 |

**Table 2.** The chemical composition of the surfaces of the shoe (alloys No. 1 and No. 2) after testing.

| Pressure | Shoe-alloy | Concentration of elements (% mass) | | | | | | | | |
|---|---|---|---|---|---|---|---|---|---|---|
| | | O | Al | Si | Cd | Sn | Fe | Cu | Pb | Bi |
| 0.5 MPa | No. 1 | 0.59 | 85.36 | 6.98 | 0.32 | 2.74 | 0.16 | 3.42 | 0.29 | 0.14 |
| | No. 2 | 16.76 | 65.54 | 4.89 | - | 2.64 | 6.53 | 3.09 | 0.35 | 0.20 |
| 1.0 MPa | No. 1 | 15.27 | 72.95 | 5.76 | 0.35 | 2.36 | 0.21 | 2.66 | 0.25 | 0.19 |
| | No. 2 | 11.19 | 75.90 | 6.49 | - | 2.61 | 0.21 | 2.97 | 0.43 | 0.20 |

IOP Conf. Series: Materials Science and Engineering **699** (2019) 012048 doi:10.1088/1757-899X/699/1/012048

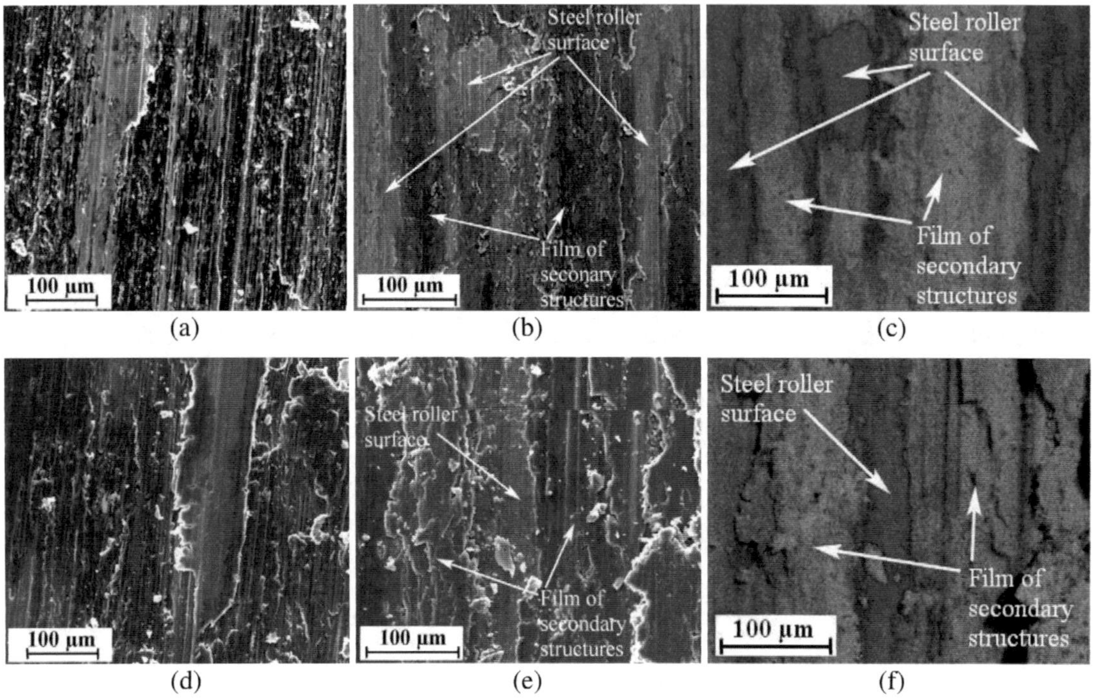


**Figure 2.** SEM image (secondary electrons) of the "shoe-roller" contact pair after tribological tests at a pressure of 1 MPa: (a) shoe from alloy No. 1, (b) roller after tests of alloy No. 1, (c) mapping the surface of the roller after testing alloy No. 1 (blue – iron, red – aluminium); (d) shoe from alloy No. 2, (e) roller after testing alloy No. 2, (f) mapping of the roller surface after testing alloy No. 2 (blue – iron, red – aluminium). (b) and (c), as well as (e) and (f) correspond to the same areas.

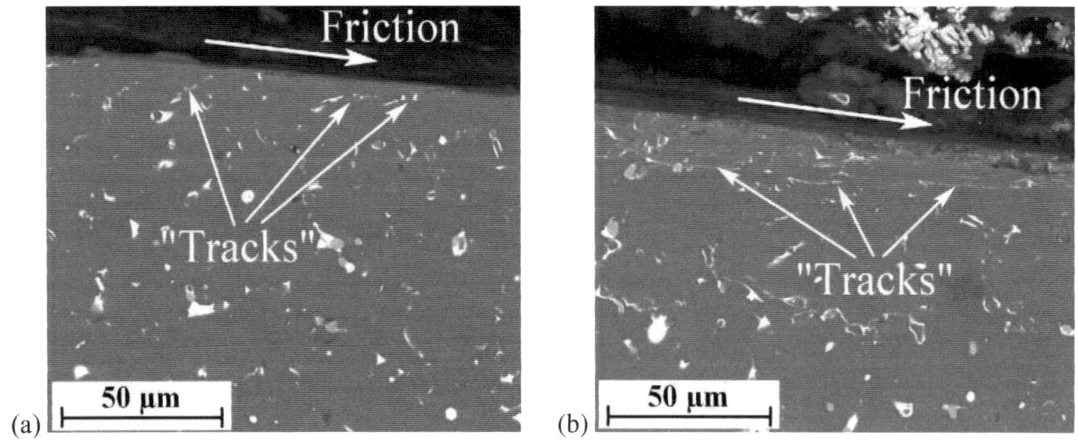


**Figure 3.** SEM image (back-scattered electrons) of the surface of an oblique cut of the shoes after tribological tests at a maximum pressure of 1 MPa: (a) alloy No. 1, (b) alloy No. 2.

of the counterbody. At the same time, mass transfer of the shoe material also occurs on the surface of the roller, resulting in the formation of a film of secondary structures (blue phase in the mapping in Figure 2c,f). With increasing pressure, an unevenly applied film of secondary structures thickens, which leads to the development of macrorelief and may contribute to a score.

*3.4. The effect of tribological tests on the change in the surface layers – study by SEM*

For a more complete study of the processes occurring in the contact zone under friction, sections of the shoes were prepared after the tests, which made it possible to investigate the surface layers of the alloys. Figure 3 shows the SEM images of slices (sections, cuts) of the No. 1 and No. 2 shoes after tribological tests at a maximum pressure of 1 MPa.

When studying sections of all samples, it can be seen that in the near-surface region the structure is noticeably different from the structure in volume. Thus, phase components traced in the direction of friction ("tracks") are traced on it. This is due to the fact that in the process of friction in the surface layers, the grains and dendritic cells of the aluminium matrix are deformed, leading to a change in the geometry of these components. The resulting images show that the differences between two alloys are small. The images allow us to estimate the thickness of the deformed layer: for both alloys, this is 20-30 µm (for pressures of 0.5 MPa) and 50-80 µm (for pressures of 1 MPa).

*3.5. The effect of tribological tests on changes in the surface layers – study by EBSD*

The use of the method of EBSD allowed obtaining additional results that demonstrate the process of deformation of the grains of the aluminium matrix in the surface layer after tribo-testing. Figure 4 shows, as an example, the results of this study for alloy No. 2 after testing at a maximum pressure of 1.0 MPa (the results of all the studied samples have a similar picture).

The images obtained can be used to estimate the average grain size in a volume of 20-80 µm. The colour of the grains in the resulting image characterizes their orientation; it can be seen that there are grains with different orientations in the volume. At the same time, analysis of EBSD paintings showed that it was not possible to identify grains and determine their orientation in a narrow (several microns) surface area and in a wider near-surface area. The reasons for this, obviously, is a strong distortion of the grains in these areas as a result of deformation during friction. At the same time, undistorted grains are clearly visible in the bulk layers of the alloy. The results obtained allowed estimating the thickness of the deformed near-surface layer, which was 50-100 µm (grey colour – indicated by an arrow – a homogeneous area, on which the grain structure is not detected). These results are in good agreement with the electron microscopy data.

*3.6. Probe microscopy*

The sections (cuts) obtained after tribological tests were studied by probe microscopy. Figure 5a,b shows the SPM images of the morphology of the cut-off surface of sample No. 2 after tribological tests (a – in the near-surface layer, b – in the volume). Figure 5c,d shows the corresponding images of the distribution map of elastic modulus.

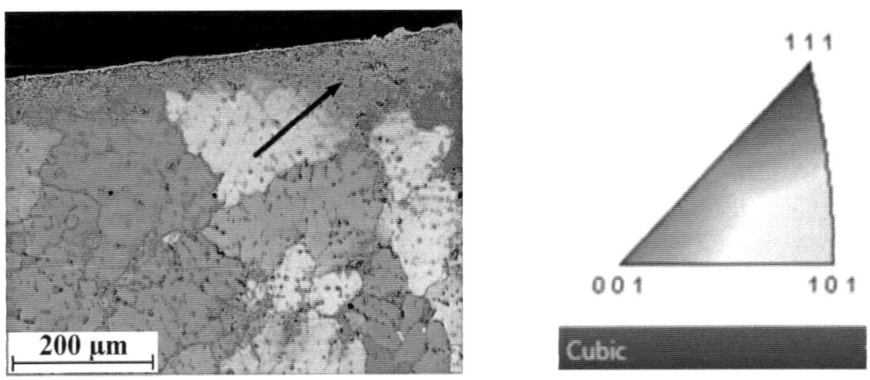


**Figure 4.** The EBSD map of the cut surface of the block (alloy No. 2) after tribological tests – areas with different colour intensities correspond to different grains. Arrow indicates narrow, deformed, near-surface area.

Corresponding to the data of Figure 5c,d, as well as intermediate areas, the numerical values of the modulus of elasticity are presented in Figure 6. Figures 5 and 6 demonstrate that the modulus of elasticity in the surface layer of the alloy is significantly higher than in the volume – 2.7 and 0.6 GPa, respectively (these values are relative and give a qualitative assessment of changes in parameters). This can be explained by the hardening of the alloy material in the process of friction. These measurements are relative and may differ significantly about the values obtained by nanoindentation. At the same time, they convincingly show surface hardening due to friction.

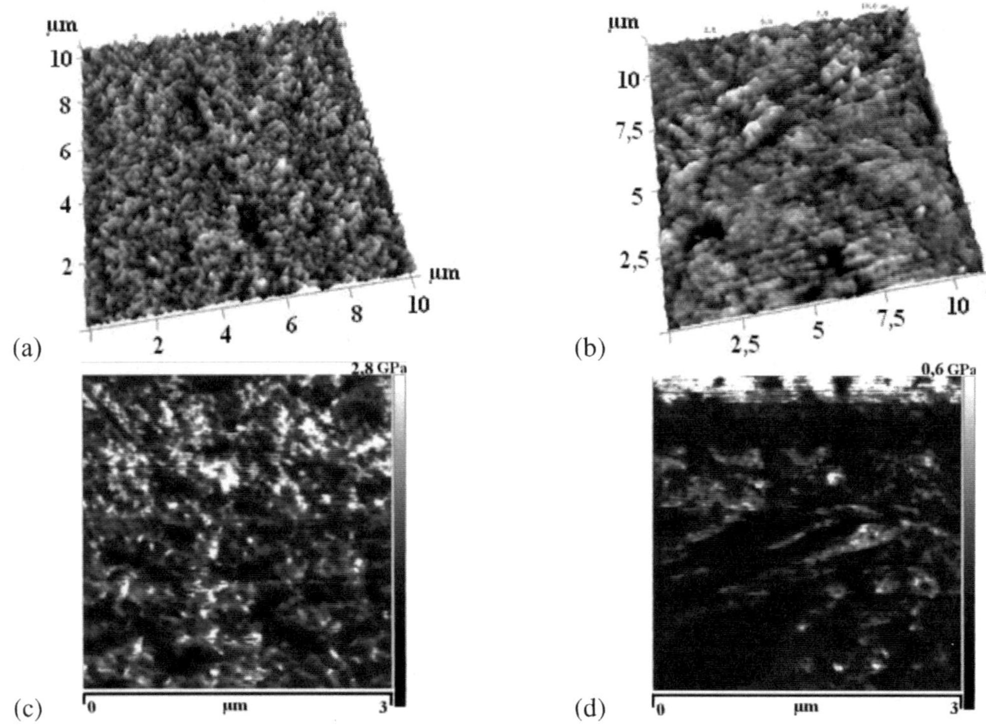


**Figure 5.** SPM image of the cut surface of the sample (shoe) from alloy No. 2 after tribological tests at a pressure of 1.0 MPa: (a) the morphology of the surface layer (the distance from the friction boundary is up to 20 µm), (b) the volume morphology (the distance from the friction boundary is 240-300 µm), (c) distribution map of the elasticity modulus of the surface layer, (d) distribution map of elastic modulus in volume.

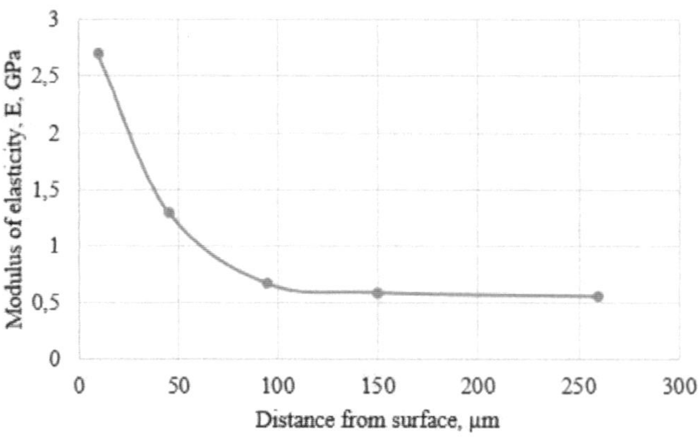


**Figure 6.** The value of the elastic modulus at the shear block of alloy No. 2 at different distances from the friction surface boundary after tribological tests 1.0 MPa.

In general, a comparison of the results obtained for the two alloys shows that the addition of cadmium practically does not change the parameters under study. Note that earlier the effect of cadmium was found for alloys of the Al–Si–Cu system. In the present work, in the presence of other elements (Sn, Bi, and Pb), cadmium did not significantly affect the properties.

## 4. Conclusion

It was found that after heat treatment in both alloys, the phase components acquired a globular shape. So, after heat treatment, the elongated phase components of the alloy – silicon and low-melting points – acquired a more compact, rounded shape. It was shown that in the studied samples, copper in the process of heat treatment dissolved in an aluminium solid solution.

Analysis of the SEM image of the shoe showed that in the process of friction, the deformation of the grains, the extraction of the soft phase to the surface, and mass transfer occurred. In the study of the surfaces of the contact pair after tribological tests, it was established that the solid phase components of the alloy acted as an abrasive, contributing to wear and mass transfer. So, there is a transfer of iron from the steel surface of the counterbody to the sample (shoe) under investigation.

The study of sections of the shoe using SEM and EBSD allowed concluding about the formation of a surface layer (30-100 μm). In this layer, low-melting elements extract on the contact surface due to the deformation of the matrix grains. On the roller surface, mass transfer of the shoe material occurs, leading to the formation of a film of secondary structures. With increasing pressure, this unevenly formed film thickens, which leads to the development of macrorelief and contributes to the score.

According to SPM studies, the value of the elastic modulus in the surface layer of the alloys is much higher than that in the volume. This can be explained by hardening the alloy material in the process of friction.

Hardness, tribological properties, and surface structure for both studied alloys are almost the same. It can be concluded that the addition of bismuth and lead contributes to the partial replacement of expensive tin. At the same time, the addition of cadmium does not change the operational parameters, i.e. the use of cadmium together with bismuth and lead for doping these alloys is impractical.

## Acknowledgements

This work was partially supported by the grant RFBR 18-38-00289 (mol-a) (sample preparation and electron microscopic studies) and the Belarusian Republican Foundation for Basic Research No. F18R-239 (research using probe microscopy). The authors are grateful to A.M. Mezrin (IPMech RAS) for conducting tribological tests.

## References

[1] Summer F, Grün F, Offenbecher M and Taylor S 2019 Tackling the problem with novel bearing materials *Tribol. Int.* **131** 238-50
[2] Ünlüa B S, Durmuşa H and Akgünb S 2009 *J. Alloy Compd.* **487** 225-30
[3] Singh J B, Cai W and Bellon P 2007 *Wear* **263** 830-41
[4] Straffelini G, Maines L, Pellizzari M and Scardi P 2005 *Wear* **259** 506-11
[5] Zhang S-C, Pan Q-L, Yan J and Huang X 2016 *Trans. Nonferr. Metal Soc.* **26** 1809-19
[6] Mironov A E, Gershman I S, Ovechkin A V and Gershman E I 2015 *J. Frict. Wear* **36** 257-61
[7] Belov N A, Stolyarova O O and Yakovleva A O 2016 *Russ. Metallurgy* **3** 198-206
[8] Sachek B Y, Mezrin A M, Muravyeva T I, Stolyarova O O, Zagorskiy D L and Belov N A 2015 *J. Frict. Wear* **36** 103-11
[9] Yakovleva A O, Belov N A, Bazlova T A and Shkalei I V 2018 *Phys. Met. Metallogr.* **119** 35-43
[10] Kurbatkin I I and Muravyeva T I 2012 *Trenie i smazka v mashinakh i mekhanizmakh* **1** 38-44 (in Russian)
[11] Stolyarova O O, Muravyeva T I, Zagorskiy D L and Belov N A 2016 *Phys. Mesomech.* **19** 104-13

IOP Conf. Series: Materials Science and Engineering **699** (2019) 012049 doi:10.1088/1757-899X/699/1/012049

# Tip-induced domain growth in the non-polar cuts of SBN:Ce single crystals

**V A Shikhova[1], E A Neradovskaia[1,2], A P Turygin[1], V V Fedorovyh[1], V A Anikin[1], L I Ivleva[3] and V Ya Shur[1]**

[1]School of Natural Sciences and Mathematics, Ural Federal University, 620000 Ekaterinburg, Russia
[2]Institute de Physique de Nice, Université Côte d'Azur – CNRS, 06180 Nice, France
[3]Prokhorov General Physics Institute, Russian Academy of Sciences, 119991 Moscow, Russia

vera@urfu.ru

**Abstract**. The local switching by conductive tip of scanning probe microscope was studied in the non-polar cuts of Ce-doped $Sr_{0.61}Ba_{0.39}Nb_2O_6$ single crystals after creation of the single-domain state. The switched domains possessed the egg-shaped heads and wedge-like tails. The dependences of lengths of the domain head and tail and width of the domain head on the voltage and pulse duration were derived. The start voltage for growth of the domain tail was revealed. The fast relaxation of the domain head and slow relaxation of the domain tail were observed. The model of the forward domain growth by step generation and kink motion was used for explanation of the experimental results. The obtained knowledge can be used for the domain engineering in ferroelectrics.

## 1. Introduction

The domain growth in ferroelectrics under the action of the local electric field produced by conductive tip of the scanning probe microscope (SPM) has been studied intensively due to applications of the ferroelectrics with engineered domain structures in optoelectronics and photonics [1-4]. Fabrication of the light frequency conversion devices based on periodically poled ferroelectric crystals [3] and piezoelectric components with improved characteristics [5] is very promising. The local switching by conductive SPM tip has been used for creation of the domain patterns in lithium niobate [6-9]. However, most papers to the date explore the lateral domain growth on the polar surfaces and have a lack of information about forward domain growth along polar direction due to some experimental difficulties.

Strontium barium niobate ($Sr_xBa_{1-x}Nb_2O_6$, SBN100x) is a uniaxial relaxor ferroelectric, which is considered to be a prospective material for domain engineering. Prominent nonlinear-optical and electro-optical properties of SBN [10] opened up new horizons for commercialization of the domain patterned crystals. Moreover, SBN crystals are very attractive for SPM recording due to their low coercive fields [10]. The initial domain state at room temperature in thermally depolarized SBN crystals represents maze-type domain structures with characteristic sizes of about a hundred nanometres, which was revealed using piezoresponse force microscopy (PFM) [11-14].

For the first time, the possibility of recording domains on the polar surface of SBN61 single crystals using conductive SPM tip was demonstrated in Ref. [15]. Afterward, the field and exposure characteristics of the SPM domain recording on the polar surface were investigated in pure and Nd, Ce,

IOP Conf. Series: Materials Science and Engineering **699** (2019) 012049 doi:10.1088/1757-899X/699/1/012049

and Ni doped SBN61 single crystals [16-19]. The influence of the initial domain state on the sizes of the domains created by the biased conductive SPM tip was studied in Ni doped SBN single crystals [20]. The relaxation of the polarized state created by conductive SPM tip was investigated in Z-cut SBN crystals at various temperatures [21].

The formation of the microdomain patterns on the nonpolar surface using conductive SPM tip was investigated in SBN61 single crystals with initial multi-domain and single-domain states [18,22].

In this paper, we present the results of experimental study of tip-induced domain growth and its relaxation in the non-polar cuts of Ce-doped SBN61 single crystals after creation of single-domain state.

## 2. Experimental

The studied samples represented plates of $Sr_{0.61}Ba_{0.39}Nb_2O_6$ single crystals doped by Ce (0.004 wt.% $CeO_2$) (SBN:Ce) grown by modified Stepanov technique [23] in the Institute of General Physics of the Russian Academy of Sciences (Moscow, Russia). The 2-mm-thick plates were cut normally to the nonpolar axis (X-cut) and carefully polished.

The single domain state was created in the studied samples before measurements using the cyclic switching at elevated temperature [24]. The sinusoidal electric field with amplitude 300 V/mm and frequency 1 Hz was applied along the polar axis during 100 s at 50°C. The switching was performed using NI-6251 multifunctional Data Acquisition board (DAQ, National Instruments, USA) and high-voltage amplifier Trek-677B (TREK Inc., USA). The silver paste electrodes were deposited for electric field application.

The PFM measurements were realized using scanning probe microscope Asylum MFP-3D (Oxford Instruments, UK) with silicon NSC-18 tips (MikroMasch, Estonia). The domain structure was created by application of dc voltage pulses perpendicular to nonpolar axis with amplitudes from 10 to 30 V and duration from 0.5 to 16 s via conductive SPM tip. The tip had a platinum conductive coating and a typical curvature radius of about 25 nm. The domains were formed at room temperature, which was below freezing temperature for the studied crystal composition [25,26].

## 3. Results and discussion

The domains formed by local switching (amplitude 20 V, duration from 0.5 to 16 s) on the nonpolar surface of the SBN are presented in Figure 1a. All created domains possessed the egg-shaped head and wedge-like tail.

The ZX distribution of the electric field, produced by SPM tip in contact with the nonpolar surface of SBN:Ce single crystals was calculated for the our experimental conditions (Fig. 2a) [27-29]. It is seen that the shape of the created domain head correlates with electric field distribution (Fig. 2b).

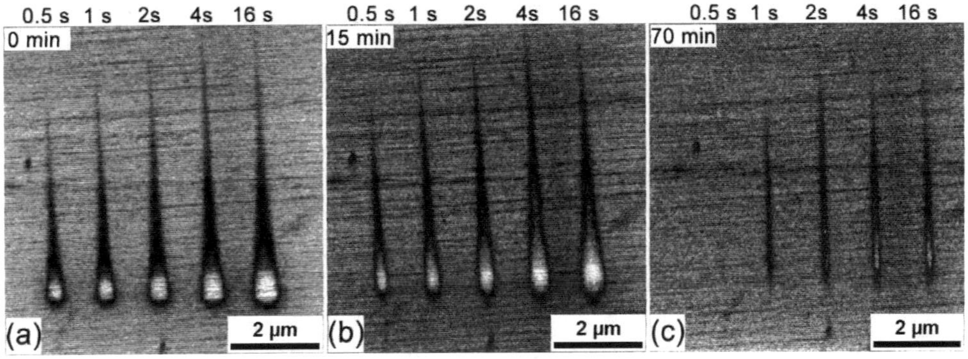


**Figure 1.** PFM images of domains formed on SBN nonpolar surface: (a) immediately after SPM local switching, (b) 15 min after switching, (c) 70 min after switching. $U_{sw}$ = 20 V. $t_{sw}$ = 0.5-16 s. Amplitude signal.

IOP Conf. Series: Materials Science and Engineering **699** (2019) 012049 doi:10.1088/1757-899X/699/1/012049

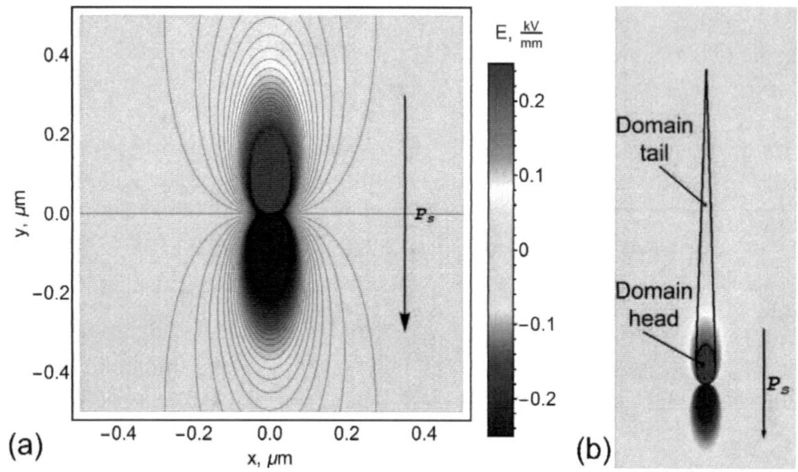


**Figure 2.** (a) Contour maps of calculated electric field $E_z$ across the ZX plane (y = 0 nm), produced by SPM tip in contact with the nonpolar surface of SBN:Ce ($\varepsilon_c$ = 2800, $\varepsilon_a$ = 450, $r_{tip}$ = 25 nm, $U_{sw}$ = 30 V). (b) Domain growth scheme.

The maximum width of the created domains corresponded to the width of the domain head (Fig. 3a,d). The length of the domain head was extracted from the measured head width according to the calculated spatial distribution of the polar component of the electric field (Fig. 3b,e). The domain lengths were divided in two parts: 1) length of the domain head and 2) length of the domain tail. So, the length of the domain tail was equal to the difference between the domain length and the domain head length (Fig. 3c,f). The growth of the domain tail in the area with negligible value of the applied field was attributed to kink motion under the field produced by neighbouring kinks [30]. Thus, we observed the forward domain growth by step generation [31] and kink motion.

All the domain sizes (lengths of the domain head and tail and width of the domain head) increased with pulse duration and voltage. The observed linear dependence of the domain sizes on time logarithm (Fig. 3a-c) and linear voltage dependence (Fig. 3d-f) are typical for ferroelectrics [29].

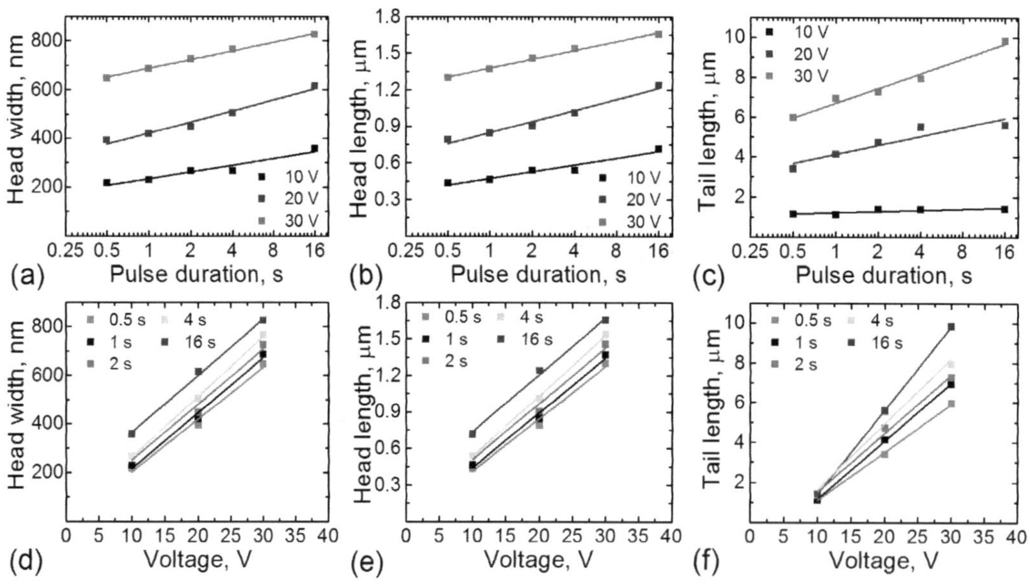


**Figure 3.** The dependences of domain head width (a, d), domain head length (b, e), and domain tail length (c, f) on pulse duration (a, b, c) and voltage (d, e, f).

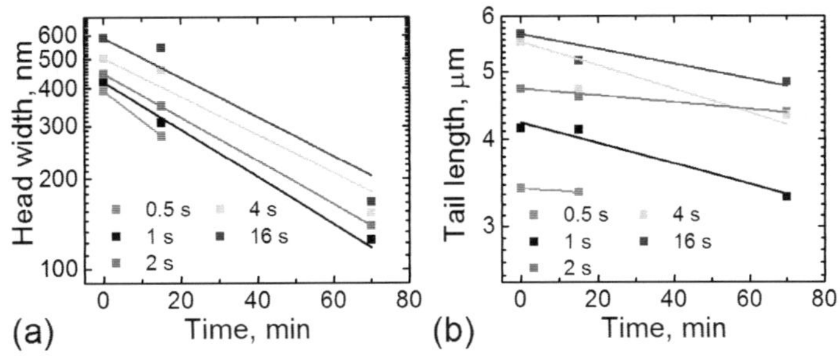


**Figure 4.** The relaxation of (a) domain head width and (b) domain tail length with time. Logarithmic scale on the Y axis. $U_{sw}$ = 20 V. $t_{sw}$ = 0.5-16 s.

The obtained voltage dependences of the domain tail length were fitted by linear function [32,33] $L_t(U) = a\ (U\text{-}U_{st})$, where $a$ is proportional to the domain wall mobility, $U_{st}$ is the start voltage. The obtained start voltage for the tail length ranged from 5 to 6.5 V.

The changes in created domains were revealed over time. The domain images created by application of the pulse with amplitude 20 V were obtained just after local switching and 15 and 70 minutes later (Fig. 1). The domain created with pulse duration 0.5 s disappeared on the third scan. The length of the other domains decreased slightly, while their width decreased significantly.

The time relaxation of the domain sizes was investigated for created domains. The time dependences of the domain head width (Fig. 4a) and domain tail length (Fig. 4b) were fitted by exponential function. The fast time relaxation of the domain head width and slow relaxation for the domain tail length were observed. The relaxation time for domain head width was in the range from 50 to 70 minutes; for domain tail length, it was in the range from 300 to 800 minutes. The fast relaxation of the domain head can be attributed to the high concentration of the kinks near domain head. The slow decrease of the length of domain tails can be attributed to electrostatic interaction of the approaching walls [3,34].

## 4. Conclusion

We studied the local switching by conductive tip of scanning probe microscope on the non-polar surface of Ce-doped $Sr_{0.61}Ba_{0.39}Nb_2O_6$ single crystals after creation of the single-domain state. It was found that created domains consisted of egg-shaped heads and wedge-like tails. The shape of the domain head correlated with spatial distribution of the polar component of electric field, produced by SPM tip in contact with the non-polar surface of SBN:Ce single crystals. The dependences of domain sizes (lengths of the domain head and tail and width of the domain head) on the voltage and pulse duration were derived. The linear dependence of domain sizes on time logarithm and linear voltage dependence were observed. The start voltage was revealed for growth of the domain tail. The fast relaxation of the domain head and slow relaxation of the domain tail were revealed. The obtained knowledge can be applied for improvement of the domain engineering methods in SBN crystals.

## Acknowledgements

The equipment of the Ural Centre for Shared Use "Modern nanotechnology" Ural Federal University was used. The research was made possible by the Russian Science Foundation (grant № 19-72-00008).

## References

[1] Shur V Ya, Rumyantsev E L, Nikolaeva E V, Shishkin E I, Batchko R G, Miller G D, Fejer M M and Byer R L 2000 *Ferroelectrics* **236** 129-44
[2] Shur V Ya, Rumyantsev E L, Nikolaeva E V, Shishkin E I, Batchko R G, Fejer M M and Byer R L 2001 *Ferroelectrics* **257** 191-202
[3] Shur V Ya 2006 *J. Mater. Sci.* **41** 199-210
[4] Balke N, Bdikin I K, Kalinin S V and Kholkin A L 2009 *J. Am. Ceram. Soc.* **92** 1629-33.

[5] Shur V Ya, Akhmatkhanov A R and Baturin I S 2015 *Appl. Phys. Rev.* **2** 040604
[6] Rodriguez B J, Nemanich R J, Kingon A, Gruverman A, Kalinin S V, Terabe K, Liu X Y and Kitamura K 2005 *Appl. Phys. Lett.* **86** 012906
[7] Rosenman G, Urenski P, Agronin A, Rosenwaks Y and Molotskii M 2003 *Appl. Phys. Lett.* **82** 103-5
[8] Ievlev A V, Morozovska A N, Shur V Ya and Kalinin S V 2014 *Appl. Phys. Lett.* **104** 092908
[9] Ievlev A V, Jesse S, Morozovska A N, Strelcov E, Eliseev E A, Pershin Y V, Kumar A, Shur V Ya and Kalinin S V 2014 *Nat. Phys.* **10** 59-66
[10] Prokhorov A M and Kuz'minov Yu S 1990 *Ferroelectric crystals for laser radiation control* (Bristol: Adam Hilger)
[11] Lehnen P, Kleemann W, Woike Th, and Pankrath R 2001 *Phys. Rev. B* **64** 224109
[12] Terabe K, Takekawa S, Nakamura M, Kitamura K, Higuchi S, Gotoh Y and Gruverman A 2002 *Appl. Phys. Lett.* **81** 2044-6
[13] Shvartsman V V, Kleemann W, Łukasiewicz T and Dec J 2008 *Phys. Rev. B* **77** 054105
[14] Shur V Ya, Shikhova V A, Pelegov D V, Ievlev A V and Ivleva L I 2011 *Phys. Solid State* **53** 2311-5
[15] Terabe K, Higuchi S, Takekawa S, Nakamura M, Gotoh Y and Kitamura K 2003 *Ferroelectrics* **292** 83-9
[16] Gainutdinov R V, Volk T R, Lysova O A, Razgonov I I, Tolstikhina A L and Ivleva L I 2009 *Appl. Phys. B: Las. Opt.* **95** 505-12
[17] Volk T R, Simagina L V, Gainutdinov R V, Tolstikhina A L and Ivleva L I 2010 *J. Appl. Phys.* **108** 042010
[18] Volk T R, Gainutdinov R V, Bodnarchuk Ya V, Simagina L V, Mishina E D, Ilyin N A, Artemov V V and Ivleva L I 2013 *Ferroelectrics* **442** 63-73
[19] Neradovskaya E A, Neradovskiy M M, Fedorovyh V V, Turygin A P, Shur V Ya, Kholkin A L and Ivleva L I 2016 *KnE Mater. Sci.* **2016** 115-21
[20] Shikhova V A, Fedorovyh V V, Turygin A P, Gimadeeva L V, Chezganov D S, Vlasov E O, Alikin D O, Ivleva L I, Kholkin A L and Shur V Ya 2018 *Ferroelectrics* **525** 100-7
[21] Shur V Ya, Shikhova V A, Alikin D O, Lebedev V A, Ivleva L I, Dec J, Lupascu D C and Shvartsman V V 2017 *Sci. Rep.* **7** 125
[22] Volk T R, Gainutdinov R V, Bodnarchuk Ya V and Ivleva L I 2013 *JETP Lett.* **97** 483-9
[23] Ivleva L I 2009 *Bull. Russ. Acad. Sci.: Phys.* **73** 1338-40
[24] Kolchina E A, Neradovskiy M M, Shikhova V A, Pelegov D V, Shur V Ya, Ivleva L I and Dec J 2016 *Ferroelectrics* **496** 149-56
[25] Kaczmarek S M, Orowski M, Skibiñski T, Jasik A and Ivleva L I 2010 *Rev. Adv. Mater. Sci.* **23** 80-7
[26] Shikhova V A, Ushakov A D, Fedorovyh V V, Anikin V A, Esin A A, Shur V Ya, Kholkin A L and Ivleva L I 2018 *IOP Conf. Ser.: Mater. Sci. Eng.* **443** 012031
[27] Pertsev N A and Kholkin A L 2013 *Phys. Rev. B* **88** 174109
[28] Ievlev A V, Alikin D O, Morozovska A N, Varenyk O V, Eliseev E A, Kholkin A L, Shur V Ya and Kalinin S V 2015 *ACS Nano* **9** 769-77
[29] Alikin D O, Ievlev A V, Turygin A P, Lobov A I, Kalinin S V and Shur V Ya 2015 *Appl. Phys. Lett.* **106** 182902
[30] Schryvers D, Salje E K H, Nishidac M, De Backer A, Idrissi H and Van Aert S 2017 *Ultramicroscopy* **176** 194-9
[31] Shur V Ya, Gruverman A L and Rumyantsev E L 1990 *Ferroelectrics* **111** 123-31
[32] Agronin A, Molotskii M, Rosenwaks Y, Rosenman G, Rodriguez B J, Kingon A I and Gruverman A 2006 *J. Appl. Phys.* **99** 104102
[33] Shur V Ya and Rumyantsev E L 1997 *Ferroelectrics* **191** 319-33
[34] Shur V Ya and Zelenovskiy P S 2014 *J. Appl. Phys.* **116** 066802

IOP Conf. Series: Materials Science and Engineering **699** (2019) 012050 doi:10.1088/1757-899X/699/1/012050

# Magnetron sputtering deposition of ultra-thin tungsten coatings onto amorphous graphite for enhancement of horseradish peroxidase adsorption

**I D Shumov[1], S L Kanashenko[1], V S Ziborov[1,2], Y D Ivanov[1], A I Archakov[1], D V Enikeev[3], N V Potoldykova[3] and T O Pleshakova[1]**

[1]Institute of Biomedical Chemistry, 119121 Moscow, Russia
[2]Joint Institute for High Temperatures RAS, 125412 Moscow, Russia
[3]Institute of Urology and Reproductive Health, Sechenov University, 119991 Moscow, Russia

shum230988@mail.ru

**Abstract.** In the present study, atomic force microscopy (AFM) has been used to investigate the adsorption of horseradish peroxidase (HRP) protein onto the surface of amorphous carbon (graphite). It has been experimentally demonstrated that modification of amorphous carbon surface by magnetron sputtering deposition of ultra-thin (1.3 nm) tungsten coatings allows one to enhance significantly the adsorption of HRP macromolecules onto this surface, as compared with bare carbon.

## 1. Introduction

Atomic force microscopy (AFM) represents a nanotechnology-based method of molecular detection, which allows visualizing various nanometre-sized objects, including viral particles [1,2], biological macromolecules [3] and their complexes [4] with up to 0.1 nm height resolution [3,5]. This makes AFM an attractive tool in biochemical research.

The macromolecules under investigation must be fixed on the surface of a solid substrate for their visualization by AFM. Regarding biological macromolecules, this can be achieved by their adsorption (non-covalent immobilization) onto the substrate surface. Accordingly, appropriate adsorption efficiency of the macromolecules of interest onto the substrate is required. Despite this efficiency is strongly affected by electrostatic interactions of the macromolecules with the support surface, hydrophobic interactions should also be taken into account, as in a number of cases they determine the possibility of macromolecule adsorption [6]. It depends on the properties of the studied objects. The substrate surface may require modification to provide adsorption of the studied objects.

Magnetron sputtering allows obtaining metal coatings of high purity with excellent adhesion to the coated surface [7], while the porosity of the so-formed coatings is low even at low coating thickness [8,9]. The deposition rate can be varied in a broad range by changing sputtering gas pressure or discharge power (or both of these parameters) [8,9]. These advantages allow employing magnetron sputtering in sample preparation for AFM [2,10]. It was demonstrated that in the case of formation of thin magnetron sputtered metal coating on the support surface, the surface roughness remained virtually unchanged, as compared with the support without the coating [2]. Retaining of atomic smoothness after surface modification is crucial in the case of supports intended for visualization of nanometre-size objects, such as protein macromolecules [3,5].

Accordingly, magnetron sputtering is suitable for modification of the surface properties of supports for AFM, which has been demonstrated herein with the example of adsorption of horseradish peroxidase (HRP) protein onto the surface of amorphous carbon (graphite). Herein, it has been experimentally demonstrated that modification of amorphous carbon surface by magnetron sputtering deposition of ultra-thin (1.3 nm) tungsten coatings significantly enhances the adsorption of HRP macromolecules onto this surface in comparison with bare carbon. That is, HRP macromolecules were visualized by AFM on the surface of the support modified with tungsten coating after its incubation in $10^{-7}$ M HRP solution. At the same time, virtually no objects were visualized on the surface of bare carbon even in the case of 10 times higher ($10^{-6}$ M) HRP concentration.

## 2. Experimental

In our experiments, peroxidase from horseradish (P2088, Sigma, USA) was adsorbed onto amorphous carbon supports (01843-F; Ted Pella, Inc., USA) according to the following procedure: (1) 5 μL of $10^{-7}$ M to $10^{-6}$ M HRP solution in Dulbecco modified phosphate buffered saline were incubated on the support surface for 5 min and then washed off with 1 mL of ultrapure deionized water; (2) the support was dried in air. In control experiments, protein-free buffer was used.

Ultra-thin tungsten coatings were formed on the surface of carbon supports analogously to the technique described elsewhere [2,11]. Briefly, Orion-3 magnetron sputtering system (AJA Inc., USA) equipped with tungsten target (thickness 0.25'', diameter 2'', 99.95% purity, Girmet, Russia) was used. Base pressure in the system was not higher than $6\times10^{-7}$ Torr. The sputtering was carried out using argon plasma in DC mode at a constant power of 70 W and 5 mTorr sputtering gas pressure during 10 s. The distance between the target and the sample was 15 cm. The rotation of the substrate with constant angle velocity of 40 rpm was provided. The thickness of the so-obtained tungsten films was 1.3 nm as measured by quartz crystal microbalance.

The HRP macromolecules adsorbed on the surface of carbon supports (either coated with tungsten or not) were imaged using NT-MDT AFM (Zelenograd, Russia) in a semi-contact mode in air with 256×256 resolution.

## 3. Results and discussion

Figure 1 displays typical AFM images of the surface of carbon supports coated with 1.3 nm tungsten film after their incubation in $10^{-7}$ M HRP solution (Fig. 1a) and in protein-free buffer (Fig. 1b).

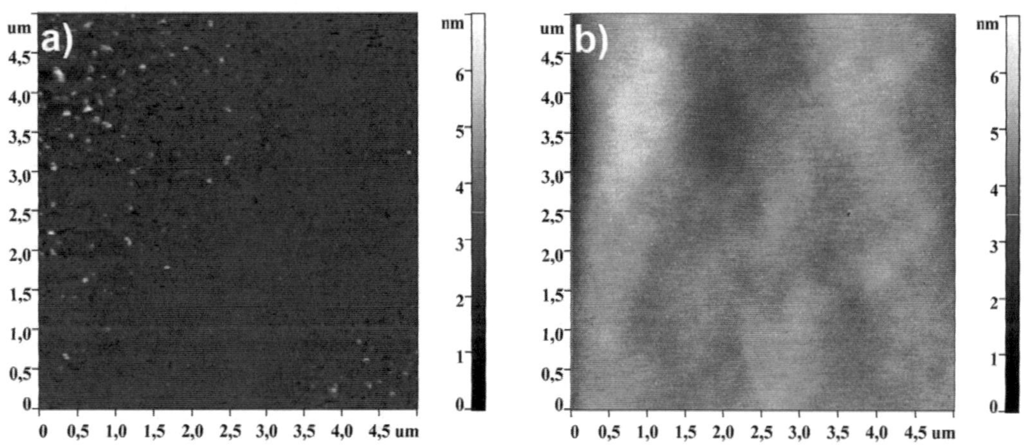


**Figure 1.** AFM image of amorphous carbon coated with 1.3-nm tungsten film (a) after adsorption of HRP from $10^{-7}$ M solution and (b) control image. Scan size 5×5 μm$^2$.

As seen from the image in Figure 1a, separate objects with heights of AFM images ($h$) from 1 to 2 nm and a number of objects with $h > 2$ nm are observed on the surface of tungsten-coated carbon after its incubation in $10^{-7}$ M HRP solution. These data are in agreement with previously reported sizes ($h$=1.5 nm [3]) of AFM images of HRP (which molecular weight $M_r$ is 40 kDa [12]), and with the data on the sizes of other proteins with similar $M_r$ (putidaredoxin reductase, $h_{max}$ = 1.8 nm [5], $M_r$=45.6 kDa [13]; adrenodoxin reductase, $h_{max}$=1.8 nm [14], $M_r$=54 kDa [15]). Objects with $h$>2 nm apparently correspond to aggregates formed on the support surface by several HRP macromolecules. It should be noted that no HRP adsorption was observed at bare amorphous carbon without tungsten coating, even for ten-times higher ($10^{-6}$ M) HRP concentration. The control image in Figure 1b indicates that artefact objects are virtually not observed on the support incubated in protein-free buffer. It indicates that the coating roughness is sufficient for AFM imaging of protein macromolecules.

The studies of HRP solutions performed by Ignatenko et al. [16] revealed that HRP forms oligomers in solutions with concentration exceeding $10^{-7}$ M. Accordingly, the compact objects with $h$<2 nm observed in Figure 1a can be attributed to HRP monomers, while objects with greater $h$ apparently correspond to HRP oligomers. Our data agree with the literature [3, 16].

## 4. Conclusion

The obtained results demonstrate the influence of surface hydrophobicity on the adsorption of heme-containing HRP protein. The adsorption of HRP onto hydrophobic surface of amorphous carbon is virtually not observed, while it readily adsorbs onto hydrophilic tungsten surface. These results are in agreement with the previously reported data [3], where the HRP adsorption onto hydrophilic chemically activated amino-mica surface was studied by AFM on the single-molecule level; in the latter case, covalent capturing of the protein onto the chemically activated surface was observed to adsorb HRP onto mica from its solution with very low ($10^{-17}$ M) concentration.

The magnetron sputtering-based technique described herein can be used for surface modification in the fabrication of analytical chips with surface-immobilized proteins. This is important for the development of novel chip-based methods of protein detection, including highly sensitive ones, intended for use in analytical proteomics as well as for diagnostic applications.

**Acknowledgements**
The present study was performed in the framework of the Program for Basic Research of State Academies of Sciences for 2013-2020.

## References

[1] Dubrovin E V, Yaminsky I V, Kirikova M, Novikov V and Drygin Y 2004 *Colloid J.* **66** 673-8
[2] Shumov I D, Kanashenko S L, Archakov A I, Ivanov Yu D and Pleshakova T O 2017 *Appl. Phys.* **4** 10-5
[3] Ivanov Yu D, Danichev V V, Pleshakova T O, Shumov I D, Ziborov V S, Krokhin N V, Zagumenniy M N, Ustinov V S, Smirnov L P and Archakov A I 2013 *Biochemistry Suppl. Series* B*: Biomedical Chemistry* **7** 46-61
[4] Ramacviciene A, Snitka V, Mieliauskiene R and Ramanavicius A 2006 *Cent. Eur. J. Chem.* **4** 194-206
[5] Ivanov Y D, Bukharina N S, Frantsuzov P A, Pleshakova T O, Kanashenko S L, Medvedeva N V, Argentova V V, Zgoda V G, Munro A W and Archakov A I 2012 *Soft Matter* **8** 4602-8
[6] Felgueiras H P 2018 Fundamentals of protein and cell interactions in biomaterials. *Peptides and Proteins as Biomaterials for Tissue Regeneration and Repair* ed M A Barbosa and M C L Martins (Duxford, UK: Woodhead Publishers/Elsevier) chapter 1 pp 1-27
[7] Swann S 1988 *Phys. Technol.* **19** 67-75
[8] Petroff P, Sheng T, Sinha A, Rozgonyi G and Alexander F 1973 *J. Appl. Phys.* **44** 2545-54
[9] Salamon K, Milat O, Radic N, Dubcek P, Jercinovic M and Bernstorff S 2013 *J. Phys.* D*: Appl. Phys.* **46** 095304

IOP Conf. Series: Materials Science and Engineering **699** (2019) 012050 doi:10.1088/1757-899X/699/1/012050

[10] Shumov I D, Kanashenko S L, Ziborov V S, Ivanov Yu D, Archakov A I and Pleshakova T O 2017 *J. Phys.: Conf. Ser.* **789** 012053
[11] Shumov I D, Kanashenko S L, Ziborov V S, Ivanov Yu D, Archakov A I and Pleshakova T O 2018 *J. Phys.: Conf. Ser.* **1058** 012048
[12] Davies P F, Rennke H G and Cotran R S 1981 *J. Cell. Sci.* **49** 69-86
[13] Unger B P, Gunsalus I C and Sligar S G 1986 *J. Biol. Chem.* **261** 1158-63
[14] Ivanov Y D, Frantsuzov P A, Zöllner A, Medvedeva N V, Archakov A I, Reinle W and Bernhardt R 2011 *Nanoscale Res. Lett.* **6** 54
[15] Chu J-W and Kimura T 1973 *J. Biol. Chem.* **248** 2089-94
[16] Ignatenko O V, Sjolander A, Hushpulian D M Kazakov S V, Ouporov I V, Chubar T A, Poloznikov A A, Ruzgas T, Tishkov V I, Gorton L, Klyachko N L and Gararyan I G 2013 *Adv. Biosens. Bioelectron.* **2** 25-34

IOP Conf. Series: Materials Science and Engineering **699** (2019) 012051 doi:10.1088/1757-899X/699/1/012051

# Creation of nanoparticles and surface nanostructures of alumina by hot water treatment

**V Ya Shur, E A Mingaliev, A V Makaev, D S Chezganov, I Y Kozheletova, and V I Pryakhina**

School of Natural Sciences and Mathematics, Ural Federal University, 620000 Ekaterinburg, Russia

andrey.makaev@urfu.ru

**Abstract**. In this report, we present a detailed study of the formation of alumina nanostructures at the surface of aluminium plate by hot water treatment (HWT) at various temperatures. The nanostructures were studied by scanning electron microscopy. The superhydrophilic property of the treated surface was revealed and its stability was investigated. It was shown that HWT could be used also for synthesis of the aqueous suspensions of alumina nanoparticles. It is proposed that the method can be applied for production of surface nanostructures and nanoparticles of various metal oxides.

## 1. Introduction

Nanomaterials have numerous applications in various areas, such as catalysis [1], photonics [2], molecular computing [3], energy storage [4,5], fuel cells [6,7], sensing [8-10], and nanomedicine [11-15]. This is due to increased reactivity compared to their micro-sized counterparts, since nanoscale materials have extremely high value of surface-to-volume ratio; therefore, their properties are no longer dominated by the bulk of the material but by the surface atoms.

There are several ways to obtain nanoscale materials: electrochemical synthesis [16-18], chemical vapour deposition [19], molecular beam epitaxy [19,20], pulsed laser ablation [21,22], and pyrolysis of volatile precursors [23-25]. Although these methods allowed producing excellent nanostructured materials of various geometry and complexity, they have their inherent limitations, such as complex and expensive equipment, toxic precursors, and scaling problems.

Metal oxide nanostructures (MONSTRs) and nanoparticles (MONPs) produced by treatment of the metal surface demonstrate unique physical and chemical properties. Recently, MONSTRs have been used in electronic and optical devices [26,27] and sensitive devices of an evidence-based concept [26,28]. However, their synthesis methods have several obstacles that limit their large-scale production.

The growth of hydrous oxide films on the surface of electropolished aluminium in hot water with temperature ranged from 50 to 70°C was discovered more than forty years ago [29]. This effect was applied for durable adhesive bonding [30] and for production of superamphiophobic aluminium alloy surface with nanoscale roughness [31].

Recently, it has been demonstrated that a simple Hot Water Treatment (HWT) process allowed producing MONSTRs on the surface of various metals, their compounds and alloys by a one-step, scalable, low-cost, and eco-friendly process [32]. The HWT process involves immersing the metal piece in hot deionized water without any chemical additives, such as a metal salt and a reducing or oxidizing agent. This method already was used for producing of copper (II) oxide nanoleaf structures on Cu sheets for optical photoconductive devices [26].

The boiling water was used to make the aluminium surfaces hydrophilic through the creation of nanostructures [33]. Water treatment of laser-ablated aluminium allowed making it superhydrophilic [33]. The surfaces with changed wettability are used in wide range of applications and industries, for example, in biomedical area as substrates for controlling protein adsorption [34,35], cell interaction [36], and bacterial growth [37,38].

It is seen that HWT is the new promising eco-friendly method for producing MONSTRs of the most metals in the periodic table. But, to date there are no systematic studies of MONSTRs formation. In this report, we present the detail study of formation of superhydrophilic alumina nanostructures and nanoparticles by HWT at various temperatures and treatment durations.

## 2. Experimental

Aluminium compound plates (Al 97%, Cu 1.2%, Fe 0.7%, Mn 0.3%) with 0.5 mm thickness and 10x10 $mm^2$ size were used. Just before HWT, the native oxide layer and organic contaminations were removed by ultrafine sandpaper and the sample was cleaned by ultra-sonication in acetone and isopropanol for 5 min each.

Treatment by hot deionized water was done at 60-95°C during 2-30 min. Treatment by boiling water was done for 30 min only. The water volume was always 50 ml. The samples were extracted from the water after HWT and dried by airflow.

Wettability of the treated surfaces was characterized by water droplet contact angle obtained by sessile drop technique. The measurements were carried out after careful sample drying with compressed air. The water droplet of 10 μl volume was placed on the surface and its profile was recorded by high resolution camera FASTCAM Mini UX100 (Photron, Germany).

The imaging of the plate surface morphology was performed by scanning electron microscopy (SEM) using dual-beam workstation Auriga CrossBeam (Carl Zeiss, Germany) operated at accelerating voltage of 3 kV and beam current about 100 pA. The secondary electron detection mode with resolution down to 2 nm was used.

The thickness of alumina nanostructure was measured by SEM imaging of the cross-sections produced by focused Ga+ ion beam (FIB) operated at accelerating voltage of 30 kV and beam current of 1 nA. The elemental composition was measured by means of energy-dispersive X-ray spectroscopy (EDS) using X-Max N detector and analysed by Aztec software (Oxford Instruments, UK).

The shape and sizes of nanoparticles were revealed by analysis of SEM images. The samples for SEM imaging were prepared by vacuum freeze drying (lyophilization) of the water suspension drop on the Si/$SiO_2$ substrate by freeze dryer Alpha 2-4 LSC (Martin Christ, Germany).

## 3. Results and discussion

### *3.1. Nanostructure morphology*

The study of nanostructures formed at various durations of HWT at fixed water temperature (80°C) allowed us to reveal three stages of nano-sheets formation. At the end of the first stage, for HWT duration below 8 min, the metal surface was covered by isolated nanowires (Fig. 1a,b). At the end of the second stage, for HWT duration about 16 min, the surface was completely covered by three-dimensional maze-like nanostructure consisting of nanowires (Fig. 1c). At the last stage, for HWT duration above 30 min, the nano-sheets mostly oriented normally to the plate surface with average thickness about 10 nm formed over the maze structure (Fig. 1d).

The element analysis of the aluminium plate surface layer was performed before and after HWT. It was obtained that the nanostructured layer appeared after HWT consisted of alumina nano-sheets.

Investigation of the obtained nanostructured layers allowed revealing significant increase in nano-sheets concentration with water temperature (Fig. 2).

IOP Conf. Series: Materials Science and Engineering **699** (2019) 012051 doi:10.1088/1757-899X/699/1/012051

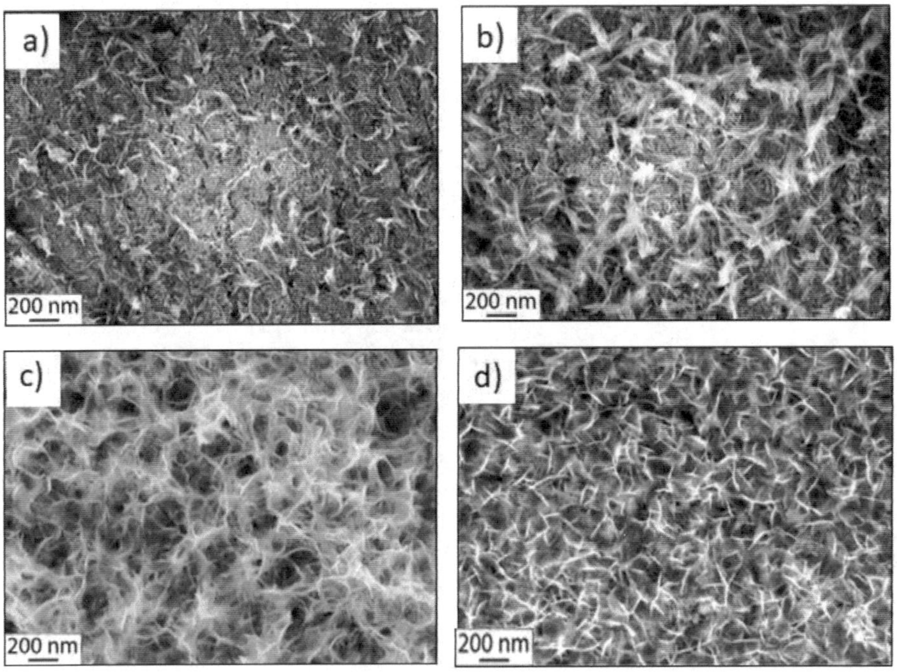


**Figure 1.** SEM images of the surface of aluminium plate after HWT at 80°C for various treatment durations, min: (a) 4, (b) 8, (c) 16, (d) 30.

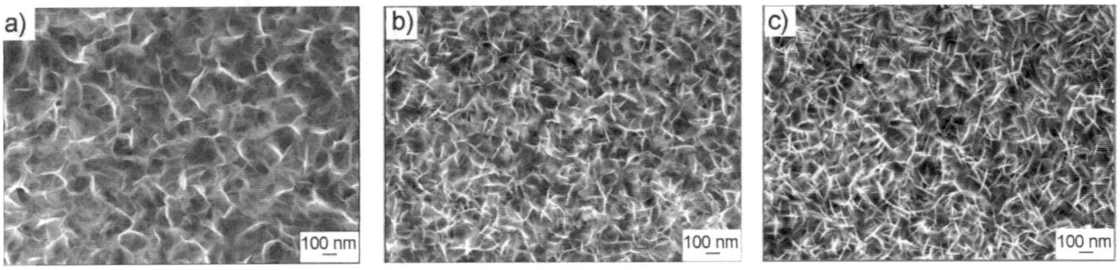


**Figure 2.** SEM images of aluminium oxide nanostructures on surface plates after HWT during 30 minutes at various temperatures, °C: (a) 70, (b) 80, (c) 95.

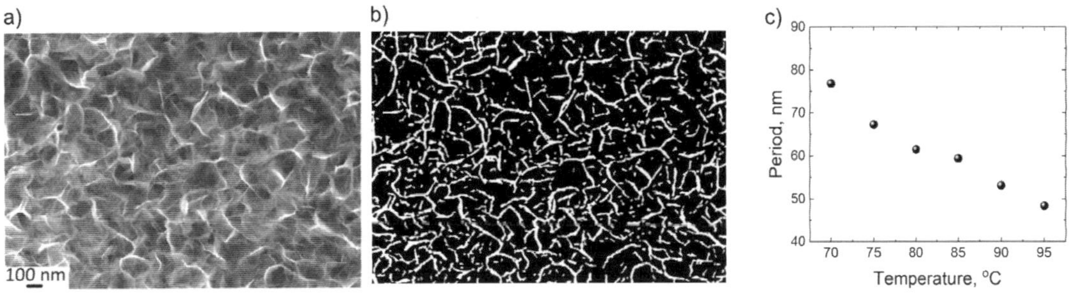


**Figure 3.** SEM image of alumina nanostructure produced by HWT at 70°C for 30 min: (a) original and (b) binarized. (c) Temperature dependence of the average effective period of the nanostructure obtained after HWT during 30 min.

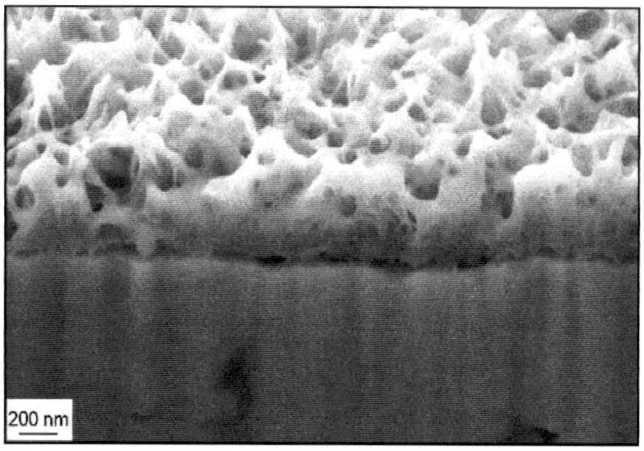


**Figure 4.** Cross-section of surface nanostructured layer by ion beam.

For quantitative characterization of the effect, we treated the SEM images. The obtained grayscale images were binarised by local adaptive binarisation function (Fig. 3a) with subsequent application of the skeletisation function (Fig. 3b). The obtained skeleton corresponds to the top edges of the nano-sheets. The average period of nanostructure was calculated as a ratio of surface area covered by it to the total length of the skeleton segments. It was obtained that the average effective period of the nanostructure decreased monotonously with water temperature from 76 to 48 nm (Fig. 3c).

The thickness of alumina nanostructured layer was measured at cross-sections produced by FIB (Fig. 4). It was revealed that the layer thickness was nonuniform over the plate surface and the difference could reach 100 nm. It can be attributed to the islet character of nanostructure formation. Places, where isolated nanowires started appearing, had a thicker alumina layer. Such nonuniformity did not allow revealing the temperature dependence of the thickness of nanostructured layer. For HWT at temperatures from 70 to 95°C during 30 min, the layer thickness varied between 350 and 550 nm.

*3.2 Change in wettability*

It was revealed that appearance of alumina nanostructure led to significant change in the surface wettability. The contact angle of untreated plate surface was about 69°. It was shown that HWT of plate during 30 min at temperature in the range from 70 to 95°C led to appearance of superhydrophilic

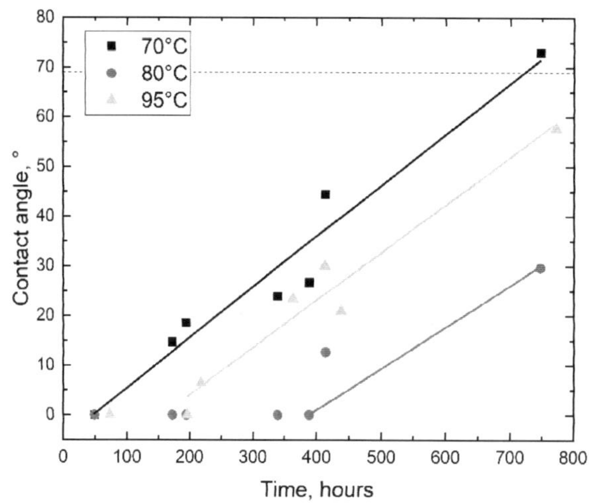


**Figure 5.** The time dependence of contact angle of alumina nanostructured surface after HWT at different temperatures during 30 min.

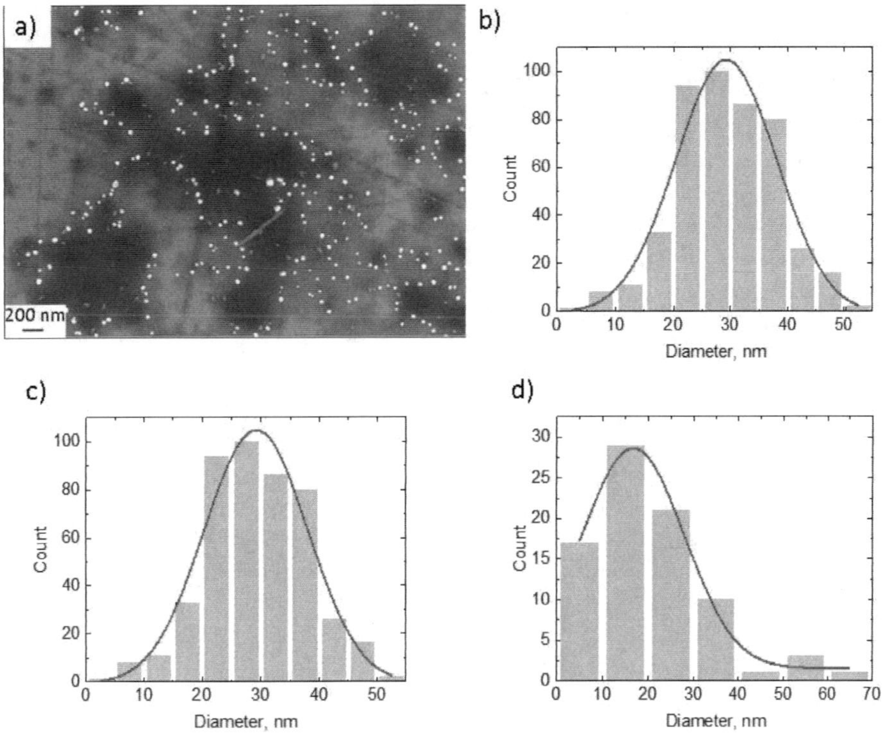


**Figure 6.** (a) SEM image of alumina nanoparticles obtained as a result of HWT at 80°C. The size distributions of nanoparticles produced at different temperatures, °C: (b) 70, (c) 80, and (d) 95.

property with wetting angle close to zero. The superhydrophilic state remained for about 400 hours after HWT at 80°C and then slowly increased to the initial value. The lifetime increased with the temperature. Nevertheless, the lifetime after HWT at 95°C was essentially shorter. This fact can be attributed to the influence of cavitation caused by boiling near the surface of aluminium plate (Fig. 5).

### *3.3 Nanoparticles*

HWT leads also to appearance of the water suspension of nanoparticles. For SEM visualization of the nanoparticle morphology, the drop of suspension was dried on the $Si/SiO_2$ substrate (Fig. 6a). It is clearly seen that the synthesized nanoparticles possess mostly spherical shape. These results are in good agreement with those previously reported [39].

The size distributions of the nanoparticles were obtained by analysis of the SEM images (Fig. 6a) using the Image J program. The nanoparticle size distributions are shown in Figure 6b-d. The average sizes of nanoparticles were 29±18 nm for 70°C, 41 ± 18 nm for 80°C, and 17 ± 22 nm for 95°C. So, we can conclude that the temperature does not affect nanoparticle size significantly.

Along with the spherical nanoparticles, the rare isolated nano-filaments with typical width about 25 nm and length above 1 μm were obtained (Fig. 7a). Moreover, the self-assembled bundles of nano-filaments were found (Fig. 7b). The similar shapes were produced by HWT of the aqueous suspensions of CuO nanoparticles produced by laser ablation in water [40].

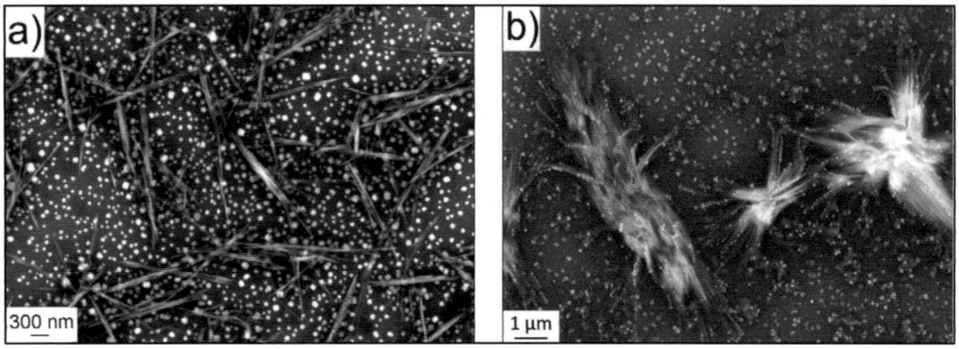


**Figure 7.** SEM images of $Al_2O_3$ nano-filaments: (a) individual, (b) bundles.

## 4. Conclusion

In this work, we have studied the appearance of alumina nanostructures and nanoparticles as a result of hot water treatment of aluminium plate for various water temperatures and treatment times. Three stages of nanostructure formation were revealed. At the beginning, the aluminium surface was covered by isolated nanowires. Later, the surface was completely covered by three-dimensional maze-like nanostructure consisting of nanowires. Finally, the nano-sheets, mostly oriented normally to the plate surface with average thickness about 10 nm, formed over the maze structure. The average period of nanostructure monotonously decreased with temperature from 76 to 48 nm. The thickness of nanostructured layer varied between 350 and 550 nm without marked temperature dependence. Hot water treatment allowed achieving the superhydrophilic surface state with contact angle close to zero. The lifetime of this state increased with temperature and achieved hundreds of hours.

The formation of the water suspension of the spherical metal oxide nanoparticles with average sizes ranging from 20 to 40 nm appeared as a result of HWT has been demonstrated. The obtained results allow applying hot water treatment for production of superhydrophilic surfaces covered by nanostructures and aqueous suspensions of nanoparticles of various metal oxides.

## Acknowledgements

E.A.M. acknowledges the support from the Ural Federal University (grant for young scientist No. 2.2.2.17.D-18/36). The work has been supported in part by the Ministry of Science and Higher Education of the Russian Federation (projects 3.9534.2017/8.9 and 3.4993.2017/6.7) and by Government of the Russian Federation (act 211, agreement 02.A03.21.0006). The equipment of the Ural Centre for Shared Use "Modern Nanotechnology" Ural Federal University was used.

## References

[1] Liu X, Iocozzia J, Wang Y, Cui X, Chen Y, Zhao S, Li Z and Lin Z 2017 *Energy Environ. Sci.* **10** 402-34
[2] Kim H, Beack S, Han S, Shin M, Lee T, Park Y, Kim K S, Yetisen A K, Yun S H, Kwon W and Hahn S K 2018 *Adv. Mater.* **30** 1-33
[3] King N P, Sheffler W, Sawaya M R, Vollmar B S, Sumida J P, Andre I, Gonen T, Yeates T O and Baker D 2012 *Science* **336** 1171-4
[4] Li F T, Ran J, Jaroniec M and Qiao S Z 2015 *Nanoscale* **7** 17590-610
[5] Wei Q, Xiong F, Tan S, Huang L, Lan E H, Dunn B and Mai L 2017 *Adv. Mater.* **29** 1602300
[6] Holade Y, Sahin N, Servat K, Napporn T and Kokoh K 2015 *Catalysts* **5** 310-48
[7] Tang J, Yuan Y, Liu T and Zhou S 2015 *J. Power Sources* **274** 170-6
[8] Kim Y, Johnson R C and Hupp J T 2001 *Nano Lett.* **1** 165-7
[9] Enoch S, Quidant R and Badenes G 2004 *Opt. Express* **12** 3422
[10] Guo S, Wen D, Zhai Y, Dong S and Wang E 2010 *ACS Nano* **4** 3959-68

[11] Lin W 2015 *Chem. Rev.* **115** 10407-9
[12] Gao Y, Lim J, Teoh S H and Xu C 2015 *Chem. Soc. Rev.* **44** 6306-29
[13] Hofmann-Amtenbrink M, Grainger D W and Hofmann H 2015 *Nanomedicine Nanotechnology, Biol. Med.* **11** 1689-94
[14] Wang Y, Sun S, Zhang Z and Shi D 2018 *Adv. Mater.* **30** 1-21
[15] Mirzaei H and Darroudi M *Ceram. Int.* **43** 907-14
[16] Hochbaum A I, Chen R, Delgado R D, Liang W, Garnett E C, Najarian M, Majumdar A and Yang P 2008 *Nature* **451** 163-7
[17] Menke E J, Brown M A, Li Q, Hemminger J C and Penner R M 2006 *Langmuir* **22** 10564-74
[18] Riley D J 2002 *Curr. Opin. Colloid Interface Sci.* **7** 186-92
[19] Swihart M T 2003 *Curr. Opin. Colloid Interface Sci.* **8** 127-33
[20] Li G, Wang W, Yang W and Wang H 2015 *Surf. Sci. Rep.* **70** 380-423
[21] Shur V Y, Gunina E V, Esin A A, Shishkina E V, Kuznetsov D K, Linker E A, Greshnyakov E D and Pryakhina V I 2019 *Appl. Surf. Sci.* **483** 835-9
[22] Nebogatikov M S, Shur V Y, Tyurnina A E, Kozin R V, Sukhanova V Y, Mingaliev E A and Zorikhin D V 2015 *Ferroelectrics* **477** 54-62
[23] Wang L P, Leconte Y, Feng Z, Wei C, Zhao Y, Ma Q, Xu W, Bourrioux S, Azais P, Srinivasan M and Xu Z J 2017 *Adv. Mater.* **29** 1603286
[24] Atienzar P, Primo A, Lavorato C, Molinari R and García H 2013 *Langmuir* **29** 6141-6
[25] Reau A, Guizard B, Mengeot C, Boulanger L and Ténégal F 2009 *Mater. Sci. Forum* **534-536** 85-8
[26] Al-Mayalee K H, Saadi N, Badradeen E, Watanabe F and Karabacak T 2018 *J. Phys. Chem. C* **122** 23312-20
[27] Yu X, Marks T J and Facchetti A 2016 *Nat. Mater.* **15** 383-96
[28] Eranna G 2011 *Metal Oxide Nanostructures as Gas Sensing Devices* (CRC Press)
[29] Alwitt R S 1974 *J. Electrochem. Soc.* **121** 1322
[30] Rider A N and Arnott D R 2000 *Int. J. Adhes. Adhes.* **20** 209-20
[31] Saifaldeen Z S, Khedir K R, Cansizoglu M F, Demirkan T and Karabacak T 2014 *J. Mater. Sci.* **49** 1839-53
[32] Saadi N S, Hassan L B and Karabacak T 2017 *Sci. Rep.* **7** 1-8
[33] Ngo C V and Chun D M 2018 *Appl. Surf. Sci.* **435** 974-82
[34] Guo S, Zhu X, Li M, Shi L, Ong J L T, Jańczewski D and Neoh K G 2016 *ACS Appl. Mater. Interfaces* **8** 30552-63
[35] Stetsyshyn Y, Raczkowska J, Lishchynskyi O, Bernasik A, Kostruba A, Harhay K, Ohar H, Marzec M M and Budkowski A 2017 *ACS Appl. Mater. Interfaces* **9** 12035-45
[36] Li Y, Wang S and Jiang L 2017 *Wettability Effect on Stem Cell Behavior* (Elsevier Inc.)
[37] Li M, Liu X, Liu N, Guo Z, Singh P K and Fu S 2018 *Colloids Surf. A Physicochem. Eng. Asp.* **554** 122-8
[38] Shateri Khalil-Abad M and Yazdanshenas M E 2010 *J. Colloid Interf. Sci.* **351** 293-8
[39] Piriyawong V, Thongpool V, Asanithi P and Limsuwan P 2012 *J. Nanomater.* **2012** 819403
[40] Liu J, Huang X, Sulieman K M, Li Y, He X and Sun F 2006 *Cryst. Growth Des.* **6** 1690-6

# Formation of the maze domain structures in lithium niobate as a result of multiple pulse irradiation by infrared laser

**V Ya Shur, E A Mingaliev, M S Kosobokov and A V Makaev**

School of Natural Sciences and Mathematics, Ural Federal University, 620000 Ekaterinburg, Russia

vladimir.shur@urfu.ru

**Abstract**. The formation of the maze domain structures in the plates of congruent lithium niobate single crystal caused by multiple irradiation of infrared pulsed laser at different plate temperatures was studied. Four stages of domain evolution with increasing pulse number were distinguished. The dependence of the formed self-organizing domain structures on pulse number and sample temperature was revealed. Suppression of the domain formation at the elevated temperatures was attributed to increasing ionic conductivity, which led to decreasing the switching field.

## 1. Introduction

The modern development of telecommunication technologies requires manufacturing submicron and nanoscale periodic domain structures for use in various types of optical components, such as electrically controlled Bragg reflectors, beam control devices, and narrow-band filters [1-3].

Backward second harmonic generation (SHG), corresponding to the propagation of the wave with the fundamental frequency in the direction opposite to its second harmonic, requires the creation of submicron gratings with nanoscale accuracy of period reproduction [4]. Lithium niobate $LiNbO_3$ (LN) crystals being a favourite ferroelectric material for domain engineering [5] is considered as one of the best candidates for nanodomain structuring. Periodically poled LN crystals are used for converting the light frequency with record efficiency based on the quasi-phasematching effect [4,6-9]. The production of tailored domain structures is usually carried out by applying an electric field using photolithographically produced electrode patterns [10]. This method is ineffective for precise domain structuring with period below 2 μm [11]. The formation of stable nanodomain structures in congruent LN (CLN) crystals was demonstrated under highly non-equilibrium switching conditions [12-13]. Recently, the surface domain inversion leading to formation of the nanodomain structures has been proposed. The intense irradiation of the single domain LN by ultraviolet pulse laser induces formation of the shallow surface nanodomain rays with width below 100 nm and depth about 2 μm [14]. In contrast, the infrared pulsed laser irradiation forms the bulk domain structure of nanodomain rays with a depth of up to 200 μm, which is enough for all above discussed applications [15]. It was shown experimentally that the polarization reversal occurred during cooling after termination of the laser heating.

In this paper, we show how multiple laser irradiation with controlled initial temperature affects morphology of the domain structure.

IOP Conf. Series: Materials Science and Engineering **699** (2019) 012052 doi:10.1088/1757-899X/699/1/012052

## 2. Experimental

The studied samples represented optically polished 0.5-mm-thick Z-cut plates of CLN (SIPAT Co., China). The pulsed $CO_2$ laser ($\lambda$ = 10.6 μm, 40 W) with Gaussian distribution of laser energy density ($\omega$ = 0.35 mm) was used for the sample irradiation by pulses with duration ($t_p$) 3 ms and energy 160 mJ. The chosen duration of the pulse period (20 s) was enough for sample cooling to the initial temperature ($T_i$). Each series of pulse irradiation was carried out at fixed $T_i$ (25, 50, 75, and 100°C). The radiation was focused on Z+ polar surface of the sample by ZnSe lens with focal distance of 50.8 mm. The power density was tuned with variation of the spot diameter by change of the distance between lens and irradiated surface.

The formed static domain structures were revealed by shallow selective chemical etching in pure HF for 13 seconds to prevent rearrangement of domain structure [16]. The surface relief corresponding to the domain structure was imaged by optical microscope Olympus BX51 (Olympus, Japan) [17].

## 3. Results and discussion

It was shown that the geometry of the domain structure formed after irradiation of CLN by series of laser pulses essentially changed with pulse number. The study of domains formed at different numbers of pulses at fixed temperature (25°C) allowed us to reveal four stages of domain formation. The first stage represented formation of the domain rays mostly oriented along Y crystallographic axis after the termination of the first pulse (Fig. 1a) [11]. During the second stage, with number of pulses below 20, the domain rays became wider with increasing number of pulses due to sideways domain wall motion (Fig. 1b). At the third stage, the flat domain walls became irregular with subsequent growth of the fingers on the wall for the number of pulses above 30 (Fig. 1c). At the last stage, for number of pulses above 70, the maze-like domain structure formed (Fig. 1d).

The statistical analysis of the domain images formed in the centre of irradiated zone allowed revealing the monotonous growth of the switched area with number of pulses (Fig. 2). The maximum fraction of the switched area, corresponding to the final domain structure, approached to 0.5. This is due to the fact that the total value of the pyroelectric field arising in periodically polarized ferroelectric crystal (e.g., of the Kittel type) is close to zero. For multiply irradiation at $T_i$ = 25°C, the fraction of the switched area stops changing after 80 pulses and the final domain structure appears (Fig. 1d).

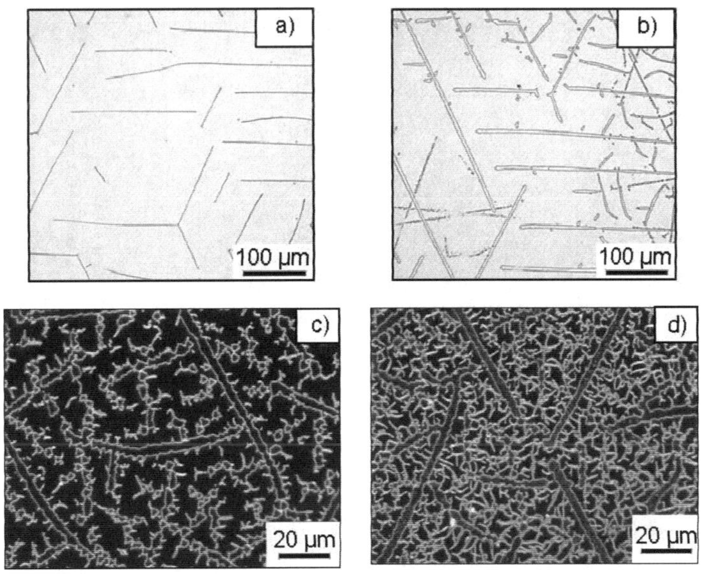


**Figure 1.** Optical visualization at (a, b) bright and (c, d) dark fields after chemical etching of domain structure after various number of pulses: (a) 1, (b) 20, (c) 50, and (d) 80.

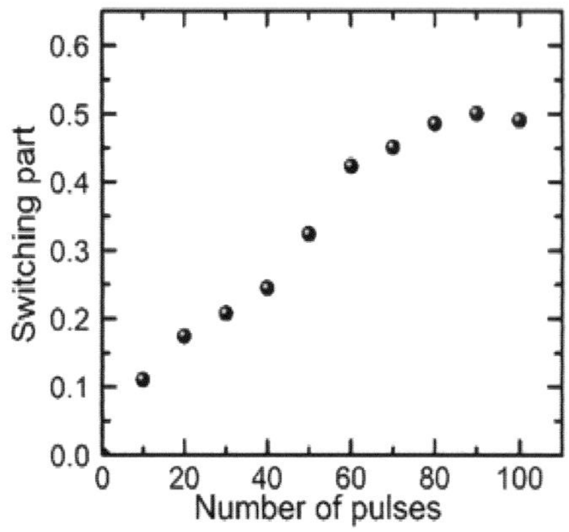


**Figure 2.** Dependence of the fraction of the switched area in the centre of irradiated zone on the number of pulses.

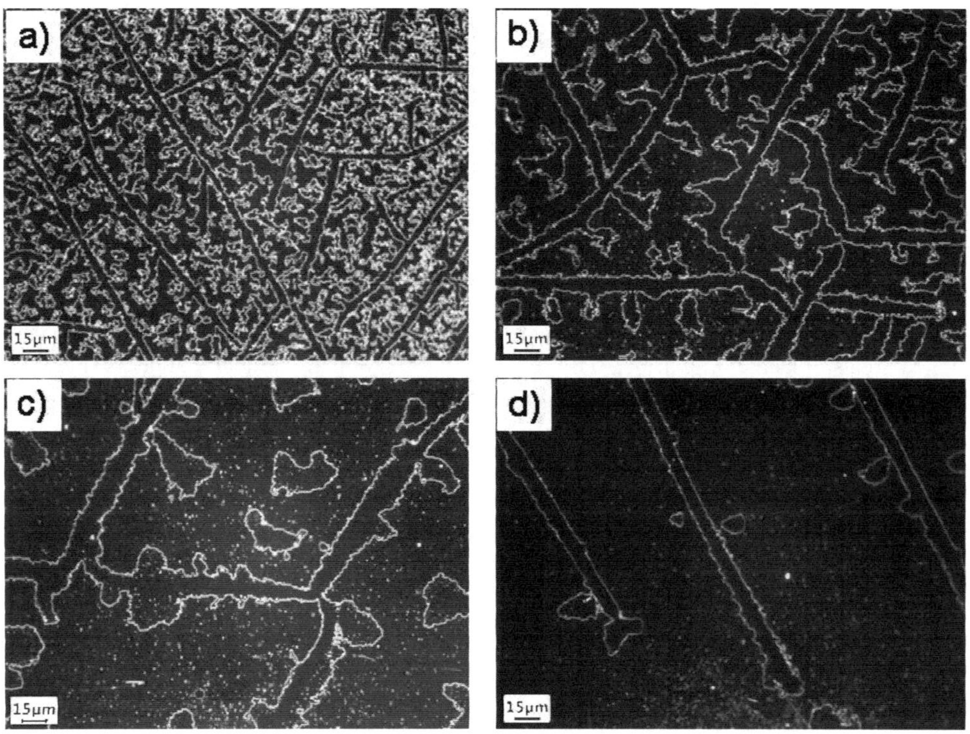


**Figure 3.** Domain structure after 100 pulses with initial temperature, °C: (a) 25, (b) 50, (c) 75, (d) 100.

It was obtained that the initial temperature of the CLN crystal plate strongly affected the final domain structure (Fig. 3). For quantitative characterization of the final domain structure at different $T_i$, two parameters were calculated: the fraction of the switched area and the domain wall density (total wall length per unit of the sample surface area). It was found that the fraction of the switched area linearly decreased with increasing $T_i$ (Fig. 4a), whereas the domain wall density decreased rapidly with increasing temperature to 50°C and slowly decreased from 50 to 100°C.

IOP Conf. Series: Materials Science and Engineering **699** (2019) 012052 doi:10.1088/1757-899X/699/1/012052

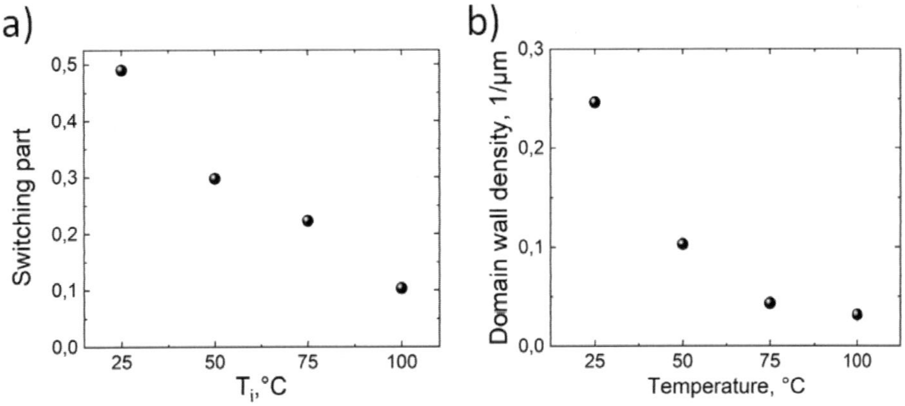


**Figure 4.** Dependence on $T_i$ of: (a) the fraction of the switched area in the centre of the irradiated zone, (b) the average domain wall density.

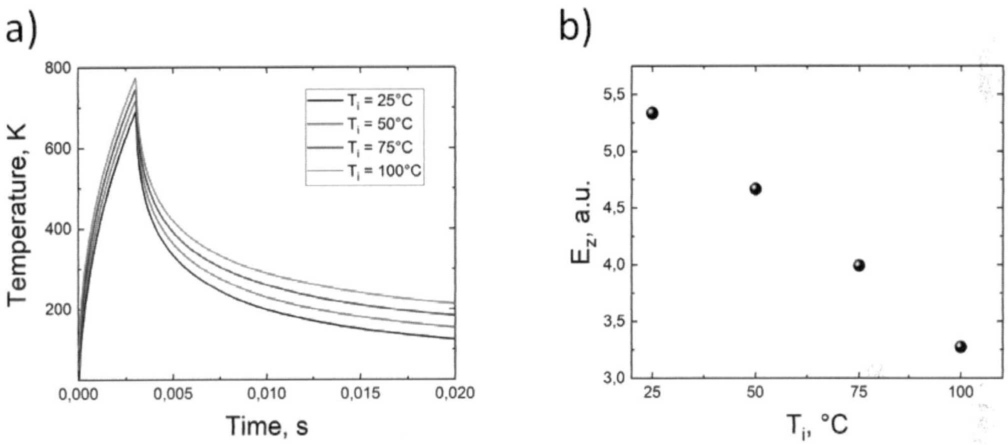


**Figure 5.** (a) The time dependences of the temperature in the centre of the irradiated zone for the first laser pulse for different $T_i$. (b) Dependence on Ti of calculated $E_z$ values at 250°C during cooling.

For explanation of the obtained results, the time dependences of spatial distributions of temperatures and local pyroelectric fields in the bulk of CLN plate were calculated by Finite Element Method (FEM) using COMSOL Multiphysics software. The detail description of the used physical model was published in Ref. [19].

The time dependence of the temperature distribution during pulse laser heating for different $T_i$ was calculated. The published temperature dependences of the heat capacity $c_p(T)$ and the thermal conductivity k(T) for CLN were used [18]. The simulated time dependences of the temperature in the centre of irradiated zone for different $T_i$ for the first laser pulse are presented in Figure 5a.

The time dependence of the pyroelectric field appeared as a result of temperature change was calculated. It was shown earlier experimentally that polarization switching started at 5-10 ms after heating termination [19], corresponding to 170-250°C. The calculated values of Z-component of the local field ($E_z$) at 250°C during cooling decreased with $T_i$ (Fig. 5b). Decreasing of the switching field is due to increasing ionic conductivity of the crystal and leads to suppression of the domain formation.

## 4. Conclusion

The domain structure formation induced by multiple pulsed infrared laser irradiation in the plates of congruent lithium niobate was investigated. The dependence of the domain structure parameters on number of pulses and initial temperature was measured. It was shown that number of laser pulses significantly affected the domain structure. Four stages of domain structure evolution were revealed: domain ray formation after the first pulse, domain widening by sideways domain wall motion, fingering, and formation of maze-like structure.

It was shown that increase in the initial plate temperature decreased the fraction of the switched area and the domain wall density. This fact can be attributed to decreasing the pyroelectric field due to growth of the ionic conductivity with initial temperature. It should be noted that the stable micro- and nanodomain structures obtained under the action of pyroelectric field attract interest for domain engineering.

**Acknowledgements**
The equipment of Ural Centre for Shared Use "Modern Nanotechnology" Ural Federal University was used. The research was made possible by Russian Science Foundation (grant No. 19-12-00210).

**References**

[1] Li F, Zhang S, Damjanovic D, Chen L Q and Shrout T R *Adv. Funct. Mater.* **28** 1-21
[2] Mhaouech I, Coda V, Montemezzani G, Chauvet M and Guilbert L 2017 *Opt. InfoBase Conf. Pap.* **Part F82-C** 4174-7
[3] Shandarov S M, Savchenkov E N, Borodin M V, Mandel A E, Akhmatkhanov A R and Shur V Y 2019 *Ferroelectrics* **542** 58-63
[4] Stivala S, Busacca A C, Curcio L, Oliveri R L, Riva-Sanseverino S and Assanto G 2010 *Appl. Phys. Lett.* **96** 2005-8
[5] Shur V Y, Gruverman A L and Rumyantsev E L 1990 *Ferroelectrics* **111** 123-131
[6] Neradovskiy M, Neradovskaia E, Richter M, Kuhl U, Aschiéri P, Tronche H, Doutre F, Baldi P and De Micheli M P 2018 *Opt. InfoBase Conf. Pap.* **Part F121** 2-3
[7] Buse K and Breunig I 2019 *Laser Resonators, Microresonators, and Beam Control* **10904** 1090403
[8] Shur V Y, Rumyantsev E L, Batchko R G, Miller G D, Fejer M M, and Byer R L 1999 *Phys. Solid State* **41** 1681-7
[9] Shur V Y, Rumyantsev E L, Nikolaeva E V, Shishkin E I, Batchko R G, Fejer M M and Byer R L 2001 *Ferroelectrics* **257** 191-202
[10] Yamada M, Nada N, Saitoh M and Watanabe K 1993 *Appl. Phys. Lett.* **62** 435-6
[11] Shur V, Rumyantsev E, Batchko R, Miller G, Fejer M and Byer R 1999 *Ferroelectrics* **221** 157-67
[12] Lobov A I, Shur V Y, Kuznetsov D K, Negashev S A, Pelegov D V, Shishkin E I and Zelenovskiy P S 2008 *Ferroelectrics* **373** 99-108
[13] Shur V Y, Kuznetsov D K, Lobov A I, Nikolaeva E V, Dolbilov M A, Orlov A N and Osipov V V 2006 *Ferroelectrics* **341** 85-93
[14] Wengler M C, Heinemeyer U, Soergel E and Buse K 2005 *J. Appl. Phys.* **98** 064104
[15] Kuznetsov D K, Shur V Y, Mingaliev E A, Negashev S A, Lobov A I, Rumyantsev E L and Novikov P A 2010 *Ferroelectrics* **398** 49-54
[16] Shur V Y, Lobov A I, Shur A G, Kurimura S, Nomura Y, Terabe K, Liu X Y and Kitamura K 2005 *Appl. Phys. Lett.* **87** 022905
[17] Shur V Y and Zelenovskiy P S 2014 *J. Appl. Phys.* **116** 066802
[18] Morgan R A, Kang K I, Hsu C C, Koliopoulos C L and Peyghambarian N 1987 *Appl. Opt.* **26** 5266
[19] Shur V Y, Kosobokov M S, Mingaliev E A and Karpov V R 2015 *AIP Adv.* **5** 107110

IOP Conf. Series: Materials Science and Engineering **699** (2019) 012053 doi:10.1088/1757-899X/699/1/012053

# Forming-free resistive switching in nanocrystalline hafnium oxide films

**V A Smirnov, R V Tominov, V I Avilov, A A Avakyan and O A Ageev**

Institute of Nanotechnologies, Electronics and Equipment Engineering, Southern Federal University, 347928 Taganrog, Russia

roman.tominov@gmail.com

**Abstract.** This work presents the results of the investigations of resistive switching effect in Si(100)/$HfO_2$ structure. It was shown that resistive switching from HRS to LRS occurred at 0.4±0.1 V, and from LRS to HRS at -0.5±0.1 V. An increase in the sweep voltage from 1 to 5 V led to a decrease in the HRS/LRS ratio from 606±36 to 204±11. Thus, it was shown that the nanocrystalline $HfO_2$ film resistance varied within two orders of magnitude at a sweep voltage of 1 V within 15 measurements. The results can be useful for manufacturing neuromorphic systems based on forming-free nanocrystalline $HfO_2$ films.

## 1. Introduction

Brain-based computing is a new area that can expand the rich of information technology beyond the von Neumann paradigm [1,2]. Every year, more and more popular solutions to the problems associated with processing heterogeneous «noisy» data appear: image recognition in real time, diagnostics of various processes, as well as in applications related to self-learning and adaptive control systems [3,4]. One of the ways to achieve this goal is the creation of integrated circuits (IC) based on memristors, which are key elements for simulating high-density biological neurons, interconnected using special channels – synapses. It should be noted that in the biological brain the number of synapses exceeds the number of neurons by 3-4 orders of magnitude, which implies that for the technical implementation of ICs based on memristor structures, memory elements must have small geometric dimensions to achieve a high degree of integration. An analysis of articles has shown that one of the most suitable elements for the technical implementation of memristors for neuromorphic structures is non-volatile resistive memory (ReRAM), which has a small cell size of several nanometres, a high integration density, a high speed and low power consumption, which allows simulating mass concurrency and low-power computing previously observed in the human brain [5-14]. The human brain has a high degree of connectivity, and any neuron can have up to 10.000 connections with other neurons. An array of ReRAM elements provides the ability to emulate such connectivity. The principle of operation of ReRAM is based on the effect of resistive switching – changing the resistance of a thin oxide film in the range of values between the resistance in the low-resistance state (LRS) and that in the high-resistance state (HRS), when an external electric field is applied. An analysis of the publications has shown that structures based on binary metal oxides are promising for creating neuromorphic systems; nanocrystalline hafnium oxide ($HfO_2$) can be particularly distinguished [5-18]. Nanocrystalline $HfO_2$ shows forming free resistive switching with high speed, low energy consumption, high HRS/LRS ratio, and also a sufficiently large number of switching cycles. However, for the manufacture of neuromorphic structures based on hafnium oxide, there are currently no systematic modes of studying the effect of sweep voltage on the HRS/LRS ratio. There are many methods for manufacturing thin films of hafnium oxide; however, it was decided

IOP Conf. Series: Materials Science and Engineering **699** (2019) 012053 doi:10.1088/1757-899X/699/1/012053

to use the pulsed laser deposition (PLD) method, which allows the formation of films in a wide range of electrophysical and morphological parameters with an acceptable growth rate. In the process of studying the effect of resistive switching, a need arises for a local study of the morphology of thin oxide films. Atomic force microscopy (AFM) is one of the promising methods that can effectively solve this problem [19-27]. Thus, the aim of this work is to study the effect of resistive switching in thin films of hafnium oxide, in particular, to study the effect of sweep voltage on the HRS/LRS ratio.

## 2. Experimental

To carry out experimental studies of the resistive switching in nanocrystalline $HfO_2$ film, the sample was prepared using pulsed laser deposition equipment Pioneer 180 (Neocera Inc., USA). Silicon substrate Si with a crystallographic orientation (100) and 10 Ω·cm resistivity was used as a wafer. Deposition was performed under the following conditions: wafer temperature 30°C, target-wafer distance 50 mm, $O_2$ pressure 1 mTorr, pulse energy 300 mJ, number of pulses 30000, and frequency 10 Hz. To provide electrical contact to the Si substrate, the $HfO_2$ film was deposited through a special mask pattern. As a result, $HfO_2$ did not precipitate on a template-protected area of the Si surface.

AFM-images of $HfO_2$ film surface were obtained and processed using atomic-force microscope Solver P47 Pro (NT-MDT, Russia) and Image Analysis 1.1, respectively. Electrical properties of obtained $HfO_2$ film were measured by Ecopia HMS-3000 equipment (Ecopia Co., Republic of Korea). Electric measurements were carried out using semiconductor characterization system Keithley 4200-SCS (Keithley, USA) with W probes. During experiment, TiN layer was grounded. To avoid electrical breakdown of the $HfO_2$ film, a current limit of 1 mA during electrical measurements was set. Current-voltage characteristics were measured at the same point (endurance test) and in different points (homogeneity test) at different voltage sweeps from 1 to 5 V. Based on the results obtained, the dependence of HRS/LRS ratio on sweep voltage was built. The curves were analysed using Origin 8.1 software.

## 3. Results and discussion

The results of experimental studies showed that nanocrystalline $HfO_2$ film surface had a granular structure with 0.25±0.08 μm grain size (Fig. 1a,c) and 12±3 nm roughness (Fig. 1b). Studies of electrophysical parameters showed that nanocrystalline $HfO_2$ film had $6.5\times10^{19}$ $cm^{-3}$ electron volume concentration, 12 $cm^2$/V·s electron mobility, and $4.2\times10^{-3}$ Ω·cm resistivity.

Figure 2a shows average current-voltage characteristics (CVC) of Si(100)/$HfO_2$ structure at –1 to +1 voltage sweep, performed from the 15 current-voltage curves at the same point. It was shown that Si(100)/$HfO_2$ structure exhibited a nonlinear bipolar behaviour, when the electric potential gradient was the dominant parameter of resistive switching effect [24]. Resistive switching from HRS to LRS occurred at 0.4±0.1 V (SET), and from LRS to HRS at -0.5±0.1 V (RESET) (Fig. 2a). Based on the

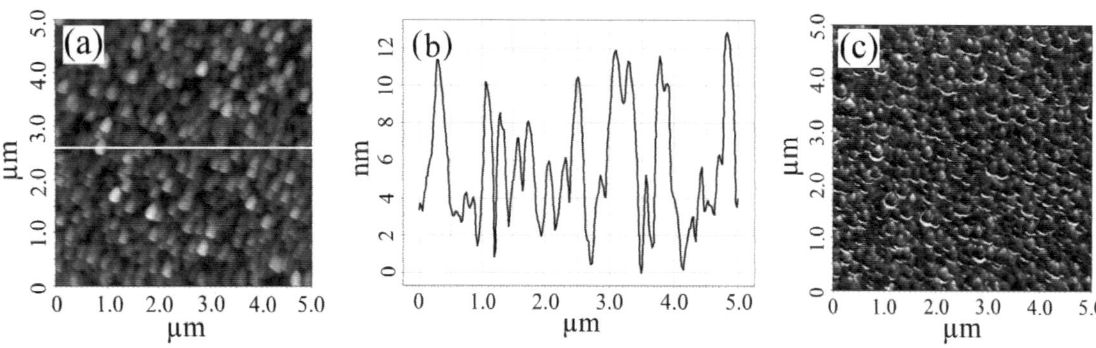


**Figure 1.** Investigation of the formed $HfO_2$ film: (a) AFM-image; (b) cross-section profile along white line on (a); (c) phase.

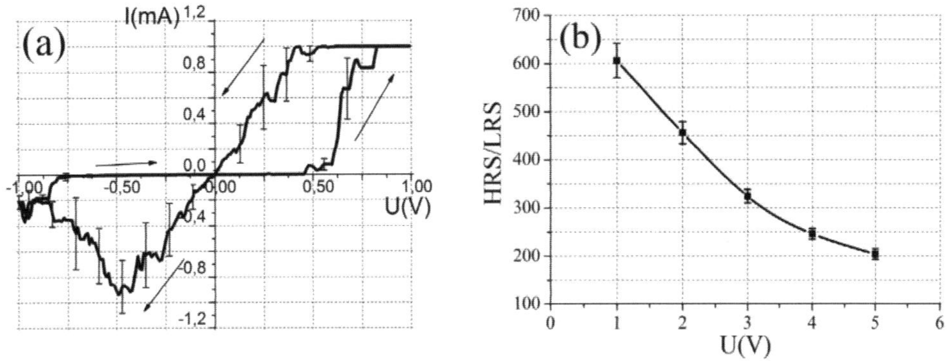


**Figure 2.** Investigation of resistive switching effect in forming-free nanocrystalline $HfO_2$ film: (a) current-voltage characteristic; (b) dependence of the HRS/LRS ratio on the sweep voltage.

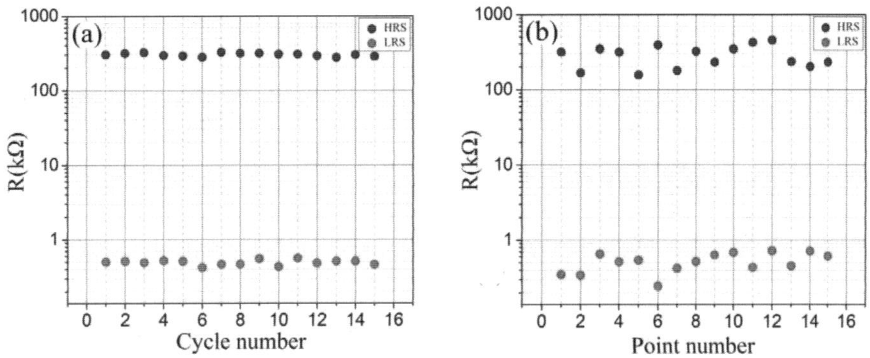


**Figure 3.** Investigation of effect of resistive switching in forming-free nanocrystalline hafnium oxide film: (a) endurance test; (b) homogeneity test.

obtained CVC for different voltage sweeps, the dependence of the HRS/LRS ratio on the sweep voltage was built. Analysis of the results showed, that an increase in the sweep voltage from 1 to 5V led to a decrease in the HRS/LRS ratio from 606±36 to 204±11 (Fig. 2b). The obtained result can be explained by a decrease in the dissolution length of the oxygen vacancy nanofilaments with an increase in the sweep voltage, and, as a consequence, an increase in the current through the $HfO_2$ film.

Investigations of effect of resistive switching uniformity in Si(100)/$HfO_2$ structure in a single point (endurance test) and in different points on the film surface (homogeneity test) were performed at voltage sweep, at which the maximum HRS/LRS ratio was observed, i.e. at 1V. Reading voltage was 0.3 V. It was shown, that for endurance test HRS was 298±25 kΩ, LRS was 0.51±0.05 kΩ (Fig. 3a), and for homogeneity test HRS was 304±147 kΩ, LRS was 0.5±0.2 kΩ (Fig. 3b). It was shown, that resistance confidence interval of endurance and homogeneity tests varied 4.3 times for HRS and 3.5 times for LRS. It can be explained by the heterogeneity of the $HfO_2$ film surface morphology and also by the different concentration of oxygen vacancies in volume at different points on the $HfO_2$ film surface. Thus, it was shown that the nanocrystalline $HfO_2$ film resistance varied within two orders of magnitude at a sweep voltage of 1 V within 15 measurements.

## 4. Conclusion

In summary, we investigated effect of resistive switching in a nanocrystalline $HfO_2$ film formed by pulse laser deposition. The results of experimental studies, obtained using AFM, showed that nanocrystalline $HfO_2$ film had a granular structure with 0.25±0.08 μm grain size and nm with roughness 12±3 nm.

Electrophysical parameters of nanocrystalline $HfO_2$ film were investigated. It was shown, that volume concentration was $6.5\times10^{19}$ cm$^{-3}$, mobility was 12 cm$^2$/V·s, and resistivity was $4.2\times10^{-3}$ Ω·cm. Resistive switching from HRS to LRS occurred at 0.4±0.1 V, and from LRS to HRS at -0.51±0.11 V. The experimental dependence of HRS/LRS ratio on sweep voltage was built. It was shown, that an increase in the sweep voltage from 1 to 5V led to a decrease in the HRS/LRS ratio from 606±36 to 204±11. The results of experimental studies of resistances for sweep voltage 1V showed that for endurance test HRS and LRS were 298±25 kΩ and 0.51±0.05 kΩ, respectively, and for homogeneity test 304±147 kΩ and 0.5±0.2 kΩ, respectively. Thus, it was shown that the nanocrystalline $HfO_2$ film resistance varied within two orders of magnitude at a sweep voltage of 1 V within 15 measurements. The results can be useful for manufacturing neuromorphic systems based on forming-free nanocrystalline $HfO_2$ films.

## Acknowledgements

This work was supported by RFBR according to the research projects № 19-29-03041 mk, № 18-37-00299 mol_a, and by Grant of the President of the Russian Federation No. MK-2721.2018.8. The results were obtained using the equipment of the Research and Education Centre and Centre for Collective Use «Nanotechnologies» of Southern Federal University.

## References

[1] Covi E, Brivio S, Serb A, Prodromakis T, Fanciulli M and Spiga S 2016 *IEEE Int. Symp. Circuits and Systems (ISCAS)* **1** 393-6
[2] Covi E, Brivio S, Fanciulli M and Spiga S 2015 *Microelectron. Eng.* **147** 41-4
[3] Jo S H, Chang T, Ebong I, Bhadviya B B, Mazumder P and Lu W 2010 *Nano Lett.* **10** 4
[4] Dongale T D, Desai N D, Khot K V, Volos C K, Bhosale P N and Kamat R K 2018 *Nanoelectron. Optoe.* **13** 1
[5] Murdoch B J, McCulloch D G, Ganesan R, McKenzie D R, Bilek M M M and Partridge J G 2016 *Appl. Phys. Lett.* **108** 143504
[6] Amer S, Sayyaparaju S, Rose G S, Beckmann K and Cady N C 2017 *IEEE Int. Symp. Circuits and Systems (ISCAS)* **1** 1-4
[7] Khakhulin D A, Vakulov Z E, Smirnov V A, Tominov R V, Yoon J G and Ageev O A 2017 *J. Phys.: Conf. Ser.* **917** 092008
[8] Waser R and Dittmann R 2009 *Adv. Mater.* **21** 2632
[9] Yang J J and Pickett M D 2008 *Nature nanotechnology* **7** 492
[10] Avilov V I, Ageev O A, Jityaev I L, Kolomiytsev A S and Smirnov V A 2016 *Proc. SPIE* **10224** 102240T
[11] Tominov R V, Zamburg E G, Khakhulin D A, Klimin V S, Smirnov V A, Chu Y H and Ageev O A 2017 *J. Phys.: Conf. Ser.* **917** 032023
[12] Avilov V I, Polupanov N V, Tominov R V, Smirnov V A and Ageev O A 2017 *IOP Conf. Ser.: Mater. Sci. Eng.* **256** 012001
[13] Prezioso M, Merrikh-Bayat F, Hoskins B D, Adam G C, Likharev K K and Strukov D B 2015 *Nature* **521** 61
[14] Tominov R V, Bespoludin V V, Klimin V S, Smirnov V A and Ageev O A 2017 *IOP Conf. Ser.: Mater. Sci. Eng.* **256** 012023
[15] Azghadi M R, Linares-Barranco B, Abbott D and Leong P H 2016 *IEEE Trans. Biomed. Circuits and Systems* **11** 434
[16] Smirnov V A 2015 *Piezoelectrics and Nanomaterials: Fundamentals, Developments and Applications* 85-103
[17] Ageev A O, Konoplev B G, Polyakov V V , Svetlichnyi A M and Smirnov V A 2007 *Russ. Microelectronics* **36** 353
[18] Gale E 2014 *Semicond. Sci. Tech.* **29** 104004
[19] Zhang L, Chen Z, Joshua Yang J, Wysocki B, McDonald N and Chen Y 2013 *Appl. Phys. Lett.* **102** 153503
[20] Yan X, Zhao J, Liu S, Zhou Z, Liu Q, Chen J and Liu X Y 2018 *Adv. Funct. Mater.* **28** 1705320

[21] Avilov V I, Ageev O A, Konoplev B G, Smirnov V A, Solodovnik M S and Tsukanova O G 2016 *Semiconductors* **50** 601
[22] Shinde S S and Dongle T D 2015 *J. Semiconductors* **36** 034001
[23] Huang C H, Huang J S, Lin, S. M., Chang, W Y, He J H and Chueh Y L 2012 *Acs Nano* **6** 8407
[24] Ageev O A, Konoplev B G, Rubashkina M V, Rukomoikin A V, Smirnov V A and Solodovnik M S 2013 *Nanotechnologies in Russia* **8** 23
[25] Park J, Lee S, Lee J and Yong K 2013 *Adv. Mater.* **25** 6423
[26] Huang C H, Huang J S, Lai C C, Huang H W, Lin S J and Chueh Y L 2013 *ACS Appl. Mater. Interfaces* **5** 6017
[27] Yang J J, Strukov D B and Stewart D R 2013 *Nature Nanotechnology* **8** 13

IOP Conf. Series: Materials Science and Engineering **699** (2019) 012054 doi:10.1088/1757-899X/699/1/012054

# Organism's responses to a long-term inhalation of silica-containing submicron particles of an industrial aerosol

**S N Solovyeva[1], B A Katsnelson[1], M P Sutunkova[1], L I Privalova[1], V B Gurvich[1], I A Minigalieva[1], T V Slyshkina[1], I E Valamina[2], V Ya Shur[3], I V Zubarev[3] and D K Kuznetsov[3]**

[1]Ekaterinburg Medical Research Centre for Prophylaxis and Health Protection in Industrial Workers, 620014 Ekaterinburg, Russia
[2]Central Research Laboratory of the Ural Medical University, 620028 Ekaterinburg, Russia
[3]School of Natural Sciences and Mathematics, Ural Federal University, 620000 Ekaterinburg, Russia

solovyevasn@ymrc.ru

**Abstract**. Female white rats were exposed during up to 6 months 5 times a week, 4 hr per day in a "nose only" inhalation device to an aerosol containing predominantly submicron (nanoscale included) particles of amorphous silica in the concentration 2.6±0.6 or 10.6±2.1 mg/m$^3$. In an auxiliary experiment with a single-shot intratracheal instillation of these particles, it was shown that they induced a pulmonary cell response comparable with that to highly cytotoxic and fibrogenic standard quartz powder $DQ_{12}$. However, in the long-term inhalation test, the studied aerosol proved to be of very low systemic toxicity and fibrogenicity. This paradox may be explained by low $SiO_2$ retention in lungs and other organs due to a relatively high *in vivo* solubility of these nanoparticles. Nevertheless, their genotoxic action and transnasal penetration into the brain should make one give a cautious overall assessment of this aerosol as an occupational or environmental hazard.

## 1. Introduction

While engineered $SiO_2$ nanoparticle toxicity is being widely investigated, mostly on cell lines or in acute animal experiments [1-9], and a lot of others, the practical importance of industrial condensation aerosols with a high $SiO_2$ particle content and the theoretical interest in it seems to be neglected. Thus, long-term inhalation exposure to nano-$SiO_2$ has not been, as far as we know, undertaken in experimental nanotoxicology research.

## 2. Experimental

Outbred female white rats were exposed for 3 or 6 months, 5 times a week, 4 hr a day to an aerosol containing 72% amorphous silica predominantly submicron particles at an exposure concentration of 2.6±0.6 or 10.6±2.1 mg/m$^3$. This material (Fig. 1a) had been collected from the flue-gas ducts of electric ore smelting furnaces producing elemental silicon, subsequently sieved through a <2 µm screen and re-dispersed in airflow (Fig. 1b) to feed a computerized "nose only" inhalation system (Fig. 2).

(a)

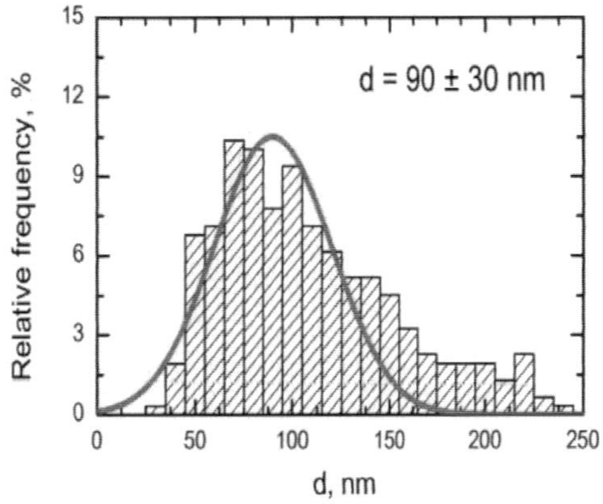


(b)

**Figure 1.** Particles of the powder collected in the flue gas duct from the hood over a silicon smelting ore-thermal furnace and sieved through a < 2 μm screen. (a) SEM image, (b) particle size distribution.

Preliminarily, we carried out an experiment to study the response of the lower airways free cells obtained from rats exposed, 24 h before bronchoalveolar lavage, to a single-shot intratracheal instillation of the same particles suspended in normal saline in comparison with the response to the highly cytotoxic and fibrogenic standard quartz dust $DQ_{12}$. A batch of each powder was incubated during 24 h at 37°C together with either the normal saline, or the Ringer-Locke's solution, or the cell-free bronchoalveolar lavage fluid (BALF) supernatant. Samples of each system were taken at several time points and analysed for the Si content with the help of atomic absorption spectrometry (AAS).

## 3. Results and discussion

Judging by a 3-4-fold increase in the total cell count of the BALF after intratracheal instillation of particles due to enhanced recruitment of alveolar macrophages (AM) and, especially, of neutrophil leukocytes (NL) to the lower airways free surface (Table 1), the cytotoxicity of Nano-Silica Containing Aerosol (NSCA) investigated and that of quartz $DQ_{12}$ are quite similar.

(a)

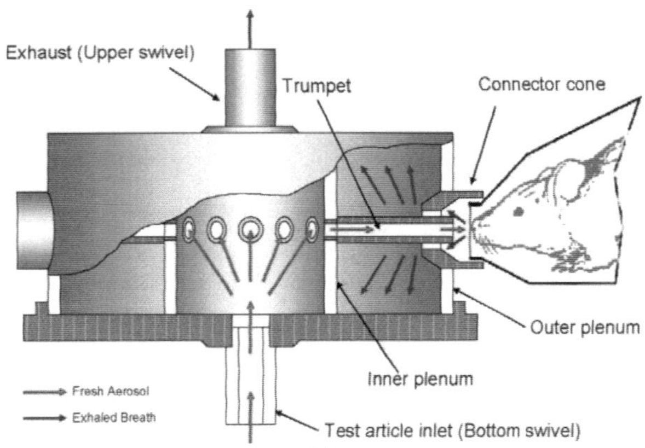


(b)

**Figure 2.** (a) General view of the experimental setup (photographed with the front chamber doors removed). Next to the exposure unit, a similar system setup for sham exposure of control rats is seen. (b) Diagram of aerosol flows in the "nose only" inhalation exposure system (courtesy CH Technology, USA).

**Table 1.** Number of cells in the bronchoalveolar lavage fluid (BALF) 24 hours after intratracheal instillation of NSCA pr $DQ_{12}$ particle suspensions administered to rats at a dose of 7 mg in 1mL normal saline (x ± s.e.).
*NB*: *statistically significant difference from the control group (P <0.05 by Student's t-test).

| What was instilled: | Number of cells * $10^6$ | | | NL/AM |
|---|---|---|---|---|
| | Total | Alveolar macrophages (AM) | Neutrophil leukocytes (NL) | |
| NSCA | 5.85±0.56* | 3.03±0.66 | 2.86±0.79* | 2.24±1.16* |
| $DQ_{12}$ | 7.44±1.30* | 3.79±0.94* | 3.10±0.80* | 1.69±0.64* |
| Normal saline (control group) | 1.96±0.29 | 1.49±0.28 | 0.52±0.24 | 0.37±0.16 |

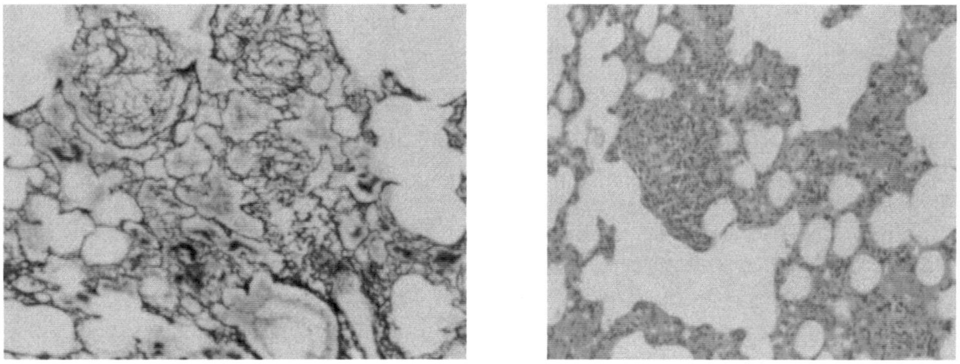

(a) (b)

**Figure 3.** Rat lungs after inhalation exposure to the NSCA at the concentration of 10.6 mg/m$^3$ during 6 months. (a) Focal coarsening of the argyrophylic framework of the lung tissue and appearance of a net of thin argyrophylic fibres within a macrophagal-lymphocytic nodule. (b) Alveolar septa are thickened, with lymphocytic infiltration, but round macrophage-lymphocytic clusters without detectable collagen fibres within or around them seen in this preparation occur as solitary observations only. Hematoxylin and eosin staining, magnification X50.

**Table 2.** Some indices for rat lungs status after inhalation exposure to particles of NSCA (X ± s.e.).
*NB*: *statistically significantly difference from corresponding control value (p<0.05 by Student's t-test); •statistically significantly difference from the low-exposure group of the same exposure period (p<0.05 by Student's t-test).

| Indices | 3 months | | | 6 months | | |
|---|---|---|---|---|---|---|
| | **Controls** | **NSCA 2.6 mg/m$^3$** | **NSCA 10.6 mg/m$^3$** | **Controls** | **NSCA 2.6 mg/m$^3$** | **NSCA 10.6 mg/m$^3$** |
| Dry lung mass, g/100g body mass | 0.08 ±0.01 | 0.08 ±0.01 | 0.09 ±0.01• | 0.17 ±0.02 | 0.17 ±0.02 | 0.16 ±0.01 |
| Total lipid content of lungs, mg | 33.82 ±2.74 | 31.84 ±2.98 | 37.66 ±4.57 | 68.02 ±5.63 | 36.98 ±4.81* | 60.64 ±4.56 |
| Hydroxyproline content of lungs, μg | 2060 ±85 | 4978 ±1187* | 4641 ±1022* | 4112 ±578 | 3048 ±533 | 6142 ±946• |
| Hydroxyproline content of lungs, μg/100 g body mass | 776 ±42 | 1898 ±460* | 1711.9 ±381* | 1653 ±209 | 1146 ±209 | 2227 ±342• |

**Table 3.** $SiO_2$ content of the rat organs, blood and excreta (x±s.e.).
*NB*: [1]Determined in each group of rats' pooled tissue, divided by the number of rats in the group;
*Statistically significantly different from the control value for the same exposure period;
+Statistically significantly different from the index obtained for the same exposure period at the lower concentration;
@Statistically significantly different from the value obtained for the previous exposure period at the same concentration ($p < 0.05$ by Student's t-test).

| Exposure period, months | Group | Organs, mg | | | | | | Blood, mg/L | Urine, mg/L | Feces, mg/g |
|---|---|---|---|---|---|---|---|---|---|---|
| | | lungs | lungs-associated lymph nodes | Liver | kidneys | spleen | brain | | | |
| **Exposure concentration of NSCA 2.6 mg/m³** | | | | | | | | | | |
| 3 | Exposed | 0.11± 0.02* | 0.01 | 0.02± 0.00* | 0.03± 0.00* | 0.02± 0.00 | 0.02± 0.00 | 0.39± 0.16* | 1.47± 0.21* | 2.18± 0.12* |
| | Control | 0.02± 0.01 | 0.00 | 0.01± 0.01 | 0.01± 0.01 | 0.02± 0.01 | 0.01± 0.00 | 0.24± 0.06 | 0.44± 0.07 | 1.36± 0.10 |
| 6 | Exposed | 0.20± 0.01*+ | 0.06 | 0.04± 0.01*+ | 0.04± 0.01*+ | 0.05± 0.01*+ | 0.02± 0.00 | 0.46± 0.04*+ | 1.68± 0.07*+ | 2.32± 0.28* |
| | Control | 0.04± 0.006 | 0.01 | 0.01± 0.002 | 0.01± 0.001 | 0.02± 0.003 | 0.01± 0.00 | 0.27± 0.04 | 0.44± 0.07 | 1.38± 0.08 |
| **Exposure concentration of NSCA 10,6 mg/m³** | | | | | | | | | | |
| 3 | Exposed | 0.63± 0.07*@ | 0.02 | 0.03± 0.00*@ | 0.04± 0.01*@ | 0.06± 0.01*@ | 0.02± 0.00 | 0.60± 0.14*@ | 4.57± 0.23*@ | 7.24± 0.30*@ |
| | Control | 0.02± 0.01 | 0.00 | 0.01± 0.01 | 0.01± 0.01 | 0.02± 0.01 | 0.01± 0.00 | 0.24± 0.06 | 0.44± 0.07 | 1.36± 0.10 |
| 6 | Exposed | 0.74± 0.05*+@ | 0.06 | 0.06± 0.01*+@ | 0.07± 0.01*+@ | 0.07± 0.00*+@ | 0.02± 0.00 | 0.71± 0.05*+@ | 3.06± 0.11*+@ | 4.39± 0.53*+@ |
| | Control | 0.04± 0.01 | 0.01 | 0.01± 0.00 | 0.01± 0.00 | 0.02± 0.00 | 0.01± 0.00 | 0.27± 0.04 | 0.44± 0.07 | 1.38± 0.08 |

However, under long-term inhalation exposures, the NSCA proved to be of a very low systemic toxicity as assessed with a lot of functional and biochemical indices, and of a negligible pulmonary fibrogenicity as assessed both morphologically (Fig. 3) and biochemically (Table 2).

As can be seen from Table 3, the AAS- determined silica content of the lungs and lungs-associated lymph nodes is substantially higher than in the controls, this increase being dependent on both the level and the duration of the inhalation exposure, which dependences confirm its causal relationship with the latter.

However, according to Ref. [10], approximately the same extent of retention of $SiO_2$ in rat lungs (0.550 mg) was observed over 6 months under a similar inhalation exposure to standard quartz dust $DQ_{12}$ at a concentration of 1 mg/m³, which is 10 times lower than in our experiment with the NSCA. In other words, if calculated per unit concentration of aerosol in the air, the accumulated mass of particles in the lungs under chronic inhalation exposure to NSCA is approximately 10 times less than that under similar inhalation exposure to quartz dust $DQ_{12}$.

The most consistent explanation of this phenomenon (which, in its turn, is a sufficient explanation of weak silica-induced pathology) is the dissolution of submicron (nanoscale, in particular) silica particles *in vivo*, especially in the lining fluid of the pulmonary area, on which they deposit when inhaled. This mechanism was modelled by us *in vitro* by a noticeable solubilisation of particles in the BALF of intact rats.

Even though the scanning transmission electron microscopy (STEM) confirmed that NSCA particles are retained in the lungs (Fig. 4), this retention was evidently not sufficient to cause any essential pathological changes in the pulmonary tissue.

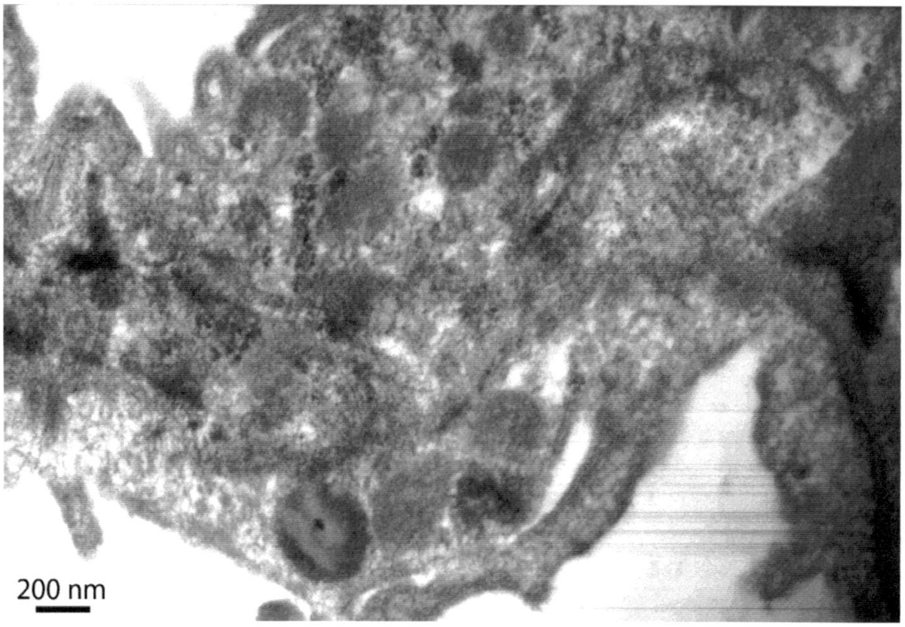


**Figure 4.** Nanoparticles within a type I alveolocytes of a rat exposed to the NSCA at the concentration of 10.6 mg/m$^3$ during 3 months. STEM image.

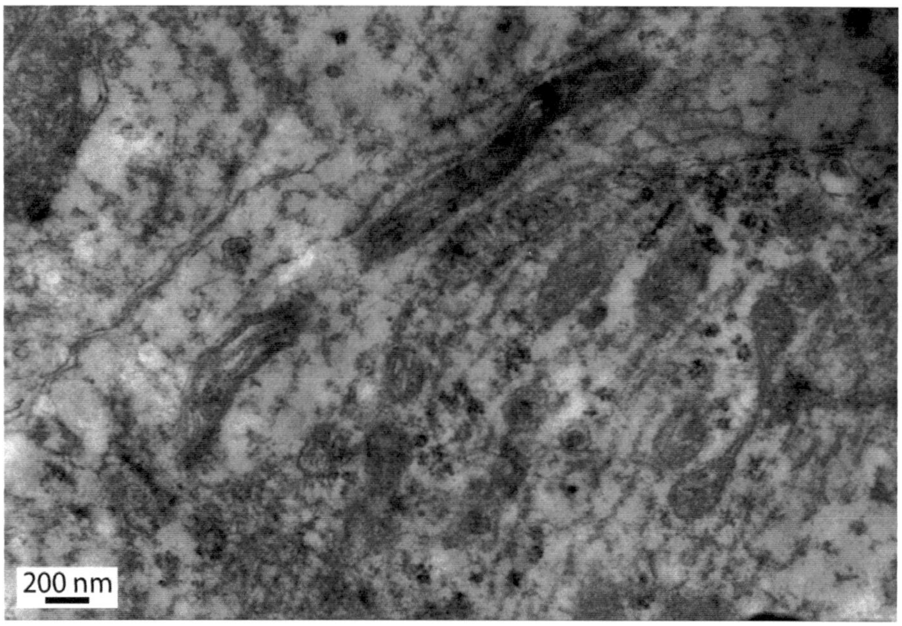


**Figure 5.** Nanoparticles in the neuron body under inhalation exposure to NSCA at the concentration of 2.6 mg/m$^3$ during 3 months. STEM image.

The ability of inhaled nanoparticles initially deposited in the nasal passages to penetrate along the olfactory nerve fibres into the olfactory area of the brain known from the literature [11-13] has been confirmed in our previous experiment with nano-$Fe_2O_3$ as well [14]. As follows from Figure 5, even in the rats that were exposed to the lowest concentration of the NSCA we revealed a considerable number of nanoscale electron-dense round formations in some neurons of this area. These formations are most likely to be nanoparticles of the aerosol under study.

**Table 4.** Genomic DNA fragmentation coefficient for blood and bone marrow cells of rats exposed to the NSCA per inhalation during 6 months (RAPD-test) (x ± s.e.)
*NB*: *Statistically significant difference from corresponding control value;
•Statistically significant difference from the low-exposure group ($p< 0.05$ by Student's t-test).

| Cells | Controls | Exposure level | |
|---|---|---|---|
| | | 2.6 mg/m³ | 10.6 mg/m³ |
| Nucleated blood cells | 0.4240±0.0005 | 0.4622±0.0004* | 0.4704±0.0005*• |
| Bone marrow | 0.3995±0.0005 | 0.4043±0.0003* | 0.4316±0.0003*• |

As can been seen from Table 4, the enhanced fragmentation of the genomic DNA was found in both nucleated blood cells and bone marrow cells, the latter clearly displaying the dependence of this effect on the exposure level.

## 4. Conclusion

The silica (mostly amorphous) containing submicron particles with a prevailing proportion of those in the nanoscale range induce, when instilled intratracheally, a pulmonary cell response comparable with that to highly cytotoxic and fibrogenic standard quartz powder $DQ_{12}$. Nevertheless, in long-term inhalation experiment at realistic concentrations, they proved to be of very low systemic toxicity and negligible pulmonary fibrogenicity. This paradox may be explained by low $SiO_2$ retention in lungs and other organs due to a relatively high solubility of these nanoparticles in relevant biological and model milieus.

However, their genotoxic action and transnasal penetration into the brain found in the same inhalation experiment should make one give a cautious overall assessment of this aerosol as an occupational or environmental hazard.

**Acknowledgements**
The research was partially supported by the Ministry of Science and Higher Education of the Russian Federation (project 3.9534.2017/8.9) and by Government of the Russian Federation (Act 211, Agreement 02.A03.21.0006). The equipment of the Ural Centre for Shared Use "Modern Nanotechnology" UrFU was used.

## References

[1] Du Z J, Zhao D L, Jing L, Cui G, Jin M, Li Y, Liu X, Liu Y, Du H, Guo C, Zhou X and Sun Z 2013 Cardiovascular toxicity of different sizes amorphous silica nanoparticles in rats after intratracheal instillation *Cardiovasc. Toxicol.* **13** 194-207
[2] Eom H J and Choi J 2009 Oxidative stress of silica nanoparticles in human bronchial epithelial cell, Beas-2B *Toxicol. In Vitro* **23** 1326-32
[3] Guo C, Xia Y, Niu P, Jiang L, Duan J, Yu Y, Zhou X, Li Y and Sun Z 2015 Silica nanoparticles induce oxidative stress, inflammation, and endothelial dysfunction in vitro via activation of the MAPK/Nrf2 pathway and nuclear factor-κB signaling *Int. J. Nanomed.* **10** 1463-77
[4] Guo C, Yang M, Jing l, Wang J, Yu Y, Li Y, Duan J, Zhou X, Li Y and Sun Z 2016 Amorphous silica nanoparticles trigger vascular endothelial cell injury through apoptosis and autophagy via reactive oxygen species-mediated MAPK/Bcl-2 and PI3K/Akt/mTOR signaling *Int. J. Nanomed.* **11** 5257-76
[5] Kim Y J, Yu M, Park H O and Yang S I 2010 Comparative study of cytotoxicity, oxidative stress and genotoxicity induced by silica nanomaterials in human neuronal cell line *Mol. Cell. Toxicol.* **6** 336-43
[6] Park E J and Park K 2009 Oxidative stress and pro-inflammatory responses induced by silica nanoparticles in vivo and in vitro *Toxicol. Lett.* **184** 18-25
[7] Petrick L, Rosenblat M, Paland N and Aviram M 2016 Silicon dioxide nanoparticles increase macrophage atherogenicity: stimulation of cellular cytotoxicity, oxidative stress, and

triglycerides accumulation *Environ. Toxicol.* **31** 713-23

[8] Sergent J A, Paget V and Chevillard S 2012 Toxicity and genotoxicity of nano-$SiO_2$ on human epithelial intestinal HT-29 cell line *Ann. Occup. Hyg.* **56** 622-30

[9] Wang J, Yu Y, Lu K, Yang M, Li Y, Zhou X and Sun Z 2017 Silica nanoparticles induce autophagy dysfunction via lysosomal impairment and inhibition of autophagosome degradation in hepatocytes *Int. J. Nanomed.* **12** 809-25

[10] Bellmann B, Creutzenberg O and Dasenbrock C 1991 Lung clearance and retention of toner utilizing a tracer technique, during chronic inhalation exposure in rats *Fund. Appl. Toxicol.* **17** 300-13

[11] Oberdörster G, Sharp Z, Atudore V, Elder A, Gelein R and Kreylin W 2004 Translocation of inhaled ultrafine particle to the brain *Inhal. Toxicol.* **16** 437-45

[12] Elder A, Gelein R, Silva V, Feikert T, Opanashuk L, Carter J, Potter R, Maynard A, Ito Y, Finkelstein J and Oberdörster G 2006 Translocation of inhaled ultrafine manganese oxide particles to the central nervous system *Environ. Health Perspect.* **114** 1172-8

[13] Kao Y-Y, Cheng T-J, Yang D-M and Liu P-Sh 2012 Demonstration of an olfactory bulb-brain translocation pathway for ZnO nanoparticles in rodent ells in vitro and in vivo *J. Mol. Neurosci.* **48** 464-71

[14] Sutunkova M P, Katsnelson B A, Privalova L I, Gurvich V B, Konysheva L K, Shur V Ya, Shishkina E V, Minigalieva I A, Solovjeva S N, Grebenkina S V and Zubarev I V 2016 On the contribution of the phagocytosis and the solubilization to the iron oxide nanoparticles retention in and elimination from lungs under long-term inhalation exposure *Toxicol.* **363** 19-28

IOP Conf. Series: Materials Science and Engineering **699** (2019) 012055 doi:10.1088/1757-899X/699/1/012055

# PFM studies of ferroelectric phase transition in superprotonic $(K_{0.43}(NH_4)_{0.57})_3H(SO_4))_2$ crystals

**A L Tolstikhina, R V Gainutdinov, E V Selezneva and I P Makarova**

Shubnikov Institute of Crystallography of Federal Scientific Research Centre "Crystallography and Photonics" of Russian Academy of Sciences", 199333, Moscow, Russia

alla@ns.crys.ras.ru

**Abstract**. The new superprotonic crystal $(K_{0.43}(NH_4)_{0.57})_3H(SO_4))_2$ is investigated by the scanning probe microscopy and X-ray diffraction. It is shown by piezoresponse force microscopy (PFM) that the crystal is in paraelectric phase at room temperature. The ferroelectric phase transition in the crystal at the temperature decreasing from 296 to 282 K is revealed by PFM. The thin nanorelief is observed at the natural cleavage surface of the crystals by studying its topology with atomic force microscopy. The pits and hills of the nanorelief demonstrate the stable depth (height) of about 0.7 nm, which corresponds to 1/3 unit cell in *c* axis direction.

## 1. Introduction

A systematic study of the salt system of $K_3H(SO_4)_2$–$(NH_4)_3H(SO_4)_2$–$H_2O$ solid solutions was performed for the first time at the Shubnikov Institute of Crystallography of the Russian Academy of Sciences [1]. It was found that using $K_3H(SO_4)_2$ and $(NH_4)_3H(SO_4)_2$ solutions with different K:$NH_4$ ratio led to the crystallization of compounds with different structure types.

The research of single crystals grown in the $K_3H(SO_4)_2$–$(NH_4)_3H(SO_4)_2$–$H_2O$ system from solutions with a K:$NH_4$ concentration ratio about 3:7 revealed that they had a composition $(K_{0.43}(NH_4)_{0.57})_3H(SO_4)_2$ with trigonal symmetry at ambient conditions [2,3] such as the superprotonic phase of $(NH_4)_3H(SO_4)$ at high temperature. Studies of dielectric properties allowed revealing the high conductivity, being typical for known superprotonic compounds at high temperatures [3]. The conductivity $\sigma_{DC}$ increased with increasing temperature from $10^{-9}$ $\Omega^{-1}$cm$^{-1}$ at 223 K to ~3× $10^{-4}$ $\Omega^{-1}$cm$^{-1}$ at 353 K. For comparison, the $(NH_4)_3H(SO_4)$ crystals have the value $\sigma_{DC}$ $10^{-4}$ $\Omega^{-1}$cm$^{-1}$ at 425 K [4]. The values obtained for the activation energies $U = 0.72$ eV below 283 K and $U = 0.52$ eV above 283 K for $(K_{0.43}(NH_4)_{0.57})_3H(SO_4)_2$ crystals correspond to the values $U = 0.76$ eV and $U = 0.47$ eV for the low- and high-temperature phases of $(NH_4)_3H(SeO_4)$ crystals, respectively [5]. Obtained results lead to the conclusion that the temperature 283 K corresponds to the structural transition of the crystals to the low-temperature phase probably with lower symmetry.

In this study, the temperature behavior of $(K_{0.43}(NH_4)_{0.57})_3H(SO_4)_2$ crystal was investigated by piezoresponse force microscopy (PFM) below room temperature to detect the phase transitions. The topography of the fresh cleavage and polished surfaces of the crystals has been studied by atomic force microscopy (AFM).

IOP Conf. Series: Materials Science and Engineering **699** (2019) 012055 doi:10.1088/1757-899X/699/1/012055

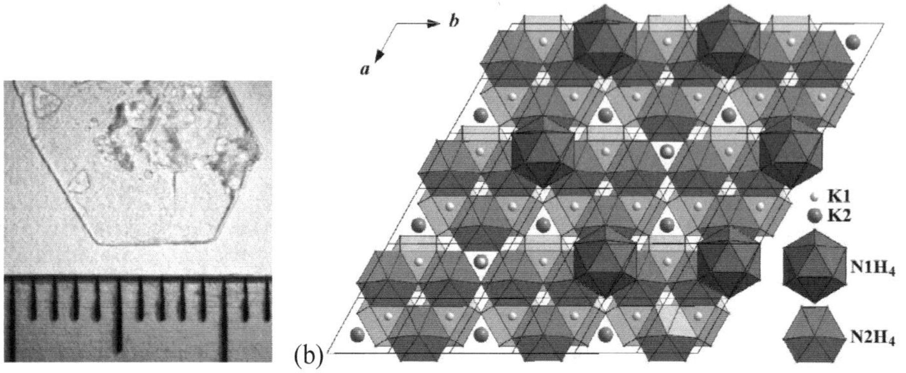


(a) (b)

**Figure 1.** (a) $(K_{0.43}(NH_4)_{0.57})_3H(SO_4)_2$ crystals. (b) Model of the arrangement of K and N atoms in $(K_{0.43}(NH_4)_{0.57})_3H(SO_4)_2$ crystals: K1 atoms and $N1H_4$ coordination polyhedra, located in the layers of $SO_4$ tetrahedra, and K2 atoms and $N_2H_4$ coordination polyhedra, located between the layers, are shown. $SO_4$ tetrahedra and hydrogen bonds are omitted.

## 2. Experimental

Single crystals of $(K_{0.43}(NH_4)_{0.57})_3H(SO_4)_2$ were grown by controlled reduction of solubility. Crystallization was carried out over the temperature range from 303 to 313 K. Solutions with a K:$NH_4$ concentration ratio of ~3:7 yielded large single crystals with high optical quality and habitus as pseudo-hexagonal thin plates with the predominant face (001) (Fig. 1a).

The structure and composition of the crystalline samples were refined by X-ray structural analysis using an Xcalibur S diffractometer (Oxford Diffraction) with a CCD area detector at room temperature. The experimental diffraction data were processed using CrysAlis PRO [6]. Structure refinement was carried out using JANA2006 [7].

The crystals surface relief was investigated with scanning probe microscope NTEGRA Prima (NT-MDT Spectrum Instruments, Russia). All experiments were carried out under controlled conditions of the TRACKPORE ROOM-05 measuring complex (purity class 5 ISO (100), the accuracy of maintaining the air temperature in the range 298± 5 K is ±0.05 K, the relative air humidity is 35 ± 1%). Before study the samples were cleaved in air along the (010) plane perpendicular to the c axis. The fresh cleavage surface was studied in tapping mode of AFM. The $(K_{0.43}(NH_4)_{0.57})_3H(SO_4)_2$ crystal was investigated with two PFM methods using various ways of recording the piezoelectric response. The first one is traditional contact PFM method, the second one is a hybrid PFM [8]. PFM measurements were carried out with the same scanning probe microscope. The silicon cantilevers (NSG01 / Pt, NT-MDT Spectrum Instruments) coated with Pt were used. The alternating voltage $V_{ac}$ was applied to the conductive cantilever with a frequency of 49 kHz and an amplitude of 8 V. The maximum interaction force between cantilever tip and surface was 160 nN.

## 3. Results and discussion

X-ray study showed that the crystals had a trigonal symmetry, the space group $R\bar{3}$, $Z = 3$, $a = b =$ 5.7768(3) Å, $c$ = 22.0983(1) Å at $T \approx 296$ K. The appearance of the threefold axis (and the trigonal symmetry) is due to the ratio K/N and the corresponding coordination of $NH_4$ groups in the $(K_{0.43}(NH_4)_{0.57})_3H(SO_4)_2$ (Fig. 1b). This symmetry leads to the disordering of O atoms involved in hydrogen bonds, and, as a consequence, to the formation of a dynamically disordered network of hydrogen bonds and an increase in conductivity. Such an increase in symmetry and the formation of a dynamically disordered network of hydrogen bonds were observed in the high-temperature phases of isostructural superprotonic compounds at elevated temperatures.

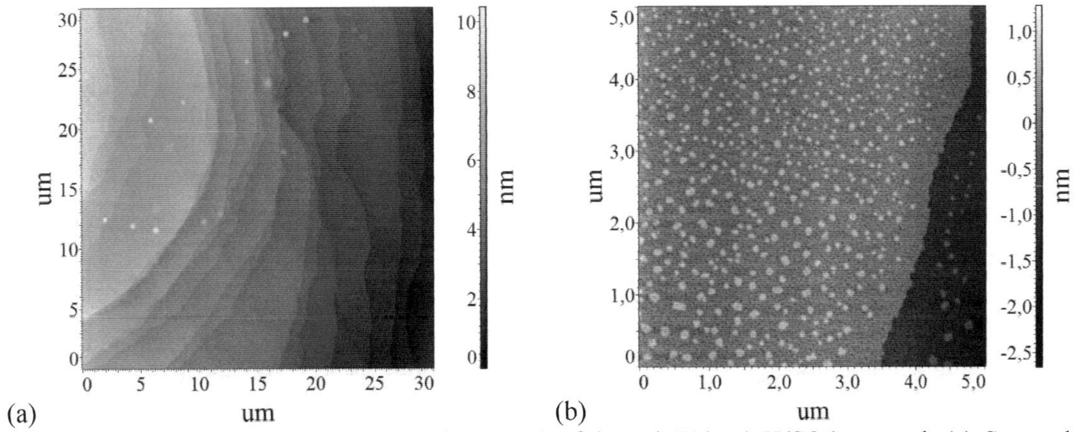


**Figure 2.** AFM images of cleavage surface (001) of $(K_{0.43}(NH_4)_{0.57})_3H(SO_4)_2$ crystal. (a) Stepped microrelief of surface region (30×30 μm²). (b) Nanorelief of one of the steps (5×5 μm²).

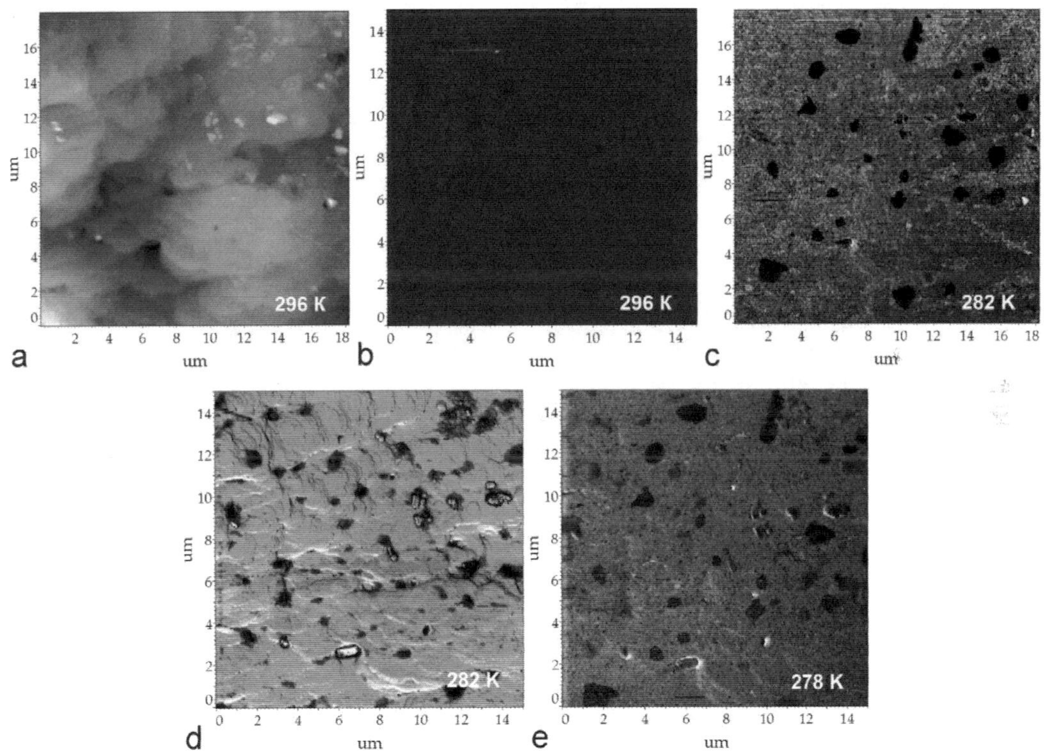


**Figure 3.** AFM images of $(K_{0.43}(NH_4)_{0.57})_3H(SO_4)_2$ crystal as a function of temperature: (a) relief, contact AFM, (b), (c), (e) contact out-of-plane PFM (phase), (d) hybrid mode, hardness (arb. un).

The typical surface area of 30×30 μm² in size with stepped microrelief is shown in Figure 2a. The minimal steps height is 0.7 nm. The RMS parameter of roughness $R_q$ is 2 nm. The AFM image of smaller size (5×5 μm²) is presented in Figure 2b. One can see the cleavage step with the islands and pits of height/depth about 0.7 nm and lateral sizes 50-150 nm and 20-100 nm, respectively. Such height/depth value is multiple to unit cell parameter and corresponds to 1/3 *c*.

As it ensues from the model of the atomic structure of $(K_{0.43}(NH_4)_{0.57})_3H(SO_4)_2$ crystal, along the c axis of the unit cell three layers of $SO_4$ tetrahedra connected by hydrogen bonds and K/NH4 positions among and between them are located. The nanorelief in the form of islands/pits observed on the surface, which height/depth is multiple 1/3 *c* axis of unit cell correlates with features of the layered structure of $(K_{0.43}(NH_4)_{0.57})_3H(SO_4)_2$ crystal. The analysis of the phase contrast AFM images showed that phase composition of the crystal cleavage surface was homogeneous.

The temperature changes (Fig. 3) were carried out with MP6LC thermostatic controller which allowed varying a temperature in the range from 243 to 393 K. The temperature changed three times from 296 ± 0.1 K to 278 ± 0.1 K and back. At room temperature PFM images demonstrated only homogeneous contrast. Decreasing temperature to 282 K allowed observing the appearance of dark contrast regions about 1 μm in size identified as ferroelectric domains. Increase temperature to 296K led to their disappearance. Therefore, the PFM investigations at decreasing temperature allowed observing the ferroelectric phase transition in $(K_{0,43}(NH_4)_{0,57})_3H(SO_4)_2$ crystal for the first time. It is shown that the phase transition is reversible. The obtained information supplements the structure and dielectric data [3].

## 4. Conclusion

In summary, we report the experimental measurements of the new superprotonic crystal $(K_{0.43}(NH_4)_{0.57})_3H(SO_4)_2$. Studies by AFM method gave the new information about morphology and piezoelectric response of natural cleavage surface and polished samples. At decreasing temperature, the PFM investigations allowed observing ferroelectric phase transition in $(K_{0,43}(NH_4)_{0,57})_3H(SO_4)_2$ crystals for the first time. The results of the PFM studies of the temperature change on the crystal surface correspond to the dielectric data for bulk samples, indicating modification of the conductivity at T~283 K, and the structural data, demonstrating changes of symmetry at superprotonic phase transitions in the crystals of the $(AO_4)_{(m+n)/2}{\cdot}yH_2O$ (M = K, Rb, $NH_4$, $AO_4 = SO_4$, $SeO_4$) system.

## Acknowledgements

We are grateful to V.V. Dolbinina (FSRC "Crystallography and Photonics" RAS, Moscow) for supplying the samples and A.S. Kalinin (NT-MDT Spectrum Instruments) for his help. This work was supported by the Ministry of Science and Higher Education within the State assignment FSRC «Crystallography and Photonics» RAS.

## References

[1] Dmitricheva E V, Makarova I P, Grebenev V V, Dolbinina V V and Verin I A 2014 *Crystallogr. Rep.* **59** 344-52
[2] Dmitricheva E V, Makarova I P and Grebenev V V 2015 *Crystallogr. Rep.* **60** 814-20
[3] Selezneva E V, Makarova I P, Malyshkina I A, Gavrilova N D, Grebenev V V, Novik V K and Komornikov V A 2017 *Acta Cryst. B* **73** 1105-13
[4] Schwalowsky L, Vinnichenko V, Baranov A, Bismayer B, Merinov B and Eckold G 1998 *J. Phys. Condens. Mat.* **10** 3019-27
[5] Lindner L, Zdanowska-Frączek M, Pawłowski A and Frączek Z J 2014 *J. Appl. Phys.* **116** 163513
[6] Oxford Diffraction. 2011 *CrysAlis PRO and SCALE3 ABSPACK.* Oxford Diffraction Ltd, Yarnton, Oxfordshire, UK
[7] Petříček V, Dušek M and Palatinus L *Jana 2006 Structure Determination Software Programs.* Institute of Physics, Praha, Czech Republic
[8] Tolstikhina A L, Gainutdinov R V, Belugina N V, Lashkova A K, Kalinin A S, Atepalikhin V V, Polyakov V V and Bykov V A 2018 *Phys. B: Condens. Mat.* **550** 332-9

# Microstructure of barium strontium titanate based glass-ceramics doped by Ce and La

**A P Turygin**[1], **D S Chezganov**[1], **I S Baturin**[1], **X Song**[2], Y **Zhang**[2] and **V Ya Shur**[1]

[1]School of Natural Sciences and Mathematics, Ural Federal University, 620000 Ekaterinburg, Russia
[2]Beijing Key Laboratory of Fine Ceramics, State Key Laboratory of New Ceramics and Fine Processing, Institute of Nuclear and New Energy Technology, Tsinghua University, 100084 Beijing, P.R. China

anton.turygin@urfu.ru

**Abstract**. The microstructure and surface morphology of $(Ba_{0.8},Sr_{0.2})TiO_3$ based glass-ceramics with different amounts of La and Ce were studied. It was shown that the addition of the 1.5% La and >2%Ce led to formation of dendrite-like structures. The isolated clusters of ferroelectric phase were found in all investigated samples. Fraction of ferroelectric phase was about 0.75% for La addition and 0.25% for Ce addition.

## 1. Introduction

Glass-ceramics representing ferroelectric grains embedded in a glass matrix are of significant interest for power electronic applications, especially for energy storage devices [1,2]. It has been shown that the application of the glass-ceramics based on the ferroelectric materials can improve energy storage properties by the utilization of a high dielectric constant of ferroelectrics and relatively high breakdown voltage of glass [2]. It is known that the aluminosilicate glass-ceramics based on $(Ba,Sr)TiO_3$ (BST) are among the most promising materials for the energy storage applications [3]. An addition of SrO is used for tuning the Curie temperature and increasing the dielectric constant at room temperature.

It has been shown that the energy storage properties of the BST glass-ceramics depend on their microstructure [4,5]. Different additives ($AlF_3$, $La_2O_5$, $MnO_2$, etc.) were used to control the microstructure, thus optimizing the dielectric properties [5].

The effect of space charges (interfacial polarization) at the interphase boundaries between crystalline grains and glass on the breakdown voltage and the dielectric properties has been discussed [6-8]. At the same time, properties of the ferroelectric materials strongly depend on the domain structure, but little is known about the domain structure of the BST glass-ceramics. The purpose of the present paper is to study the microstructure of the BST glass-ceramics with different concentrations of La and Ce additives.

## 2. Experimental

The samples of BST based glass-ceramics with different concentrations of La and Ce additives were prepared from well-mixed powders by a melt technique. The glass samples with La addition have the composition of $m$BaO–$n$SrO–$29TiO_2$–$22SiO_2$–$12Al_2O_3$–$2.4BaF_2$–$x La_2O_3$ (x=0, 0.5, 1.0, and 1.5 mol%, $m=4n$, $m+n+x=34.6$ mol%) (BST-La$x$%). The ball-milled wet powder was melted in a platinum crucible at 1550°C for 4 h and then poured rapidly into a container filled with distilled water to obtain glass frits. The glass frits were subsequently ball-milled for 4 h to produce glass powder. This glass powder was mixed with PVA and then pressed into small disk with 10 mm in diameter and 0.6 mm in thickness using uniaxial press. The pellets were heat-treated by two-step process for nucleation and subsequent crystal growth. The green pellets were first heated at 850°C for 1 h and then sintered at 950°C for 2 h to form the La-doped BST glass-ceramics. Glass samples with Ce addition have composition $m$BaO–$n$SrO–29$TiO_2$–22$SiO_2$–12$Al_2O_3$–$x$$CeO_2$ (x = 0, 1, 2, and 3 mol%, $m = 4n$, $m+n+x = 37$ mol%) (BST-Ce$x$%). The powders were melted in a platinum crucible at 1550°C for 2-3 h and then poured rapidly into a container filled with distilled water to obtain glass frits. The glass frits were subsequently ball-milled for 4 h to produce glass powder. This glass powder was mixed with PVA and then pressed into small disks with 8 mm in diameter and 0.25 mm in thickness using uniaxial press. The pellets were heat-treated by two-step process for nucleation and subsequent crystal growth. The green pellets were first heated at 700°C for 4 h and then sintered at 950°C for 2 h to form the Ce-doped BST glass-ceramics.

The sample surface was polished by diamond paste with decreasing grain sizes from 6 to 0.25 μm. The final polishing was done by colloidal silica (SF1 Polishing Suspension, Logitech, UK). The sample thickness after polishing was about 200 μm.

The visualization of the domain structure with high spatial resolution was performed using piezoresponse force microscopy (PFM). The samples were mounted by conductive silver paste to the metal disks. PFM measurements were realized by a scanning probe microscope Asylum MFP-3D (Asylum Research, Oxford Instruments, UK). The silicon HA_NC/W2C tips (ScanSens, Russia) with a tungsten carbide conductive coating, radius of curvature R = 35 nm, resonance frequency f = 380 kHz, and spring constant k = 35 N/m were used. Piezoresponse measurements were realized by the application of 5 V AC far from the contact resonance frequency to the SPM tip.

The microstructure of the glass-ceramic samples was studied by a scanning electron microscope (SEM, Merlin, Carl Zeiss, Germany) with field-emission using secondary electron imaging with charge compensation.

## 3. Results and discussion

Investigation of the BST-La glass-ceramics microstructure by SEM demonstrated irregular-shaped grains with size below 100 nm for all samples (Fig. 1). It should be noted that the grain boundaries have low contrast, which makes it difficult to distinguish individual grains. All the grains can be separated by contrast in two groups: corresponding to glass matrix and ferroelectric phase (Fig. 1a-d). For structure analysis, auto-correlation function (ACF) technique can be used [6]. 2D ACF image is defined as:

$$C(r_1, r_2) = \sum_{x,y} D(x,y) \times D(x + r_1, y + r_2) \quad (1)$$

where $C(r_1, r_2)$ is an intensity of ACF in ($r_1$, $r_2$) point, D(x, y) is an intensity of SEM image in (x, y) one.

SEM images (Fig. 1a-d) were analysed after binarization. Angle averaged profile of ACF image was fitted by function:

$$C(r) \sim exp\left[-\left(\frac{r}{\xi}\right)^{2h}\right] \quad (2)$$

where $r$ is the distance from the central maximum, $C(r)$ is ACF amplitude, $\xi$ is the correlation radius, and $h$ is a parameter.

Correlation radius for BST-La samples is about 55 nm for all samples and weakly depends on La concentration.

Formation of dendrite-like structures was found for high La concentration (>1%) (Fig. 2a). The formation of dendrite structures in BST-based glass-ceramics with addition of $AlF_3$, $MnO_2$ and rare earth elements were reported [1,7-11]. In general, the dendrites violate the uniform microstructure of ceramics, which leads to degradation of dielectric breakdown properties. However, such structures can increase dielectric permittivity due to large surface of phase boundaries.

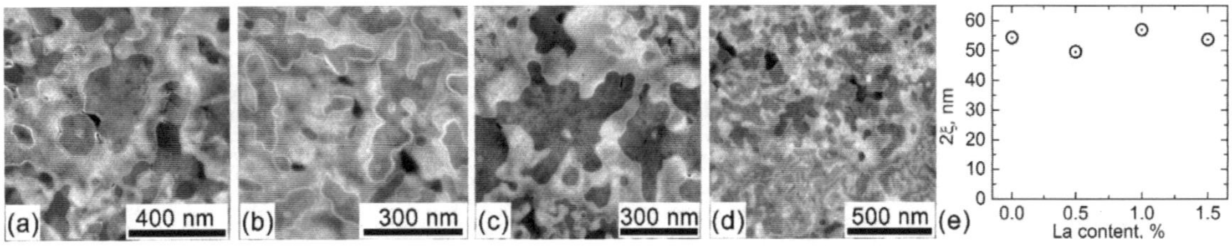


**Figure 1.** SEM images of BST-La glass-ceramics microstructure with various La concentration: (a) 0%, (b) 0.5%, (c) 1%, (d) 1.5%, (e) dependence of correlation radius on La content.

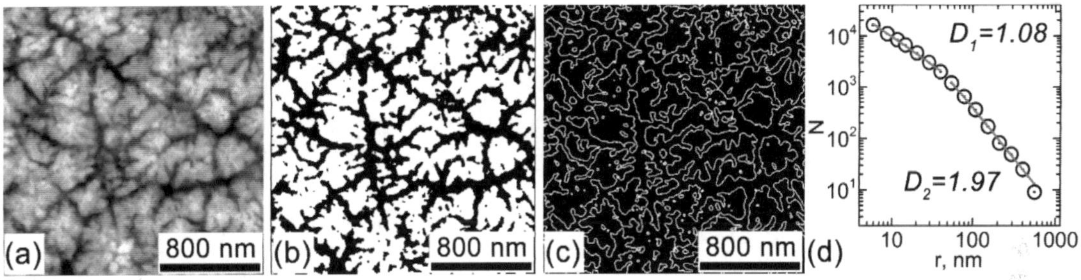


**Figure 2.** The dendrite structures in BST-La1.5%: (a) SEM image, (b) binarized image, (c) edge detected image, (d) dependence of box number on box size.

The analysis of fractal dimension of structure was done using box counting method after image binarization and edge detection (Fig. 2b,c). Two-level fractal structure was obtained (Fig. 2d) with fractal dimensions $D_1$ = 1.08 at scale below 100 nm and $D_2$ = 1.97 above 100 nm. Such large value of $D_2$ can indicate the compact dendrite structure [12]

Study of BST-Ce glass-ceramics did not reveal individual grains. The obtained microstructure slightly depended on Ce concentration (Fig. 3a-d). In the samples with Ce content below 2%, as-grown lamellar structures with period below 30 nm were found. Correlation radius decreased with increasing Ce content (Fig. 4e).

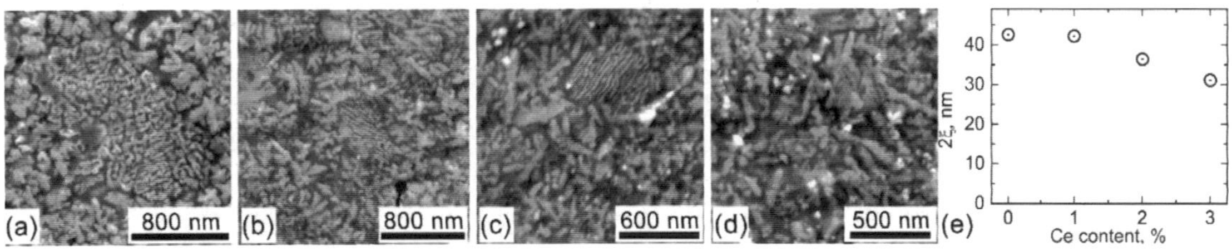


**Figure 3.** SEM images of BST-Ce glass-ceramics microstructure with various Ce concentrations: (a) 0%, (b) 1%, (c) 2%, (d) 3%, (e) dependence of correlation radius on Ce content.

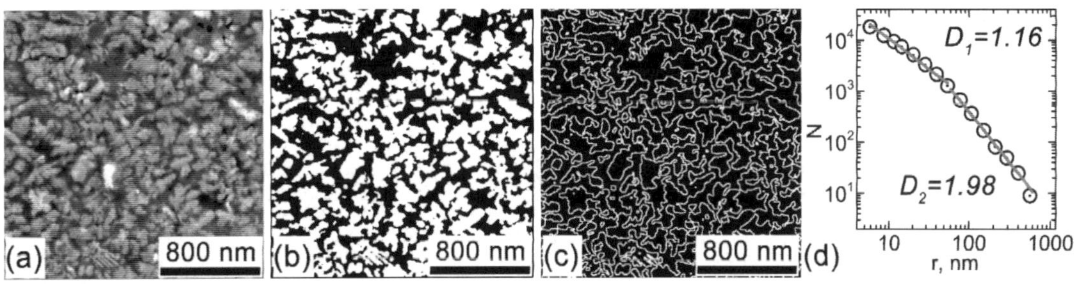


**Figure 4.** The dendrite structures in BST-Ce0%: (a) SEM image, (b) binarized image, (c) edge detected image, (d) dependence of box number on box size.

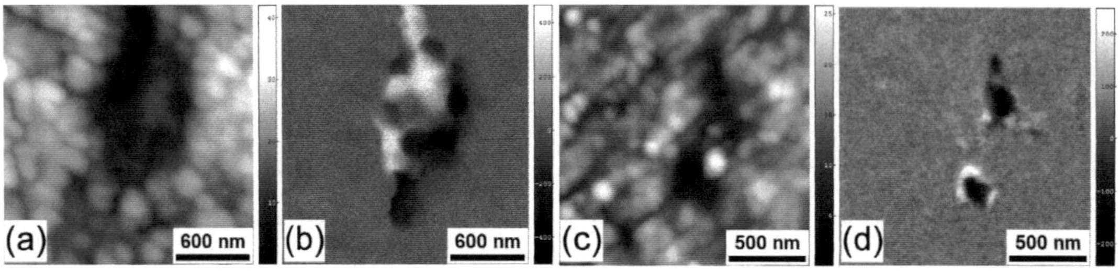


**Figure 5.** Domain structure in BST-La1%: (a) topography, (b) PFM image. Domain structure in BST-Ce2%: (c) topography, (d) PFM image.

In all BST-Ce samples, dendrite-like structures with size about several hundred nanometres were found (Fig. 3a-d). Application of box counting method for structure analysis (Fig. 4) allowed obtaining fractal dimension $D_2$ of about 1.98 for all BST-Ce samples, which indicates the compact structure.

In BST-La samples isolated clusters of polar phase with size from 0.2 to 3 μm were found by PFM (Fig. 5a,b). Fraction of polar phase was about 0.75% for all the samples and did not depend on La concentration. PFM studies in BST-Ce demonstrated presence of the same polar phase clusters with size from 0.2 to 1.5 μm and fraction about 0.25%, which was also independent on Ce concentration.

Application of the single voltage pulse to the tip with amplitude up to ±200 V and duration 10 s did not allow obtaining local polarization reversal even at the elevated temperatures up to 90°C. This fact can be attributed to a strong clamping of the domain walls by glass matrix [13].

## 4. Conclusion

The microstructure and surface morphology of the BST glass-ceramics with La and Ce additions were studied. It was found that the addition of 1.5% of La led to formation of the compact dendrite structure. Ceramics with Ce addition possessed dendrite-like structures, decreasing in size with Ce concentration. Fraction of ferroelectric phase is below 1% in all the studied samples. The obtained information about microstructure can be used for further improvement of the dielectric properties of BST glass-ceramics.

### Acknowledgements

The research was made possible by Russian Foundation for Basic Research (grant 18-52-53032-NNSF_a). The equipment of the Ural Centre for Shared Use "Modern Nanotechnology" UrFU was used.

### References

[1] Gorzkowski E P, Pan M, Bender B and Wu C C M 2007 *J. Electroceramics* **18** 269-76
[2] Herczog A 1973 *IEEE Trans. Parts, Hybrids, Packag.* **9** 247-56
[3] Fletcher N H, Hilton A D and Ricketts B W 1996 *J. Phys. D Appl. Phys.* **29** 253-8
[4] Chen J, Zhang Y, Deng C, Dai X and Li L 2009 *J. Am. Ceram. Soc.* **92** 1350-3
[5] Chen J, Zhang Y, Deng C and Dai X 2009 *J. Am. Ceram. Soc.* **92** 1863-6
[6] Shvartsman V and Kholkin A 2007 *J. Appl. Phys.* **101** 064108
[7] Baturin I S, Chezganov D S, Neradovskiy M M, Naumova N A, Shur V Y, Zhang Y and Wang X 2013 *Ferroelectrics* **442** 131-6
[8] Xiu S, Xiao S, Zhang W, Xue S, Shen B and Zhai J 2016 *J. Alloys Compd.* **670** 217-21
[9] Zhang W, Xue S, Liu S and Zhai J 2015 *Ferroelectrics* **488** 71-8
[10] Wang X, Zhang Y, Ma T, Yuan Z and Zhang Q 2013 *Int. J. Appl. Ceram. Technol.* **10** 301-6
[11] Turygin A P, Abramov A S, Alikin D O, Chezganov D S, Esin A A, Baturin I S, Song X, Zhang T, Zhang Y, Hu K, Zhao Z and Shur V Y 2018 *IOP Conf. Ser.: Mater. Sci. Eng.* **443** 012037
[12] Ihle T and Müller-Krumbhaar H 1994 *Phys. Rev. E* **49** 2972-91
[13] Patel S, Chauhan A, Kundu S, Madhar N A, Ilahi B, Vaish R and Varma K B R 2015 *AIP Adv.* **5** 087145

IOP Conf. Series: Materials Science and Engineering **699** (2019) 012057 doi:10.1088/1757-899X/699/1/012057

# Fabrication of superhydrophobic and superoleophilic teflon surfaces using irradiation by nanosecond infrared laser

**K V Volchetskaya, D K Kuznetsov and V Ya Shur**

School of Natural Sciences and Mathematics, Ural Federal University, 620000 Ekaterinburg, Russia

vladimir.shur@urfu.ru

**Abstract**. Polytetrafluoroethylene (PTFE, teflon) possesses excellent stability and low surface energy, which enables its extensive use in the manufacture of biomimetic superhydrophobic surfaces. We proposed a simple and fast method for producing superhydrophobic surfaces by treating the teflon plate by nanosecond pulse infrared irradiation by means of $CO_2$ laser. The surface relief was visualized by scanning electron microscope. The surface wettability was estimated by measuring the contact angle. The treated teflon superhydrophobic surfaces with static contact angle up to 154°demonstrated self-cleaning properties. The creation of regular two-dimensional structures leading to super-adhesive property of the teflon surface was shown. Thus, it was shown that the surface treating of teflon by nanosecond infrared laser could be used for creation of the super-hydrophobic surfaces for various applications such as self-cleaning, microfluidic devices, and the laboratory-on-chip systems.

## 1. Introduction

Inspired by nature, superhydrophobic surfaces with a contact angle above 150° have attracted great attention and research efforts in recent years. Lotus leaves and rose petals are two types of superhydrophobic surfaces in nature. Lotus leaves are well-known for the “lotus effect”. These surfaces demonstrate low adhesion to water. The water droplets cannot remain stable on such surfaces, but spontaneously roll off thus removing the usually obtained dust particles from the surfaces [1]. On the other hand, the “petal effect”, being typical for the rose petals, represents a very high adhesion to water resulting in fixation of the water droplets on the surface at any tilted angles [2].

Artificial low-adhesive and high-adhesive superhydrophobic materials find their application in many fields: microfluidics, cell engineering, biosensors, etc. [3,4]. Various technologies have been developed to fabricate such materials. In recent years, laser technologies using picosecond laser [5], femtosecond laser [6], and excimer laser [7] become widely used to prepare superhydrophobic biomimetic materials because of rapid surface micromachining of various patterns with controllable parameters. The femtosecond laser microfabrication has been successfully applied for producing strong non-wetting surfaces because of its negligible heat-affected zone, non-contact process, precise ablation threshold, and high spatial resolution. It has been studied how the irradiation by femtosecond laser can change the surface morphology and wettability of teflon surfaces [8-10].

In this work, we demonstrate that the surface microstructuring of the teflon plates by nanosecond infra-red (IR) laser irradiation leads to essential modification of the wetting characteristics and allows realizing superhydrophobic and superoleophilic properties. It has been shown that the laser treatment

IOP Conf. Series: Materials Science and Engineering **699** (2019) 012057 doi:10.1088/1757-899X/699/1/012057

increases essentially the initial wetting angle, which allows demonstrating the lotus and rose petal effects and effective self-cleaning.

## 2. Experimental

### *2.1. Materials*

The study of the polymer surface modification by laser radiation was carried out on 2-mm-thick teflon (fluoroplastic-4) samples. Before irradiation, the samples were cleaned by isopropanol or acetone in an ultrasound bath for 7 min and in deionized water for 5 min. The cleaned samples were dried by compressed dry air. The liquids used for testing of the wetting conditions represent: deionized water, petroleum and mineral aviation oil (MS-8P), and turbine oil (TP-22S).

### *2.2. Surface treatment*

The free-standing platform VersaLASER VLS 3.60 (Universal Laser System, US) equipped with 40 W infrared $CO_2$ laser source (10.6 μm wavelength) was used. The surface treatment was made with the following parameters: pulse frequency (repetition rate) 450 Hz, scanning speed 270 mm/s, and pulse power 6 W. The size of the spot on the teflon surface after single pulse irradiation was about 100 μm.

The imaging of the surface morphology before and after laser treatment was made by field emission scanning electron microscope (SEM) Merlin (CarlZeiss, Germany) with spatial resolution about 1 nm. The chemical compositions of initial and treated teflon surfaces were measured by X-ray fluorescence spectrometer K-ALPHA (Thermo Fisher Scientific, UK). The contact angle was measured by the lying drop method [4]. A drop of water or oil with a volume of 5 μL was dropped from a height about 5 mm onto a horizontally lying sample at room temperature. The contact angle θ was calculated from the measured droplet sizes:

$$\theta = arccos\frac{(d/2)^2 - h^2}{(d/2)^2 + h^2} \quad (1)$$

where $d$ is the width of the droplet base and $h$ is the droplet height (Fig. 1).

The droplet on the teflon surface was imaged with a digital camera Sony DSC-H9 (Sony Corporation, Japan) with a CCD size of 1 x 2.5 inch. The droplet width and height were measured using the Adobe Illustrator vector graphics editor.

## 3. Results and discussion

### *3.1. Initial surface*

The surface morphology of the initial teflon surface before laser radiation has been studied by SEM (Fig. 2). It has been shown that it is relatively smooth with irregular graininess and roughness (Ra) about 600 nm measured by optical profilometer WYKO NT 1100 (Veeco, USA). The contact angle for water is about 101°. The well-known teflon hydrophobicity before any treatment can be attributed by the chemical composition only without any significant contribution of the surface topography to the wetting.

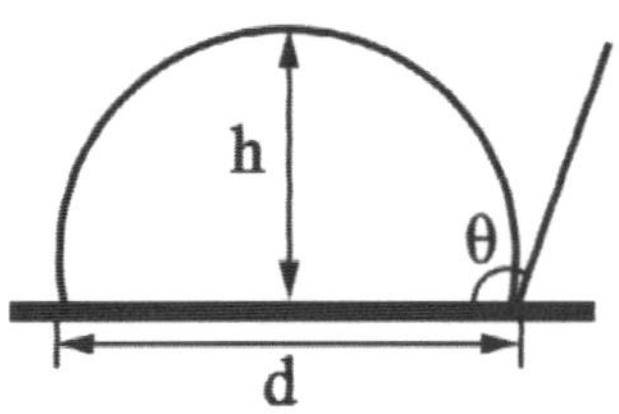


**Figure 1.** Scheme for measuring the contact angle.

(a) 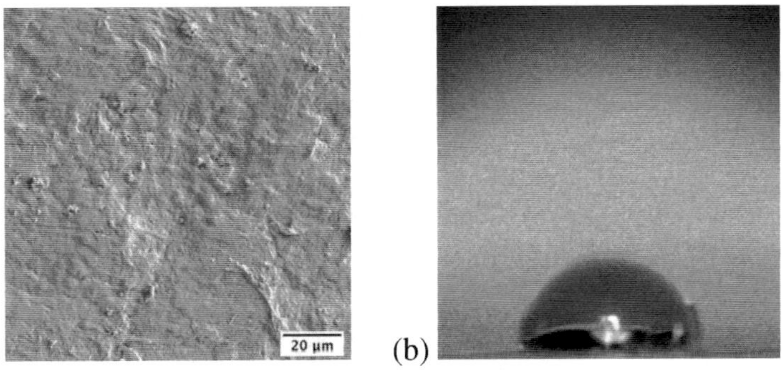 (b)

**Figure 2.** (a) SEM image of the initial teflon surface, (b) water droplet on it. Droplet volume 5 μL.

*3.2. Hydrophobicity of teflon surfaces after laser treatment*

The infrared laser treatment of the teflon surface leads to appearance of random intertwined fibrous structure resembling a thick “sponge”. Along with the formation of fibrous teflon in the irradiated zone, polymer destruction occurs with the formation of monomer tetrafluoroethylene [11,12]. Secondary polymerization of teflon leads to the deposition of irregular shaped particles on the polymer fibres with the size range from 20 to 100 nm, which can be assembled into agglomerates (Fig. 3).

The local polymer melting and foaming during laser processing leads to formation of the fibres, which are carried away by the ablation flow. The ablated particles are deposited on the surface along the propagation path. Being in a state of viscous flowing melt, the fibres are fused at the points of contact forming a "spongy" structure.

The contact angle on such a surface is 140.5°. The results of X-ray fluorescence analysis showed that the chemical composition of the polymer was not changed by laser treatment. The obtained increase of hydrophobic properties can be attributed to essential increase of the surface roughness resulting in a decrease of the area of the liquid contact with the surface and increase of the contact angle.

(a)  (b)



**Figure 3.** SEM images of the treated teflon surfaces: (a) the fibrous structure; (b) nanoparticle agglomerates.

IOP Conf. Series: Materials Science and Engineering **699** (2019) 012057 doi:10.1088/1757-899X/699/1/012057

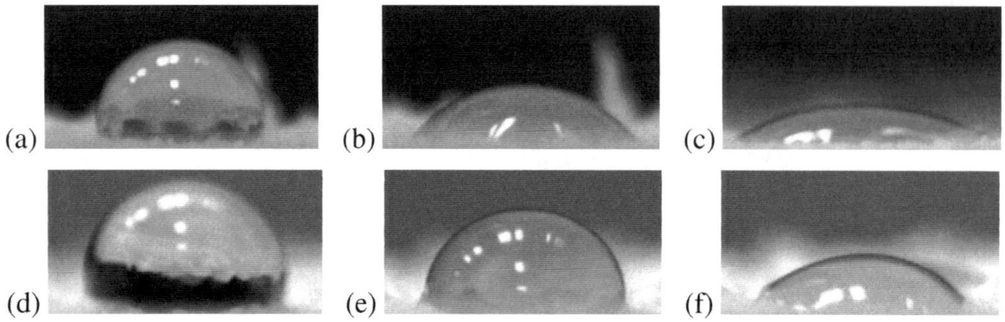

**Figure 4.** Time evolution of the droplets of (a-c) aviation oil and (d-f) turbine oil on the surface of the modified teflon after: (a, d) 3 min, (b, e) 8 min, and (c, f) 15 min. Droplet volume 5 μL.

### *3.3. Oleophilicity of teflon surfaces after laser treatment*
The wetting of the treated surface by petroleum, turbine and aviation oils was studied. The initial teflon surfaces show strong oleophilicity with contact angle about 15°. The droplets were placed on the surface, and the contact angle was measured after 3, 8, and 15 min.

It was shown that the contact angles of the aviation and turbine oil droplets on the treated teflon surface changed with time (Fig. 4). The oil droplets on the modified teflon surface maintains a spherical shape for 3 min with subsequent decreasing of the contact angle both for aviation (Fig. 4 a-c) and turbine (Fig. 4e-g) oils. Contact angle decreases from 100° to 42° for aviation oil and from 112° to 58° for turbine oil after 15 min demonstrating strong oleophilicity.

It is known that the contact angle change depends on the liquid viscosity and the droplet volume. The observed temporary oleophobicity can be attributed to the high oil viscosity, which affects the rate of oil penetration into existing voids of modified teflon surface.

The oleophilic properties are even more pronounced for petroleum. The contact angle on the initial surface is 18°, while the modified teflon surface exhibits superoleophilic properties with zero contact angle. A petroleum drop placed on the sample spreads over the surface immediately.

The creation of the porous fibrous structures on the teflon surface makes it possible to describe the system under consideration by the Wenzel model [4], according to which the increase of the surface roughness leads to increase of the oleophilic properties up to superoleophilicity observed for the petroleum.

### *3.4. Self-cleaning surfaces*
It is necessary to point out that the described fibrous structures can be easily removed by the mechanical impact. This is the reason to study the effect of the layer thickness of the formed fibres on the wetting effects. The unstable initial fibrous layer with thickness ranged from 420 to 480 μm was mechanically removed from the part of the treated surface (Fig. 5). The thickness of the stable residual layer is about 40 μm. The decrease of the layer thickness leads to noticeable increase of the contact angle. On the surface with initial fibrous layer the contact angle was 134°, while on the residual layer – about 140°.

It was also revealed that the teflon surface modified with IR laser radiation, after removal of the excess fibrous layer, exhibited the pronounced self-cleaning property. For testing, the graphite shavings were placed on the initial and treated teflon plates (Fig. 6). The plates were tilted by 20-30° and water droplets were placed on the surface. On the initial surface, droplets were collected, and water slowly drained from the plate, which led to the formation of spots on

**Figure 5.** The cross-section of the surface layer of the teflon plate just after laser processing (right) and after mechanical removal of the fibrous layer (left).

the surface. Water droplets placed on the surface with increased hydrophobicity instantly rolled off from an inclined plane, taking away graphite particles. This demonstrates the "lotus effect" on the modified teflon surface.

Also, the self-cleaning process was carried out by dipping the plates graphite shavings in the water. After several dipping cycles, the initial teflon surface remained dirty, while the modified surface was cleaned due to its superhydrophobic properties (Fig. 6).

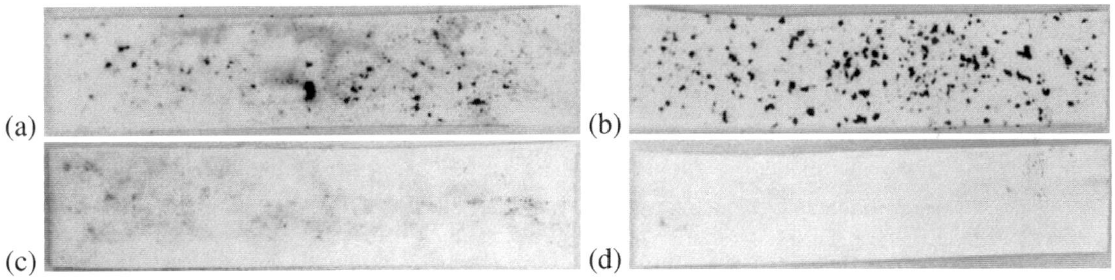


**Figure 6.** Optical images of (a) initial and (b) modified teflon surfaces contaminated with graphite particles; (c) initial and (d) modified surfaces after dipping into the water.

**Figure 7.** SEM images of the two-dimensional structure on teflon surface: (a) an array of hills; (b) the individual hill.

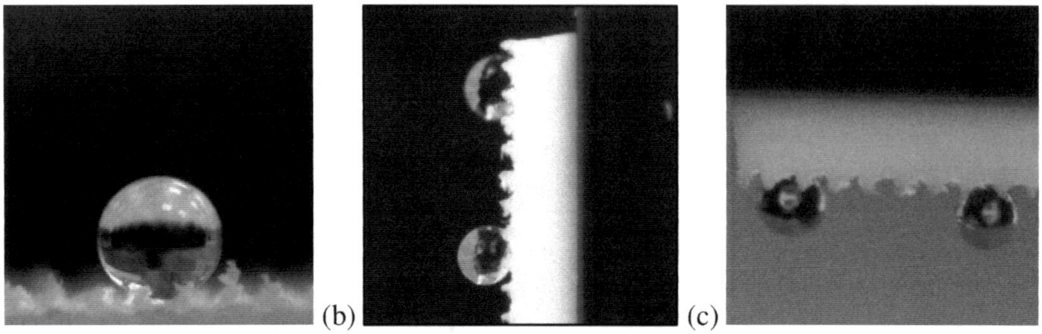
(a) (b) (c)

**Figure 8.** Water droplets placed on the highly adhesive superhydrophobic surface: (a) horizontal surface; (b) surface tilted on 90°; (c) surface tilted on 180°. Droplet volume (a) 6 μL, (b,c) 3 μL.

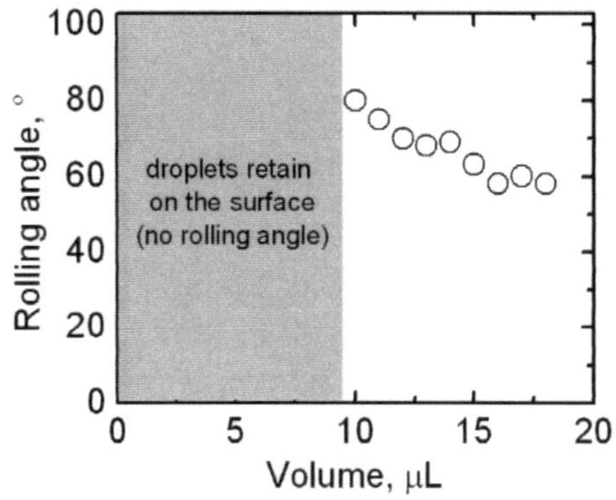


**Figure 9.** Dependency of the rolling angle on the droplet volume for teflon surfaces with two-dimensional structure.

### *3.5. Rose petal effect*

It was shown that the samples with two-dimensional structures on the surface demonstrated super adhesive properties – "rose petal effect". The structures consisting of hills were produced by laser scanning and subsequent mechanical removal of the formed fibrous layer. SEM imaging showed that the produced hills had a pyramidal shape with height about 500 μm, width about 205 μm, and length about 295 μm (Fig. 7). The period of the structure was 500 μm in one direction and 380 μm in another one.

It was revealed that the surfaces with two-dimensional structure were super-hydrophobic with the maximal value of the contact angle 154.5° in our research (Fig. 8a).

For the obtained highly adhesive superhydrophobic surface, the sliding (rolling) angles depended on the droplet volume. As in the case of rose petal effect, the surfaces held the droplets with volume up to 9 μL (Fig. 9). The maximal volume of the retained droplet was about 18 μL. The observed effects can be attributed to the surface relief of the two-dimensional structure. The further research will be done for detail explanation of the obtained results.

## 4. Conclusion

The teflon surface modification by nanosecond $CO_2$ laser treatment to improve hydrophobic properties to superhydrophobicity has been demonstrated. The modified teflon surfaces also have exhibited superoleophilic properties. It has been revealed that the modified teflon surfaces exhibit the self-cleaning properties or "lotus effect". The highest value of the contact angle

($\theta$ = 154.5°) has been obtained for a sample with two-dimensional structures created using two-stage surface treatment: structure formation by $CO_2$ laser and mechanical removal of the fibrous layer. Such two-dimensional structures with morphology similar to rose petal have demonstrated super adhesive properties ("rose petal effect"), which allowed retaining water droplets with volume up to 18 μL.

**Acknowledgements**
The work has been supported in part by the Ministry of Science and Higher Education of the Russian Federation (projects 3.9534.2017/8.9 and 3.7046.2017/7.8) and by Government of the Russian Federation (Act 211, Agreement 02.A03.21.0006). The equipment of the Ural Centre for Shared Use "Modern Nanotechnology" Ural Federal University was used.

**References**

[1] Neinhuis C and Barthlott W 1997 *Annals of Botany* **79** 667-77
[2] Feng L, Zhang Y, Xi J, Zhu Y, Xia F and Jiang L 2008 *Langmuir* **18** 4114-9
[3] Ueda E and Levkin P A 2013 *Adv. Mater.* **25** 1234-47
[4] Yong J, Chen F, Yang Q, Jiang Z and Hou X 2018 *Adv. Mater. Interfaces* **5** 1701370
[5] Long J, Fan P, Zhong M, Zhang H, Xie Y, Lin C 2014 *Appl. Surf. Sci.* **311** 461
[6] Wang Z B, Hong M H, Lu Y F, Wu D J, Lan B and Chong T C 2003 *J. Appl. Phys.* **93** 6375-80
[7] Adhi K P, Owings R L, Railkar T A, Brown W D and Malshe A P 2003 *Appl. Surf. Sci.* **218** 17-23
[8] Yong J, Huo J, Chen F, Yang Q and Hou X 2018 *Phys. Chem. Chem. Phys.* **39** 25140-63
[9] Liang F, Lehr J, Danielczak L, Leask R and Kietzig A M 2014 *Int. J. Mol. Sci.* **15** 13681-96
[10] Falah Toosi S, Moradi S, Kamal S, Hatzikiriakos S G 2015 *Appl. Surf. Sci.* **349** 715-23
[11] Diskinson J T, Shin J, Jiang W and Norton G 1993 *J. Appl. Phys.* **74** 4729-36
[12] Serafetinides A A, Makropoulou M I, Skordoulis C D and Kar A K 2001 *Appl. Surf. Sci.* **180** 42-56

IOP Conf. Series: Materials Science and Engineering **699** (2019) 012058 doi:10.1088/1757-899X/699/1/012058

# Features of Raman spectra in barium titanate pressed powder

**I V Zaytseva[1], A M Pugachev[1], N V Surovtsev[1] and A S Krylov[2]**

[1]Institute of Automation and Electrometry SB RAS, 630090 Novosibirsk, Russia
[2]Kirensky Institute of Physics Federal Research Centre KSC SB RAS, 660036 Krasnoyarsk, Russia

ZaytsevaIV@iae.sbras.ru

**Abstract.** Raman spectra in barium titanate powder under non-hydrostatic pressures are presented. The relationship between the pressure (up to 11 GPa) and the shift and the width of E(TO)-phonon line has been established. It has been found that under the action of uniaxial pressure a new Raman peak arises in the vicinity of 100 $cm^{-1}$. It is shown that the Raman shift and width of both peaks increase depending on the applied pressure. These Raman lines broadening may be associated with an increase of the concentration of defect states in the volume of the sample.

## 1. Introduction

The primary method used in the manufacture of piezoelectric ceramics is a non-hydrostatic pressure treatment of the ferroelectric powders [1]. Studies of pressure treatment of ferroelectric powders are usually devoted to a case of hydrostatic pressures [2,3]. However, one can note that the effect of the non-hydrostatic pressure treatment may differ significantly from hydrostatic one. The phase transition temperature in $BaTiO_3$ powders is known to be decreased under the hydrostatic pressure [4] in contrast to the non-hydrostatic one. It was shown that residual mechanical stresses inside the pressure treated samples (mechanical stresses after removal of external non-hydrostatic ones) led to increase in the temperature and width of the ferroelectric phase transition from cubic to tetragonal phase. These pressures, caused by the formation of defects in the crystal cell, were studied in detail in Refs. [5-8]. The behaviour of the phase transition temperature (from tetragonal to cubic phase) in $BaTiO_3$ powders on the non-hydrostatic pressure is similar to analogous one in strained $BaTiO_3$ films grown on different substrates due to the existence of enormous strains [9-11].

It is known that Raman spectroscopy allows measuring the mechanical stress. In the case of barium titanate ($BaTiO_3$), mechanical stresses can be determined by the shift of E(TO)-phonon line (near 307 $cm^{-1}$) in Raman spectra [6,12]. However, a detailed study of the Raman spectra when non-hydrostatic pressures are applied directly to the powder was not carried out up to now.

The barium titanate crystal at room temperature passes into a cubic phase under applied hydrostatic pressure above 2 GPa. Therefore, it is important to investigate the position and width of E(TO) line in order to determine the phase of barium titanate at high non-hydrostatic mechanical stresses. In the present paper, we investigate the effect of heterogeneity or anharmonism, which should be reflected on the dependence of the E(TO) peak width under the mechanical pressure.

It was demonstrated that a new Raman line appeared under high pressure. In the present paper, this line is investigated in detail.

## 2. Experimental

High-pressure experiments were carried out in nominally pure $BaTiO_3$ powder (Aldrich) with a grain size less than 2 μm. Raman spectra of $BaTiO_3$ powder under high pressure were recorded at room temperature (296 K) in the EasyLab μScope DAC-HT(G) (UK). In the present experiments, the non-hydrostatic pressing was realised, due to the fact that no liquid pressure transmittance media was used. The value of mechanical stresses was measured by the shift of the Raman line in a diamond anvil near the surface of the sample. Details of this technique are described in Ref. [6].

Raman spectra were measured in backscattering geometry. The laser beam was focused on the sample, and scattered light was collected by with a 50× Olympus LMPlanFl objective lens with N.A.=0.35. Raman spectra were excited by the $Ar^+$ ion laser Spectra-Physics Stabilite 2017 with $\lambda$=514.5 nm and power of 5 mW. Spectra of scattered light were recorded by the triple T64000 Horiba Jobin Yvon spectrometer.

## 3. Results and discussion

Typical Raman spectra under different non-hydrostatic pressures are demonstrated in Figure 1. It is clearly seen that:

1. E(TO) line is present at all applied mechanical stresses.
2. At non-hydrostatic pressures above 2 GPa a new line in the vicinity of 100 $cm^{-1}$ appears in the spectra.
3. When mechanical stresses are applied, the position of mentioned above lines shifts toward the higher frequencies, and the width increases. The dependences of the position and width of lines under study are shown in Figures 2 and 3.

It is known that the line near 307 $cm^{-1}$ is a characteristic feature of the tetragonal (ferroelectric) phase in barium titanate [13]. The experimental fact that this line is manifested for the whole pressure range points out that the $BaTiO_3$ powder does not undergo the ferroelectric phase transition from tetragonal phase to cubic one under non-hydrostatic pressures up to 11 GPa. On the contrary, the

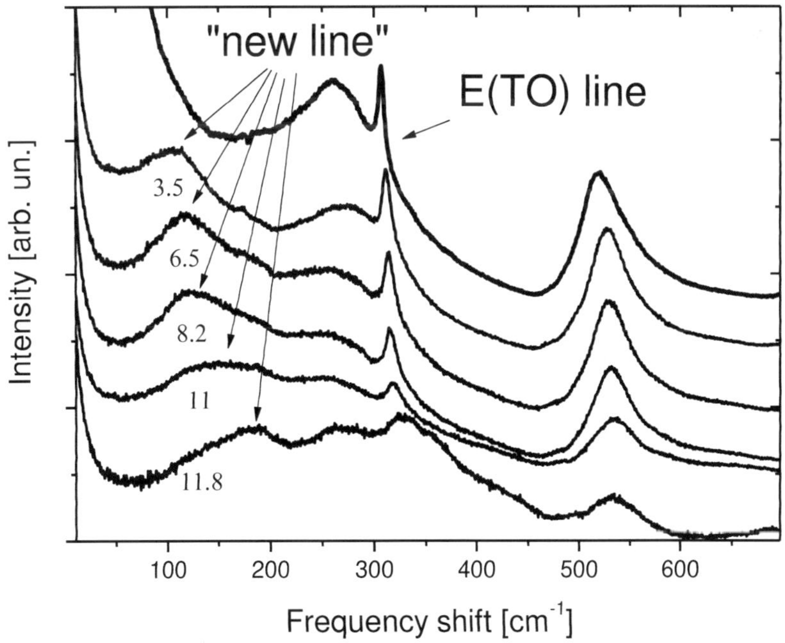


**Figure 1.** Typical Raman spectra under different non-hydrostatic pressures. The numbers indicate the values of non-hydrostatic pressures in GPa. Upper curve is a Raman spectrum at zero pressure.

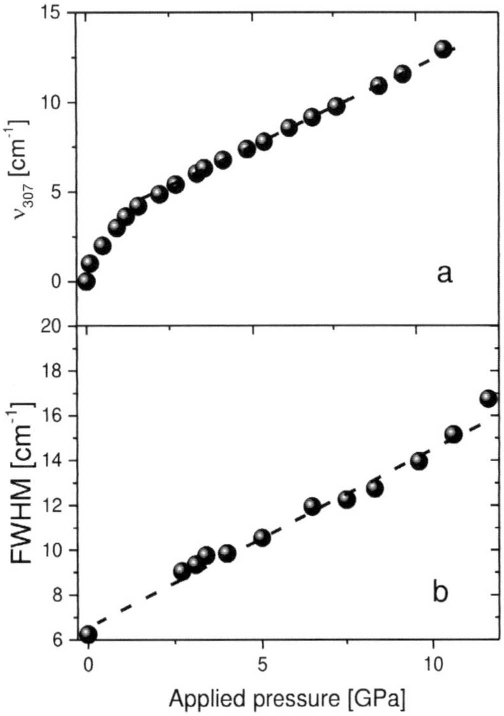


**Figure 2.** Dependence of the (a) position and (b) FWHM of E(TO) line on the applied non-hydrostatic pressure.

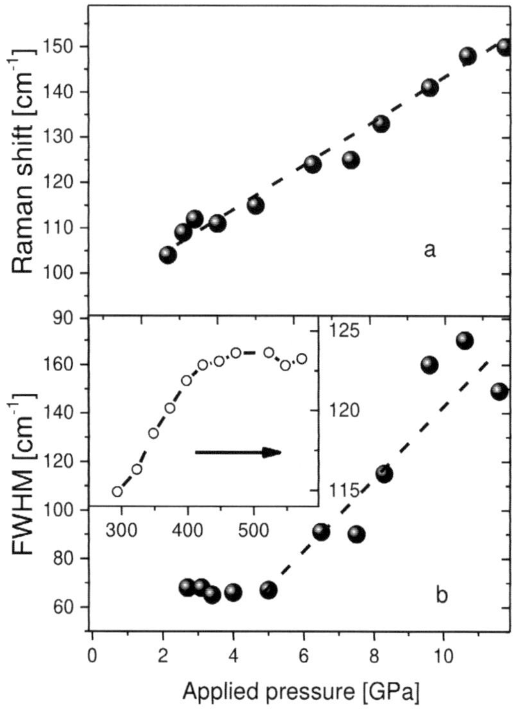


**Figure 3.** Dependence of the (a) shift and (b) FWHM of the "new peak" on the applied non-hydrostatic pressure. The temperature dependence of the Raman shift of this peak at 5 GPa is shown in the inset.

hydrostatic pressure above 2 GPa applied to $BaTiO_3$ powder leads to the transition into the cubic phase [3]. We can propose that uniaxial mechanical stress causes a reorientation of domains by the ferroelastic effect. In this way, non-hydrostatic mechanical stress increases the density of domain (and dipole momentum, respectively) in a certain direction in originally unpolarized ceramics, which leads to the "conservation" of the ferroelectric phase.

Increasing the frequency of the E(TO) line shift presented in Figure 2a has been discussed earlier [12]. It is shown that the dependence of the line shift on the applied pressure can serve as an effective tool for determining mechanical stresses in ceramics obtained by pressing of barium titanate powder.

The full width of high maximum (FWHM) of the E(TO) Raman line is presented in Figure 2b. For this purpose, this line was fitted by the Voigt contour (the procedure of this fitting was described in Ref. [6] in detail). The Raman line width (Lorentz part of the contour) increases with increasing the pressure.

For a uniformly broadened line, one would expect a decrease in its width. Therefore, it should be noted that inhomogeneous broadening is observed in this experiment. The inhomogeneous broadening of this Raman line may be associated with a shortening of the corresponding lifetimes of vibration states. On the other hand, it could be caused by the appearance of the defect in the sample. Such inhomogeneous broadening can occur on much smaller scale, down to atomic one [14].

An increase of the non-hydrostatic pressure leads to the appearance of a new peak in the Raman spectra. It was first discovered in Ref. [15]. The dependence of the position of this peak on non-hydrostatic stresses is demonstrated in Figure 3a. In turn, the increase of FWHM of this line upon the pressure increase indicates an inhomogeneous broadening due to the anharmonicity of oscillations

(Fig. 3b). The temperature dependence of the Raman shift of this peak is presented in the inset in Figure 3b.

It is known that without the application of mechanical stresses, the soft mode in barium titanate is overdamped [16]. The question of whether a new peak is a manifestation of a soft mode can be answered by the temperature dependence of this peak. The temperature dependence of this line under the non-hydrostatic pressure of 5GPa excludes its behaviour, which could be interpreted as a soft mode.

## 4. Conclusion

The dependence of the Raman shift and width of E(TO) Raman line (307 $cm^{-1}$) on applied non-hydrostatic pressures up to 11 GPa in the $BaTiO_3$ powder has been established. It is found that under the action of uniaxial pressure, a new peak appears in the vicinity of 100 $cm^{-1}$. Its position, width, and temperature dependence, are measured. It is shown that the positions and widths of both peaks mentioned above increase depending on the applied non-hydrostatic pressure. These Raman lines' broadening may be associated with an increase of the concentration of defect states in the volume of the sample. The temperature dependence of this line under the non-hydrostatic pressure of 5 GPa excludes its soft mode behaviour.

**Acknowledgements**

The reported study was funded by RFBR according to the research projects No. 19-42-543-016 and 18-02-00399 and State assignment No AAAA-A17-117052410033-9. The experiments were performed in the multiple-access centre "High-Resolution Spectroscopy of Gases and Condensed Matter" in IA&E SBRAS (Novosibirsk, Russia) and Centre for Common Use of the Krasnoyarsk Scientific Centre, SBRAS (Krasnoyarsk, Russia).

**References**

[1] Samara G A 1996 Pressure and temperature dependences of the dielectric properties of the perovskites $BaTiO_3$ and $SrTiO_3$ *Phys. Rev.* **151** 378-86

[2] Merz W J 1950 The effect of hydrostatic pressure on the Curie point of barium titanate single crystals *Phys. Rev.* **77** 52-4

[3] Iniguez J and Vanderbilt D 2002 First-principles study of the temperature-pressure phase diagram of $BaTiO_3$ *Phys. Rev. Lett.* **89** 115503

[4] Samara G A 1966 Pressure and temperature dependences of the dielectric properties of the perovskites $BaTiO_3$ and $SrTiO_3$ *Phys. Rev.* **151** 378

[5] Pugachev A M, Zaytseva I V, Kovalevskii V I, Malinovsky V K, Surovtsev N V, Borzdov Yu, Raevski I, Raevskaya S and Malitskaya M 2016 Local residual stresses in pressure-treated barium titanate powders probed by second harmonic generation *Ferroelectrics* **501** 9-14

[6] Zaytseva I V, Pugachev A M, Okotrub K A, Surovtsev N V, Mikerin S L and Krylov A S 2019 Residual mechanical stresses in pressure treated $BaTiO_3$ powder *Ceram. Int.* **45** 12455-60

[7] Pugachev A M, Zaytseva I V, Krylov A S, Malinovsky V K, Surovtsev N V, Borzdov Yu M and Kovalevsky V 2017 Uniaxial mechanical stresses and their influence on the parameters of the ferroelectric phase transition in pressure-treated barium titanate *Ferroelectrics* **508** 161-6

[8] Pugachev A M, Kovalevsky V I, Malinovsky V K, Borzdov Yu M and Surovtsev N V 2015 Relaxor-like features in pressure-treated barium titanate powder *Appl. Phys. Lett.* **107** 102902

[9] Tenne D A, Turner P, Schmidt J D, Biegalski M, Li Y L, Chen L Q, Soukiassian A, Trolier-McKinstry S, Schlom D G, Xi X X, Fong D D, Fuoss P H, Eastman J A, Stephenson G B, Thompson C and Streiffer S K 2009 Ferroelectricity in ultrathin $BaTiO_3$ films: Probing the size effect by ultraviolet Raman spectroscopy *Phys. Rev. Lett.* **103** 177601

[10] Choi K J, Biegalski M, Li Y L, Sharan A, Schubert J, Uecker R, Reiche P, Chen Y B, Pan X Q,

Gopalan V, Chen L Q, Schlom D G and Eom C B 2004 Enhancement of ferroelectricity in strained $BaTiO_3$ thin films *Science* **306** 1005

[11] Diéguez O, Tinte S, Antons A, Bungaro C, Neaton J B, Rabe K M and Vanderbilt D 2004 Ab initio study of the phase diagram of epitaxial $BaTiO_3$ *Phys. Rev. B* **69** 212101

[12] Pugachev A M, Kovalevsky V I, Malinovsky V K, Borzdov Yu M and Surovtsev N V 2015 Relaxor-like features in pressure-treated barium titanate powder *Appl. Phys. Lett.* **107** 102902

[13] Jiang Y J, Zeng L Z, Wang R P, Zu Y and Liu Y L 1996 Fundamental and second order Raman spectra of $BaTiO_3$ *J. Raman Spectrosc.* **27** 31-4

[14] Surovtsev N V, Kupriyanov I N, Malinovsky V K, Gusev V A and Pal'yanov Yu N 1999 Effect of nitrogen impurities on the Raman line width in diamonds *J. Phys.: Condens. Mat.* **11** 4767-74

[15] Zhang W, Chen L, Jin C, Deng X, Wang X and Li L 2008 High pressure Raman studies of dense nanocrystalline $BaTiO_3$ ceramic *J. Electroceram.* **21** 859-62

[16] Shirane Q, Frazer B C, Minkiewicz V J and Leake J A 1967 Soft optic mode in barium titanate *Phys. Rev. Lett.* **5** 234-5

# AUTHOR INDEX